Smart Nanomaterials Technology

Series Editors

Azamal Husen, Wolaita Sodo University, Wolaita, Ethiopia

Mohammad Jawaid, Department of Chemical and Petroleum Engineering, United Arab Emirates University, Al Ain, United Arab Emirates

Nanotechnology is a rapidly growing scientific field and has attracted a great interest over the last few years because of its abundant applications in different fields like biology, physics and chemistry. This science deals with the production of minute particles called nanomaterials having dimensions between 1 and 100 nm which may serve as building blocks for various physical and biological systems. On the other hand, there is the class of smart materials where the material that can stimuli by external factors and results a new kind of functional properties. The combination of these two classes forms a new class of smart nanomaterials, which produces unique functional material properties and a great opportunity to larger span of application. Smart nanomaterials have been employed by researchers to use it effectively in agricultural production, soil improvement, disease management, energy and environment, medical science, pharmaceuticals, engineering, food, animal husbandry and forestry sectors.

This book series in Smart Nanomaterials Technology (SNT) publishes cutting edge research monographs and edited volumes by top researchers and professionals across the globe. This book series aims to comprehensively cover topics in the fabrication, synthesis and application of these materials for applications in the following fields:

- Energy Systems—Renewable energy, energy storage (supercapacitors and electrochemical cells), hydrogen storage, photocatalytic water splitting for hydrogen production
- Biomedical—controlled release of drugs, treatment of various diseases, biosensors
- Agricultural—agricultural production, soil improvement, disease management, animal feed, egg, milk and meat production/processing
- Forestry—wood preservation, protection, disease management
- Environment—wastewater treatment, separation of hazardous contaminants from wastewater, indoor air filters

Review Process

The proposal for each volume is reviewed by the main editor and/or the advisory board. The chapters in each volume are individually reviewed single blind by expert reviewers (at least two reviews per chapter) and the main editor.

Ethics Statement for this series can be found in the Springer standard guidelines here - https://www.springer.com/us/authors-editors/journal-author/journal-author-helpdesk/before-you-start/before-you-start/1330#c14214

Azamal Husen
Editor

Sustainable Uses of Aromatic Plants in Nanotechnology

Springer

Editor
Azamal Husen
Wolaita Sodo University
Wolaita, Ethiopia

ISSN 3004-8273 ISSN 3004-8281 (electronic)
Smart Nanomaterials Technology
ISBN 978-981-95-7408-7 ISBN 978-981-95-7409-4 (eBook)
https://doi.org/10.1007/978-981-95-7409-4

This Springer imprint is published by the registered company Springer Nature Singapore Pte Ltd.
The registered company address is: 152 Beach Road, #21-01/04 Gateway East, Singapore 189721, Singapore

Preface

The convergence of botanical sciences and nanotechnology represents one of the most exciting frontiers in sustainable research. Among the diverse plant groups, aromatic plants renowned for their distinctive fragrances, essential oils, and therapeutic properties are increasingly recognized as a valuable resource for the eco-friendly synthesis of nanoparticles. The book *Sustainable Uses of Aromatic Plants in Nanotechnology* explores this emerging synergy, highlighting its potential applications in medicine, agriculture, environmental remediation, cosmetics, and food technology. Aromatic plants have been cherished for centuries in traditional medicine, culinary practices, and perfumery, primarily due to their essential oils and volatile bioactive compounds. These phytochemicals are hydrophobic, highly concentrated, and rich in functional groups, making them ideal candidates for nanoparticle synthesis through green, sustainable routes. Unlike conventional chemical or physical methods of nanoparticle production often criticized for being costly, energy-intensive, and environmentally hazardous, the use of aromatic plants provides an eco-friendly, cost-effective, and biocompatible alternative. This method leverages nature's inherent chemical diversity to design nanomaterials with precise biological activities, reduced toxicity, and improved performance.

The overarching theme of this book is sustainability. Nanotechnology, while offering tremendous promise, must evolve beyond traditional synthesis routes to minimize its environmental impact. Aromatic plants provide a renewable, biodegradable, and versatile source of biomolecules for fabricating nanoparticles that meet this challenge. Through their diverse chemical repertoire, including terpenoids, flavonoids, alkaloids, and phenolic compounds, these plants act as reducing, capping, and stabilizing agents, thereby eliminating the need for harsh chemicals and promoting green nanotechnology practices.

One of the key areas explored in this book is the biomedical potential of aromatic plant-based nanoparticles. These nanosystems have demonstrated remarkable antimicrobial, antifungal, antiviral, anticancer, and wound-healing properties. For instance, nanoparticles synthesized from essential oils can enhance the efficacy of therapeutic formulations by improving drug stability, bioavailability, and targeted delivery. Furthermore, nanocarrier systems such as nanoemulsions, nanogels, and nanocomposites are being developed to encapsulate active compounds, ensuring controlled release and prolonged shelf-life, an advancement critical for pharmaceutical and cosmetic applications. The chapters dedicated to wound healing and skin restoration underscore the transformative role of nanotechnology in improving recovery rates and reducing infection risks.

Beyond healthcare, this book delves into agricultural and environmental applications. Aromatic plants, when combined with nanotechnology, can promote sustainable farming practices through nanopriming, biofertilizers, and natural nanopesticides. Green-synthesized nanoparticles can boost plant growth, improve resistance to pathogens, and minimize the use of synthetic chemicals, thereby reducing ecological harm. Additionally, in environmental sciences, aromatic plant-based nanomaterials are being explored for their ability to absorb and neutralize pollutants, including heavy metals, thus contributing to environmental detoxification and restoration. The structural organization of this book reflects a systematic approach to this interdisciplinary field. It begins with foundational concepts of aromatic plants, their chemical profiles, and their unique potential in nanotechnology. Specific plant families such as Lauraceae, Umbelliferae, Myrtaceae, and Labiatae are highlighted for their phytochemical richness and bioactivity, followed by detailed discussions on the synthesis, characterization, and applications of various nanoparticle types, including metallic, metal-oxide, carbon-based, and composite nanostructures. Further chapters are dedicated to biomedical applications, categorizing them by therapeutic function: antimicrobial, antiviral, antifungal, anticancer, wound healing, and skin restoration, thereby providing readers with a clear understanding of how specific nanoparticles address distinct medical challenges. Subsequent sections explore nanoemulsions and their multifaceted roles in the food industry, pharmaceuticals, cosmetics, and insect repellents, emphasizing how these advanced formulations enhance stability, efficacy, and safety. A crucial part of this book also examines how nanoparticles affect the growth and performance of aromatic plants themselves. Studies on carbon-based, metal-based, and metal-oxide nanoparticles reveal insights into their influence on physiological, biochemical, and morphological processes, offering potential strategies for precision agriculture. Finally, this book addresses the regulatory and safety aspects of using aromatic plants in nanotechnology, an area of growing importance as industries move toward large-scale applications. Issues such as toxicity, environmental impact, standardization of synthesis protocols, and compliance with international safety guidelines are critically discussed to ensure responsible scientific advancement.

This book is intended for a wide audience, including researchers, academicians, industry professionals, and students working in the fields of botany, nanotechnology, pharmacology, food science, agriculture, and environmental sciences. It aims to serve as both a reference and a guide, providing a blend of theoretical knowledge, experimental findings, and future perspectives. Readers will gain insights into not only the underlying science but also the potential for innovation across industries through sustainable practices. The interdisciplinary nature of this work reflects the collaborative efforts of experts from diverse scientific backgrounds. Chemists, botanists, pharmacologists, materials scientists, and environmental researchers have collectively contributed to developing the chapters, ensuring a holistic and authoritative treatment of the subject. Each section integrates fundamental principles with cutting-edge applications, supported by case studies and experimental data where applicable. The promise of integrating aromatic plants into nanotechnology extends beyond laboratory research; it represents a paradigm shift toward green innovation. By leveraging the natural wealth of plant-derived compounds, we can develop biocompatible, biodegradable, and efficient nanomaterials that address some of the most pressing challenges of our time, ranging from infectious diseases to environmental degradation. Importantly, this approach aligns with global sustainability goals, emphasizing renewable resources, reduced chemical waste, and environmentally responsible manufacturing. In curating this book, great care has been taken to maintain a balance between scientific rigor and accessibility. While technical concepts such as nanoparticle synthesis, characterization techniques, and bioactivity assays are presented in detail, the language and organization aim to ensure clarity for readers from diverse academic and professional backgrounds. Moreover, each chapter concludes with a forward-looking perspective, highlighting gaps in knowledge and potential avenues for future research.

As global demand for sustainable technologies grows, integrating aromatic plants with nanotechnology opens a visionary path for innovation. From advanced biomedical solutions to safer food packaging, enhanced agricultural productivity, and environmental remediation, the possibilities are vast. This book aims to inspire researchers and industries to refine and adopt these eco-friendly approaches, fostering a greener and more resource-conscious future. I extend heartfelt gratitude to all contributors for sharing their expertise and incorporating suggestions to enhance their chapters. Special thanks to Shagufta, Zaara, Mehwish, and Huzaifa for their time and invaluable support in bringing this work to completion.

Wolaita, Ethiopia Azamal Husen

Contents

About the Editor

Azamal Husen has served as a Full Professor and Head of the Department of Biology at the University of Gondar, Ethiopia. He is currently affiliated as a Foreign Delegate with Wolaita Sodo University, Ethiopia, and holds an Adjunct Professorship in the Department of Biotechnology, Graphic Era (Deemed to be University) in Dehradun, Uttarakhand, India. His earlier academic roles include Full Professor at Sankalchand Patel University in Visnagar, India, and Visiting Faculty at the Forest Research Institute and Doon College of Agriculture and Forest in Dehradun, India. With more than 25 years of experience in academia and research, he is highly regarded for his expertise in nanobiotechnology, plant physiology, environmental botany, medicinal plants, and plant propagation. His research has been supported by prominent national and international organizations, including the World Bank, the National Agricultural Technology Project, the Indian Council of Agricultural Research, the Indian Council of Forestry Research and Education, and the Japan Bank for International Cooperation. He is a prolific scholar, having published more than 350 academic works, including research articles, books, and book chapters. His editorial contributions are equally noteworthy; he serves on the editorial boards and peer review panels of numerous prestigious journals published by Elsevier, Frontiers Media, Springer Nature, Taylor & Francis, RSC, Oxford University Press, the Royal Society, CSIRO, MDPI, PLOS, Sciendo, John Wiley & Sons, and UPM Journals. He is also a member of the advisory board of Cambridge Scholars Publishing, United Kingdom. He is a Fellow of the Plantae group under the American Society of Plant Biologists and holds membership in several scientific organizations, including the International Society of Root Research, the Asian Council of Science Editors, and the International Natural Product Sciences Taskforce. Throughout his career, he has delivered invited lectures, keynote addresses, and technical presentations at numerous international conferences, symposia, and workshops. In the field of academic development, he has made significant contributions through the design and delivery of orientation programs for university faculty, the development of educational modules, and the formulation of curricula at the undergraduate, postgraduate, and doctoral levels. In recognition of his contributions to teaching, research, and community service, he was honored with a special award from the University of Gondar. He currently serves as the Editor-in-Chief of the *American Journal of Plant Physiology* and is the Series Editor for several major academic collections, including *Exploring Medicinal Plants* (Taylor & Francis, USA), *Plant Biology, Sustainability, and Climate Change* (Elsevier, USA), and *Smart Nanomaterials Technology* (Springer Nature, Singapore). He also leads the *Handbook Series on Nanotechnology and Its Applications* (Springer Nature, Singapore). His outstanding contributions to science have earned him international acclaim, including the prestigious distinction of being named among the "World's Top 2% Scientists" for three consecutive years (2022 onward) by Stanford University, a recognition published in the Elsevier Data Repository. In addition to his regular academic responsibilities, he has served as a Guest Editor for several high-impact special issues, including *Advances in Plant Metabolomics: Insights, Applications, and Integration* (*Plant Physiology and Biochemistry*), *Exploring Nanotechnology in Medicinal Plants* (*Journal of Herbal Medicine*), and *Integrating Nanotechnology with Plant Compounds: Enhancing Abiotic Stress Resilience and Crop Improvement via Signaling Molecules* (*Frontiers in Plant Science*), among others. His work continues to shape contemporary research at the intersection of plant science, biotechnology, and sustainable development. His academic leadership, research excellence, and international collaborations have positioned him as a prominent figure in global scientific circles.

Sustainable Use of Aromatic Plants in Nanotechnology

Shifa Shaffique, Muhammad Saleem Khan, Muhammad Aaqil Khan, Sang-Mo Kang, and In-Jung Lee

Abstract Aromatic plants, characterized by their rich content of phenolic and phenylpropanoid compounds, possess diverse biological activities and strong aromatic properties that make them valuable in sustainable nanotechnology. Their phytochemical profile enables them to act as reducing, stabilizing, and functionalizing agents in the green synthesis of metallic and metal oxide nanoparticles. This eco-friendly approach eliminates the reliance on toxic reagents, minimizes energy consumption, and reduces environmental pollution. Beyond serving as green synthesis agents, aromatic plants impart their intrinsic bioactive properties to nanomaterials, thereby enhancing their functional value. Such bio-inspired nanomaterials show significant potential for applications in medicine, agriculture, food safety, and cosmetics. This chapter explores the sustainable use of aromatic plants in nanotechnology, highlighting their role in nanoparticle synthesis and the diverse applications of the resulting nanomaterials. It further examines current trends, prospects, and challenges in integrating aromatic plant-mediated nanotechnology into agricultural, pharmaceutical, and industrial value chains. By emphasizing both environmental and economic benefits, this chapter positions aromatic plants as key contributors to the development of circular, low-carbon economies.

Keywords Aromatic compounds · Green synthesis · Nanomaterials · Sustainable use

---

S. Shaffique · S.-M. Kang · I.-J. Lee (✉)
Department of Applied Biosciences, Kyungpook National University, Daegu, Republic of Korea
e-mail: ijlee@knu.ac.kr

M. S. Khan
Department of Zoology, University of Okara, Okara, Pakistan

M. A. Khan
International Joint Research Laboratory for Global Change Ecology, School of Life Sciences, Henan University, Kaifeng, Henan, China

A. Husen (ed.), *Sustainable Uses of Aromatic Plants in Nanotechnology*, Smart Nanomaterials Technology, https://doi.org/10.1007/978-981-95-7409-4_1

1 Introduction

Aromatic plants are present naturally in the environment, which contain aromatic compounds, a source of bioactive compounds with a distinctive essence and a diverse range of biological activities [1–8]. They are also known as herbs and spices and have been used for traditional medicine since approximately 5000 BC. They are organic molecules that contain one or more aromatic rings (benzene) in their chemical structure. The rings are composed of six carbon atoms with double bonds. These double bonds are alternating and delocalized, which gives them a distinctive essence, stability, and reactivity in a specific manner [9, 10].

These compounds are used in several industries, such as cosmetics, medicine, etc., and are considered safe and eco-friendly [11, 12]. In plants, these compounds are secondary metabolites that are involved in the defense mechanism. Usually, these compounds are present in essential oils, extracts, and resins of aromatic plants such as clove, basil, lavender, etc. [13–15]. Aromatic plants contain volatile, highly concentrated compounds known as essential oils. It is present in several parts of plants, such as flowers, buds, wood, fruits, etc. It is composed of secondary metabolites and terpenes. The detailed metabolomics revealed that it contains terpenoids, α-terpineol, hydrocarbons, phenolic compounds, and aromatic aldehyde. These essential oils have a wide range of pharmacological properties [1, 16]. The studies reported a diverse range of their application, i.e., animal feeding [17], feed additive [18], industrial interest, etc. [19].

Nanotechnology is a branch of science that involves the synthesis and application of nanomaterials. It has revolutionized all sectors of applied sciences, from health care to agricultural management [20, 21]. The conventional effect of nanomaterials involves substances that are even lethal to the environment, high energy, and temperature input, which further affect sustainability. Green synthesis of nanomaterials is an innovative and emerging field that involves natural compounds, such as aromatic compounds. The nanomaterials that are prepared from the green synthesis approach are biodegradable, cost-effective, and energy efficient [22–25]. de Cristo Soares Alves et al. [26] synthesized the nanoencapsulation of gallic acid (GA) and evaluated its toxicity and antioxidant potential in an in vitro platform. Gallic acid (GA, 3,4,5-trihydroxylbenzoic acid) is an important phenolic aromatic compound. It is present in several fruits, nuts, green tea, etc. It has the potential to scavenge the free radicals [27, 28]. Several studies reported on its therapeutic and biological potential. However, the lower bioavailability remains a challenge for gallic acid application. A study reported that synthesizing silver NPs by using the clove bud extract showed enhanced antibacterial activity against the marine bacterium and diatoms. The study showed that using the clove bud extract as a capping or reducing agent is an eco-friendly approach to using the clove extract in the synthesis of silver NPs. The obtained NPs were 10–27 nm cubic crystals; the FTIR analysis confirmed the presence of organic functional compounds in them. The results confirmed that the biological synthesis of the NPs using the aromatic plant extract is a sustainable way [29]. In another study, the clove extract (CE) was used to synthesize the silver nanoparticles, and the antibacterial and fungal activities were compared. The results suggested that application of the CE-mediated silver nanoparticles showed enhanced antibacterial and fungal activities as compared to the other non-mediated NPs. The results further indicate that the presence of eugenol acts as a reducing and capping agent. It enhances the biological activities of NPs. As aromatic plants, such as clove buds, clover extracts are nontoxic in nature, so it is sustainable to use them [30]. Gold NPs were synthesized by using clove oil, which acts as a reducing as well as stabilizing agent. This study further confirms that the addition of aromatic plants has a significant role in stabilizing and enhancing the biological activities of NPs [31].

Basil is also considered an important aromatic plant that is rich in essential oils and phenolic compounds, such as linalool, eugenol, methyl chavicol, etc. Various previous studies reported that it has strong biological activities, such as antimicrobial, antifungal, and antioxidant [32–34]. Copper (Cu) and zinc (Zn) NPs were prepared with green synthesis by using the basil extract and applied to the basil plant via foliar application. The results revealed that the potential application of synthesizing Cu and Zn NPs enhanced the physiological, biochemical, and antioxidant traits in plants, reducing the need for agrochemicals [35]. Another study reported that basil-loaded ZnO NPs had strong anticancer and anti-inflammatory activities. For this experiment, the two cellular lines HEPG-2 and HCT-116, hepatocellular and colorectal, respectively. The results were interesting, as these NPs had the strongest anticancer activities [36]. de Cristo Soares Alves et al. [26] reported that nanoencapsulation of GA was successfully synthesized within a range of 220 nm, with 25% approximately drug encapsulation efficiency. The NPs were stable even at storage and room temperature. They contain significant antioxidant potential and negligible hemolysis as compared to the non-coated NPs.

Caffeic acid is an aromatic compound that is present in several aromatic plants. It has a strong antioxidant potential, and it acts as a precursor for other natural compounds [37, 38]. In a study conducted in India, where caffeic acid silver nanoformulation was prepared and its effectiveness against cytotoxicity and cell cycle was monitored, the results suggested that it works as a reducing agent; the presence of the hydroxyl group in the caffeic acid provides stabilization as well as activates the reduction process. The outcome of flow cytometric analysis revealed that it is the best anticancer therapeutic alternative. One study also reported that caffeic acid extracted from the nutmeg when it is green is synthesized to form the silver nanoparticles, which have the strongest antibacterial activity [39].

Mint (*Mentha*) is an aromatic plant that is used as traditional folk medicine. The trend to replace the traditional metallic NPs with green synthesis of NPs is a more sustainable approach. In several reported studies, mint was used to synthesize the silver NPs. In a study, mint (*Mentha piperita*) leaf extract was used as a reducing agent to synthesize the silver NPs. The results declared that it has the strongest antibacterial activity compared to the control group [40]. Similarly, in another study, it was reported that the mint leaf extract has the strongest antibacterial activity [41]. Mint leaves (*Mentha arvensis*) were used to synthesize the calcium oxide CaO NPs. The results suggested it has the strongest anti-inflammatory and anticancer activity

[42]. In 2023, Saudi mint was used to prepare the silver nanoparticles. The results are in favor of the previous findings, and the usage of aromatic plants for green synthesis of NPs enhances their biological activity [43].

2 Sustainable Use of Aromatic Plants in Nanotechnology

The sustainable synthesis of nanomaterials from aromatic plants offers a versatile platform for applications that extend across health care, agriculture, food preservation, and cosmetic industries. The integration of renewable plant resources with nanotechnology not only reduces dependence on hazardous synthetic methods but also enhances the functionality of the resulting nanomaterials through phytochemical activity [44, 45]. The sustainable use of aromatic compounds in nanotechnology and a few examples are given in Table 1, and the properties are given in Fig. 1.

Table 1 The potential use of aromatic plant-mediated nanotechnology and their sustainable usage

Aromatic plants/ compounds	Nanotechnology	Sustainable use	Key references
Clove buds	Silver nanoparticles	Enhanced antibacterial potential against bacteria and diatoms	Lakhan et al. [29]
Clove extract/ eugenol	Silver nanoparticles	Enhanced antibacterial activity and fungal activity	Parlinska-Wojtan et al. [30]
Basil	Nano-fertilizer	Enhanced antioxidant, physiological, and biochemical traits	Abbasifar et al. [35]
Basil	Loaded nanoparticles	Anticancer activities	Barakat et al. [36]
Gallic acid	Nanoencapsulation	Increase antioxidant potential Increase cytotoxicity Increase hemolysis	de Cristo Soares Alves et al. [26]
Caffeic acid	Bio-nano-formulation	Highly stable anticancer	Pushpanathan et al. [46]
Caffeic acid	Silver nanoparticles	Antibacterial potential	Thummaneni et al. [39]
Mint leaf extract	Silver nanoparticles	Antibacterial activity	Gabriela et al. [40]
Mint leaves	Silver nanoparticles	Antimicrobial activities	Aziz and Jassim [41]
Mint leaves	Calcium oxide nanoparticles	Anti-inflammatory anticancer	Shanmuganathan et al. [42]
Saudi mint	Silver nanoparticles	Antimicrobial antioxidant	Qanash et al. [43]

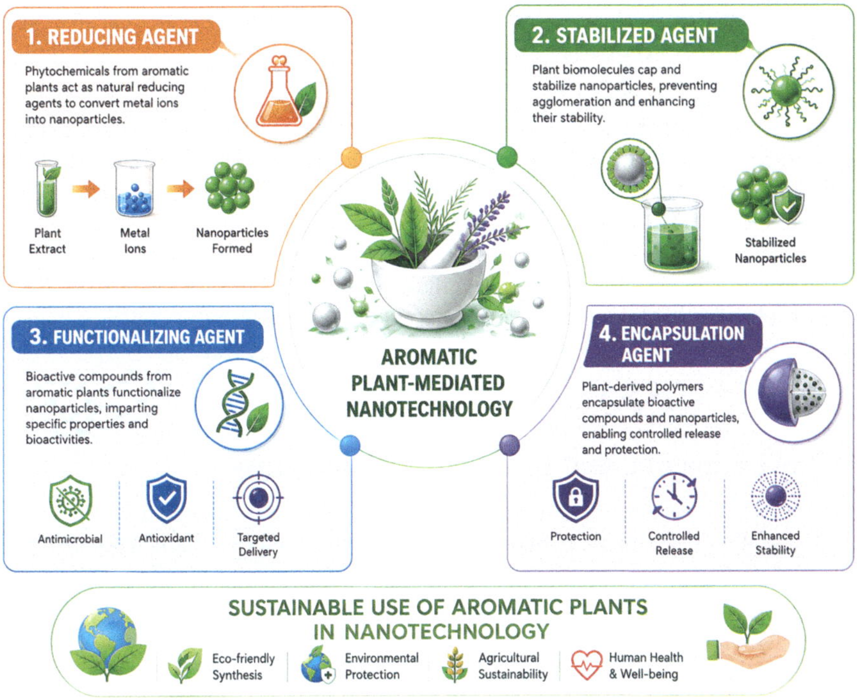

Fig. 1 Aromatic plant-mediated nanotechnology provides the four key functions for enhancing sustainability

3 Antimicrobial, Anti-Inflammatory, and Antioxidant Activities

Aromatic plant-derived silver nanoparticles (AgNPs) demonstrate remarkable efficacy against multidrug-resistant bacterial and fungal pathogens [47]. The synergistic interaction between metal NPs and plant-based capping agents such as thymol, eugenol, and rosmarinic acid enhances membrane disruption, oxidative stress induction, and biofilm inhibition. AgNP-loaded wound dressings synthesized from *Thymus vulgaris* or *Rosmarinus officinalis* extracts have been shown to accelerate epithelialization and reduce infection rates while minimizing cytotoxicity to mammalian cells [48, 49]. NPs stabilized by polyphenol-rich aromatic plant extracts exhibit strong radical scavenging capacity. Zinc oxide nanoparticles (ZnONPs) synthesized from *Lavandula angustifolia* or *Ocimum basilicum* can be incorporated into topical formulations for skin protection, leveraging both UV-blocking and anti-inflammatory properties [50].

4 Sustainable Use of Aromatic Plant-Mediated Nanomaterials in Agriculture

Plant-derived nanoparticles offer a promising route for eco-friendly pest and pathogen management. For example, ginger oil-based nanomaterials are being used to control tropical plant diseases [51]. Metal oxide NPs, particularly zinc oxide (ZnONPs) derived from *Aloe barbadensis* Mill, can act as slow-release micronutrient sources, improving nutrient use efficiency [52].

Selenium nanoparticles (SeNPs) synthesized from *Nigella sativa* extracts have been observed to modulate plant cadmium (Cd) stress responses, as well as enhance antioxidant enzyme activity and promote plant growth and protection [53].

5 Environmental and Economic Sustainability

The integration of aromatic plants into nanomaterial synthesis aligns with multiple pillars of sustainability: environmental preservation, economic viability, and social benefit. The approach reduces the reliance on hazardous chemicals traditionally used in nanoparticle production, thereby lowering the risk of environmental contamination and occupational hazards [51]. The future of this field lies in interdisciplinary collaboration combining the expertise of botanists, chemists, nanotechnologists, engineers, and policymakers to fully realize the potential of aromatic plants in shaping the next generation of eco-friendly nanotechnological solutions. By harnessing the synergy between nature's molecular diversity and cutting-edge nanoscience, aromatic plant-based nanotechnology can play a pivotal role in addressing global challenges in health, food security, and environmental stewardship. The convergence of aromatic plant biotechnology and nanotechnology is poised to expand into new frontiers driven by advances in synthetic biology, materials engineering, and sustainability science. Future research is expected to focus on several key directions, such as the biodiversity of aromatic plants and toxicity of nanomaterials.

6 Conclusion

The diverse phytochemical profile of aromatic plants enables the sustainable low-energy synthesis of metallic and metal oxide NPs, eliminating the need for toxic reagents and reducing environmental pollution. Beyond green synthesis, the endogenous bioactivity of these plants imparts additional functional value to the resulting nanomaterials, enabling applications across medicine, agriculture, food safety, and cosmetics. The integration of aromatic plant-mediated NPs into existing agricultural and industrial value chains offers both environmental and economic benefits, contributing to the development of resilient, circular, and low-carbon economies. While technical and regulatory challenges remain, advances in precision phytochemistry, process engineering, and sustainability assessment are likely to accelerate the transition from laboratory-scale innovation to large-scale industrial adoption.

Competing Interests The authors declare that they have no competing interest.

Declaration of Generative AI and AI-Assisted Technologies in the Writing Process During the preparation of this work, the book chapter author(s) have not taken any assistance from AI, and the book chapter author(s) reviewed and edited the content as needed and take(s) full responsibility for the content of the publication.

References

1. Christaki E, Bonos E, Giannenas I, Florou-Paneri P (2012) Aromatic plants as a source of bioactive compounds. Agriculture 2(3):228–243
2. Husen A (2022) Herbs, shrubs and trees of potential medicinal benefits. CRC Press, Boca Raton, pp 33487–32742. https://doi.org/10.1201/9781003205067
3. Husen A (2023) Antidiabetic medicinal plants and herbal treatments. CRC Press, Boca Raton, pp 33487–32742. https://doi.org/10.1201/b23347
4. Husen A (2023) Exploring poisonous plants: medicinal values, toxicity response and other potential uses. CRC Press, Boca Raton, p 33487. https://doi.org/10.1201/b23017
5. Husen A (2024) Antimalarial medicinal plants. Taylor & Francis Group, New York. https://doi.org/10.1201/9781003378396
6. Husen A (2024) Medicinal spices and condiment crops. Taylor & Francis Group, New York. https://doi.org/10.1201/9781003387046
7. Máthé Á, Hassan F, Kader AA (2015) Medicinal and aromatic plants of the world. In: Medicinal and aromatic plants world. Springer, Cham
8. Skaria BP (2007) Aromatic plants, vol vol. 1. New India Publishing, New Delhi
9. Blumer M (1976) Polycyclic aromatic compounds in nature. Sci Am 234(3):34–45
10. McCague R (1986) Aromatic compounds. Ann Rep Sect B 83:147–169
11. Doğan MS, Çelik H (2023) Organic compounds containing aromatic structure used in pharmaceutical production. J Biochem Technol 14(2–2023):102–111
12. Lubbe A, Verpoorte R (2011) Cultivation of medicinal and aromatic plants for specialty industrial materials. Ind Crop Prod 34(1):785–801
13. Manjunath MS, Koshariya AK, Sharma N, Rajput A, Pandey SK, Singh S, Singh BV (2023) Exploring the use of aromatic compounds in crop growth and protection. Int J Plant Soil Sci 35(21):78–89
14. Şenkal BC (2020) The role of secondary metabolites obtained from medicinal and aromatic plants in our lives. ISPEC J Agric Sci 4(4):1071–1079
15. Shelepova OV, Baranova EN, Tkacheva EV, Evdokimenkova YB, Ivanovskii AA, Konovalova LN, Gulevich AA (2022) Aromatic plants metabolic engineering: a review. Agronomy 12(12):3131
16. Reichling J, Schnitzler P, Suschke U, Saller R (2009) Essential oils of aromatic plants with antibacterial, antifungal, antiviral, and cytotoxic properties–an overview. Res Complement Med 16(2):79–90
17. Franz C, Baser K, Windisch W (2010) Essential oils and aromatic plants in animal feeding–a European perspective. A review. Flavour Fragr J 25(5):327–340

18. Zeng Z, Zhang S, Wang H, Piao X (2015) Essential oil and aromatic plants as feed additives in non-ruminant nutrition: a review. J Animal Sci Biotechnol 6(1):7
19. Guillén MD, Cabo N, Burillo, Jesus (1996) Characterisation of the essential oils of some cultivated aromatic plants of industrial interest. J Sci Food Agric 70(3):359–363
20. Hulla JE, Sahu SC, Hayes AW (2015) Nanotechnology: history and future. Hum Exp Toxicol 34(12):1318–1321
21. McNeil SE (2005) Nanotechnology for the biologist. J Leukoc Biol 78(3):585–594
22. Husen A (2023) Secondary metabolites based green synthesis of nanomaterials and their applications. Springer, Singapore. https://doi.org/10.1007/978-981-99-0927-8
23. Husen A (2023) Nanomaterials from agricultural and horticultural products. Springer, Singapore. https://link.springer.com/book/9789819934348
24. Joy J, George N, Wilson R (2025) Chemical synthesis versus green synthesis. In: Nanomaterial green synthesis. Springer, Cham, pp 483–512
25. Parveen K, Banse V, Ledwani L (2016) Green synthesis of nanoparticles: their advantages and disadvantages. AIP Conf Proc 1724:020048
26. de Cristo Soares Alves A, Mainardes RM, Khalil NM (2016) Nanoencapsulation of gallic acid and evaluation of its cytotoxicity and antioxidant activity. Mater Sci Eng C 60:126–134. https://doi.org/10.1016/j.msec.2015.11.014
27. Badhani B, Sharma N, Kakkar R (2015) Gallic acid: A versatile antioxidant with promising therapeutic and industrial applications. RSC Adv 5(35):27540–27557
28. Zhao J, Khan IA, Fronczek FR (2011) Gallic acid. Struct Rep 67(2):316–317
29. Lakhan MN, Chen R, Shar AH, Chand K, Shah AH, Ahmed M, Wang J (2020) Eco-friendly green synthesis of clove buds extract functionalized silver nanoparticles and evaluation of antibacterial and antidiatom activity. J Microbiol Methods 173:105934. https://doi.org/10.1016/j.mimet.2020.105934
30. Parlinska-Wojtan M, Depciuch J, Fryc B, Kus-Liskiewicz M (2018) Green synthesis and antibacterial effects of aqueous colloidal solutions of silver nanoparticles using clove eugenol. Appl Organomet Chem 32(4):e4276
31. Fizer MM, Mariychuk RT, Fizer OI (2022) Gold nanoparticles green synthesis with clove oil: spectroscopic and theoretical study. Appl Nanosci 12(3):611–620. https://doi.org/10.1007/s13204-021-01726-6
32. Ch M, Naz S, Sharif A, Akram M, Saeed M (2015) Biological and pharmacological properties of the sweet basil (Ocimum basilicum). Br J Pharm Res 7(5):330–339
33. Juliani H, Simon J (2002) Antioxidant activity of basil. Trends New Crops New Uses 9:575
34. Pushpangadan P, George V (2012) Basil. In: Handbook of herbs and spices. Elsevier, Amsterdam, pp 55–72
35. Abbasifar A, Shahrabadi F, ValizadehKaji B (2020) Effects of green synthesized zinc and copper nano-fertilizers on the morphological and biochemical attributes of basil plant. J Plant Nutr 43(8):1104–1118
36. Barakat LAA, Rady S, El Dossoki FI, Hosny NM (2024) Green synthesis of basil loaded with ZnO NPs and assessing it's anti-cancer activities on hepatocellular (HEPG-2) and colorectal (HCT-116). Paper presented at the Proceedings of the First International Conference on Green Sciences, Cham
37. Gülçin İ (2006) Antioxidant activity of caffeic acid (3, 4-dihydroxycinnamic acid). Toxicology 217(2–3):213–220
38. Magnani C, Isaac VLB, Correa MA, Salgado HRN (2014) Caffeic acid: a review of its potential use in medications and cosmetics. Anal Methods 6(10):3203–3210
39. Thummaneni C, Surya Prakash DV, Golli R, Vangalapati M (2022) Green synthesis of silver nanoparticles and characterization of caffeic acid from Myristica fragrans (nutmeg) against antibacterial activity. Mater Today Proc 62:4001–4005. https://doi.org/10.1016/j.matpr.2022.04.586

40. Gabriela Á-M, Gabriela MDOV, Luis A-M, Reinaldo P-R, Michael H-M, Rodolfo G-P, Roberto V-BJ (2017) Biosynthesis of silver nanoparticles using mint leaf extract (Mentha piperita) and their antibacterial activity. Adv Sci Eng Med 9(11):914–923
41. Aziz WJ, Jassim HA (2018) Green chemistry for the preparation of silver nanoparticles using mint leaf leaves extracts and evaluation of their antimicrobial potential. World News Na Sci 18(2):163–170
42. Shanmuganathan R, Le QH, Devanesan S, Sayed SR, Rajeswari VD, Liu X, Jhanani G (2023) Mint leaves (Mentha arvensis) mediated CaO nanoparticles in dye degradation and their role in anti-inflammatory, anti-cancer properties. Environ Res 236:116718
43. Qanash H, Bazaid AS, Binsaleh NK, Alharbi B, Alshammari N, Qahl SH, Bagher AA (2023) Phytochemical characterization of Saudi mint and its mediating effect on the production of silver nanoparticles and its antimicrobial and antioxidant activities. Plants 12(11):2177
44. Kralova K, Jampilek J (2021) Responses of medicinal and aromatic plants to engineered nanoparticles. Appl Sci 11(4):1813
45. Singh S, Panwar N, Kumar SS, Singh R, Anand G, Kumar A, Malyan SK (2023) Aromatic oils from medicinal plants and their role in nanoparticles synthesis, characterization, and applications. In: Secondary metabolites from medicinal plants. CRC Press, Boca Raton, pp 305–314
46. Pushpanathan S, Yahya S, Gunasekaran A, Natarajan SR, Kannan K, Krishnan K (2025) Caffeic acid functionalized silver nanoparticles: a bionanoformulation and its assessment of cell cycle and in vitro cytotoxicity. Next Nanotechnol 7:100105. https://doi.org/10.1016/j.nxnano.2024.100105
47. Adil M, Alam S, Amin U, Ullah I, Muhammad M, Ullah M, Khan T (2023) Efficient green silver nanoparticles-antibiotic combinations against antibiotic-resistant bacteria. AMB Express 13(1):115
48. Abdellatif AA, Alhathloul SS, Aljohani AS, Maswadeh H, Abdallah EM, Hamid Musa K, El Hamd MA (2022) Green synthesis of silver nanoparticles incorporated aromatherapies utilized for their antioxidant and antimicrobial activities against some clinical bacterial isolates. Bioinorg Chem Appl 2022(1):2432758
49. Parlinska-Wojtan M, Kus-Liskiewicz M, Depciuch J, Sadik O (2016) Green synthesis and antibacterial effects of aqueous colloidal solutions of silver nanoparticles using camomile terpenoids as a combined reducing and capping agent. Bioprocess Biosyst Eng 39(8):1213–1223
50. Saadatmand M, Al-Awsi GRL, Alanazi AD, Sepahvand A, Shakibaie M, Shojaee S, Mahmoudvand H (2021) Green synthesis of zinc nanoparticles using Lavandula angustifolia Vera. Extract by microwave method and its prophylactic effects on Toxoplasma gondii infection. Saudi J Biol Sci 28(11):6454–6460
51. Kumari P, Luqman S, Meena A (2019) Application of the combinatorial approaches of medicinal and aromatic plants with nanotechnology and its impacts on healthcare. DARU J Pharm Sci 27(1):475–489. https://doi.org/10.1007/s40199-019-00271-6
52. Singh J, Kumar S, Alok A, Upadhyay SK, Rawat M, Tsang DCW, Kim K-H (2019) The potential of green synthesized zinc oxide nanoparticles as nutrient source for plant growth. J Clean Prod 214:1061–1070. https://doi.org/10.1016/j.jclepro.2019.01.018
53. Ahmad A, Javad S, Iqbal S, Shahzadi K, Gatasheh MK, Javed T (2024) Alleviation potential of green-synthesized selenium nanoparticles for cadmium stress in Solanum lycopersicum L: modulation of secondary metabolites and physiochemical attributes. Plant Cell Rep 43(4):113. https://doi.org/10.1007/s00299-024-03197-9

Harnessing the Potential of Lauraceae, Apiaceae, Myrtaceae, and Lamiaceae Families in Nanotechnology

Mine Koçyiğit

Abstract Active ingredients from the Lauraceae, Apiaceae, Myrtaceae, and Lamiaceae families demonstrate great potential in nanotechnology. These natural phytochemicals act as stabilizing and reducing agents, offering biocompatible and environmentally friendly alternatives to conventional chemical synthesis. The content and efficacy of essential oils are significantly influenced by ontogenetic stages and environmental factors. These factors also influence the potential for their use in nanotechnology. The increase in linalool and linalyl acetate concentrations in *Lavandula angustifolia* during flowering is an example of the evolution of the secondary metabolite profile with plant development. Essential oils from the Lauraceae family, rich in monoterpenes and sesquiterpenes, have been shown to be effective in nanotechnological production. Secondary metabolites obtained from *Ocotea indecora*, in particular, exhibit larvicidal activity against *Aedes aegypti* while simultaneously protecting the health of pollinators. *Coriandrum sativum*, a member of the Apiaceae family, stands out with its approaches used to produce biocidal nanoparticles, which have yielded positive results in the treatment of diabetes. However, concerns regarding the phytotoxicity of nanomaterials persist. Despite these positive results, there are still challenges to address, for example, the difficulty of achieving consistency in production and transitioning to industrial scale, the lack of standardization of plant extracts, and concerns about the toxicity of nanoparticles and their long-term environmental impact. The development of plant-based nanotechnologies requires the integration of biochemistry, toxicology, and engineering and the support of unified global regulatory frameworks to ensure safety and efficacy.

Keywords Aromatic plants · Essential oils · Nanomaterials · Phytochemicals · Sustainable alternatives

---

M. Koçyiğit (✉)
Faculty of Pharmacy, Department of Pharmaceutical Botany, Istanbul University, Istanbul, Turkey
e-mail: mkocyigit@istanbul.edu.tr

A. Husen (ed.), *Sustainable Uses of Aromatic Plants in Nanotechnology*, Smart Nanomaterials Technology, https://doi.org/10.1007/978-981-95-7409-4_2

1 Introduction

This section explains how the Lauraceae, Apiaceae (Umbelliferae), Myrtaceae, and Lamiaceae (Labiatae) families' rich phytochemical makeup can be used in nanotechnology. The development of biocompatible and ecologically friendly nanotechnology solutions is facilitated by the growing application of these plant families in fields including targeted medication delivery systems, green nanoparticle synthesis, nanoencapsulation, and antimicrobial nanoformulations. This framework assesses phytochemicals, the processes that produce nanoparticles, agricultural and biological uses, and prospects for the future.

2 Factors Influencing the Phytochemical Properties of Aromatic Plants

Natural substances with a variety of biological actions, essential oils are vital to the defense and attraction systems of plants. Essential oil production and accumulation are influenced by three primary causes (Fig. 1):

2.1 *Ontogenetic Changes*

The amount and chemical makeup of a plant's essential oils are greatly influenced by its growth stages. Changes in the quantity and makeup of key components can occur, especially during the pre- and post-flowering stages. Some plants attain their

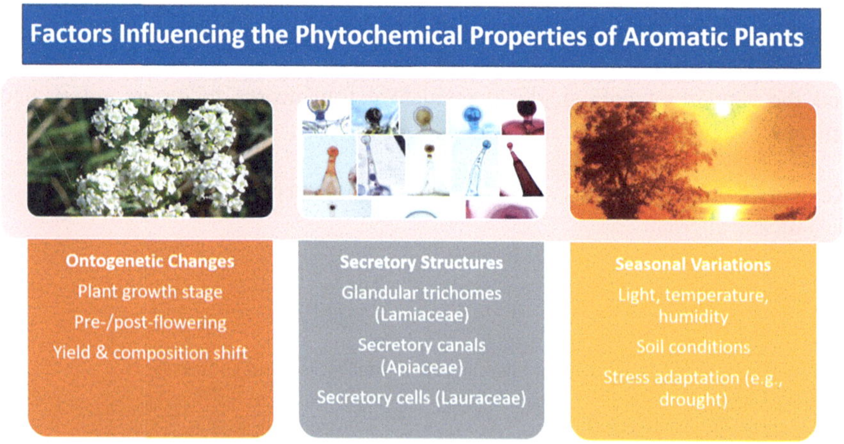

Fig. 1 Schematic summary of factors affecting phytochemical properties of aromatic plants

greatest essential oil concentration during the vegetative growth phase, whereas others reach their peak during flowering [1]. For instance, lavender (*Lavandula angustifolia* Mill.), a member of the Lamiaceae family, often produces the most essential oils and the best smell profile during the early or full flowering period [2]. Full blooming is when the essential oil content of rosemary (*Salvia rosmarinus* Spenn.) reaches its peak [3, 4]. On the other hand, essential oil production is highest in mint species (*Mentha* L. spp.) during the vegetative development phase before flowering. The maximum concentrations of menthol and other important components are found in leaves that are collected at this point [5, 6].

2.2 Secretory Structures

Essential oils are synthesized and stored in specialized secretory cells or structures. These include glandular trichomes in the Lamiaceae family [7], secretory ducts in the Apiaceae family [8, 9], and secretory cells in the Lauraceae family [10]. The morphology of these structures plays a decisive role in determining both the chemical profile and the overall yield of essential oils [1].

2.3 Seasonal Variations

Environmental factors, especially seasonal variations, can cause significant changes in both the composition and yield of essential oils. Factors such as light intensity, temperature, humidity, and soil conditions regulate plant metabolism by modulating the activity of enzymes and the expression of genes involved in essential oil biosynthesis. This phenomenon can also be interpreted as an adaptive strategy of plants to cope with stress conditions. For example, the essential oil of oregano *(Origanum vulgare* L.) contains major components such as thymol and carvacrol, the proportions of which vary considerably depending on altitude and climate. Plants grown in hot and sunny summers tend to accumulate higher levels of carvacrol than those grown in cooler and shaded environments [11]. This may be explained as a protective mechanism against high light intensity and heat stress. Similarly, the essential oil composition of mint *(Mentha* spp.) is highly sensitive to seasonal changes in temperature and humidity. Under hot and dry summer conditions, the proportion of menthol—the major component of mint essential oil—may decrease, reflecting a shift in metabolic pathways in response to water stress. Conversely, mint grown in cooler and wetter environments tends to contain higher levels of menthol [12, 13]. Rosemary *(Salvia rosmarinus*), a plant characteristic of the Mediterranean climate, also shows seasonal and climatic variability in its essential oil constituents. Under drought stress, rosemary plants often show increased levels of compounds with antioxidant activity, such as 1,8-cineole, reflecting a defense mechanism through increased antioxidant production [14, 15].

3 Lauraceae

3.1 Botanical Characteristics of the Lauraceae Family

About 20 *Cassytha* species are obligatory parasitic climbers, which is an exception to the Lauraceae family's rule of evergreen trees. The fruits are drupes with a hard endocarp around them. They are usually meaty and single-seeded. However, the fruit physically resembles a single-seeded berry because the endocarp is typically relatively thin in this family [16]. Some genera, most notably *Ocotea*, have fruit that resembles an acorn because it is partially or entirely encased in a thick, deep cupule that is formed from the calyx tube. At the base of the fruit of *Lindera* species, a structural development called a hypocarpium is seen. Typically, the flowers are protogynous, a characteristic of a complicated reproductive system that inhibits self-pollination. Birds use the fruits as a major source of food. Certain mite species use the domatia that some species have in the axils of their leaf veins as microhabitats. Ants in the genus *Pleurothyrium* have been known to engage in mutualistic interactions with plants, defending them, but some *Ocotea* species build colonies inside trees by nesting in leaf sacs or hollow stems [17]. Herbivore-deterring or herbivore-poisoning tissues and exudates are examples of additional defense systems. The Lauraceae family's medicinal significance is based on several compounds.

3.1.1 Phytochemical Characteristics of the Lauraceae Family

Around 3384 species and 55 recognized genera make up the Lauraceae family globally (https://wfoplantlist.org/taxon/wfo-7000000321-2025-06). Essential oils are abundant in these woody plants and are stored in their leaves, stems, and seeds. Because of their antibacterial, antioxidant, insect-repellent, preservative, and medicinal qualities, Lauraceae essential oils find extensive use in the food, cosmetic, biopesticide, and pharmaceutical industries [18].

The main chemical constituents are monoterpenes (like linalool, 1,8-cineole, camphor, and limonene) and sesquiterpenes (like caryophyllene, germacrene D, and nerolidol) and lactone derivatives and phenylpropanoids (like safrole, eugenol, and cinnamaldehyde) [19, 20]. While more basic groupings like *Beilschmiedia* Nees, *Cassytha* Mill., and *Cryptocarya* R.Br. show lower yields but more chemical variety, genera like *Cinnamomum* Schaeff, *Litsea* Lam., *Laurus* L., and *Lindera* Thunb. stand out for their large essential oil yields. These variations are chemotaxonomic indicators and are linked to the family's evolutionary processes [21].

In Lauraceae, specialized secretory cells—oil and mucilage cells—are responsible for the manufacture of essential oils. The Golgi apparatus, vacuoles, endoplasmic reticulum, and plastid all play a dynamic role in terpene production during the four stages of oil cell formation. The terpene synthase (TPS) gene family is vital to the generation of essential oils in the Lauraceae family at the genetic level. TPS gene families are notably more numerous in the Lauraceae family; 93 TPS genes have been found in *Camphora officinarum* Boerh. ex Fabr. This growth has accelerated chemical differentiation by increasing the diversity of monoterpenes and sesquiterpenes. While gene clusters (such the MYB44-ADH cluster involved in citral biosynthesis in *Litsea cubeba* (Lour.) Pers.) offer adaptive advantages under environmental constraints, gene duplications, segmental repeats, and tandem repeats contribute to TPS diversity. Additionally, transcription factors control the expression of TPS, which raises the synthesis of certain terpenes [21].

The composition of essential oils is also influenced by environmental factors. While high light intensity encourages the production of cineoles, low temperatures may raise camphor levels. The quality and output of essential oils are further impacted by altitude, soil nutrients, and extraction methods (such as steam distillation and supercritical CO_2 extraction) [22].

3.1.2 Nanotechnological Potential of the Lauraceae Family

The extensive use of synthetic pesticides has harmed nontarget creatures, such as people and animals, and caused mosquitoes to develop resistance. In order to solve this problem, a nanotechnology-based strategy has been investigated that uses *Ocotea indecora* (Schott) Mez essential oils to create a nanoemulsion that has larvicidal activity against the dengue vector mosquito, *Aedes aegypti*, while maintaining ecotoxicological safety for nontarget pollinators like honeybees (*Apis mellifera*). Sesquirosefuran (81.4%) was the primary ingredient identified by chemical characterization of *O. indecora* essential oil in this investigation [23]. Using a low-energy phase inversion technique, the nanoemulsion was created. An optimized surfactant mix produced a droplet size of 122.8 nm, a polydispersity index (PdI) of 0.262, and an HLB value of 14.22. The formulation was optimized using a 2^3 factorial design, resulting in spherical nanoemulsions with a PdI of 0.263 and an average size of 105.3 nm. Although samples kept at 40 °C showed signs of deterioration, long-term stability testing showed that the nanoemulsion stayed stable for up to a year at ambient temperature and under refrigeration.

An optimized surfactant blend produced an HLB value of 14.22, a droplet size of 122.8 nm, and a polydispersity index (PdI) of 0.262 for the nanoemulsion, which was made by a low-energy phase inversion technique. After optimizing the formulation using a 2^3 factorial design, spherical nanoemulsions with an average size of 105.3 nm and a PdI of 0.263 were produced. At room temperature and under refrigeration, the nanoemulsion remained stable for up to a year, according to long-term stability studies; however, samples kept at 40 °C showed signs of disintegration. These results support the application of biorational solutions that combine the effectiveness of vector management with pollinator protection by highlighting *O. indecora* essential oil-based nanoemulsion as an environmentally benign and efficient alternative for mosquito control.

In another study, the bioactivity and substance composition of essential oils from four plants in the Lauraceae family, namely, *Cinnamomum burmanni* (Nees and T.Nees) Blume, *Cinnamomum migao* (H.W.Li) Y.Yang, Bing Liu, and Zhi Yang,

Litsea pungens Hemsl., and *Laurus nobilis* L., were tested against pests linked to stored goods, such as the booklouse (*Liposcelis bostrychophila*) and the red flour beetle (*Tribolium castaneum*) [24]. The oils were extracted by hydrodistillation and analyzed by GC-MS. The bioassays employed three modes of action: fumigant, contact, and repellent. The median lethal concentrations (LC_{50}) and lethal doses (LD_{50}) were determined using probit analysis. All essential oils had a lot of monoterpenoids, although the primary constituents differed. For example, *C. migao* oil showed high levels of limonene (22.32%) and linalool (51.96%), but *L. nobilis* oil had notable levels of eucalyptol (54.38%) and α-terpinyl acetate (15.33%). Both bug species were toxically affected by all four oils, but *C. burmannii* and *L. nobilis* showed the strongest fumigant toxicity. All oils exhibited repellent properties as well. Of the various chemicals that were examined, borneol was effective against *L. bostrychophila*, whereas sabinene and α-terpineol have shown strong repellency against *T. castaneum*. These results also show that plant-based treatments are environmentally friendly substitutes for chemical pesticides, with great promise for controlling pests in grain storage facilities.

4 Apiaceae (Umbelliferae)

4.1 Botanical Characteristics of the Apiaceae Family

Although some members of the Apiaceae family have woody shrub or small tree forms, the majority of the family's plants are herbaceous and can be annual, biennial, or perennial. The plant's base is often where the majority of the leaves are found. Though opposite phyllotaxis may occur in the higher portions, they are often alternating and vary in size. Petioles are frequently sheathing and, in certain species, completely wrap the stem, resulting in a perfoliate morphology. Leaves can be sessile or petiolate. Although leaves in some genera, like *Bupleurum* L., are simple and whole, the leaf blade is usually split, ternate, or deeply pinnatifid. When crushed, the majority of species emit a pleasant or disagreeable smell [25]. Terminal umbels, which can be simple or compound and sometimes create umbelliform cymes, are nearly usually used to arrange flowers. Although the majority of the flowers are hermaphrodite and actinomorphic, some species, like coriander (*Coriandrum sativum* L.) and carrot (*Daucus carota* L.), have zygomorphic marginal flowers, meaning that the outside petals are larger than the inner ones. Some species have flowering systems that are andromonoecious, polygamomonoecious, or dioecious. Both the corolla and the calyx are present; however, the calyx is frequently blurry and somewhat diminished in most species. The corolla might be purple, pink, yellow, or white. Pentamerous flowers are distinguished by their five petals and five stamens [25].

It is common to see geitonogamy among the same individual's many flowers. An inferior ovary and a bicarpellate pistil make up the gynoecium. The stylopodium has

Fig. 2 Fruit types of some Apiaceae species most commonly used in research: (**a**) *Cuminum cyminum*, (**b**) *Coriandrum sativum*, (**c**) *Apium graveolens*, (**d**) *Pimpinella anisum*, (**e**) *Anethum graveolens*, and (**f**) *Ammi visnaga* (scale bar 1 mm)

two styles that draw pollinators by secreting nectar. The fruit is a schizocarp made up of two carpels that split into two mericarps with a single seed each when they reach maturity (Fig. 2). The distinctive scents of economically significant species like anise, cumin, and coriander are caused by the volatile oils found in seeds, which also contain an oily endosperm. At the species level, ripe fruits' shape and surface decoration are essential diagnostic characteristics [26].

4.1.1 Phytochemical Characteristics of the Apiaceae Family

With 451 recognized genera and 3940 species worldwide (https://wfoplantlist.org/taxon/wfo-7000000036-2025-06), the Apiaceae family is a rich source of chemicals with potential medical applications and antioxidants. In the food and cosmetic industries, the phenolic components of Apiaceae species which contribute to flavor, color, and sensory qualities offer sustainable substitutes [27–29]. For example, celery (*Apium graveolens*) has high levels of luteolin and apigenin, coriander (*Coriandrum sativum*) has high levels of quercetin, and dill (*Anethum graveolens*) also has high levels of quercetin [30]. These anti-inflammatory, antiaging, anti-atherosclerotic, antidiabetic, anticancer, and neuroprotective properties are among the many biological impacts of these antioxidant substances [31–34].

Antioxidant concentration and activity are greatly influenced by the type of plant, the organ employed (seeds, leaves, roots, etc.), and environmental or geographic conditions [35]. Strong toxicity against bacteria, fungus, and mammalian cells is demonstrated by polyacetylenes that have been isolated from Apiaceae species. Carrots, celery, and parsley contain C17 aliphatic polyacetylenes, which have been shown to have anticancer and anti-inflammatory properties. The primary components of essential oils are terpenoids, a significant class of secondary metabolites in the Apiaceae family that also have antibacterial and antioxidant qualities and contribute scent [36, 37]. The bioactive chemicals of Apiaceae are interesting candidates for natural biocidal applications in the management of weeds, fungi, and pests, given the detrimental effects that synthetic pesticides have on the environment and human health.

4.1.2 Nanotechnological Potential of the Apiaceae Family

The cytogenetic effects of two frequently used nanomaterials, copper oxide (CuO) and cadmium sulfide (CdS), were assessed in a study on the possible ecological hazards of nanomaterials employing coriander (*Coriandrum sativum*), a significant Apiaceae plant. The study found harmful effects on chromosome architecture, seed germination, and seedling growth that were dose-dependent. Significant decreases in seedling length and germination rate were seen with increasing nanoparticle concentrations. Both CuO and CdS nanoparticles caused a number of cytogenetic aberrations and decreased the mitotic index, which suggests that cell division was disrupted. Additionally, exposure to nanoparticles enhanced pollen sterility. These results demonstrate the cytotoxic and genotoxic hazards that CdS and CuO nanoparticles pose to *C. sativum*, indicating the necessity of additional research to evaluate the ecological consequences of nanoparticle spread in food chains [38]. Nanoparticles can cause genetic changes in plants, simulating the effects of traditional mutagens, according to another study on the same species [39].

Essential oils' stability and bioavailability have been improved by the use of nanotechnological techniques including nanoemulsions, which increase their effectiveness at lower concentrations. This highlights the potential of Apiaceae plants as eco-friendly and natural sources of biocidal chemicals for agricultural applications such as food preservation, weed control, and pest management. However, additional toxicological and field-based evaluations are necessary to guarantee their safe commercial use [40].

Another study used *Apium graveolens* extracts to biosynthesize gold nanoparticles (AG-AuNPs), which were then tested for their impact on glucose uptake in rat hemidiaphragm tissue that had been extracted. The nanoparticles even improved the vitality of fibroblast cells and showed noncytotoxic and biocompatible characteristics. Interestingly, rat muscle tissue exposed to AG-AuNP showed a substantial increase in glucose absorption when compared to controls, indicating an insulin-like action that facilitates cellular glucose transport. Mechanistic insight into the reported impact was provided by molecular docking research, which identified possible interactions between important nanoparticle ingredients and proteins including GLUT4 and PI3K, which are essential for glucose transport. These results demonstrate AG-AuNPs' potential as natural therapeutic agents for the treatment of diabetes [41].

The effects of conventional and nano-formulated NPK fertilizers on growth and seed oil content were examined in a comparative study involving *Cuminum cyminum* L., *Pimpinella anisum* L., and *Anethum graveolens*. Both fertilizer types improved plant growth and oil yield; according to the results, however, nano-NPK formulations outperformed traditional ones. The best results were obtained with mixed formulations, highlighting the potential of nanofertilizers in Apiaceae crop development [42].

Ammi visnaga L. is a prominent member of the pharmacologically useful Apiaceae family, producing furanocoumarins including visnagin and khellin that are used to treat asthma, kidney stones, and coronary heart disease. According to Khalil et al. [43] and Ahmed et al. [44], these substances also have antiviral, antifungal, antibacterial, antioxidant, and anticancer qualities. The elicitor effects of graphene oxide-coated SiO_2 nanoparticles (GO-SiO_2 NPs) and SiO_2 nanoparticles (NPs) at several doses (0, 5, 15, 25 mg/L) on *A. visnaga* callus cultures were examined in a recent study. Ferric reducing antioxidant power (FRAP), hydrogen peroxide (H_2O_2) content, total phenolic content (TPC), total flavonoid content (TFC), callus morphology, and antioxidant enzyme activities (catalase, superoxide dismutase, guaiacol peroxidase, ascorbate peroxidase, and polyphenol oxidase) were among the parameters evaluated. The findings demonstrated that, at 15 mg/L, GO-SiO_2 NPs considerably boosted phenolic accumulation (193.3 mg GAE/g FW) and catalase, guaiacol peroxidase, and ascorbate peroxidase activity. The maximum levels of H_2O_2 and FRAP activity were generated by SiO_2 NPs at 25 mg/L, although total flavonoids peaked at 15 mg/L (62.27 mg QE/g FW). Necrotic lesions and morphological damage resulted from the inhibition of callus formation caused by high concentrations (25 mg/L) of both nanoparticles. Positive correlations between phenolic/flavonoid contents and enzyme activity, especially APX and GPX, were found via correlation analysis. These findings suggest that whereas high doses of GO-SiO_2 NPs are phytotoxic, low-to-moderate dosages (15 mg/L) are efficient elicitors for increasing the production of secondary metabolites in *A. visnaga* callus. These results highlight the role of nanotechnology-based elicitors in the biotechnological synthesis of pharmacologically active compounds [45].

To sum up, the Apiaceae family is a valuable resource for both agriculture and health because of its abundance of phenolics, flavonoids, terpenoids, and polyacetylenes. These substances serve as natural insecticides and biocidal agents in addition to their medicinal and antioxidant properties. The bioavailability and efficacy of metabolites generated from Apiaceae have been further enhanced by recent developments in nanotechnology, which have also made it possible to employ nanoparticles for innovative medicinal techniques. However, a thorough assessment of the possible cytogenetic and ecological hazards of nanoparticles on plant growth and ecosystems is necessary. Therefore, if their safe and efficient usage is confirmed by thorough field and toxicological research, the natural phytochemicals of the Apiaceae family and their nanotechnological derivatives constitute a viable frontier for sustainable agriculture, food security, and contemporary medicine.

5 Myrtaceae

5.1 Botanical Characteristics of the Myrtaceae Family

This family's leaves are often whole and fragrant. Some species adapt to dry areas and strong sunshine by taking on an upright posture and needle-like shape. Most flowers are epigynous, actinomorphic, and hermaphrodite. Usually, four to five lobes and gamosepalous, the calyx can be shortened, caducous as a cap during anthesis (as in *Eucalyptus* L'Hér.), or absent altogether in some species. Four to five petals, which can be free or gamopetalous, typically make up the corolla; in certain eucalyptus species, the petals are joined to form an operculum. Many (polyandrous) stamens are grouped in whorls along the edge of the container. A monadelphous arrangement is seen in certain taxa. The stamens of *Melaleuca* L. are arranged in five antipetalous bundles, whereas those of *Melaleuca leucadendra* (L.) L. in numerous stamens are organized into five antipetalous fascicles. The carpels are two or more in number and syncarpous. The ovary is epigynous, or inferior. Inflorescences are usually cymose; however, variations such as dichasial cymes (*Syzygium*), paniculate cymes (*Eucalyptus*), racemose–cymose, and corymbose–cymose types may also occur. In some taxa, flowers are borne in the axils (e.g., *Psidium* L., *Myrtus communis* L.). Dense pendulous spikes are a defining feature of *Melaleuca*. Species-specific fruit types include drupe-like (*Eugenia*), capsule (*Eucalyptus, Melaleuca*), and berry (*Psidium*). Exalbuminous seeds exist [46]. Heteropyxidoideae, which has only four species, and Myrtoideae, which has 6543 species, make up the two subfamilies of the family, which has 6547 species in total (https://wfoplantlist.org/taxon/wfo-7000000404-2025-06).

5.1.1 Phytochemical Characteristics of the Myrtaceae Family

The phytochemicals, nutritional makeup, and bioactive characteristics of *Syzygium aromaticum* L. (clove) differ depending on the portion of the plant. According to Liu and Liu [47], fruits have the maximum amount of total carbohydrates, flower buds are the richest in minerals and crude oil, and leaves have the highest quantities of calcium and protein. Volatile components including eugenol, eugenyl acetate, β-caryophyllene, and α-humulene dominate the phytochemical profile. Flavonoids, chromones, tannins, and triterpenoids are examples of nonvolatile components. These substances are responsible for the plant's wide range of pharmacological actions.

Clove's potent antioxidant action, which is mediated by scavenging free radicals, increasing the activity of antioxidant enzymes, and preventing lipid peroxidation, is one of its most notable qualities. Notably, according to one study, clove extract had antioxidant activity that was ten times more than that of vitamin E [48]. Its effectiveness against germs and fungi is another significant action. The respiratory pathogens *Pseudomonas aeruginosa*, *Staphylococcus aureus*, *Klebsiella pneumoniae*, and *Streptococcus pneumoniae* are all well combatted by clove bioactive. Numerous mechanisms, such as bacterial cell wall destruction, enzymatic function disruption, and suppression of biofilm formation, contribute to these effects [49].

By reducing pain signals, eugenol has analgesic and anti-inflammatory properties, such as in osteoarthritis. By lowering oxidative stress, clove extract shields neuronal cells in neurodegenerative conditions like Parkinson's and Alzheimer's. Furthermore, clove and eugenol have antiviral qualities against the Ebola virus, influenza A virus, and herpes simplex virus (HSV) [50]. Additional reports point to possible insecticidal action and the capacity to quicken the body's removal of acetaldehyde, which would lessen the negative effects of alcohol [51].

5.2 Nanotechnological Potential of the Myrtaceae Family

The production of silver (AgNPs) and iron oxide nanoparticles (FeO_3NPs) using aqueous leaf extracts and their antifungal efficacy when applied to pulp made from oak wood and *Imperata cylindrica* were demonstrated in a study on *Syzygium cumini* (L.) Skeels. In the creation of nanoparticles, plant extracts served as stabilizing and reducing agents. AgNPs have shown greater antifungal activity than Fe_2O_3NPs against *Aspergillus fumigatus*, *Aspergillus niger*, and *Fusarium culmorum* at doses of 25, 50, and 75 μg/mL. AgNPs were suggested to interfere with fungal metabolism by entering the cell and preventing growth [52].

In a different work, AgNPs were green synthesized using fruit extract from *Syzygium alternifolium* (Wight) Walp., an endangered plant of the Eastern Ghats. A surface plasmon resonance peak at 442 nm was detected by UV-Vis spectroscopy to confirm the production of nanoparticles, while FTIR identified primary amines and phenolics as stabilizing and capping agents. AFM, SEM, and TEM microscopic investigations revealed spherical, polydisperse nanoparticles with a diameter of 5–68 nm. Strong antibacterial action against a variety of bacteria and fungi was demonstrated by these AgNPs [53].

When *Eucalyptus globulus* L. leaf extract and its ZnO nanoparticles (ZnO-NPs) were compared against the grain pest *Rhyzopertha dominica*, the nanoparticles demonstrated superior insecticidal effectiveness. After 15 days, at 600 ppm, ZnO-NPs caused 80.5% mortality with an LC_{50} of 202.11 ppm, while crude extract caused 1043.06 ppm. Additionally, ZnO-NPs and extract administration showed 87.0% and 75.7% suppression of larval development after 30 days, respectively, indicating the increased potency of nanoparticle formulations [54].

In contrast to chemical approaches, plant-based biosynthesis is safer, easier, and more environmentally friendly, according to a review of eucalyptus-mediated production of metallic nanoparticles (MNPs). Eucalyptus leaves, which are abundant in polyphenols, terpenoids, and flavonoids, effectively reduce and stabilize metal ions into nanoparticles. The most prevalent type of reported particles was silver nanoparticles, which had a spherical or cubic shape and ranged in size from 1.25 nm to 468.7 nm. These biosynthesized nanoparticles can be used in medical (as antibacterial, antioxidant, anticancer, and wound-healing agents), agriculture (as pesticides and fungicides), and water purification (as lead, chromium, and arsenic removal agents). There have also been indications of antiaging effects [55–58].

Additionally, the synthesis of AgNP, which stimulated the degradation of methylene blue (MB) dye, was accomplished with the help of eucalyptus leaf extracts. In comparison to their non-microwave-treated counterparts (39.27%), microwave-assisted hot extract AgNPs demonstrated the highest degrading efficiency (71.34% within 24 h). This underlines their potential as low-cost, fast, and eco-friendly catalysts for wastewater treatment [59].

In order to circumvent the low bioavailability of plant flavonoids with antiviral potential, including against SARS-CoV-2, nanotechnological solutions have also been investigated. Flavonoids work by targeting viral proteins including hACE-2 and Mpro and preventing viral entrance and replication. The absorption and targeted administration of polymeric micelles and nanoemulsions are greatly improved [60].

Furthermore, green-synthesized AgNPs from leaf extracts of *Lawsonia inermis* L. (henna) and *Psidium guajava* L. (guava) showed cytotoxic action against the colon cancer cell line HCT-116 and the breast cancer cell line MCF-7. Interestingly, compared to *P. guajava*, AgNPs produced from *L. inermis* had around double the cytotoxic potency. Flavonoids and phenolics' reducing and capping functions were emphasized as important factors influencing these biogenic nanoparticles' bioactivity [61].

Similarly, silver oxide nanoparticles were created using an aqueous leaf extract of *Melaleuca citrina* (Curtis) Dum. Cours, producing spherical to hexagonal particles with a size range of 3–30 nm. Promising pharmacological potential was indicated by these biogenic nanoparticles' strong antioxidant activity and dose-dependent lethal effects against *Artemia* larvae [62].

In conclusion, the Myrtaceae family is a valuable source for green nanotechnology because of its abundance of phytochemicals. The production of AgNPs, ZnONPs, and Fe_2O_3NPs, which exhibit a wide range of bioactivities, including antifungal, antibacterial, antiviral, insecticidal, and anticancer effects, is efficiently mediated by extracts of *Syzygium*, *Psidium*, *Eucalyptus*, and *Melaleuca* species. In addition to their use in biomedicine, these nanoparticles aid in sustainable pest control and environmental cleanup (such as dye degradation and heavy metal removal). The Myrtaceae are positioned as a vital botanical resource for the advancement of pharmacological, environmental, and industrial nanotechnology because biogenic nanoparticles provide a safe, economical, and environmentally friendly alternative to conventional chemical synthesis.

6 Lamiaceae (Labiatae)

6.1 *Botanical Characteristics of the Lamiaceae Family*

The Lamiaceae family is one of the largest plant families with 8091 species and 232 recognized genera worldwide (https://wfoplantlist.org/taxon/wfo-7000000318-2025-06). The opposite-decussate leaf arrangement, verticillaster inflorescence, bilabiate calyx and corolla, gynobasic pistil style, and fruit that splits into four nutlets set this family apart from others. The presence of capitate glandular trichomes with eight cells at the apex sets this family apart.

The number of stamens, the corolla's form and color, the stamens' length in relation to the corolla's upper lip, the indumentum (hairiness), the calyx length, the nutlet structure, and the plant's annual or perennial status are the factors used to distinguish genera within the Lamiaceae. Most species in the Lamiaceae family are hermaphrodites. Nevertheless, the male organs of about half of the species in the genera *Nepeta* L., *Ziziphora* L., and *Mentha* L. are sterile and reduced, making the blooms functionally female. The indumentum is primarily multicellular-uniseriate, simple, branching, trifid, or stellate, and it can be glandular or nonglandular, long and thick, or sparsely short. Short-stalked trichomes are nearly nonexistent, but occasionally trichomes have 13-celled apices with lengthy basal cells [63].

6.2 *Phytochemical Characteristics of the Lamiaceae Family*

Because of its rich and varied phytochemical makeup, members of the Lamiaceae family are widely used in traditional medicine, the food industry, and cosmetics. The following are the main classes of chemicals that make up this family's phytochemical profile (Fig. 3):

6.2.1 Essential Oils

One of the most characteristic features of the Lamiaceae family is the high production of essential oils (EOs). These are complex mixtures of compounds responsible for the distinctive aroma and biological activities of the plants. Essential oils play a crucial role in plant defense mechanisms and demonstrate various therapeutic effects on human health. Owing to their antimicrobial activity, they hold potential as natural food preservatives and agents in combating infections. The principal constituents include:

- Monoterpenes: Compounds such as limonene, pinene, camphene, myrcene, and terpinolene are abundant in the essential oils of many Lamiaceae species and exhibit antimicrobial and anti-inflammatory properties.

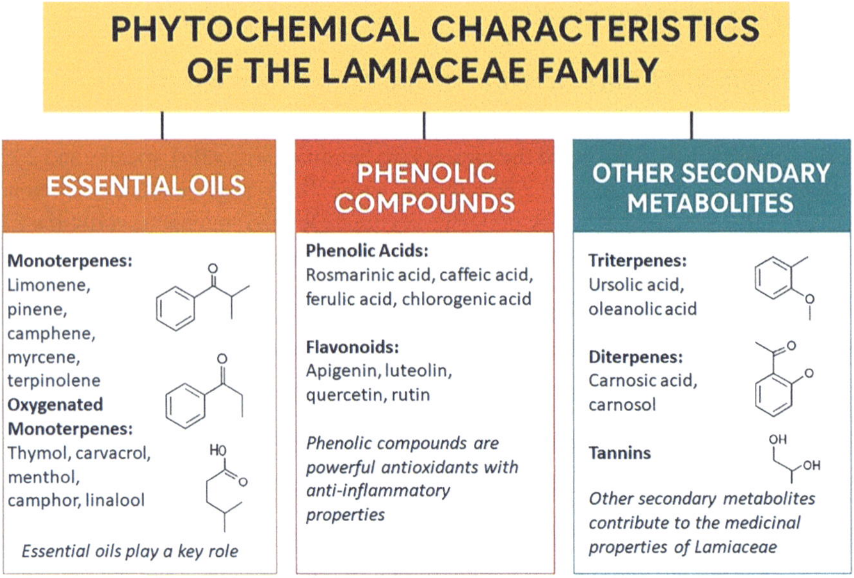

Fig. 3 Schematic summary of phytochemical characteristics of the Lamiaceae family

- Oxygenated monoterpenes: Thymol and carvacrol (in thyme and oregano), menthol (in mint), camphor (in rosemary), and linalool (in lavender) are well known for their strong antimicrobial, antioxidant, and analgesic effects.
- Sesquiterpenes: Compounds such as caryophyllene and humulene are also present in the essential oils of some Lamiaceae species and contribute to anti-inflammatory activity.

6.2.2 Phenolic Compounds

The Lamiaceae family is also particularly rich in phenolic compounds, which exhibit strong antioxidant and anti-inflammatory properties. These compounds can be divided into two main categories:

- Phenolic acids: Rosmarinic acid, caffeic acid, ferulic acid, and chlorogenic acid are found in high concentrations, especially in rosemary, sage, and thyme. Among these, rosmarinic acid is notable for its potent antioxidant and anti-inflammatory effects.
- Flavonoids: Apigenin, luteolin, quercetin, and rutin are common in many Lamiaceae species. These flavonoids neutralize free radicals, reduce oxidative stress, and contribute to the prevention of various chronic diseases.

Phenolic compounds are essential for plant protection against UV radiation, pathogens, and herbivores. In the human diet, they also serve as important sources of natural antioxidants.

6.2.3 Other Secondary Metabolites

In addition to essential oils and phenolic compounds, members of the Lamiaceae family contain other significant secondary metabolites:

- Triterpenes: Compounds such as ursolic acid and oleanolic acid exhibit anti-inflammatory, anticancer, and hepatoprotective activities.
- Diterpenes: Carnosic acid and carnosol, predominantly found in rosemary, are characterized by strong antioxidant and anticancer properties.
- Tannins: Certain Lamiaceae species contain tannins with astringent and antimicrobial activities.

This phytochemical diversity has contributed to the broad application of the Lamiaceae family in traditional medicine. Modern research highlights the great potential of these compounds in the development of new pharmaceutical, nutraceutical, and cosmetic products. Particularly when integrated with nanotechnology, these natural metabolites can offer enhanced bioavailability and targeted delivery to specific tissues, thereby enabling more effective therapeutic strategies [64].

6.3 Nanotechnological Potential of the Lamiaceae Family

A review examining the effect of foliar micronutrient application on the accumulation of secondary metabolites in Lamiaceae plants analyzed studies conducted between 2010 and 2023. The study highlighted the critical role of micronutrients in plant metabolism and secondary metabolite production. Key findings indicated that foliar application of relatively immobile micronutrients within the plant, such as Fe, Zn, Mn, Cu, B, and Si, significantly enhanced the biosynthesis of various secondary metabolites, including flavonoids, phenolic compounds, and terpenoids, in Lamiaceae species. Moreover, the foliar application of nano-micronutrients was reported to be more effective than traditional soil treatments. The study emphasized the need for further research on cerium, titanium, and boron and suggested that genera such as *Lavandula*, *Satureja*, *Origanum*, and *Hyssopus* warrant more detailed investigation in future studies [65].

An academic study explored the use of *Lavandula angustifolia* (lavender) extracts in the green synthesis of silver nanoparticles (AgNPs) and assessed the antioxidant and antimicrobial potential of the synthesized nanoparticles. Addressing the need for eco-friendly and sustainable nanoparticle synthesis methods, the study utilized plant-derived compounds for their reducing and stabilizing properties. AgNPs were synthesized using lavender extracts and characterized via UV-Vis spectroscopy, Fourier-transform infrared spectroscopy (FTIR), scanning electron microscopy (SEM), X-ray diffraction (XRD), and dynamic light scattering (DLS). The characterization results revealed the morphological, structural, and optical properties of the synthesized AgNPs. Their biomedical potential was assessed through antioxidant activity assays and antimicrobial evaluations against various microorganisms. The study demonstrated that AgNPs synthesized with lavender extracts possess significant antioxidant and antimicrobial properties, suggesting potential applications in pharmaceutical, cosmetic, and food industries. It highlighted that *Lavandula angustifolia* extracts provide an effective platform for green synthesis, producing biologically active nanoparticles suitable for biomedical applications and offering a safer, less toxic, and more sustainable alternative to conventional chemical methods [66].

Another study investigated the green synthesis of silver-doped zinc oxide nanoparticles (Ag-ZnO NPs) using *Lavandula stoechas* extracts and examined their potential environmental and biomedical applications. The plant extract acted as both a reducing and stabilizing agent. Synthesized nanoparticles were characterized in detail using UV-Vis spectroscopy, FTIR, SEM, XRD, and DLS, revealing their morphological, structural, optical, and crystallographic features. Environmental applications, such as the photocatalytic degradation of dyes, were explored, while biomedical potential was assessed via antibacterial, antifungal, and cytotoxic activities against cancer cell lines. The results demonstrated that Ag-ZnO NPs synthesized with *Lavandula stoechas* extracts hold promise for pollutant remediation, antimicrobial applications, and potential anticancer therapies, highlighting the extract as an effective green synthesis platform with versatile applications [67].

Additionally, studies have extensively examined the potential of Lamiaceae essential oils (EOs) in combination with nanotechnology for medical applications. Lamiaceae species are known for their rich secondary metabolite profiles and broad biological activity. Research has emphasized the antioxidant, anti-inflammatory, wound-healing, antiaging, antimelanogenic, and anticancer properties of Lamiaceae EOs, attributing these activities to terpenes, phenolic compounds, and other bioactive molecules. Integrating nanotechnology has the potential to enhance EO bioavailability, stability, and targeted delivery. Controlled release and efficacy improvement through nanoparticles, liposomes, and other nanocarrier systems have been highlighted, enabling more effective therapeutic strategies than traditional applications. Consequently, combining Lamiaceae EOs with nanotechnology demonstrates significant promise for developing novel therapeutic approaches and maximizing the pharmacological potential of natural compounds [68].

Another study investigated the effect of metallic nanoparticles (CuO, ZnO, Fe_2O_3) on in vitro cultures of two Mediterranean Lamiaceae species, *Lavandula viridis* L'Her. and *Thymus latifolius* G. López and R. Morales, focusing on phenolic profiles and biological activities. Different nanoparticle concentrations were added to culture media, and plant morphological development, biomass production, and physiological responses were evaluated. Results indicated that nanoparticles enhanced antioxidant enzyme activity and influenced secondary metabolite production, suggesting potential for strengthening plant defense mechanisms under stress. Changes in phenolic profiles and their impact on antioxidant and other biological activities were analyzed, demonstrating that metallic nanoparticles can modulate secondary metabolite synthesis and enhance the bioactivity of Lamiaceae species in vitro [69].

A study in Iran examined the effect of precipitation levels on the size of silver nanoparticles (AgNPs) synthesized using Lamiaceae plants. The research focused on the biosynthetic capacity of Lamiaceae species from different regions and how environmental factors, especially rainfall, influence nanoparticle properties. Statistical analyses, including t-tests, revealed that plants from low-rainfall areas exhibited higher phenolic content than those from high-rainfall regions. This suggests that stress adaptation mechanisms increase phenolic compound production, which plays a key role as reducing and stabilizing agents in AgNP synthesis. The study concluded that environmental conditions, particularly rainfall, significantly affect the efficiency of plant-based AgNP synthesis and the properties of the resulting nanoparticles [70].

Collectively, these studies indicate that species from the four families considered (Lauraceae, Apiaceae, Myrtaceae, and Lamiaceae) are valuable resources for green nanoparticle synthesis due to their rich content of secondary metabolites, particularly phenolics, terpenes, and essential oils. These findings can be considered strategic resources for nanotechnological innovations in sustainable environmental, biomedical, and industrial applications (Table 1).

Table 1 Prominent phytochemicals, nanotechnological applications, and plants by compiled families

Plant family	Key phytochemicals	Nanotechnological potential and applications	Examples
Lauraceae	Monoterpenes (linalool, 1,8-cineole, limonene), sesquiterpenes (caryophyllene), and phenylpropanoids (cinnamaldehyde, eugenol)	Nanoemulsions and natural pesticides for mosquito control	*Ocotea indecora*, *Laurus nobilis*, *Cinnamomum* species
Apiaceae	Phenolic compounds (luteolin, quercetin), terpenoids, and polyacetylenes	Potential as bionanomaterials for diabetes treatment and natural biocidal agents; some nanoparticles have shown phytotoxic effects	*Coriandrum sativum*, *Apium graveolens*, *Ammi visnaga*, *Cuminum cyminum*
Myrtaceae	Eugenol, eugenyl acetate, β-caryophyllene; also, flavonoids, chromones, tannins, and triterpenoids	Detailed information on nanotechnological potential is not provided in the document	*Syzygium aromaticum*, *Eucalyptus*, *Melaleuca*
Lamiaceae	Essential oils	Nanotechnological applications are not mentioned; the focus is primarily on factors affecting phytochemical properties	*Lavandula angustifolia*, *Salvia rosmarinus*, *Mentha* species, *Origanum vulgare*

7 Challenges

7.1 *Standardization of Plant Extracts and Production Continuity*

The chemical composition of plant extracts varies depending on numerous factors, including species, geographical region, climate, phenological stage, harvest time, processing, and drying conditions. This variability affects the reducing and stabilizing capacity during nanoparticle synthesis, leading to inconsistencies in morphology, size, stability, and biological activity of the final product [71]. Protocols developed at the laboratory scale are often difficult to replicate at an industrial scale. Ensuring sufficient quantities of plant material with consistent quality for large-scale production remains a significant challenge [72]. If additional components, such as essential oils, phenolic compounds, or terpenoids, are not standardized in terms of metabolite profile, the synthesis process may yield different results each time, posing problems for quality control and routine production [73].

7.2 *Nanoparticle Toxicity on Organisms and the Environment*

Due to their small size and high surface-to-volume ratio, nanoparticles have been reported to exert toxic effects on plants (e.g., germination, growth, physiological stress responses) and soil microorganisms. High concentrations, prolonged exposure, or inappropriate particle size/shape further increase these toxic risks [74]. Nanoparticles can disperse through water and air, potentially entering soil and aquatic ecosystems, where they may bioaccumulate and enter the food chain, posing long-term risks to human and animal health [75].

Phytochemicals present on plant-extract-derived nanoparticles may also influence toxicity; for example, stabilizing agents themselves may be toxic, or compounds bound to the nanoparticle surface may disrupt plant or environmental physiology. Therefore, toxicological evaluation should include both the metal ions and any capping or coating agents [76].

Nanoparticles (1–100 nm) with large surface areas, widely used across scientific disciplines, have been shown to cause ecosystem toxicity following nano-waste release [77, 78]. Nanotoxicity is a global concern. Due to cost-effectiveness and operational simplicity, investigating nanotoxicity using root plants as reservoirs for nano-waste is essential for environmental safety [38].

7.3 Compliance with International Regulations and Ethical Considerations

Present regulatory frameworks are primarily designed for chemical drugs, pesticides, or agricultural inputs. Plant-extract-based nanoparticles may fall into multiple categories—pharmaceutical, biopesticide, or cosmetic—creating ambiguity and prolonging approval processes [76]. Ethical responsibilities include labeling, product safety, consumer information, and sustainable sourcing for applications in food, cosmetics, or agriculture. Transparency, clear disclosure of contents, presentation of toxicity data, and sustainable resource use must be ensured. Regional regulatory differences (e.g., EU, USA, Asian countries) create challenges for international trade and market entry, as testing standards, safety limits, and approval procedures vary across regions [79].

7.4 Integration into Bioeconomy and Circular Agriculture Systems

Green nanoparticle synthesis aligns with circular economy principles through low energy and chemical input and the use of biological resources. However, practical implementation in agricultural systems faces gaps in supply chain, logistics, and waste management. The valorization of plant residues from extract processing, waste reuse, and recovery of by-products is not yet widespread. Within circular agriculture, plant-based biogenic nanoparticle production should be economically sustainable and environmentally low impact. Access to nanoparticles or nanomicronutrients for farmers depends on cost, technical knowledge, infrastructure, and support policies. Field-scale efficiency requires pilot studies, adaptation with local varieties, and stakeholder involvement [76].

8 Future Perspectives

- More systematic studies should be conducted on factors influencing the metabolite profiles of plant extracts, such as species variation, genetic differences, cultivation conditions, season, and harvest time, which could serve as a basis for standardization.

- Nanoparticle toxicity thresholds, especially under real agricultural conditions, should be established; long-term ecosystem effects, sequential exposures, and bioaccumulation must be evaluated, along with chronic effects on human and animal health.
- International regulations should include specific guidelines for plant-extract-derived nanomaterials, with clearly defined testing standards, labeling, quality control, and safety limits. Ethical principles should guide consumer information, biodiversity protection, and sustainable resource use.
- Nanoparticle synthesis processes should be integrated into a bioeconomy framework, emphasizing circularity: utilization of extract residues, wastewater and energy efficiency, and sustainable cultivation of plant resources.
- Field and pilot studies are needed to further characterize the potential of unique species within the Lauraceae, Apiaceae, Myrtaceae, and Lamiaceae families, particularly in terms of phytochemistry, resilience, and regional adaptation. This will accelerate the translation of these technologies into agricultural production systems.

9 Conclusion

Plants from the Lauraceae, Apiaceae, Myrtaceae, and Lamiaceae families offer significant potential for nanotechnology-based applications due to their rich phytochemical composition. Extracts from these families serve as valuable biological reducing and stabilizing agents for environmentally friendly and sustainable nanoparticle synthesis. However, the present literature highlights considerable challenges regarding the standardization of plant extracts, production continuity, and scalability to industrial levels. Comprehensive studies on nanoparticle toxicity profiles and long-term environmental effects are still required.

Gaps in international regulations, product safety, ethical responsibilities, and commercial integration remain. Nevertheless, integration into bioeconomy and circular agricultural systems can enhance the economic and ecological sustainability of these technologies. Future research should focus on better characterization of phytochemical compositions, assessment of nanoparticle toxicity across biological systems, and harmonization of regulatory frameworks.

In conclusion, these four plant families are strong candidates to play a strategic role in the development of green nanotechnology. Realizing this potential effectively will require multidisciplinary approaches, standardized production protocols, and globally harmonized regulatory mechanisms.

Competing Interests The authors declare that they have no competing interest.

Declaration of Generative AI and AI-Assisted Technologies in the Writing Process During the preparation of this work, the book chapter author(s) have not taken any assistance from AI, and the book chapter author(s) reviewed and edited the content as needed and take(s) full responsibility for the content of the publication.

References

1. Lee YL, Ding P (2016) Production of essential oil in plants: ontogeny, secretory structures, and seasonal variations. PJSRR 2(1):1–10
2. Crişan I, Ona A, Vârban D, Muntean L, Vârban R, Stoie A, Morea A (2023) Current trends for lavender (Lavandula angustifolia mill.) crops and products with emphasis on essential oil quality. Plants 12(2):357
3. Schmiderer C, Steinborn R, Novak J (2023) Monoterpene synthases of three closely related sage species (*Salvia officinalis, S. Fruticosa,* and *S. Pomifera*, Lamiaceae). Plant Physiol Biochem 196:318–327
4. Stafiniak M, Ślusarczyk S, Pencakowski B, Matkowski A, Rahimmalek M, Bielecka M (2021) Seasonal variations of rosmarinic acid and its glucoside and expression of genes related to their biosynthesis in two medicinal and aromatic species of *salvia* subg. Perovskia Biol 10(6):458
5. Çelik C, Tuğlu Ü, Telci İ, Karakurt Y, Özek T, Özek G (2025) Determination of gene expression of some essential oil components and some genes responsible for the synthesis of these components in *Mentha* species harvested in different phenological periods. Acta Physiol Plant 47(6):68
6. Yousefian S, Esmaeili F, Lohrasebi T (2023) A comprehensive review of the key characteristics of the genus *Mentha*, natural compounds, and biotechnological approaches for the production of secondary metabolites. Iran J Biotechnol 21(4):e3605
7. Aobuliaikemu N, Koçyiğit M (2020) Anatomical studies of *Dracocephalum moldavica* L., *Ocimum basilicum* L., and *Agastache rugosa* (Fisch. & CA Mey) Kuntze (Lamiaceae) used for cardiovascular diseases in traditional Uyghur medicine. EMU J Pharm Sci 3(1):16–28
8. Tuncay HO, Akalın E (2023) *Angelica remotiserrata* (Apiaceae), a new species from Turkey based on morphological and molecular data. Pl Biosystems 157(5):950–957
9. Tuncay HO, Yeşil Y (2019) Comparative fruit anatomy and morphology of four species known as cumin (Kimyon) in Turkey. KSÜ Tar Doga Derg 22(4):547–556
10. de Deus Bento KB, Canaveze Y, Machado SR (2024) Oil and mucilage idioblasts co-occur in the vegetative organs of *Ocotea pulchella* (Lauraceae): comparative development, ultrastructure and secretions. Protoplasma 261(5):877–895
11. Moradi M, Hassani A, Sefidkon F, Maroofi H (2021) Qualitative and quantitative changes in the essential oil of *Origanum vulgare* ssp. gracile as affected by different harvesting times. J Agric Sci Technol 23(1):179–186
12. Alvarenga JP, Braga AF, Pacheco FV, Carvalho AAD, Pinto JEBP, Bertolucci SKV (2021) Seasonal variation in essential oil content and chemical profile of mint in the southeast of Brazil. Cienc Rural 51:e20200979
13. Stefanakis MK, Papaioannou C, Lianopoulou V, Philotheou-Panou E, Giannakoula AE, Lazari DM (2022) Seasonal variation of aromatic plants under cultivation conditions. Plants 11(16):2083
14. Ben Arfa A, Gouja H, Hannachi H, Isoda H, Neffati M, Najjaa H (2022) Seasonal changes in rosemary species: a chemotaxonomic assessment of two varieties based on essential oil compounds, antioxidant, and antibacterial activities. PLoS One 17(8):e0273367
15. Rathore S, Mukhia S, Kapoor S, Bhatt V, Kumar R, Kumar R (2022) Seasonal variability in essential oil composition and biological activity of *Rosmarinus officinalis* L. accessions in the western Himalaya. Sci Rep 12(1):3305
16. Khodja YK, Bachir-Bey M, Belmouhoub M, Ladjouzi R, Dahmoune F, Khettal B (2023) The botanical study, phytochemical composition, and biological activities of *Laurus nobilis* L. leaves: a review. Int J Second Metab 10(2):269–296
17. de Moraes P, Cardoso D, Gregório B, Baitello J (2022) A remarkable new gynodioecious species of *Ocotea* (Lauraceae) from the Brazilian Atlantic rain forest and its phylogenetic placement in the *Ocotea* complex. Edinb J Bot 79:1–23
18. Almeida KCR, Silva BB, Alves CCF, Vieira TM, Crotti AEM, Souza JM, Martins CHG, Ribeiro AB, Squarisi IS, Tavares DC (2020) Biological properties and chemical composition

of essential oil from *Nectandra megapotamica* (Spreng.) Mez. Leaves (Lauraceae). Nat Prod Res 34:3149–3153
19. Damasceno CSB, Higaki NTF, Dias JDFG, Miguel MD, Miguel OG (2019) Chemical composition and biological activities of essential oils in the family Lauraceae: a systematic review of the literature. Planta Med 85(13):1054–1072
20. Du Y, Wang Y, Xue W, Fan Y, Jiang Z (2023) Research progress of essential oils from Lauraceae plants. Food Sci 44:346–359
21. Wu F, Chen Y, Gao M, Li W, Zhao Y, Wang Y (2025) An updated review on essential oils from Lauraceae plants: chemical composition and genetic characteristics of biosynthesis. Int J Mol Sci 26:5690
22. Wang F, Lin J, Shan S, Pan J, Wang M, Deng G (2020) Extraction of the essential oil from *Cinnamomum camphora* leaves by CO_2-expanded. Ethanol J Chin Cereals Oils Assoc 35:125–130
23. Machado FP, Folly D, Enriquez JJS, Mello CB, Esteves R, Araujo RS, Pedro FS, Toledo JG, Mantilla-Afanador MG, Santos EE, Oliveira ER-J, Rocha L (2023) Nanoemulsion of *Ocotea indecora* (Shott) Mez essential oil: Larvicidal effects against *Aedes aegypti*. Ind Crop Prod 192:116031
24. Wang Y, Zhang LT, Feng YX, Zhang D, Guo SS, Pang X, Zhu-Feng G, Chao X, Du SS (2019) Comparative evaluation of the chemical composition and bioactivities of essential oils from four spice plants (Lauraceae) against stored-product insects. Ind Crop Prod 140:111640
25. Kljuykov EV, Zakharova EA, Ostroumova TA, Tilney PM (2021) Most important carpological anatomical characters in the taxonomy of Apiaceae. Bot J Linn Soc 195(3):532–544
26. Birjees M, Ahmad M, Zafar M, Khan AS, Ullah I (2022) Palyno-anatomical characters and their systematic significance in the family Apiaceae from Chitral, eastern Hindu Kush. Pakistan Microsc Res Tech 85(3):980–995
27. Kalia P, Selvakumar R (2023) Advanced breeding approaches for quality enhancement in the Apiaceae family. Int J Seed Spices 13(1 and 2):1–14
28. Kozhanova K, Alimanova A, Datkhayev U, Serikbayeva E, Kayupova F, Zhumalina K, Zhakipbekov K (2025) Unlocking the potential: exploring pharmacological properties from the Apiaceae family. Pharmacia 72:1–14
29. Thiviya P, Gamage A, Piumali D, Merah O, Madhujith T (2021) Apiaceae as an important source of antioxidants and their applications. Cosmetics 8:111
30. Yashin A, Yashin Y, Xia X, Nemzer B (2017) Antioxidant activity of spices and their impact on human health: a review. Antioxidants 6:70
31. Bagheri SM, Hakimizadeh E, Allahtavakoli M (2024) A review on the antidiabetic potential of the genus ferula (Apiaceae). Curr Tradit Med 10(1):114–122
32. Salem MA, Khalil HM, Manaa EG, Bass AK, Osama N, Samaka RM, Magda T, Ibrahim Hamdan DI (2025) Antioxidant potential of selected Apiaceae plant extracts: a study focused on the chemical composition and neuroprotective effect of *Coriandrum sativum* L. extract against lead (Pb)-induced neurotoxicity in rats. Biol Trace Elem Res 203:6093
33. Shariatzadeh M, Karami A, Moghadam A, Lotfi M, Maggi F, Ebrahimie E (2022) The essential oil from *Oliveria decumbens* vent. (Apiaceae) as an inhibitor of breast cancer cell (MCF-7) growth. Pharmaceuticals 16(1):59
34. Taibi M, Elbouzidi A, Haddou M, Yahyaoui MI, Baraich A, Allali A, Khoulati A, Bellaouchi R, Asehraou A, Addi M, El Guerrouj B, Qamar MMU, Wondmie GGF, El-Sheikh MMA, Chaabane K (2024) Exploring dermatological applications of *Ammoides pusilla* essential oil: evaluation of anti-aging, dermatoprotective, and antimicrobial properties against dermatopathogenic microorganisms. Discov Life 54(1):19
35. Jayakodi Y, Thiviya P, Gamage A, Evon P, Madhujith T, Merah O (2024) Antioxidant activity of essential oils extracted from Apiaceae family plants. Agrochemicals 3(1):57–69
36. Umar AW, Hussain H, Ahmad N (2025) The known unknowns: an enigmatic pathway of C17-Polyacetylenic Oxylipins in carrot (*Daucus carota* L.). CIMB 47(6):471

37. Wang XJ, Luo Q, Li T, Meng PH, Pu YT, Liu JX, Zhang J, Liu H, Tan GF, Xiong AS (2022) Origin, evolution, breeding, and omics of Apiaceae: a family of vegetables and medicinal plants. Hortic Res 9:uhac076
38. Pramanik A, Datta AK, Das D, Kumbhakar DV, Ghosh B, Mandal A, Gupta S, Saha A, Sengupta S (2018) Assessment of nanotoxicity (cadmium sulphide and copper oxide) using cytogenetical parameters in *Coriandrum sativum* L. (Apiaceae) Cytol Genet 52(4):299–308
39. Pramanik A, Datta AK, Gupta S, Ghosh B (2018) Copper oxide nanoparticles induced a fertile desynaptic mutant line in *Coriandrum sativum* L. (Apiaceae). Cytologia 83(1):103–107
40. Thiviya P, Gunawardena N, Gamage A, Madhujith T, Merah O (2022) Apiaceae family as a valuable source of biocidal components and their potential uses in agriculture. Horticulturae 8(7):614
41. Panchamoorthy R, Mohan U, Muniyan A (2022) *Apium graveolens* reduced phytofabricated gold nanoparticles and their impacts on the glucose utilization pattern of the isolated rat hemidiaphragm. Heliyon 8(1):e08805
42. Jameel DA, Al-Tai AAS (2018) Effect of normal and nano NPK compound fertilizer on mineral and protein content of three species of Apiaceae plant. J Glob Pharma Technol 9(12):341–348
43. Khalil N, Bishr M, Desouky S, Salama O (2020) *Ammi visnaga* L., a potential medicinal plant: a review. Molecules 25(2):301
44. Ahmed SS, Fahim JR, Youssif KA, AboulMagd AM, Amin MN, Abdelmohsen UR, Hamed ANE (2022) Metabolomics of the secondary metabolites of *Ammi visnaga* L. roots (family Apiaceae) and evaluation of their biological potential. South Afr J Bot 149:860–869
45. Golkar P, Akbari R, Bazarganipour M, Javed R (2023) Biochemical and phytochemical responses of Ammi visnaga L. (Apiaceae) callus culture elicited by SiO_2 and graphene oxide-SiO_2 nanoparticles. Plant Physiol Biochem 200:107741
46. Wilson PG (2011) Myrtaceae. In the families and genera of vascular plants. In: Kubitzki K (ed) Sapindales, Cucurbitales, Myrtaceae, vol X. Springer-Verlag, Heidelberg, pp 212–271
47. Liu JG, Liu HP (2021) Total nutrient analysis of various parts of *Eugenia caryophyllata*. Biotic Res 43:357–362
48. Marangoni F, Agostoni C, Borghi C, Catapano AL, Cena H, Ghiselli A (2020) Dietary linoleic acid and human health: focus on cardiovascular and cardiometabolic effects. Atherosclerosis 292:90–98
49. Garcia MO, Allend SO, Cunha KFD, Hartwig DD (2023) Myrtaceae family: an update on plant-derived bioactive compounds against bacteria that affect the respiratory system. Rodriguésia 74:e00822022
50. Diyya ASM (2022) Thomas NV (2022) multiple micronutrient supplementation as a supportive therapy in the treatment of covid-19. Biomed Res Int 1:3323825
51. Xue Q, Xiang Z, Wang S, Cong Z, Gao P, Liu X (2022) Recent advances in nutritional composition, phytochemistry, bioactivity, and potential applications of *Syzygium aromaticum* L. (Myrtaceae) Front Nutr 9:1002147
52. Elshaer MA, Abd-Elraheem MA, Taha AS, Abo-Elgat WA, Abdel-Megeed A, Salem MZ (2024) Green synthesis of silver and ferric oxide nanoparticles using *Syzygium cumini* leaf extract and their antifungal activity when applied to oak wood and paper pulp from *Imperata cylindrica* grass biomass. Waste Biomass Valor 15(11):6191–6211
53. Yugandhar P, Savithramma N (2016) Biosynthesis, characterization, and antimicrobial studies of green synthesized silver nanoparticles from fruit extract of *Syzygium alternifolium* (Wt.) Walp., an endemic, endangered medicinal tree taxon. Appl Nanosci 6(2):223–233
54. Siddique MA, Hasan MU, Sagheer M, Sahi ST (2022) Comparative toxic effects of *Eucalyptus globulus* L. (Myrtales: Myrtaceae) and its green synthesized zinc oxide nanoparticles (ZnONPs) against *Rhyzopertha dominica* (F.) (Coleoptera: Bostrichidae). Int J Trop Insect Sci 42(2):1697–1706
55. Abdelkhalek A, Salem MZM, Kordy AM, Salem AZM, Behiry SI (2020) Antiviral, antifungal, and insecticidal activities of eucalyptus bark extract: HPLC analysis of polyphenolic compounds. Microb Pathog 147:104383

56. Abiri R, Atabaki N, Sanusi R, Malik S, Abiri R, Safa P, Ab Shukor NA, Abdul-Hamid H (2021) New insights into the biological properties of eucalyptus-derived essential oil: a promising green anti-cancer drug. Food Rev Int 38(sup1):598–633
57. Rana A, Yadav K, Jagadevan S (2020) A comprehensive review on green synthesis of nature-inspired metal nanoparticles: mechanism, application, and toxicity. J Clean Prod 272:122880
58. Sawalha H, Abiri R, Sanusi R, Shaharuddin NA, Noor AAM, Ab Shukor NA, Abdul-Hamid H, Ahmad SA (2021) Toward a better understanding of metal nanoparticles: a novel strategy from eucalyptus plants. Plants 10:929
59. Rabeea MA, Owaid MN, Muslim RF (2021) Synthesis and characterization of silver nanoparticles by natural organic compounds extracted from eucalyptus leaves and their role in the catalytic degradation of methylene blue dye. Songklanakarin J Sci Technol 43(1):14–23
60. Jannat K, Paul AK, Bondhon TA, Hasan A, Nawaz M, Jahan R, Rahmatullah M (2021) Nanotechnology applications of flavonoids for viral diseases. Pharmaceutics 13(11):1895
61. Bar FMA, Habib MMA, El-Senduny FF, Badria FA (2021) Cytotoxic activity of silver nanoparticles prepared from *Psidium guajava* L. (Myrtaceae) and *Lawsonia inermis* L. (Lythraceae) extracts. Trop J Pharm Res 20(9):1791–1799
62. Ravichandran S, Paluri V, Kumar G, Loganathan K, Kokati Venkata BR (2016) A novel approach for the biosynthesis of silver oxide nanoparticles using aqueous leaf extract of *Callistemon lanceolatus* (Myrtaceae) and their therapeutic potential. J Exp Nanosci 11(6):445–458
63. Rattray RD, Van Wyk BE (2021) The botanical, chemical and ethnobotanical diversity of southern African Lamiaceae. Molecules 26(12):3712
64. Napoli E, Siracusa L, Ruberto G (2020) New tricks for old guys: recent developments in the chemistry, biochemistry, applications, and exploitation of selected species from the Lamiaceae family. Chem Biodivers 17(3):e1900677
65. Shahhat IM, Elsheikh SYS (2024) Accumulation of secondary metabolites in the family Lamiaceae as influenced by foliar micronutrients. Bol Latinoam Caribe Plantas Med Aromát 23(6):823–838
66. Şuică-Bunghez IR, Senin RM, Sorescu AA, Ganciarov M, Răut I, Firincă C, Fierăscu RC (2024) Application of *Lavandula angustifolia* mill. Extracts for the phytosynthesis of silver nanoparticles: characterization and biomedical potential. Plants 13(3):333
67. Kiani Z, Mirjalili S, Heydaryan K, Mohammadparast P, Aramjoo H, Bahraini F, Mortazavi-Derazkola S (2024) Harmonizing nature and nanotechnology: Phytoextract-mediated synthesis of ag-doped ZnO nanoparticles using *Lavandula stoechas* extract for environmental and biomedical applications. J Drug Deliv Sci Technol 96:105708
68. Kowalczyk T, Merecz-Sadowska A, Ghorbanpour M, Szemraj J, Piekarski J, Bijak M, Sitarek P (2023) Enhanced natural strength: Lamiaceae essential oils and nanotechnology in in vitro and in vivo medical research. Int J Mol Sci 24(20):15279
69. Gonçalves S, Mansinhos I, Rodríguez-Solana R, Pereira-Caro G, Moreno-Rojas JM, Romano A (2021) Impact of metallic nanoparticles on in vitro culture, phenolic profile and biological activity of two Mediterranean Lamiaceae species: *Lavandula viridis* L'Hér and *thymus lotocephalus* G. López and R Morales Molecules 26(21):6427
70. Saberiyan Sani MN, Sayadi Anari MH (2018) Study on the effect of rainfall level on the size of silver nanoparticles synthesized by plants of the Lamiaceae Family in different regions of Iran. Int J New Chem 5(3):103–112
71. Altammar KA (2023) A review on nanoparticles: characteristics, synthesis, applications, and challenges. Front Microbiol 14:1155622
72. Abdi G, Tendulkar R, Thatte C, Mishra S, Desai B, Surve S, Tarighat MA (2024) Scaling up nature's chemistry: a guide to industrial production of valuable metabolites. In: Advances in metabolomics. Springer, Singapore, pp 331–375
73. Antunes Filho S, Dos Santos MS, Dos Santos OAL, Backx BP, Soran ML, Opriş O, Bououdina M (2023) Biosynthesis of nanoparticles using plant extracts and essential oils. Molecules 28(7):3060

74. Wohlmuth J, Tekielska D, Čechová J, Baránek M (2022) Interaction of the nanoparticles and plants in selective growth stages—usual effects and resulting impact on usage perspectives. Plants 11(18):2405
75. Djanaguiraman M, Anbazhagan V, Dhankher OP, Prasad PV (2024) Uptake, translocation, toxicity, and impact of nanoparticles on plant physiological processes. Plants 13(22):3137
76. Karnwal A, Jassim AY, Mohammed AA, Sharma V, Al-Tawaha ARMS, Sivanesan I (2024) Nanotechnology for healthcare: plant-derived nanoparticles in disease treatment and regenerative medicine. Pharmaceuticals 17(12):1711
77. Masarovičová E, Král'ová K (2013) Metal nanoparticles and plants. Ecol Chem Eng Soc 20(1):9–22
78. Rico CM, Majumdar S, Duarte-Gardea M, Peralta-Videa JR, Gardea-Torresdey JL (2011) Interaction of nanoparticles with edible plants and their possible implications in the food chain. J Agric Food Chem 59(8):3485–3498
79. Lakhani KG, Hamid R, Motamedi E, Marviya GV (2025) A review on plant metabolite-mediated nanoparticle synthesis: sustainable applications in horticultural crops. Front Nanotechnol 7:1545413

Harnessing the Potential of Aromatic Oils in Nanotechnology

Shoaeb Mohammad Syed, Priyanka Thore, and Muhammad Afzal

Abstract Essential oils, and in particular aromatic oils, have attracted an ever-growing interest in their pharmacological properties, such as antimicrobial, antioxidant, anti-inflammatory, and anticancer effects. Their translation into clinical use though is commonly hampered by the nature of their inherent pitfalls of a high volatility, lack of aqueous solubility, instability, and degradation under environmental conditions. Recent breakthroughs in nanotechnology, in particular the nanostructured lipid carriers (NLCs), solid lipid nanoparticles (SLNs), polymeric nanoparticles, nanoemulsions, and liposomes, have led to valuable solutions on how to circumvent these disadvantages. The nanocarrier systems are also able to shield the aromatic oils against oxidative deteriorations and volatilization and, furthermore, improve their solubility, stability, bioavailability, and targeted delivery. There has been a rising number of studies since 2020 and continuing to 2025, which have shown that oils like lavender, rosemary, cinnamon, clove, and eucalyptus are successfully encapsulated into nanocarriers and then used in applications focusing on dermatology, wound healing, food preservation, and cancer treatment. Although this indicates encouraging results, issues still exist in scaled-up production, sustained shelf life, regulatory acceptance, and safety certification. This review provides an overview of recent advancements in the nanoformulation of aromatic oils, critically examines formulation strategies and therapeutic uses and limitations, and lays out how these natural products can be taken to the commercial and clinical through research processes. To capitalize on the benefits of aromatic oils, it is possible to utilize nanotechnology to optimize the capabilities of aromatic oils since it

S. M. Syed (✉)
Department of Pharmaceutics, Dayanand Education Society's, Dayanand College of Pharmacy Latur, Latur, Maharashtra, India

P. Thore
MET's Institute of Pharmacy Nashik, Nashik, Maharashtra, India

M. Afzal
Department of Pharmaceutical Sciences, Pharmacy Program, Batterjee Medical College, Jeddah, Saudi Arabia

A. Husen (ed.), *Sustainable Uses of Aromatic Plants in Nanotechnology*, Smart Nanomaterials Technology, https://doi.org/10.1007/978-981-95-7409-4_3

is a more safe, effective, and sustainable development in therapeutics and functional products.

Keywords Aromatic oils · Essential oils · Nanotechnology · Nanocarriers

1 Introduction

1.1 Overview of Aromatic (Essential) Oils

Aromatic oils, also referred to as essential oils, are at the heart of natural medicine, and their use dates back to thousands of years in different cultures such as those of Egypt, India, and China. They are highly concentrated, volatile liquids derived out of various parts of plants, e.g., flowers, leaves, barks, and roots. Essential simply means the essence or soul of the plant, which reproduces the characteristic fragrance and therapeutic qualities of the plant [1].

These oils have high therapeutic values as are directly related to their complex composition. They are not pure compounds but complex mixtures of tens of chemical components. Two broad categories of these compounds could be singled out: terpenoids (such as monoterpene and sesquiterpene, e.g., limonene and pinene) and phenylpropanoids (e.g., eugenol and carvacrol). The combination of these components individual to each oil determines its profile and a large spectrum of biological effects. These actions involve potent antimicrobial effects including resistance to bacteria, fungi, and viruses, strong antioxidant properties to guard cells against destruction, and potent anti-inflammatory effects. As a result of such versatility, essential oils are applied in many different ways, in everything as diverse as perfume and food additives to medicines [2].

1.2 Potential Challenges in Conventional Use

Although they have great advantages, the traditional application of essentials oils is shrouded with difficulties that restrain their potentialities. These constraints are caused directly by this physicochemical character [2].

High Volatility and Degradation: Essential oils have a volatile nature and are unstable, as the effective constituents are easily lost to evaporation when exposed to air, light, or heat. This fast destruction of active ingredients lowers the efficacy of these ingredients as time goes by and decreases their shelf life. Moreover, most of these substances are affected by oxidation which is a phenomenon that may cause their degradation and the occurrence of ineffective or even detrimental byproducts [2].

Low Bioavailability and Poor Water Solubility: The lipophilic essential oils have poor water solubility. This presents a serious issue to formulation, in that it renders them incompatible with a broad volume of water-based products such as gels, drinks, and aqueous solutions. This low solubility also equates to low bioavailability, where, when in a dietary context, much of the oil is not incorporated by the body due to low absorption and therefore results in diminished effects in the body.

Safety and Formulation Instability: Formulation, safe even at very high concentration of bioactive compounds in essential oils, may lead to skin irritation, allergies, or other undesirable effects in direct application. This requires dilution, but stabilization and homogeneous formulation is not easily achieved. The oil does not mix well with the water phase resulting in an inconsistent product with an inconsistent dosing approach [2].

1.3 *Introduction to Nanotechnology and Its Advantages for Aromatic Oils*

Nanotechnology offers a breakthrough to some of these challenges through the application of materials science at the nanoscale (1–100 nm). Essential oils can be encapsulated and have their properties radically improved by utilizing the properties of nanocarriers, i.e., nanoemulsions, liposomes, and polymeric nanoparticles. These nanocarriers serve as protective vesicles that protect the oils against the external surroundings. Figure 1 provides an insight to incorporation of essential oil into nanoformulation with enhanced functioning.

The most important benefits of nanotechnology essential oil delivery are:

Enhanced stability: The nanocarrier shell acts as a protective shell that prevents the volatile components from evaporating, thus preventing the degradation of the

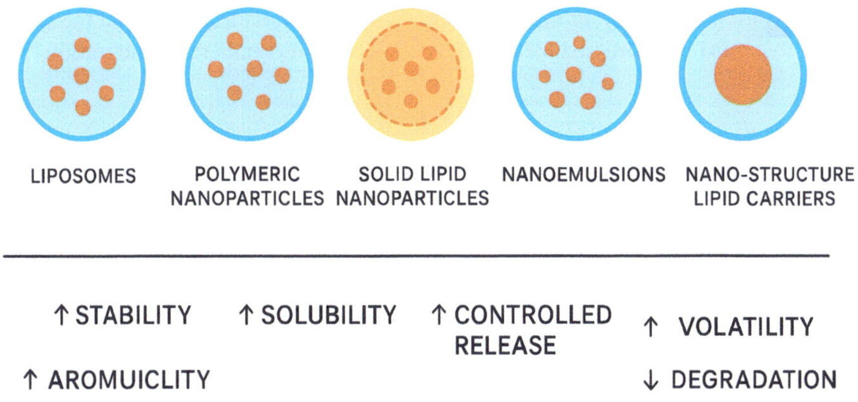

Fig. 1 Nanotechnology aromatic oil formulations

components by light and oxygen. This greatly increases the shelf life of the essential oil and keeps its medicinal strength intact over time.

Improved water solubility: Nanocarriers are able to disperse the hydrophobic essential oils to stable aqueous medium, which results in translucent or transparent nanoemulsions. This enables an improved flexibility in their incorporation into *water-based* products and enhances their facility in a great deal.

The efficacy and increase in bioavailability: Nanoparticles are 100 times smaller than conventional formulations and therefore can penetrate biological barriers such as the skin and cell membranes more readily. That results in increased rates of absorption, target drug distribution, and, finally, more effective treatment. The nanocarriers may also be tailored to have controlled release which results in a *long-term* delivery of the oil [3].

1.4 Scope of the Study

This critical review would attempt to give an in-depth analysis of the new studies concerning the use of nanotechnology in aromatic oils. The goal is to summarize the state of the available knowledge on the development and use of nanocarrier systems to address the impediments of essential oils inherently present based on rationales. This review has a complex scope in which the following statements are provided [2]:

A detailed description of the various types of nanocarrier systems (e.g., nanoemulsions, liposomes, SLNs, NLCs) and their specific advantages for essential oil delivery.

An analysis of the key formulation methods used to create these nano-based systems, such as high-pressure homogenization and ultrasonication.

A critical discussion of their diverse therapeutic and commercial applications in areas like dermocosmetics, food preservation, and medicine.

A thorough evaluation of the critical issues surrounding safety, nanotoxicity, and the existing regulatory frameworks that govern these products.

An exploration of current market trends, future research directions, and the challenges that must be addressed for successful clinical and commercial integration [3].

2 Aromatic Oils: Composition and Biological Activities

2.1 Major Constituents

The primary components of aromatic oils fall into two main groups: terpenes and aromatic/aliphatic compounds. Terpenes, which include monoterpenes and sesquiterpenes, are the most abundant and diverse group. Phenolics and aldehydes are key examples of aromatic compounds. The unique biological properties of an oil are often determined by the dominant compounds present. For example, some common components include [4]:

Terpenes: Found in citrus and pine oils, these include d-limonene, α-pinene, and γ-terpinene. They contribute to the oil's scent and often have antimicrobial and anti-inflammatory properties.

Phenolics: Compounds like thymol (in thyme oil) and carvacrol (in oregano oil) are known for their potent antimicrobial and antioxidant effects.

Aldehydes: Cinnamaldehyde (in cinnamon oil) and citral (in lemongrass oil) are powerful antimicrobial agents [3, 4].

2.2 *Pharmacological Properties*

The bioactive compounds in aromatic oils give them a broad range of pharmacological effects:

Antimicrobial: Aromatic oils are effective against various bacteria, fungi, and viruses. Components like thymol and carvacrol damage the cell membranes of microorganisms, leading to leakage of cellular contents and cell death.

Anti-inflammatory: Many oils, such as those from chamomile and thyme, can inhibit pro-inflammatory enzymes like cyclooxygenase and lipoxygenase, reducing inflammation.

Antioxidant: Phenolic compounds in aromatic oils can neutralize free radicals, which helps to prevent oxidative damage to cells and tissues.

Anticancer: Some studies suggest that certain aromatic oil components can induce apoptosis (programmed cell death) and inhibit the proliferation of cancer cells [4].

2.3 *Applications*

The high properties of aromatic oils have earned their usefulness in a variety of industries:

Food: They are used as natural flavoring agent and as preservatives. They have antimicrobial and antioxidant effects that can increase the food products' shelf life by preventing the growth of the microbes and food spoilage [5].

Cosmetics: Aromatic oils are commonly added in skin care, hair care, and perfumes. They are also appreciated because of the scent but also because of the medicinal effects, like calming one down and being anti-inflammatory or antibacterial [5].

Medicine: Aromatic oils play an important role in aromatherapy to help with many conditions such as stress, pain, sleep problems, etc. They are also under examination as possible alternatives to conventional antibiotics because of the increasing antimicrobial resistance [6].

3 Limitations of Aromatic Oils in Conventional Delivery

Aromatic oils are known to have impressive therapeutic uses, but there are serious limitations to aromatic oils when they are administered using normal ways. Such limitations are mainly due to their intrinsic chemical and physical properties that govern their stability, bioavailability, and their applicability to be used in different formulations [7].

3.1 Volatility and Degradation

One major drawback of the aromatic oils is they are highly volatile. These oils contain very active compounds with low boiling points, meaning they will evaporate easily when the air, heat, or light touches them. This loss of active ingredients causes a drop in therapeutic efficacy and low shelf life. Moreover, they are strongly unstable to oxidize, which goes faster in sunlight and in the presence of oxygen. This degradation has the potential to produce less viable or even potentially harmful byproducts, which compromise safety as well as potency of the oil [8].

3.2 Limited Bioavailability

Bioavailability is the ability of a substance (an active substance used to treat the symptoms of a disease) to be absorbed by the body and achieve its target. Aromatic oils also do not usually dissolve efficiently in water, which means that the body has a hard time absorbing them, particularly when given enterally. Most of this oil can be excreted without being absorbed, given this low bioavailability, and so the therapeutic effect may be reduced. The hydrophobic characteristics of aromatic oils also render them difficult to incorporate in *water-based* formulations, which is the case in many other drug and cosmetic products [9].

3.3 Irritation and Instability in Formulations

It has been reported that concentrated aromatic oils directly applied to the skin may be irritating and/or cause allergic reactions in some people. This poses a great safety issue that restricts most of their topical applications and requires dilution. Also, they have low water solubility, and consequently it is challenging to develop stable formulations such as lotions, creams, and gels. In formal study, the oil may separate out of the water phase in an unstable formulation resulting in a variable concentration of the active compounds. This does not only influence the quality and the shelf life of the product but also creates an obstacle in producing a predictable dose of therapy [9].

4 Role of Nanotechnology in Enhancing Aromatic Oil Utility

The technological application of nanotechnology in the delivery of aromatic (essential) oils has progressed beyond theoretical concepts and is now an expanding area of scientific research and development. Recent evidence just points out once again that nanocarriers are revolutionizing the way essential oils are applied [1]. Recent research has highlighted on optimizing the design of nanocarriers to enhance the delivery of aromatic oils.

4.1 Nanoemulsions and Nanocapsules

Studies show that nanoemulsions and other nanocarriers with extremely small droplet sizes (e.g., 10–20 nm) exhibit superior stability and bioactivity. This is because a smaller size increases the surface area, which enhances the interaction of the active components with the target system. For example, specific nanoemulsions of peppermint oil have shown exceptional larvicidal activity against mosquitoes, highlighting their effectiveness in pest management. Similarly, a study on thyme oil nanoemulsions reported robust antimicrobial properties, with inhibition zones against pathogens up to 36 mm [10].

Encapsulation Efficiency: The development of advanced encapsulation techniques has led to high encapsulation efficiency, ensuring that a greater percentage of the essential oil is protected from degradation and delivered effectively. For instance, co-encapsulating certain essential oils with other compounds has achieved encapsulation efficiencies of over 80%, significantly enhancing their anti-inflammatory and analgesic effects in medical applications [10].

Smart and Triggered Release: Recent research highlights the mechanism that could be applied, where the nanocarrier is made intelligent and the release of payload can be stimulated by a particular environmental parameter that may be pH or temperature. It enables very localized and specific delivery, and this has potential benefit in applications such as cancer treatment or targeted antimicrobial therapy [11].

Dermal and Transdermal: Nanotechnology can enhance the dermal delivery of medication in a number of *ways. The* small size of nanocarriers, such as solid lipid nanoparticles (SLNs) and nanoemulsions, can help nanotechnology bypass the skin barrier more than traditional forms. It has been recently demonstrated that the oil-loaded SLNs can considerably enhance the oil deposition in the skin. The increased penetration leads to more effective treatment of dermatological conditions and transdermal drug delivery [12].

Dermocosmetic Applications: Nanotechnology and essential oils are synergetic and hence another big trend in dermocosmetics. Nanocarriers are utilized to transport the key oil ingredients deep into the skin, tailoring therapeutic options to the antiaging, tissue rebuilding, and other skin care uses. As an illustration, peppermint oil nanocarriers and rosemary oil nanocarriers have increased efficacy in promoting hair growth because of their unique capability of better targeting the skin [13, 14].

4.2 Targeted Delivery and Bioefficacy

Nanotechnology presents the possibility of advancing the level of bioefficacy of aromatic oils by concentrating them to the desired point of action. Figure 2 demonstrates application of aromatic oils in nanocarrier system.

Antimicrobial Resistance: In view of the worldwide emergency of antimicrobial resistance, there is an ongoing study on the potential of essential oil-loaded nanocarriers as a very potent tool. The encapsulation of the essential oils means that the system can attack the microbial cells and disrupt its membranes to provide a new form of fighting with drug-resistant bacteria [14].

Anticancer Therapies: Recent studies are measuring the efficacy of nanocarriers loaded with essential oils in the treatment of cancer. Such systems may be customized to target tumor cells maximizing cellular uptake and delivering an effective dose of the anticancer compounds present in the essential oil with minimal side effects and systemic toxicity. Review articles in this field suggest the promise of different types of nanocarriers—polymeric nanoparticle, nanoemulsion, and liposome—to transform cancer treatment [15].

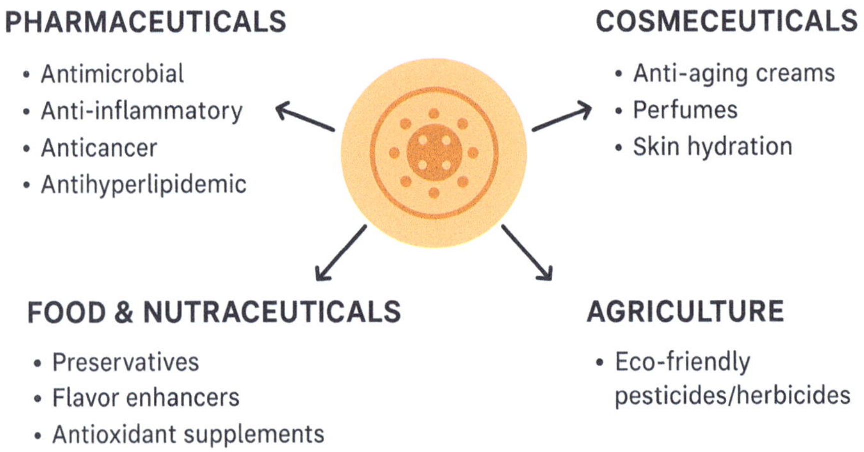

Fig. 2 Applications of aromatic oil nanocarriers

5 Types of Nanocarrier Systems for Aromatic Oils

5.1 Nanoemulsions

Nanoemulsions are thermodynamically stable formulations of oil, water, and surfactants. The aromatic oil phase is then emulsified into minute drops (20–200 nm) in an aqueous phase which acts as a continuous phase. Their minute size means that they have a high surface area, thus increasing the amounts of oils that dissolve in a body and become accessible to the body. These have found major application in the food, cosmetics, and medical areas, where they are used to deliver antimicrobial and antioxidant agents [16].

5.2 Liposomes and Niosomes

Liposomes are sphere-shaped membranes which consist of one or more lipid layers with an aqueous center. The aromatic oil is generally embedded into the lipid bilayer with the lipophilic character. Niosomes are analogous, with only the nonionic surfactants forming them rather than phospholipids, thus being more stable and cheaper in general. The two systems are biocompatible and could be fashioned to have prolonged release; thus, they can carefully work in the conveyance of aromatic oils in prescription drugs and cosmetics [17].

5.2.1 Solid Lipid Nanoparticles (SLNs) and Nanostructured Lipid Carriers (NLCs)

SLNs are soft lipid carriers at room temperature. The aromatic oil is entrapped or dissolved in this solid lipid mass. A second generation of lipid nanoparticles, NLCs, is a combination of both solid and liquid lipids. The combination results in the formation of an amorphous, less organized matrix, capable of carrying increased drug loading and one that inhibits the forces that cause oil to be discharged during storage, a frequent problem with SLNs. SLNs and NLCs are not only very stable, controlled-release form, useful in both topical and transdermal applications [18].

5.3 Polymeric Nanoparticles

Polymeric nanoparticles are solid cluster structures made of biodegradable and biocompatible polymers. They may be nanocapsules (where the oil is contained in a core enclosed by a polymer shell) or nanospheres (where the oil is dispersed in various parts of the polymer). They are very versatile, can be customized to suit certain applications, and provide strong protection and long-term release of encapsulated aromatic oils [19].

5.4 *Cyclodextrin Complexes*

Cyclodextrins (CDs) are cyclic oligosaccharides whose outer region has water-loving behavior, whereas the inside is water-repellent. Such a special T structure enables them to create inclusion complexes with aromatic oil molecules, which are totally comfortable inside the opening. This packaging guards against the degradation of the oils, their volatilization, and oxidation. During the complexation process, water solubility is enhanced immensely, thereby allowing the oil to be used in aqueous preparations [20].

6 Methods of Formulating Nano-Based Aromatic Oil Systems

6.1 *High-Pressure Homogenization*

A top-down method of manufacturing nanoemulsions and other nanosuspensions is high-pressure homogenization method. It entails the pressurization of oil and water mixture using a very small orifice (generally 50–200 MPa). The shear stress, turbulence, and cavitation effect created inside the valve disintegrates the oil droplets to a highly dispersed fine, uniform, and stable nano-sized dispersion. The process is tremendously scalable and repeatable; it found wide use in manufacture [21, 22].

6.2 *Ultrasonication*

High-intensity ultrasound, called ultrasonication, is also a popular means of developing nanoemulsions. A probe of high-frequency sound is inserted into the oil-water mixture; it creates compression-decompression in associated rapid succession. This forms microbubbles which implode violently (referred to as cavitation) producing high local shearing forces and shock waves that in effect disintegrate the oil droplets down to the nano-sized regime. This is a relatively straightforward and inexpensive method but less scalable than that of high-pressure homogenization [23].

6.3 Solvent Evaporation

The common types of materials that are produced by the solvent evaporation method are polymeric nanoparticles and solid lipid nanoparticles (SLNs). In this method, the aromatic oil and the carrier material (e.g., a polymer or a lipid) are dissolved in an immiscible organic solvent to water. This organic phase is then emulsified into aqueous phase that has a stabilizer present in it. The organic solvent is then usually removed, usually under reduced pressure, so that the dissolved polymer or lipid acrylic precipitates and forms nanoparticles in which the aromatic oil is entrapped [24].

6.4 Thin-Film Hydration (for Liposomes)

Thin-film hydration is the classic method for preparing liposomes and niosomes. The process begins by dissolving the lipid (e.g., phospholipids) and the aromatic oil in a volatile organic solvent in a round-bottom flask. The solvent is then evaporated under vacuum, leaving a thin, dry lipid film on the inner wall of the flask. This film is subsequently hydrated with an aqueous solution (e.g., buffer) at a temperature above the lipid's phase transition temperature. Gentle agitation causes the lipid film to swell and detach from the flask, spontaneously forming multi-lamellar liposomes that encapsulate the aqueous solution and the aromatic oil [25].

7 Therapeutic and Commercial Applications

7.1 Dermatological Applications

Nanocarriers are highly effective in delivering aromatic oils for skin conditions such as acne and wound healing. The small size of nanocarriers like nanoemulsions and solid lipid nanoparticles allows for deep penetration into the skin layers and hair follicles, which are the primary sites of acne development. They can deliver antibacterial and anti-inflammatory compounds from oils like tea tree and peppermint directly to the target area. For wound healing, these nanocarriers can be integrated into hydrogels or topical creams to provide a sustained release of antimicrobial and tissue-regenerating agents, protecting the wound from infection and promoting faster healing [26].

7.2 Antimicrobial and Antiviral Potential

There is wide antimicrobial activity of aromatic oils. Their stability and high volatility can lower their effectiveness, however. This potential is maximized by nanocarriers which prevent degradation of the oil. Nano-based formulation of such oils as oregano and thyme can also be used to fight multidrug-resistant bacteria due to their powerful compounds, like thymol and carvacrol, which can dissolve microbial cell membranes. The systems are also being studied with regard to their antiviral activity with certain studies showing it inhibits viral infection through interfering with viral replication [2].

7.3 Cancer Therapy

Nanotechnology is used in delivering aromatic oils as adjuvants or as cytotoxic material in the treatment of cancer. Frankincense or lemon oil encapsulation into nanocarriers enables its transfer to the tumor cell in a controlled way. The nanocarriers may be designed to localize in the tumor tissues through the enhanced permeability and retention (EPR) mechanism. This can more than double the concentration of the active components contained in an oil at a cancerous site, maximizing its cytotoxic action (e.g., apoptosis) and minimizing the damage of healthy cells in other areas of the body, theoretically preventing some damage otherwise possible due to systemic toxicity [15].

7.4 Food Preservation and Packaging

Nanotechnology is also changing the food sector by the use of aromatic oils which creates active intelligent packaging. Films and coatings nano-impregnated with essential oils can be used on foodstuffs or packaging material. These movies slowly liberate antimicrobial compounds, and this forestalls the development of food-spoiling bacteria and fungi. This does not only lengthen the shelf life of fresh produce as well as processed food but also minimizes the artificial preservatives to address the consumer pressure which increases the demand of natural products [27].

8 Safety, Toxicity, and Regulatory Concerns

8.1 Nanotoxicity Considerations

The primary concern with nanocarriers is their potential for nanotoxicity, which refers to the adverse effects of nanomaterials on living organisms and the environment. While the essential oils themselves may be safe, their encapsulation in a nanocarrier can alter their properties and biological interactions. Key factors contributing to nanotoxicity include:

Particle size and surface area: The small size and high surface area of nanoparticles can increase their reactivity and allow them to penetrate biological barriers, such as cell membranes, that larger particles cannot. This can lead to increased cellular uptake and potential interference with cellular processes.

Carrier material toxicity: The nanocarrier material itself (e.g., polymers, lipids, surfactants) must be biocompatible and biodegradable. However, some materials may accumulate in organs or cause a toxic response, especially with long-term exposure.

Reactive oxygen species (ROS): Nanoparticles can induce the production of ROS within cells, leading to oxidative stress, DNA damage, and cell death. The encapsulated essential oil may have antioxidant properties, but the carrier material's interaction with the biological system is a critical safety consideration [5, 15, 19].

8.2 Regulatory Framework

The regulatory landscape for nano-based products is complex and still evolving. There are no specific, dedicated regulatory frameworks for nanomedicines or nano-based cosmetics in many parts of the world. Instead, regulatory bodies like the U.S. Food and Drug Administration (FDA) and the European Medicines Agency (EMA) regulate these products based on their intended use, under existing rules for drugs, food additives, or cosmetics [28].

FDA and EMA: These agencies evaluate nano-based products on a case-by-case, science-based approach. The key challenge for manufacturers is to provide sufficient data demonstrating the safety and efficacy of the nanoformulation, including its unique properties (e.g., size, surface chemistry), potential for toxicity, and metabolic fate.

AYUSH (India): In India, the AYUSH ministry oversees traditional and alternative medicine systems. While it provides a framework for herbal products, specific guidelines for nano-based aromatic oil systems are still developing. The focus remains on demonstrating the safety and quality of the product in accordance with existing AYUSH standards [28].

8.3 Challenges in Clinical Translation

Moving a promising nano-based aromatic oil system from the lab to a clinical product is a significant challenge.

Manufacturing and Scalability: Reproducibly manufacturing nanoparticles on a large, industrial scale is difficult. Maintaining consistent particle size, stability, and encapsulation efficiency is a major hurdle.

Pharmacokinetic and Pharmacodynamic Variability: The behavior of nanocarriers can vary significantly in different biological systems. Factors such as protein binding, opsonization, and clearance by the reticuloendothelial system can affect the nanocarrier's circulation time and ultimate therapeutic effect.

Cost-Effectiveness: The complex manufacturing processes for many nanocarriers can be expensive, which may make the final product unaffordable compared to conventional treatments [29].

9 Current Market Products and Trends

The market already features a variety of nano-based essential oil products, particularly in the fields of cosmetics, food, and medicine. Some notable examples are provided in Table 1.

9.1 Market Potential and Consumer Trends

The nano-based essential oil products section of the market is presently showing substantial growth on the back of several important consumer trends.

Taste or Flavor Profile: More people want to eat natural and clean label foods. This tendency pushes the demand of products that comprise natural-origin compounds, and nano-based essential oils can be an alluring answer.

Wellness and Health Consciousness: There's a growing focus on health, wellness, and self-care. The therapeutic benefits of essential oils, such as their antioxidant, anti-inflammatory, and mood-enhancing properties, are now more accessible through effective nanoformulations. This is fueling market growth in aromatherapy, dietary supplements, and personal care.

Technological Innovation: Advances in nanotechnology have made it possible to produce stable, safe, and effective nanoformulations on a larger scale. This has

Table 1 Commercially available nanoformulation of essential oils

Brand/product	Description and active ingredient	Manufacturer	References
Nanorepair Q10™ (SLN-based)	Solid lipid nanoparticle (SLN) delivering coenzyme Q_{10} for antiaging benefits	Dr. Rimpler, Germany	López et al. [30]
Lavender essential oil SLN gel	Topical nanohybrid gel containing lavender essential oil loaded into SLNs for enhanced antibacterial effect	Research formulation (no commercial brand provided)	Silva et al. [31]
Nanorepair Q10™ (SLN-based)	Solid lipid nanoparticle (SLN) delivering coenzyme Q_{10} for antiaging benefits	Dr. Rimpler, Germany	Silva et al. [31]

allowed companies to bring these innovative products to market, expanding their applications and reaching new consumer bases [1, 2, 13].

10 Challenges and Future Perspectives

10.1 Scale-Up and Manufacturing Hurdles

One of the most significant challenges is the difficulty in scaling up laboratory-based production methods to industrial levels. Many of the techniques for creating nanocarriers, such as high-pressure homogenization and ultrasonication, are effective at small scales but face issues with consistency, energy consumption, and cost when scaled up. Maintaining a uniform particle size distribution and encapsulation efficiency across large batches is a major manufacturing hurdle. Without robust, reproducible manufacturing protocols, it is difficult to ensure product quality and *consistency, which is a prerequisite for regulatory approval.*

10.2 Need for Standardization and Quality Control

There is a pressing need for standardization and quality control (QC) protocols specific to nano-based essential oil products. Unlike conventional essential oils, which have established standards for composition (e.g., GC-MS analysis), nanoformulations require additional QC measures. These include:

Characterization of nanoparticles: Verifying the particle size, zeta potential, and morphology of the nanocarriers is crucial for predicting their stability and biological activity [5, 15, 23].

Encapsulation efficiency: How much of the essential oil encapsulated is of key importance to guarantee consistency in effective dose [2].

Stability testing: Stability test over time, to be conducted to establish the shelf life of the nanoformulation and to verify that the encapsulated oil should not degrade with time [2].

10.3 Opportunities for Synergistic Formulations

Aromatic oil-based nano-systems have a future in the making of compatible formulations. This means that essential oils are being mixed with other active compounds in order to come up with products with improved effects. An example may include a nanoformulation that entails the combination of an essential oil and an antibiotic that would provide better response against drug-resistant bacteria. The essence oil would interrupt the cell membrane of the bacteria, rendering the antibiotic more active and the required amount would reduce. Similarly, loading an anticancer antioxidant essential oil onto an experienced chemotherapeutic agent in the same platform could increase the therapeutic effect against the cancerous cells as well as prevent the peroxidation wreckage on healthy cells [10, 17, 24–26, 30].

10.4 Future Research Directions

There are few important areas that need to be addressed in the futuristic research to develop further in this field:

Green and sustainable nanofabrication: There is a need to have less environmentally unfriendly and expensive ways of creating nanocarriers. These include making use of biodegradable polymers and green solvents.

Smart nanocarriers: Nanocarriers that have the ability to release their therapeutic contents in response to a trigger system, such as pH, temperature, or others, would allow more effective and targeted delivery.

In vivo studies and clinical trials: There is a significant requirement of more in-depth in vivo studies and clinical trials to clearly define the long-term safety, efficiency, and pharmacokinetics of nano-based essential oil products.

Although this potential is so substantial, there are numerous factors to be taken into consideration in order to achieve adequate integration into clinical practice and commercial markets. Reproducing and standardizing nanocarrier manufacturing is the most difficult challenge. To scale up to practical production, a greater standardization and cost-effective production process is urgently required where the integrity of the product can be assured, by producing physiologically similar sized particles and being able to deliver high quantities of such a product. The safety and regulatory regime remain to be harmonized in the field. Regulatory agencies such as US FDA and the EMA are assessing nano-based products, on a case-by-case basis, under the existing guidelines. The future contributions of nanotechnology depend on detailed nanotoxicity investigations that would demonstrate the long-term safety of nanocarrier itself and the addition of nanotoxicity experiments on the oil-containing nanoparticles. This plays a paramount role in developing consumer and physician confidence. Future activity is likely to be in synergistic formulations, i.e., combinations of essential oils and other therapeutic agents within a common nanocarrier. Such a strategy can increase the efficacy of the two compounds. In another example, nano-based delivery system may also be used to blend an essential oil and an antibiotic with drastic results in the fight against drug-resistant bacteria. The potential of nano-based aromatic oils remains attractive, and research continues on the enlightenment, by creating so-called smart nanocarriers that can only release the payload when being triggered by predetermined factors such as pH changes or temperatures. The combination of these systems has potential applications in personalized medicine, environmentally friendly food preservation, and sophisticated/high-end dermocosmetic products [1, 7, 20, 29, 31, 32].

11 Conclusion

Nanotechnology has the potential of overcoming the main disadvantages associated with aromatic oils that include volatility, low-level water solubility, and low bioavailability. Oils are encapsulated in nanocarriers (nanoemulsions, liposomes, polymeric nanoparticles), protecting them against environmental perturbation (light, oxygen, and heat). This not only makes them last longer in shelves but also guarantees that they are potent. Included in the basic chemical process of flow-through encapsulation is that the solubility of oils in water is increased considerably, broadening their utility in food, cosmetics, and medicine. Moreover, owing to the nanoscale size of these carriers, it operates through an increase in bioavailability since it can penetrate biological barriers such as the skin and intestine wall of the body easier. This ability is also capable of being built into the nanocarriers giving them controlled release and a system to be selective about where they release their contents and increase the area which the active substances that they contain will work and reduce the instance of side effects.

Competing Interests The authors declare that they have no competing interest.

Declaration of Generative AI and AI-Assisted Technologies in the Writing Process During the preparation of this work, the book chapter author(s) have not taken any assistance from AI, and the book chapter author(s) reviewed and edited the content as needed and take(s) full responsibility for the content of the publication.

References

1. Zahra AA, Susanti I (2023) Review development of essential oil nano preparation formulations to pharmacological activity. J Pharm Sci 6:381–387. https://doi.org/10.36490/journal-jps.com.v6i2.95
2. Dupuis V, Cerbu C, Witkowski L et al (2022) Nanodelivery of essential oils as efficient tools against antimicrobial resistance: a review of the type and physical-chemical properties of the delivery systems and applications. Drug Deliv 29:1007–1024. https://doi.org/10.1080/10717544.2022.2056663
3. Sharifi-Rad J, Sureda A, Tenore GC et al (2017) Biological activities of essential oils: from plant chemoecology to traditional healing systems. Molecules 22. https://doi.org/10.3390/molecules22010070
4. Dhifi W, Bellili S, Jazi S et al (2016) Essential oils' chemical characterization and investigation of some biological activities: a critical review. Medicine 3:25
5. Bakkali F, Averbeck S, Averbeck D et al (2008) Biological effects of essential oils—a review. Food Chem Toxicol 46:446–475. https://doi.org/10.1016/j.fct.2007.09.106
6. Sakkas H, Papadopoulou C (2017) Antimicrobial activity of basil, oregano, and thyme essential oils. J Microbiol Biotechnol 27:429–438. https://doi.org/10.4014/jmb.1608.08024
7. Sousa VI, Parente JF, Marques JF et al (2022) Microencapsulation of essential oils: a review. Polymers 14. https://doi.org/10.3390/polym14091730

8. Cimino C, Maurel OM, Musumeci T et al (2021) Essential oils: pharmaceutical applications and encapsulation strategies into lipid-based delivery systems. Pharmaceutics 13. https://doi.org/10.3390/pharmaceutics13030327
9. Sánchez-Osorno DM, López-Jaramillo MC, Caicedo Paz AV et al (2023) Recent advances in the microencapsulation of essential oils, lipids, and compound lipids through spray drying: a review. Pharmaceutics 15. https://doi.org/10.3390/pharmaceutics15051490
10. Lammari N, Louaer O, Meniai AH et al (2020) Encapsulation of essential oils via nanoprecipitation process: overview. Progr Challenges Prosp Pharm 12:431
11. Helal MA, Abdel-Gawad AM, Kandil OM et al (2022) Microfluidic-based formulation of essential oils-loaded chitosan coated PLGA particles enhances their bioavailability and nematocidal activity. Pharmaceutics 14:2030
12. Rohilla S, Rohilla A, Narwal S et al (2023) Global trends of cosmeceutical in nanotechnology: a review. Pharm Nanotechnol 11:410–424. https://doi.org/10.2174/2211738511666230508161611
13. Achagar R, Ait-Touchente Z, El Ati R et al (2024) A comprehensive review of essential oil–nanotechnology synergy for advanced dermocosmetic delivery. Cosmetics 11:48
14. Carvalho IT, Estevinho BN, Santos L (2016) Application of microencapsulated essential oils in cosmetic and personal healthcare products - a review. Int J Cosmet Sci 38:109–119. https://doi.org/10.1111/ics.12232
15. Alabrahim OAA, Lababidi JM, Fritzsche W et al (2024) Beyond aromatherapy: can essential oil loaded nanocarriers revolutionize cancer treatment? Nanoscale Adv 6:5511–5562. https://doi.org/10.1039/d4na00678j
16. Jawaid T, Alaseem AM, Khan MM et al (2023) Preparation and evaluation of nanoemulsion of citronella essential oil with improved antimicrobial and anti-cancer properties. Antibiotics 12:478
17. Majdizadeh M, Akbarzadeh S, Al-Turnachy HH et al (2025) Synthesis of nano-liposomes containing Mentha piperita essential oil and investigating their stability, antimicrobial, antioxidant, and antiproliferative properties. Nanomed J 12:122–139. https://doi.org/10.22038/nmj.2024.79435.1960
18. Naseri N, Valizadeh H, Zakeri-Milani P (2015) Solid lipid nanoparticles and nanostructured lipid carriers: structure. Prep Appl Adv Pharma Bull 5:305–313. https://doi.org/10.15171/apb.2015.043
19. Adeyemi SB, Akere AM, Orege JI et al (2023) Polymeric nanoparticles for enhanced delivery and improved bioactivity of essential oils. Heliyon 9:e16543. https://doi.org/10.1016/j.heliyon.2023.e16543
20. Stancu AI, Mititelu M, Ficai A et al (2025) Comparative evaluation of β-cyclodextrin inclusion complexes with eugenol, eucalyptol, and clove essential oil: characterisation and antimicrobial activity assessment for pharmaceutical applications. Pharmaceutics 17:852
21. Ferreira M, Ogren M, Dias JNR et al (2021) Liposomes as antibiotic delivery systems: a promising nanotechnological strategy against antimicrobial resistance. Molecules 26. https://doi.org/10.3390/molecules26072047
22. Mushtaq A, Mohd Wani S, Malik AR et al (2023) Recent insights into nanoemulsions: their preparation, properties and applications. Food Chem: X 18:100684. https://doi.org/10.1016/j.fochx.2023.100684
23. Bilapatte A, More A, Satpute K et al (2025) Formulation and evaluation of carbamazepine loaded ethosomal nasal in-situ gel for brain targeted drug delivery. J Holist Integr Pharm 6:57–63. https://doi.org/10.1016/j.jhip.2025.03.002
24. Mendoza-Muñoz N, Alcalá-Alcalá S, Quintanar-Guerrero D (2016) Preparation of polymer nanoparticles by the emulsification-solvent evaporation method: from Vanderhoff's pioneer approach to recent adaptations. In: Vauthier C, Ponchel G (eds) Polymer nanoparticles for nanomedicines: a guide for their design, preparation and development. Springer, Cham, pp 87–121

25. Kyadalwar S, Waghmare R, Satpute K et al (2025) Development and characterization of a nanocomposed pioglitazone hydrochloride-loaded niosomal suspension for targeted drug delivery. J Bio-X Res 8:0053. https://doi.org/10.34133/jbioxresearch.0053
26. Modarresi M, Farahpour MR, Baradaran B (2019) Topical application of Mentha piperita essential oil accelerates wound healing in infected mice model. Inflammopharmacology 27:531–537. https://doi.org/10.1007/s10787-018-0510-0
27. Pattnaik R, Panda SK, Biswas S et al (2024) Prospects and challenges of nanomaterials in sustainable food preservation and packaging: a review. Discover Nano 19:178. https://doi.org/10.1186/s11671-024-04142-1
28. Ivanova S, Ivanov K, Gvozdeva Y et al (2025) Essential oils—a review of the regulatory framework in the European Union. Pharmacia 72:1–12
29. Younis MA, Tawfeek HM, Abdellatif AAH et al (2022) Clinical translation of nanomedicines: challenges, opportunities, and keys. Adv Drug Deliv Rev 181:114083. https://doi.org/10.1016/j.addr.2021.114083
30. López KL, Ravasio A, González-Aramundiz JV et al (2023) Solid lipid nanoparticles (SLN) and nanostructured lipid carriers (NLC) prepared by microwave and ultrasound-assisted synthesis: promising green strategies for the nanoworld. Pharmaceutics 15. https://doi.org/10.3390/pharmaceutics15051333
31. Silva EFD, Santos F, Pires HM et al (2025) Lipid nanoparticles carrying essential oils for multiple applications as antimicrobials. Pharmaceutics 17. https://doi.org/10.3390/pharmaceutics17020178
32. Omidian H, Cubeddu LX, Gill EJ (2025) Harnessing nanotechnology to enhance essential oil applications. Molecules 30:520

Aromatic Plants and Their Role in Metal Nanoparticle Synthesis, Characterisation, and Application

Maharudra Pratap Singh, Akhilesh Kumar Singh, and Anand Prakash

Abstract The green synthesis of nanoparticles using plant biomolecules has gained attention as a cost-effective and environmentally friendly alternative to conventional chemical and physical processes. Aromatic plants offer unique advantages due to their abundance of secondary metabolites, such as phenolic compounds, flavonoids, terpenoids, aldehydes, ketones, and essential oils, which work together as reducing, capping, and stabilising agents during nanoparticle biosynthesis. This chapter delves into the role of these phytochemicals in the nucleation, growth, and stabilisation phases of metal and metal oxide nanoparticle formation, using comprehensive characterisation techniques like spectroscopy, electron microscopy, X-ray diffraction, and surface charge analysis. The chapter emphasises the ecological compatibility and enhanced biocompatibility of aromatic plant-mediated nanoparticles, making them attractive for various applications, including biomedicine, agriculture, environmental remediation, and industrial catalysis. However, challenges remain in achieving reproducible standardisation, scalability of production processes, and navigating regulatory frameworks to ensure safety and efficacy. The chapter concludes by discussing future directions, including integrating green nanotechnology with cutting-edge fields like green chemistry, nanotechnology, bioinformatics, toxicology, and policy, to advance the development of sustainable and efficacious nanoparticle-based solutions.

Keywords Aromatic plants · Green nanotechnology · Metal nanoparticles · Spectroscopy · Electron microscopy · X-ray diffraction

M. P. Singh · A. K. Singh (✉) · A. Prakash
Department of Biotechnology, School of Life Sciences, Mahatma Gandhi Central University, Motihari, Bihar, India
e-mail: akhileshsingh@mgcub.ac.in

A. Husen (ed.), *Sustainable Uses of Aromatic Plants in Nanotechnology*, Smart Nanomaterials Technology, https://doi.org/10.1007/978-981-95-7409-4_4

1 Introduction

Plants are the foundation of life on Earth, acting as the primary autotrophic organisms responsible for transforming solar energy into chemical energy via photosynthesis. This crucial role not only supports terrestrial ecosystems and atmospheric balance, but it also serves as a huge and dynamic repository of primary and secondary metabolites. The development of plants and ecological interactions depends heavily on these metabolites. Humans have used them for centuries in a variety of applications, spanning medicines, food, cosmetics, dyes, and other industrial products [1–6]. The ability of aromatic plants to biosynthesise and store volatile organic compounds (VOCs) makes them a specialised category within the varied botanical repertoire with significant ecological, medicinal, and industrial significance. Aromatic plants are a type of medicinal plant that can produce essential oils, which are complex combinations of volatile, lipophilic secondary metabolites with low molecular weights. These substances are made and stored in specialised plant structures, including glandular trichomes, secretory ducts, and resin canals. They include monoterpenes, sesquiterpenes, phenylpropanoids, aldehydes, ketones, alcohols, and esters. These bioactives play important physiological and ecological functions, including abiotic stress reduction, pollinator attraction, herbivore deterrent, and antimicrobial defence [7]. In contrast to primary metabolites, which are widely used in growth and metabolism, secondary metabolites, like essential oils, are species- or lineage-specific and are influenced by edaphic factors, temperature, light intensity, humidity, photoperiod, altitude, and other environmental factors. Chemotypic differences, in which plants of the same species exhibit different chemical profiles, are caused by this environmental sensitivity [8]. For example, the essential oil profiles of *Ocimum sanctum* and *Mentha arvensis* grown in tropical temperatures differ significantly from those grown in temperate regions, affecting therapeutic efficacy and commercial value. Aromatic herbs have long been included in the pharmacopoeias of traditional medical systems, such as Greco-Roman medicine, Ayurveda, traditional Chinese medicine, and Unani [9]. With the advancement of present scientific validation, the phytochemical richness of these plants has been increasingly connected to a wide range of biological activities, including antibacterial, antiviral, antifungal, antioxidant, anti-inflammatory, and anticancer properties. According to the World Health Organization (WHO), traditional medicines are used for primary health care by around 80% of the world's population, with the majority of them coming from medicinal and aromatic plants (MAPs) [10]. Furthermore, around 30% of all presently available medications originate from plants or compounds associated with them. More than 7500 plant species, or almost 50% of India's native flora, are used in traditional medicine. Similar diversity and use are documented in China, Africa, and Europe, where thousands of aromatic plants are still in use, many of which have not been explored for their potential in terms of phytochemistry, pharmacogenomics, and biotechnology. To streamline ethnobotanical documentation and enhance the industrial exploitation of aromatic plants, these species are systematically classified based on a number of parameters, such as botanical features, physical qualities, ecological preferences, and chemical composition [11]. Classification by botanical family is an important approach; members of the Lamiaceae (*Ocimum*, *Rosmarinus*, *Mentha*), Apiaceae (*Coriandrum*, *Anethum*), Myrtaceae (*Eucalyptus*, *Melaleuca*), Zingiberaceae (*Zingiber*, *Curcuma*), and Rutaceae (*Citrus*, *Ruta*) are notable examples (Table 1). Another common classification approach is based on plant parts used, such as leaves (*Mentha*, *Eucalyptus*), flowers (*Jasminum*, *Rosa*), roots (*Vetiveria*), seeds (*Carum carvi*), and bark or wood (*Cinnamomum*, *Santalum*). Aromatic plants are also classed according to their growth habits, such as herbs (*Thymus vulgaris*), shrubs (*Pogostemon cablin*), grasses (*Cymbopogon* spp.), or trees (*Melaleuca* and *Camphor*). They are further classified by their means of propagation, which might be vegetative (*Pelargonium graveolens*), seed-based (*Eucalyptus globulus*), or dual-mode (*Cymbopogon flexuosus*). Furthermore, classification by climatic and biological niche indicates distinct preferences, with tropical species like *Cymbopogon citratus*, *Ocimum sanctum*, and *Santalum album*, as well as temperate species like *Matricaria chamomilla* and *Mentha piperita*. These classification systems are critical for understanding plant characteristics and agricultural processes. Additionally, it facilitates the identification of plant chemical profiles and allows for a number of uses, such as standardising essential oils, altering metabolism, and producing nanoparticles in an eco-friendly manner [13].

Nanotechnology is the fabrication and modification of materials with different physicochemical characteristics from their bulk counterparts at a nanometre scale (1–100 nm). Metallic and metal oxide nanoparticles (MNPs) are particularly useful for their antibacterial, optical, electrical, magnetic, and catalytic properties. However, conventional methods for creating metal nanoparticles typically involve energy-intensive processes, organic solvents (such as ethanol and toluene), and hazardous substances (such as sodium borohydride and hydrazine), all of which pose serious issues for environmental, economic, and safety issues. As a result, green synthesis technologies have arisen as a cost-effective, nontoxic, and environmentally friendly alternative [15]. Aromatic plants serve as natural nanofactories due to their high phytochemical content, providing surface modifiers, reducing agents, and stabilising agents for the production of nanoparticles (Fig. 1). Metal nanoparticles are produced by reducing metal ions with plant-derived compounds such as polyphenols, terpenoids, flavonoids, sugars, and proteins. These molecules reduce the ions to their zero-valent or oxide states, adhere to the surface of the nanoparticles to prevent aggregation, and provide colloidal stability and biological activity [16]. For instance, silver nanoparticles (Ag NPs) derived from Mentha piperita extract exhibit remarkable antibacterial activity when applied to drug-resistant microbes. Gold nanoparticles (Au NPs) produced from Ocimum sanctum extract exhibit potent anti-proliferative capabilities against cancer cells. In order to manage the size, shape, and distribution of the nanoparticles, the synthesis process is heavily controlled by the amount of the plant extract, pH levels, temperature, and reaction duration. In contrast to wet chemical approaches, the green synthesis of FeO_4 nanoparticles using aromatic plant extracts has produced particle sizes as small as 2–80 nm, which are significantly smaller and uniform [17, 18]. This chapter explores

Table 1 Botanical classification of aromatic plants [12–14]

Division	Subdivision	Class	Family	Genus
Embryophyta	Gymnospermae	Coniferae	Cupressaceae	*Callitris, Chamaecyparis, Cupressus, Juniperus, Thuja, Thujopsis*
			Pinaceae	*Picea, Tsuga, Pseudotsuga, Abies, Cedrus, Pinus*
			Podocarpaceae	*Dacrydium*
			Taxodiaceae	*Cryptomeria, Sciadopitys*
	Angiospermae	Monocotyledonae	Gramineae (Poaceae)	*Elionurus, Vetiveria, Cymbopogon, Andropogon*
			Cyperaceae	*Cyperus*
			Palmae (Palmaceae)	*Cocos*
			Araceae	*Acorus, Allium, Lilium, Hyacinthus, Convallaria*
			Amaryllidaceae	*Narcissus, Polianthes*
			Iridaceae	*Crocus, Iris*
			Zingiberaceae	*Hedychium, Kaempferia, Curcuma, Zingiber, Amomum, Elettaria*
		Dicotyledonae	Piperaceae	*Piper*
			Betulaceae	*Betula*
			Moraceae	*Humulus*
			Santalaceae	*Osyris, Santalum, Fusanus*
			Aristolochiaceae	*Asarum*
			Chenopodiaceae	*Chenopodium*
			Caryophyllaceae	*Dianthus*
			Ranunculaceae	*Nigella*
			Magnoliaceae	*Magnolia, Michelia, Illicium*
			Annonaceae	*Cananga*
			Myristicaceae	*Myristica*
			Lauraceae	*Cinnamomum, Ocotea, Sassafras, Cryptocarya, Laurus, Umbellularia, Aniba*
			Cruciferae	*Cochlearia, Brassica, Raphanus*
			Resedaceae	*Reseda*
			Saxifragaceae	*Philadelphus*
			Hamamelidaceae	*Hamamelis, Liquidambar*
			Rosaceae	*Spiraea, Rosa, Prunus*

			Leguminosae	*Acacia, Copaifera, Myroxylon, Lupinus, Genista, Spartium, Wistaria, Hardwickia, Myrocarpus*
			Geraniaceae	*Geranium, Pelargonium*
			Zygophyllaceae	*Bulnesia*
			Rutaceae	*Xanthoxylum, Ruta, Pilocarpus, Cusparia, Boronia, Barosma, Amyris, Clausena, Citrus*
			Burseraceae	*Boswellia, Bursera, Commiphora, Canarium*
			Euphorbiaceae	*Croton*
			Anacardiaceae	*Pistacia, Schinus*
			Tiliaceae	*Tilia*
			Malvaceae	*Hibiscus*
			Dipterocarpaceae	*Dryobalanops, Dipterocarpus*
			Cistaceae	*Cistus*
			Violaceae	*Viola*
			Myrtaceae	*Myrtus, Pimenta, Eugenia, Leptospermum, Melaleuca, Eucalyptus*
			Umbelliferae (Apiaceae)	*Coriandrum, Cuminum, Apium, Petroselinum, Carum, Pimpinella, Foeniculum, Anethum, Oenanthe, Levisticum, Angelica, Ferula, Peucedanum, Daucus, Crithmum*
			Ericaceae	*Gaultheria*
			Primulaceae	*Cyclamen*
			Oleaceae	*Syringa, Jasminum*
			Verbenaceae	*Lippia/Aloysia*
			Labiatae (Lamiaceae)	*Rosmarinus, Lavandula, Nepeta, Salvia, Monarda, Melissa, Hedeoma, Satureia, Hyssopus, Origanum marjorana, Thymus, Mentha, Perilla, Pogostemon, Ocimum, Mosla/Orthodon, Pycnanthemum, Coridothymus*
			Myoporaceae	*Eremophila*
			Rubiaceae	*Gardenia, Leptactina*
			Caprifoliaceae	*Lonicera*
			Valerianaceae	*Valeriana*
			Compositae	*Solidago*
			Asteraceae	*Erigeron, Blumea, Helichrysum, Inula, Tagetes, Santolina, Anthemis, Achillea, Matricaria, Artemisia, Arnica, Saussurea, Tanacetum*

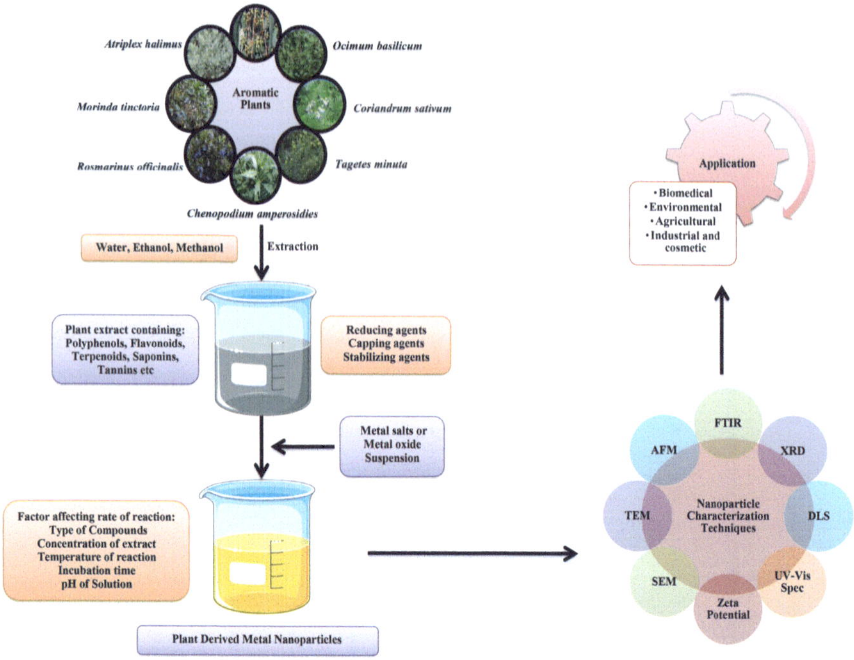

Fig. 1 Schematic representation of aromatic plants in metal nanoparticle synthesis, characterisation, and their applications

aromatic plants, their classification and chemical composition, and green synthesis principles in nanotechnology. It discusses how phytochemicals in these plants create metal nanoparticles and their potential applications in medicine, agriculture, and environmental cleanup. Green synthesis combines natural biochemical diversity with advanced nanoscience, offering a sustainable and effective method for producing safer, multifunctional nanomaterials.

2 Aromatic Plants in Metal Nanoparticle Synthesis

Aromatic plants like *Mentha piperita*, *Azadirachta indica*, *Cymbopogon citratus*, and *Ocimum sanctum* are being utilised to make nanoparticles as a safer, more environmentally friendly, and more sustainable substitute for traditional chemical and physical processes. This method uses naturally occurring phytochemicals that stabilise metal ions and reduce them to nanoparticles, including terpenoids, flavonoids, phenolics, and alkaloids. These bioactive substances improve the functionality and durability of nanoparticles whilst expediting the production process by doing away with the requirement for extra chemical reagents [19]. Aromatic plants' inherent metabolic complexity gives them a clear advantage when it comes to producing nanoparticles. Numerous hydroxyl and carboxyl groups found in phytochemicals, including ursolic acid, citral, limonene, and eugenol, enable surface binding for stabilising nanoparticles as well as electron donation for metal ion reduction. This multifunctionality offers a number of benefits, including simplifying the synthesis process, improving the nanoparticles' shelf life and dispersibility, modifying the surface chemistry for specific applications, and providing therapeutic benefits like antioxidant, antimicrobial, or anti-inflammatory properties (Table 2) [44]. From a

Table 2 Aromatic plants in metal nanoparticle synthesis and their applications

Aromatic plant	Metal nanoparticle	Size (nm)	Application/ remarks	References
Atriplex halimus	Platinum NP	1–3	Antimicrobial, antioxidant, MB dye removal	Eltaweil et al. [20]
Hibiscus sabdariffa	Platinum NP	4.40	Photodegradation of MB dye, antibacterial (gram-positive)	Seckin et al. [21]
Areca catechu	Zinc oxide NP	–	Antibacterial and photodegradation of textile dyes	Raghavendra et al. [22]
Ziziphus spina-christi	Gold NP	1–10	Removal of malachite green dye	Hosny et al. [23]
Polygonum salicifolium	Platinum NP	1–3	100% removal of 100 ppm MB dye	Hosny et al. [24]
Crataegus monogyna	Zinc oxide NP	55–70	High antibacterial activity	Fouladi-Fard et al. [25]
Ruellia tuberosa	Zinc oxide NP	40–50	92–94% degradation of MG and MB dyes	Vasantharaj et al. [26]
Justicia gendarussa	Copper oxide NP	10–15	Antibacterial and dye degradation	Vasantharaj et al. [27]
Ricinus communis	Iron NP	–	96.8% MB dye removal at 25 °C and pH 6	Abdelfatah et al. [28]
Ricinus communis	Iron NP	–	Tetracycline removal from wastewater	Abdelfatah et al. [29]
Phragmites australis	Gold NP	18	MB dye removal in 1 min	El-Borady et al. [30]
Convolvulus arvensis	Silver NP	–	Removal of copper ions	Al-Senani and Al-Kadhi [31]
Ruellia tuberosa	Iron NP	52.78	Antibacterial and violet dye degradation	Vasantharaj et al. [32]

(continued)

Table 2 (continued)

Aromatic plant	Metal nanoparticle	Size (nm)	Application/ remarks	References
Phoenix dactylifera	Silver NP	15–40	100% efficacy against *C. Albicans* and *E. Coli* at 40 μg/L	Oves et al. [33]
Atriplex halimus	Platinum NP	1–3	Antimicrobial and antioxidant	Eltaweil et al. [20]
Ziziphus spina-christi	Gold NP	1–10	Anticancer	Hosny et al. [23]
Morinda tinctoria	Zinc oxide NP	8–10	Anticancer	Anitha et al. [34]
Aristolochia bracteolata	Silver NP	41.43–60.51	Antioxidant	Thanh et al. [35]
Ferula pseudalliacea	Silver NP	–	Antibacterial and antifungal	Kocak et al. [36]
Atriplex halimus/ Chenopodium amperosidies	Gold NP	2–10/40	Anticancer	Hosny et al. [37]
Phragmites australis	Gold NP	18	Anticancer	El-Borady et al. [30]
Persicaria salicifolia	Gold NP	5.23	Anticancer	Hosny and Fawzy [38]
Persicaria salicifolia	Gold NP	5–23	Anticancer	Hosny and Fawzy [38]
Ziziphus jujuba	Silver NP	25–35	Antibacterial and anticancer	Naghizadeh et al. [39]
Lawsonia inermis	Gold NP	20	Dye and nitrophenol degradation	Kumari and Meena [40]
Anchusa italica	Zinc oxide NP	–	Antimicrobial and antifungal	Ouassil et al. [41]
Mentha pulegium	Silver NP	18–35	Antioxidant, antimicrobial	Ouassil et al. [41]
Mentha pulegium	Copper NP	–	Antimicrobial and antioxidant	Ouassil et al. [41]
Cinnamomum zeylanicum	Gold, silver NP	20–50	Antimicrobial, eugenol-driven bioreduction	Elmitwalli et al. [42]
Anchusa officinalis	Silver NP	Not stated	Antimicrobial and cytotoxic; alternative for antibiotic and anticancer therapy	Keskin et al. [43]

biomedical perspective, aromatic plant extract-derived nanoparticles are remarkably biocompatible, with naturally occurring coatings that are often physiologically active and more resilient to living systems. Along with reducing cytotoxicity, these phytochemical layers may improve drug delivery efficiency, boost cellular absorption, and enable targeted therapeutic action. Aromatic plants are already cultivated for their essential oils all over the world, which results in a large volume of agricultural waste in the shape of leaves and stems. These by-products can be used as low-cost feedstock to create nanoparticles, which will help advance the circular economy [45]. The biosynthesis process uses minimal energy and simple equipment, and it is frequently conducted at ambient temperature or slightly elevated conditions, which saves a lot of money on manufacturing. Recent developments in continuous flow reactors and bioreactors have greatly improved plant-mediated nanoparticle synthesis, enabling controlled and dependable large-scale manufacturing. Aromatic plant chemical diversity allows for precise control over nanoparticle size, shape, surface charge, and reactivity. The phytochemical profile, which is impacted by plant species, seasonal variation, geographical origin, and extraction procedures, has a direct impact on synthesis results. This method is cost-effective and ecologically friendly since it enables the production of tailored nanoparticles with particular characteristics. As evidence of the promise of this strategy, it also makes it possible to create novel nanomaterials for a range of industrial and biological applications [46].

3 Plant Metabolites in the Binding and Reduction of Metal Ions

Green synthesis of MNPs using aromatic plant extracts is an environmentally responsible and sustainable method in nanobiotechnology. This process uses metabolites from plants, which makes plants safe, affordable, and accessible biofactories for the production of MNPs. Because plant extracts contain a high concentration of phytochemicals, they can act as both reducing agents, promoting the conversion of metal ions into their zero-valent state, and stabilisers, avoiding MNP aggregation [47]. Although the entire process of plant-mediated nanoparticle synthesis is still unclear, a number of potential routes have been proposed. The phytochemical profiles of plants are diverse and complicated, making it difficult to identify the specific chemicals that stabilise and reduce MNPs [48, 49]. However, research that is presently published suggests that several chemicals derived from plants work in concert to mediate the creation and morphological regulation of MNPs [50]. The biosynthesis process has been linked to a number of kinds of secondary metabolites, including proteins, carbohydrates, vitamins, terpenoids, alkaloids, tannins, saponins, phenolic acids, and flavonoids. These metabolites, which are naturally produced by plants for both therapeutic and defence reasons, are multifunctional in the creation of MNPs (Table 3). These substances improve the stability of the nanoparticle in

Table 3 Roles of plant metabolite classes in green synthesis of metal nanoparticles

Metabolite class	Chemical nature	Mechanism of action	Role in NP synthesis	Impact on NP characteristics	References
Flavonoids	Polyphenolic compounds with C6–C3–C6 structure	Electron donation via hydroxyl groups; metal chelation	Reducing agent, mild stabiliser	Controls size and morphology; prevents aggregation	Sysak et al. [51]
Alkaloids	Nitrogen-containing heterocycles	Electron donation via nitrogen lone pairs; metal ion complexation	Reducing and capping agent	Unique geometries, high dispersion	Chaachouay et al. [52]
Tannins	High MW polyphenols with multiple –OH /–COOH groups	Strong metal binding and redox activity	Reducing and strong capping agent	Uniform particle size; enhanced stability	Sun et al. [53]
Saponins	Glycosidic compounds with hydrophobic aglycone	Surface activity; micelle formation	Mild reducing agent; strong stabiliser	Enhanced dispersion and solubility	Adnan et al. [54]
Phenolic compounds	Simple and polyphenolic acids	Phenolic –OH oxidation; metal chelation	Efficient reducing agent and capping molecule	Size-controlled synthesis; antioxidant activity	Scroccarello et al. [55]
Terpenoids	Isoprenoid units (mono-, di-, triterpenes)	Functional groups (aldehydes, alcohols) act in redox reactions	Reducing and capping agent	Affects shape and size; improves biocompatibility	Ilavenil et al. [56]
Sugars/carbohydrates	Mono-, di-, and polysaccharides	Reducing action via aldehyde/ketone groups	Mild reducing agent; viscous capping medium	Uniform particle distribution; long-term stability	Falsafi et al. [57]
Proteins/amino acids	Chains of amino acids with $-NH_2$, –COOH, –SH	Metal binding via side chains; electron donation	Reducing and stabilising agent	Forms protein corona; useful for biofunctionalisation	Theodosiou et al. [58]

solution by attaching to its surface and reducing metal ions. These active chemicals are abundant in plants, which makes them ideal for the synthesis of ecologically friendly MNPs. It has been shown that extracts from a variety of aromatic plants can effectively facilitate the creation of MNPs whilst providing benefits in terms of cost-effectiveness, environmental safety, and biocompatibility [59]. Numerous plant species have distinct phytochemical fingerprints that have a direct impact on the stability, size, shape, and yield of the produced MNPs. This metabolic variability leads to the morphological diversity of MNPs, which includes spheres, cubes, ellipsoids, triangles, hexagons, pentagons, nanorods, nanowires, nanoprisms, and nanohexahedrons. Plant metabolites' role in reducing and stabilising metal ions has been confirmed using analytical approaches such as Fourier transform infrared spectroscopy (FTIR). It has been demonstrated that metabolites such as proteins, sugars, polyphenols, alkaloids, terpenoids, and phenolic acids interact directly with metal ions to aid in their reduction and aid in the surface capping of the MNPs. These interactions also affect the nanostructures' final shape, growth dynamics, and nucleation rate. Additionally, it seems that the steric effects and binding affinity of these biomolecules are linked to the control of particle size and shape. Some substances may stimulate anisotropic growth, resulting in nanorods or triangular forms, whilst others may inhibit the growth of MNPs, resulting in smaller particles [60]. Three separate steps are typically involved in the production of nanoparticles by plant-mediated synthesis. Phytochemicals in the plant extract reduce metal ions like Ag^{+}, Au^{3+}, or Cu^{2+} to their zero-valent state during the first activation phase. Small metallic clusters are formed as a result of the nucleation that is started by this reduction. This stage is crucial because it establishes the quantity of nucleation sites and, in turn, affects the MNP size distribution [61]. The growth phase, which follows nucleation, is when neighbouring MNPs group together and develop via Ostwald ripening, coalescence, and heterogeneous nucleation. By dissolving and redepositing onto bigger particles, smaller MNPs reduce surface energy and improve thermodynamic stability. Particle development during this phase is also influenced by the ongoing reduction of remaining metal ions [62]. Finally, in the termination phase, the system reaches a thermodynamically stable state, and the MNPs achieve their final form. During this phase, surface energy is important. In the lack of appropriate capping agents, unstable morphologies may convert into more stable forms in order to reduce Gibbs free energy. The capping ability and binding strength of the plant metabolites involved play a major role in how much of the final form is preserved. Longer reaction times or extended growth phases can also lead to the formation of complex and irregular nanostructures, particularly when the plant extract supports prolonged stabilisation and particle development [63]. Plant metabolites that interact with metal precursors increase synthesis pace and yield, allowing for more precise control over MNP properties. This approach produces custom-designed MNPs with specified properties, broadening the applications of plant-based green nanotechnology. Furthermore, compared to conventional chemical procedures, the synthesis of nanoparticles utilising plant extracts offers a sustainable and environmentally benign alternative. This strategy not only reduces environmental impact but also promotes the utilisation of natural resources in nanotechnology applications [60].

4 Techniques for Characterising Metal Nanoparticles

In order to assess the functionality and possible uses of MNPs, it is crucial to comprehend their important properties, including their size, shape, surface charge, and optical activity. These qualities are examined using a range of analytical methods (Table 4). Spectroscopy is widely employed to analyse their optical properties. The molecular structure and chemical makeup of the nanoparticles can be ascertained using methods such as Fourier transform infrared spectroscopy (FTIR), X-ray diffraction (XRD), energy-dispersive X-ray spectroscopy (EDX), and fluorescence correlation spectroscopy (FCS). Scanning electron microscopy (SEM), transmission electron microscopy (TEM), scanning tunneling microscopy (STM), environmental SEM (ESEM), and tip-enhanced Raman spectroscopy (TERS) are often used instruments for detailed imaging and surface analysis. These techniques facilitate the high-resolution visualisation of the surface topography and morphology of nanoparticles [87]. In this section, we will look at each of these strategies in terms of their basic concepts, common uses, benefits, and potential restrictions.

4.1 Ultraviolet–Visible (UV–vis) Spectroscopy and Surface Plasmon Resonance

UV–Vis spectroscopy has proven to be an important and non-destructive analytical method for characterising MNPs produced from aromatic plant extracts. This technique monitors electromagnetic radiation absorption and emission in the visible (400–700 nm) and ultraviolet (185–400 nm) ranges, which makes it especially useful for detecting optical alterations linked to the production of nanoparticles. One significant feature of UV–Vis spectroscopy is its capacity to estimate the concentration of nanoparticles in solution using Beer–Lambert's equation, which connects light absorption to solute concentration. This makes the method extremely sensitive and effective for monitoring the creation of nanoparticles in real time. When it comes to green synthesis with aromatic plants, UV–Vis spectroscopy is essential for identifying and verifying the synthesis of nanoparticles. The colour of the reaction mixture usually changes noticeably when phytochemicals found in plant extracts reduce metal salts, such as those of gold, silver, or iron. Surface plasmon resonance (SPR), a phenomenon where conduction electrons on the surface of the nanoparticle oscillate in response to particular light wavelengths, is responsible for this colour shift. Each metal has a distinct colour and absorbance profile, allowing researchers to determine not only the type of nanoparticle produced but also preliminary information on its size and distribution [88]. Additionally, UV–Vis spectroscopy is useful for examining SPR changes brought on by modifications in the shape, aggregation, or surface functionalisation of nanoparticles. Furthermore, by analysing changes in the SPR band, UV–Vis spectroscopy has been used to track functionalisation events, such as the binding of molecules to the surfaces of nanoparticles.

Table 4 Characterisation techniques involved in aromatic plant-based metal nanoparticle synthesis

Characterisation technique	Principle	Analytical utility	Advantage	Limitation	References
Ultraviolet–visible (UV–vis) spectroscopy	Measures absorption of UV and visible light by nanoparticles, relating to electronic transitions	Optical properties: Size, shape, concentration, agglomeration, shape hints, electronic transitions	Rapid, non-destructive, simple; analyses colloids	Only optically active materials; indirect size info; low spatial resolution	Rocha et al. [64]
X-ray diffraction (XRD)	X-rays diffract through crystal lattice; peak patterns reveal crystal structure	Crystal structure, composition, crystallite size, phase ID, strain, size estimation	Direct crystal/phase info; quantitative size	Crystalline samples needed; insensitive to amorphous materials; misses organics	Holder and Schaak [65]
X-ray absorption (XAS) spectroscopy	Measures X-ray absorption as function of energy, providing local structure and oxidation state	Absorption coefficient, chemical state, interatomic distances, Debye–Waller	Sensitive to local environment; dilute samples; element-specific	Requires synchrotron; complex analysis; not routine	Wattanawikkam et al. [66]
Small-angle X-ray (SAXS) scattering	X-rays scattered at small angles reveal structure at nanometre scale	Particle size, size distribution, kinetics, ordering	Non-destructive; bulk size/ordering	Low spatial resolution; indirect structure; synchrotron best	Li et al. [67]
X-ray photoelectron (XPS) spectroscopy	Measures kinetic energy of electrons emitted by X-ray irradiation; reveals surface chemistry	Electronic structure, elemental composition, oxidation, ligand, surface chemistry	Surface-sensitive; chemical state; quantitative	Shallow depth; best for flat/conductive; vacuum needed	Greczynski and Hultman [68]
Fourier transform infrared (FTIR) spectroscopy	Absorption of IR light causes molecular vibrations, identifying functional groups	Functional groups, ligand binding, chemical composition, capping agents	Rapid; minimal prep; identifies organics	Overlapping bands; less sensitive to inorganics; poor spatial resolution	Pasieczna-Patkowska et al. [69]
Nuclear magnetic resonance (NMR) spectroscopy	Nuclear spins in magnetic field interact with radio waves, revealing chemical environment	Ligand density, arrangement, core structure, composition, ligand influence	Detailed ligand info; solution analysis; quantitative	Large sample mass; less metal sensitivity; costly	Wolff et al. [70]
Thermogravimetric analysis (TGA)	Monitors sample mass as temperature changes, revealing stability and composition	Thermal stability, mass loss, stabiliser/capping composition	Quantifies composition change; detects volatiles	No molecular ID; cannot probe structure; thermal only	Loganathan et al. [71]

(continued)

Table 4 (continued)

Characterisation technique	Principle	Analytical utility	Advantage	Limitation	References
Low-energy ion scattering (LEIS)	Bombards surface with low-energy ions; scattered ions reveal the outermost layers	Thickness, composition of monolayers and surface films	Extremely surface-sensitive; monolayer analysis	Shallow penetration; ultrahigh vacuum (UHV) needed; only top atomic layers	Rafati et al. [72]
Photoluminescence (PL) spectroscopy	Measures light emitted after excitation, revealing electronic and defect states	Optical emission, defect states, transitions (structure, size, composition)	Sensitive to defects/energy states; non-destructive	Only optically active materials; info is indirect; quenching	Zhang et al. [73]
Dynamic light scattering (DLS)	Measures light scattered by moving particles; analyses time-dependent fluctuations	Hydrodynamic size, polydispersity, agglomeration in suspensions	Quick size estimate in solution; broad distributions	Sensitive to contaminants; biased to larger particles; assumes spheres	Rodriguez-Loya et al. [74]
Inductively coupled plasma mass spectroscopy (ICP–MS)	Vaporises sample by plasma and then measures ion mass and concentration	Elemental composition, nanoparticle size/distribution, concentration	Highly sensitive; multielement detection	Destructive; needs digestion; size info indirect unless combined	Montoro Bustos [75]
Secondary ion mass spectrometry (SIMS/TOF–SIMS/MALDI)	Bombards sample surface; measures ejected secondary ions for surface chemistry	Surface groups, molecular orientation, conformation; MALDI: Size	Surface-sensitive; molecular/elemental detail; good spatial resolution	Destructive; matrix effects; complex data	Shen et al. [76]
X-ray magnetic circular dichroism (XMCD)	Measures differential absorption of left/right circularly polarised X-rays for magnetic moment info	Site symmetry, magnetic moments (element-specific) in transition metals	Element-specific magnetic info; highly sensitive	Needs synchrotron; magnetics only; complex interpretation	Yamamoto et al. [77]
Atom probe microscopy (APM)	Atom-by-atom field evaporation and time-of-flight analysis for 3D composition/imaging	3D atomic-scale imaging, compositional mapping	Atomic resolution; isotope/elemental analysis	Needs difficult prep; destructive; costly	Zand et al. [78]
Chemical force microscopy (CFM)	AFM with chemically functionalised tips probes chemical interactions at nanoscale	Maps chemical interactions/surface chemistry in nanodomains	Nanoscale chemical mapping; direct force measure	Needs functionalised tips; limited specificity; complex	Sarkar [79]
Atomic absorption spectroscopy (AAS)	Atoms absorb particular wavelengths of light; decrease in intensity enables	Quantitative elemental analysis via absorption	Sensitive, quantitative, standard method	No structure info; destructive prep; few elements at once	Brandt et al. [80]

Transmission electron microscopy (TEM)/ high-resolution TEM (HR–TEM)	High-energy electrons transmitted through thin specimen reveal inner structure	Size, shape, dispersion, aggregation, structure, kinetics	High res.; direct imaging Atomic lattice res.; distinguishes crystal types	Difficult prep; beam damage; vacuum needed	Malatesta [81]
Scanning transmission electron microscopy (STEM)	Electron probe scans sample; detectors reveal structure and composition at atomic scale	Atomic resolution imaging; compositional analysis	Atomic imaging plus element mapping	Beam damage; complex; costly	Ponce et al. [82]
Aberration-corrected TEM/ STEM	Advanced electron optics correct lens aberrations for higher spatial resolution	Resolves atomic structure, especially bimetallics	Highest spatial res.; advanced imaging	Expensive; must be precisely aligned; contamination sensitive	Esparza et al. [83]
Electron tomography	Sample imaged at multiple angles in TEM; data reconstructs 3D nanostructure	3D reconstruction at nano–/ atomic resolution	3D visualisation; combines multiple angles	Time-consuming; expensive; artefacts possible	Jenkinson et al. [84]
Scanning electron microscopy (HRSEM, SEM–EDX)	Electron beam scans sample surface; secondary/EDX signals analysed for structure and composition	Morphology, dispersion, surface features; EDX: Element analysis	High-res imaging; composition with EDX	Sample charging/drift; vacuum, coating needed; lower res. Than TEM	Rades et al. [85]
Electron backscatter diffraction (EBSD)	Backscattered electrons diffracted from crystal planes give orientation/grain info	Crystal orientation, grain boundaries, phase map	Microstructure mapping; orientation quantification	Polished samples needed; surface-sensitive; needs SEM	DeMott et al. [86]
Atomic force microscopy (AFM)	Nanoscopic cantilever tip scans surface; deflection gives topography	3D topography, size, shape, surface coverage	High-res; various substrates; minimal prep	Scanning slow; tip artefacts; only surface	Sarkar [79]

In colloidal nanoparticle systems, aggregation can also be identified early by widened or shifted absorption peaks, which cause instability. There are numerous benefits to using UV–Vis spectroscopy. It is perfect for routine biosynthetic process monitoring because it is quick, affordable, and non-invasive and requires little sample preparation. Although UV–Vis spectroscopy can provide general information about the optical behaviour of nanoparticles and validate their presence, it is unable to provide precise structural or morphological information. This is why advanced characterisation methods like FTIR, XRD, and TEM are frequently used in conjunction with UV–Vis analysis to get detailed information on particle size, crystallinity, and surface chemistry [89].

4.2 Fourier Transform Infrared (FTIR) Spectroscopy

Fourier transform infrared spectroscopy is critical for understanding the chemistry involved in the green synthesis of MNPs from aromatic plants. This method is mostly used to analyse the infrared absorption or emission spectra of the plant extracts and the produced nanoparticles in order to pinpoint the precise functional groups that are present in both. FTIR uses a large frequency range (5000–400 cm^{-1}) to detect the vibrational modes of bonds inside molecules, identifying biomolecules including flavonoids, terpenoids, and phenolic chemicals that contribute to nanoparticle reduction and stability. One of the most significant advantages of FTIR is its non-destructive nature, which allows for speedy and accurate analysis without affecting the material. It provides direct identification of ligands or surface functional groups affixed to nanoparticles, providing information on their surface chemistry and structural makeup. Nevertheless, FTIR has some drawbacks in spite of its efficiency. It frequently offers little quantitative information and can have trouble telling apart overlapping vibrational bands, particularly in intricate biological matrices like plant extracts. Furthermore, supplementary techniques—such as UV–Vis, XRD, or TEM—are often required to properly characterise the nanoparticles' size, shape, and crystalline nature [90].

4.3 Energy-Dispersive X-Ray (EDX) Spectroscopy

Energy-dispersive X-ray spectroscopy is critical in determining the elemental composition and confirming the successful production of nanoparticles. In order to provide simultaneous viewing and elemental analysis, EDX is frequently used in conjunction with imaging methods such as scanning electron microscopy (SEM) or transmission electron microscopy (TEM). When an electron beam interacts with a sample, it emits distinctive X-rays, which are collected and translated into a spectrum that shows the presence and percentage of various components. This is especially significant in plant-mediated synthesis, where phytochemicals are responsible for metal ion reduction and nanoparticle stabilisation [91]. One of the most significant advantages of EDX is that it is non-destructive, allowing the sample to be tested again. In addition to providing quantitative, semi-quantitative, and qualitative data, it may also produce elemental mapping, which shows the distribution of particular elements inside the sample. These characteristics aid in evaluating the effectiveness, uniformity, and purity of biosynthesised nanoparticles—elements essential to their secure deployment in industrial, pharmaceutical, or biological settings. But EDX is not without its drawbacks. It might have trouble distinguishing between elements with identical atomic numbers or detecting trace elements. Additionally, its surface-level examination might not provide enough molecular detail. Thus, methods such as FTIR, Raman microscopy, NMR, X-ray photoelectron spectroscopy (XPS), or time-of-flight secondary ion mass spectrometry (TOF-SIMS) can be used in conjunction with EDX when more detailed or specific data is required [92].

4.4 Fluorescence Correlation Spectroscopy (FCS)

One very useful and non-invasive method for describing metal nanoparticles generated from plants is fluorescence correlation spectroscopy. Because it can clarify dynamic features at the molecular level, it is a vital method for investigating the functional behaviours of green-synthesised nanoparticles. FCS was first developed by Magde and others in the early 1970s to investigate the diffusion and binding dynamics of tiny molecules. Since then, it has experienced significant improvement, most notably when coupled with confocal microscopy, which greatly increased its resolution and usefulness [93]. When it comes to metal nanoparticles made from aromatic plants, FCS has a number of clear benefits. It makes it possible to analyse nanoparticle dynamics in real time at the single-particle level, enabling accurate observations of surface ligand behaviour in solution, translational and rotational mobility, and interaction kinetics. Such measurements are particularly useful when investigating functionalisation processes involving complex phytochemical mixtures from plant extracts. For example, in a variety of application settings, FCS can identify the desorption or exchange of naturally attached ligands on the surface of nanoparticles, which has a direct effect on the particles' stability, aggregation behaviour, and biological activity. FCS offers a greater degree of sensitivity and requires only small sample volumes, usually in the sub-nanomolar to nanomolar concentration range, when compared to traditional techniques like dynamic light scattering (DLS). This reduces the interference frequently found in intricate plant extract matrices whilst also conserving essential biological content. Furthermore, FCS is especially well suited for assessing size fluctuations frequently observed in biogenically generated nanoparticles since it can probe particle diameters ranging from a few nanometres to several hundred nanometres [94]. Despite its advantages, FCS has certain drawbacks. Its accuracy can be influenced by fluorescence quenching, photobleaching, and optical heterogeneity in the sample matrix. These considerations might be particularly important when working with NPs generated from plants because plant components often have naturally occurring fluorescent compounds that might influence results. Furthermore, comprehensive structural characterisation of intricate nanoparticle complexes may be hampered by the poor spatial resolution of conventional FCS systems. These shortcomings have led to the development of improved variants of the technology, such as FCS combined with near-field scanning optical microscopy (FCS-NSOM). This hybrid approach expands the use of FCS for molecular imaging and nanobiotechnology applications by providing enhanced spatial resolution and analytical precision [95].

4.5 Surface-Enhanced Raman Spectroscopy (SERS)

An especially useful analytical technology that has gained popularity is SERS, which can identify and detect molecules with great sensitivity without requiring any chemical modification or labelling. This technique extends the capabilities of conventional Raman spectroscopy, which offers rich molecular-level information based on vibrational modes, by using metal nanoparticles that amplify electromagnetic fields at their surfaces. Even at very low concentrations, analyte identification is made possible by these improvements. SERS uses localised surface plasmon resonances (LSPRs), which are strong electron oscillations at the surface of MNPs [96]. Plant-derived metal nanoparticles, when used as substrates in SERS, retain excellent plasmonic properties and offer unique surface chemistries due to plant-based capping agents, enhancing analyte binding and signal intensity. SERS has several advantages, including its non-destructive nature, short processing time, and high sensitivity—even at single-molecule detection levels. This makes it an especially appealing tool for bioimaging, protein–ligand interaction research, real-time intracellular drug monitoring, and biomolecule structural characterisation. Despite its adaptability and benefits, SERS has some practical disadvantages. The efficacy of a SERS experiment is highly influenced by numerous critical aspects, including the metallic substrate used, the size and shape of the nanoparticles, and the excitation wavelength [97].

4.6 Tip-Enhanced Raman Spectroscopy (TERS)

In order to overcome the shortcomings of conventional Raman spectroscopy, specifically its resolution constraint brought on by the diffraction limit of light, a state-of-the-art analytical method known as TERS has been developed. In contrast to traditional Raman systems, TERS concentrates and enhances the Raman signal at the nanoscale using a sharp metallic tip instead of an optical fibre. By fusing the great spatial accuracy of scanning probe microscopy with the principles of SERS, this method makes it possible to detect even the smallest vibrational changes at the molecular level. TERS has great potential for the synthesis of metal nanoparticles based on aromatic plants. Its main advantage is that it may simultaneously provide topographical and chemical information, providing in-depth knowledge of the molecular makeup, electrical properties, and surface morphology of nanoparticles. This dual purpose is especially helpful when researching green-synthesised nanomaterials, as determining stability, reactivity, and bioactivity requires an understanding of surface interactions and functional groups [98]. Among TERS's main benefits is its extremely high spatial resolution, which makes it possible to map the distribution of biomolecules and capping agents made from plant extracts on individual nanoparticles. This information is sometimes unavailable using conventional Raman or even SERS techniques. Furthermore, TERS enables label-free, non-destructive examination, which makes it perfect for studying fragile nanostructures generated from plants without changing their original state. Nevertheless, TERS has drawbacks in spite of its advantages. The procedure necessitates highly specialised equipment and knowledge, and obtaining consistent findings can be difficult due to the sensitive nature of tip–sample interactions. Furthermore, the complexity and cost of TERS setup may limit its utility in routine laboratory use, particularly in developing countries where green nanotechnology is being actively investigated [99].

4.7 Mass Spectrometry (MS)

An essential analytical tool for characterising MNPs, especially those produced utilising environmentally friendly methods with aromatic plant extracts, is mass spectrometry. Its ability to provide highly sensitive and precise information about the elemental composition, molecular structure, chemical state, and surface modifications of nanoparticles has made it an essential element of modern nanotechnology. MS is critical in determining the physical and biochemical properties of nanoparticles, particularly when they are functionalised for specific industrial, environmental, or therapeutic applications. Notable characteristics of MS include its versatility and compatibility with a wide variety of sample types, as well as its smooth integration with separation techniques like liquid chromatography, ion mobility, field-flow fractionation, and electrophoresis. Because MS offers better resolution and more in-depth understanding of nanoparticle characteristics, it can be used for both fundamental research and real-time monitoring in practical settings. For elemental analysis, inductively coupled plasma mass spectrometry (ICP-MS) is a popular MS-based method. It offers useful data on mass, elemental concentration, particle size, size distribution, and the existence of trace-level contaminants. This method has been successful in quickly profiling metal-based nanoparticles whilst guaranteeing their composition and purity [100]. Electrothermal vaporisation ICP-MS (ETV-ICP-MS), another sophisticated technique, is especially useful for measuring metal nanoparticles in colloidal dispersions of any size. This is particularly helpful when evaluating nanoparticles in complicated matrices, like environmental materials. Single-particle ICP-MS (SP-ICP-MS) has also been developed to detect individual nanoparticles and to ascertain their isotopic composition, elemental content, and number concentrations. It has shown great promise in environmental toxicology and biomedicine, particularly for detecting and measuring nanoparticles in living systems. Molecular MS techniques like matrix-assisted laser desorption/ionisation (MALDI-MS) and electrospray ionisation (ESI-MS) are crucial for examining nanoparticle surface functionalisation, especially when biological materials like proteins, nucleic acids, or small ligands are coupled with nanoparticles for therapeutic or diagnostic purposes [101]. MALDI-MS uses metal nanoparticles' strong laser absorption capability to enable effective desorption and ionisation without extra organic matrices. It has been used to identify surface capping agents on nanoparticles. ESI-MS offers a major advantage due to its soft ionisation capabilities, making it highly effective for analysing complex liquid-phase samples. Although MS has many uses and benefits, there are certain drawbacks to using it to characterise plant-based metal nanoparticles. Its accessibility and effectiveness in regular investigations may be limited by issues including the requirement for precise sample preparation, possible matrix interferences, expensive instrumentation, and the difficulty of interpreting data, especially in multifunctional or hybrid nanoparticle systems. Furthermore, the precise identification of specific biomolecules or fragile organic coatings on nanoparticles may be hampered by ionisation inefficiencies in particular MS techniques [102].

4.8 Circular Dichroism (CD)

A sophisticated optical method called circular dichroism spectroscopy may identify the differential absorption of left- and right-handed circularly polarised light by optically active (chiral) molecules. This effect is particularly significant when one or more of the interacting molecular species—such as proteins, peptides, nucleic acids, or specific metabolites produced from plants—display chirality. In the green synthesis of MNPs, especially those made using bioactive chemicals from aromatic plants, CD has become a well-known non-invasive, extremely sensitive analytical technique in recent years. In the synthesis of plant-based metal nanoparticles, CD spectroscopy provides important information about the conformational and structural alterations of the biomolecules involved in the capping and bioreduction processes [103]. In the creation of nanoparticles, aromatic plant extracts that are abundant in secondary metabolites such as flavonoids, alkaloids, terpenoids, and phenolic acids frequently serve as both stabilising and reducing agents. When these molecules bond with metal ions, CD allows researchers to track the structural rearrangement that occurs, allowing for the creation of nanoparticles with particular chiral or optoelectronic characteristics. Additionally, CD has been successfully utilised to investigate how nanoparticles interact with biological macromolecules, such as the conformational dynamics of proteins and peptides, interactions between DNA and nanoparticles, and the assembly behaviour of nanocarriers used in drug delivery systems. One prominent advantage of CD is its capacity to detect induced chirality, in which achiral molecules exhibit optical activity only when they interact with chiral species—an important trait when researching plant-derived compounds that bind to metal surfaces. CD is also very selective and sensitive to changes in molecular structure, making it an excellent tool for studying protein folding, conformational transitions, and stereochemical configurations in nanoparticle–biomolecule complexes [104]. Additionally, CD's non-destructive properties enable complementary techniques to be used for additional downstream investigation of the same sample. Advanced CD spectroscopy variants, including vibrational circular dichroism (VCD), magnetic CD (MCD), near-infrared CD, and fluorescence-detected CD, have broadened its uses in nanomaterials research. Despite its extensive use, CD spectroscopy has some drawbacks. It is less helpful for the investigation of solely achiral molecules unless they are involved in chiral interactions because its efficiency is mostly limited to chiral systems. Additionally, it can be difficult to understand CD spectra, particularly in systems that are heterogeneous or multicomponent and frequently exhibit overlapping signals. Strict experimental control is required since sample parameters, including temperature, pH, ionic strength, and solvent composition, can have a substantial impact on spectral results. To accomplish thorough structural elucidation in these situations, CD is frequently combined with additional methods such as NMR, X-ray diffraction, or fluorescence spectroscopy [105].

4.9 Dynamic Light Scattering (DLS)

An extensively used method for assessing particle size distribution, particularly at the nanoscale, is dynamic light scattering. It provides a quick, sensitive, and non-invasive method of evaluating the physicochemical characteristics of MNPs made from aromatic plants, making it especially useful in this process. This method examines the scattering of monochromatic light—typically from a laser—caused by the random Brownian motion of particles suspended in a liquid medium. The hydrodynamic diameter of the nanoparticles is calculated by analytically analysing the variations in scattered light intensity over time using the Stokes–Einstein equation. DLS is useful for revealing essential properties such as particle size, size distribution, aggregation state, and even changes in biomolecular structure on the nanoparticle surface. It is extremely helpful in the quality control and standardisation of plant-mediated nanoparticle production because of its capacity to operate with low-concentration samples and identify tiny amounts of larger aggregates. DLS has several noteworthy benefits, including excellent sensitivity, quick analytical times, and reproducibility, particularly when assessing suspensions of monodisperse nanoparticles [106]. This is especially helpful for researchers who employ aromatic plant-based synthesis techniques to optimise synthesis methodologies and guarantee batch consistency of nanoparticles. DLS is also regarded as a non-destructive technique, which preserves the sample for additional analysis. This is a crucial aspect when working with expensive or scarce samples. DLS has significant drawbacks in addition to its many advantages. To improve colloidal stability, metal nanoparticles are commonly coated with biological macromolecules obtained from plant extracts. These coatings have the potential to falsely exaggerate the observed hydrodynamic radius (RH), resulting in overestimates of particle size. But there is also a benefit to this: the thickness of the organic layer encircling the nanoparticles may be estimated using the elevated RH. Interparticle interactions present another difficulty, particularly for non-superparamagnetic MNPs such as iron, where measurement accuracy may be distorted by magnetic dipole–dipole interactions. Furthermore, polydisperse or aggregated samples may result in confusing results since DLS assumes spherical particle geometry and struggles to resolve multimodal distributions adequately [74].

4.10 Zeta Potential (ZP)

A crucial analytical technique frequently used to evaluate the surface charge of MNPs suspended in colloidal systems is zeta potential. This value is crucial for forecasting the stability of nanoparticles, particularly those made with aromatic plants and green synthesis techniques. The Stern layer is a firmly bonded layer of counterions that is drawn to charged particles during the creation of nanoparticles. A zone known as the diffuse layer, which is loosely linked, surrounds this. They come together to form the electrical double layer. The electrical potential at the sliding plane in this double layer is known as the zeta potential, and it is commonly determined by monitoring the movement of nanoparticles under an external electric field. Zeta potential becomes essential for assessing colloidal stability in green nanotechnology, particularly when plant extracts are used as stabilising and reducing agents. Because of interparticle interactions like van der Waals forces, nanoparticles with zeta potential values below ±25 mV may cause undesirable consequences, including aggregation, flocculation, or sedimentation. In contrast, nanoparticles with zeta potential values above this range often show outstanding stability [107]. Understanding zeta potential in plant-mediated metal nanoparticle synthesis helps in surface chemistry customisation for subsequent uses as well as long-term dispersion stability prediction. Studies on zinc oxide nanoparticles, for example, have demonstrated that the dispersant selection—such as ethylene glycol or water—can have a substantial impact on the particle size distribution and zeta potential. Additionally, determining the zeta potential of metal nanoparticles before functionalising them aids in predicting their behaviour in intricate biological or environmental systems. There are some restrictions to take into account, though. Colloidal behaviour is not exclusively caused by the zeta potential of nanoparticles, especially when biological molecules are present. Hydrophobic, hydrophilic, steric, and structural interactions are some of the forces that proteins and other biomolecules can use to interact with nanoparticles; these forces may be stronger than electrostatic repulsion or attraction. For example, these additional interactions might cause nanoparticles to assemble when exposed to biological fluids, even if they have a favourable zeta potential [108]. Furthermore, zeta potential can be significantly changed by environmental factors like pH. According to research, the zeta potential of noble metal nanoparticles, such as gold and silver, is typically highest at neutral or slightly alkaline pH values and lowest in acidic environments. Strongly acidic media can actually accelerate aggregation, prevent nanoparticle dispersion, or encourage undesired adherence to surfaces like glass or polydimethylsiloxane (PDMS). Furthermore, the zeta potential of metal nanoparticles can be greatly affected by surface changes, such as the application of thiol-containing self-assembled monolayers. This is especially important when creating nanoparticles for specific industrial or medicinal uses where performance is determined by surface interactions [109].

4.11 X-Ray Diffraction (XRD)

A popular analytical method for determining the crystalline nature of materials, such as metal nanoparticles made from aromatic plant extracts, is X-ray diffraction. Bragg's law, which describes how X-rays interact with the ordered arrangement of atoms in a crystal lattice, is the basis of its fundamental idea. Depending on the distance between atomic planes, monochromatic X-rays are diffracted at certain angles when they are directed at a crystalline sample. Important structural details like the crystalline phase, lattice constants, and degree of crystallinity are revealed by this diffraction pattern. Because of its many benefits, XRD is a great tool for describing biogenically produced nanoparticles. It can identify and quantify various crystal phases, determine particle size, and even detect minute impurities [110]. It is non-destructive, reasonably simple to execute, and versatile. Notably, XRD has been crucial in describing the crystal structures of nanoparticles made from plant extracts. Apart from verifying the manufacturing of nanoparticles, these studies facilitate the understanding of their physicochemical properties, which are crucial for applications in medicine, cosmetics, and environmental remediation. Nevertheless, XRD has several drawbacks in spite of its benefits. The material frequently needs to be homogenised and finely milled for accurate findings, which is not always possible with suspensions of plant-derived nanoparticles. Furthermore, it cannot disclose particle morphology or surface chemistry, which is equally significant in applications based on nanoparticles, and it offers little information on amorphous or non-crystalline components [111].

4.12 Scanning Electron Microscopy (SEM)

Scanning electron microscopy is essential for examining metal nanoparticles made from aromatic plants because it can create incredibly fine surface images. In contrast to conventional optical microscopes, which employ glass lenses and visible light, SEM scans a sample's surface using a concentrated electron beam. This enables highly accurate topographical, compositional, and morphological information to be obtained at the nanoscale. One of the main advantages of SEM is its high-resolution imaging, especially in the secondary electron (SE) mode, which is extensively utilised for capturing detailed surface structures with resolutions down to less than 1 nm. Furthermore, SEM may function in several imaging modes, including energy-dispersive X-ray spectroscopy (EDS) and backscattered electrons (BSE), which further permits elemental analysis of nanoparticles, offering insights into their distribution and composition. SEM is especially useful for visualising the surface morphology of the particles and verifying their homogeneity and dispersion in the context of the synthesis of metal nanoparticles produced from plants [112]. Since SEM does not require ultrathin samples and is less sensitive to exposure to ambient air, its preparation is simpler than that of transmission electron microscopy (TEM). There are certain restrictions, though. Since many of the biomolecules found in aromatic plant extracts are nonconductive, charging effects may cause visual artefacts. A tiny layer of conductive material, such as gold or platinum, must frequently be applied to samples in order to counteract this; this extra step may somewhat change the nanoparticles' inherent properties. Furthermore, SEM is not appropriate for imaging hydrated or cryogenic materials, such as surface-functionalised nanoparticles in their natural aqueous form. In such instances, environmental SEM (ESEM) is favoured because it allows imaging in more natural, hydrated settings without requiring considerable sample modification. Notwithstanding its advantages, SEM has drawbacks as well, such as the possibility of sample degradation during preparation and a low degree of precision in estimating the size distribution of nanoparticles, which might result in skewed statistical analysis. However, SEM is a vital tool for characterising metal nanoparticles generated from aromatic plants due to its high resolution, adaptability, and relative ease of use [113].

4.13 Environmental Scanning Electron Microscopy (ESEM) and Wet-SEM

An enhanced version of conventional SEM, environmental scanning electron microscopy (ESEM) has the significant benefit of enabling sample visualisation in a controlled water vapour atmosphere as opposed to a high vacuum. This special feature removes the need for intensive sample preparation or coating by maintaining the natural state of fragile or moistened samples. As a result, ESEM is especially well suited for analysing biogenically produced MNPs from aromatic plants since it preserves the organic matter formed from plants whilst imaging. ESEM is particularly helpful for studying nanomaterials made of larger metals, such as platinum (Pt), gold (Au), and silver (Ag), even when these metals are very small or present in trace amounts. The use of backscattered electron (BSE) detection improves contrast for such high atomic number elements, making it easier to discover and characterise nanoparticles inside complicated matrices. The capacity of ESEM to perform elemental analysis in addition to imaging is another significant benefit [114]. Wet-SEM is a different but comparable method that uses a transparent electron membrane to image nanoparticles in liquid environments. Because of this, scientists may see NPs floating in their native matrices, like water, without being totally dehydrated. Additionally, Wet-SEM can be used in conjunction with energy-dispersive X-ray spectroscopy (EDX) to provide in situ elemental characterisation. This is very helpful when monitoring the behaviour of NPs produced by plants in biological and environmental samples. Both methods have drawbacks in spite of these advantages. It might be technically difficult to maintain the precise balance between sample moisture and chamber pressure in an ESEM. Notwithstanding its strength, Wet-SEM is constrained by the Brownian motion of nanoparticles, which can obstruct exact localisation or result in picture blurring, particularly for smaller NPs in suspension. Additionally, when examining ultrasmall or low-density NPs in extremely dynamic systems, both approaches could have trouble with resolution [111].

4.14 Transmission Electron Microscopy (TEM)

Transmission electron microscopy is essential for characterising metal nanoparticles synthesised from aromatic plant extracts. In contrast to SEM, which records surface features, TEM uses an electron beam to pass through an incredibly thin sample, enabling researchers to see the inside structure of nanoparticles. After interacting with the sample and being focussed by objective lenses, this beam creates high-resolution images that highlight important features of the morphology and crystallinity of the nanoparticles. One of the main benefits of TEM is its high resolution, which allows for the study of individual nanoparticles down to the atomic level. This is especially helpful for researching distinct crystalline structures of metal nanoparticles made from plant-based synthesis processes, such as iron, copper, silver, and selenium. TEM can offer precise information regarding nanoparticle size, shape, dispersion, and aggregation state—all of which are important for understanding their functional capabilities in biomedical and industrial applications [115]. The samples are usually prepared for TEM by placing the nanoparticles on a thin supporting sheet or grid. The instrument's necessity for vacuum conditions frequently necessitates fixing or stabilising the samples using methods like phosphotungstic acid or uranyl acetate negative staining. As an alternative, the sample can be preserved in its natural aqueous form by embedding it in vitreous ice and keeping it at liquid nitrogen temperatures for cryo-TEM. When examining nanoparticles in solution, such as gold nanoparticles suspended in water, this is particularly helpful for tracking their stability and dynamic behaviour. Furthermore, TEM becomes an effective technique for elemental composition research and structural imaging when used with EDX. This dual ability is crucial for verifying that particular metal components are present in nanoparticles made from aromatic plants. TEM has certain drawbacks in spite of its advantages. The process of preparing samples can be both technically complex and time-consuming. Additionally, there is a chance of structural artefacts or damage during preparation because the approach calls for ultrathin and vacuum-stable samples. In certain research contexts, the equipment's accessibility may be restricted due to its high cost and complexity [116].

4.15 Scanning Tunneling Microscopy (STM)

In nanoscience, scanning tunnelling microscopy is a potent analytical method that enables researchers to see surfaces at the atomic level. It works on the basis of quantum mechanical principles, namely, electron tunnelling, which happens when a conductive, sharp tip is pushed very close to the surface of a conducting or semiconductive sample, usually less than a nanometre. A current is created as electrons tunnel via this little opening. STM creates incredibly detailed images of surface topography and structure by carefully altering the tip's height to maintain a constant tunnelling current whilst scanning across the surface. STM is essential for characterising important physicochemical properties in the field of green nanotechnology, especially for metal nanoparticles made from aromatic plants [117]. It is particularly helpful in determining particle size, shape, surface morphology, and aggregation condition. In order to maximise synthesis conditions and guarantee the stability and homogeneity of nanoparticles made from biological extracts, such atomic-scale detail is crucial. STM is a helpful tool for analysing nanoparticles that are frequently synthesised from plant sources because of its remarkable spatial resolution, which is frequently less than 1 nm. Furthermore, STM's usefulness in biosynthetic systems can be increased by adapting it to image biological or nonconductive materials by putting an ultrathin conductive covering, like a layer of gold. Nevertheless, STM has drawbacks in spite of its advantages. Working with plant-based materials can be complicated because it requires samples to be either naturally conductive or adjusted properly. Its application is also limited in dynamic or large-scale situations since STM is sensitive to vibrations in the environment and usually needs a controlled atmosphere [118].

4.16 Atomic Force Microscopy (AFM)

A popular physical scanning method for examining the surface structure and particle size of metal nanoparticles at the nanoscale is atomic force microscopy, which is extremely sensitive. It works by scanning a sample's surface with a small probe tip and then exploiting the forces of interaction between the tip and the sample surface to create a detailed topographical map. Depending on the sample properties, AFM can be employed in either contact mode, where the probe barely taps or touches the surface, or noncontact mode, where it hovers just above the surface without direct contact. In surface mapping, contact mode is frequently chosen due to its higher resolution; nonetheless, the decision is mostly influenced by the sample's conductivity and delicacy. AFM is an essential tool for describing the size distribution, surface roughness, and shape of metal nanoparticles in the environmentally friendly synthesis process that uses aromatic plants. The capacity of AFM to image biological samples that are not conductive, like those made from plant extracts, without the need for conductive coatings or intensive sample preparation, which can change the natural structure of delicate samples, is one of its main advantages [119]. Moreover, AFM eliminates the need for intricate modelling or mathematical estimation by offering immediate, real-space imaging of nanoparticle size and form. This real-time visualisation improves our comprehension of how biological factors, such plant metabolites, impact the properties of nanoparticles. AFM has been widely utilised to analyse and visualise nanoparticles synthesised using plant-based techniques, including metals such as Ag, Au, and cadmium sulphide. In addition to identifying their dimensions and morphology, it can evaluate how they interact with biomolecules, which is crucial for applications in the environment and biomedicine. AFM does have certain restrictions, though. In contact mode, soft or extremely sensitive biological samples may experience tip-induced deformation, and the scanning process might be comparatively slow, particularly when examining vast sample areas. Furthermore, the quality and sharpness of the probe tip can affect the resolution and accuracy [120].

5 Applications of Aromatic Plant-Derived Metal Nanoparticles

5.1 Biomedical Applications

MNPs generated from aromatic plants have sparked substantial interest in the biomedical field due to their environmentally friendly synthesis, biocompatibility, and distinct physicochemical characteristics. The bioactive substances found in aromatic plants provide these green-synthesised nanoparticles great potential for use in a variety of therapeutic fields, such as cancer treatment, drug delivery systems, and antibacterial and antioxidant therapies [44]. One of the most well-known biomedical applications for these nanoparticles is their antimicrobial properties. With growing global concern about antibiotic resistance as a result of overuse and misuse of conventional antibiotics, innovative antimicrobial techniques are urgently needed [121]. Biogenically manufactured nanoparticles, particularly Ag NPs and Cu NPs derived from aromatic plant extracts rich in essential oils, have emerged as effective alternatives. These nanoparticles have antimicrobial effects through various mechanisms: they adhere to microbial membranes, causing structural disruption, producing reactive oxygen species (ROS), resulting in oxidative stress, and interfering with microbial DNA, proteins, and signalling cascades [122]. For example, Ag NPs obtained from *Ocimum sanctum* (holy basil) have demonstrated strong bactericidal activity against *Staphylococcus aureus* and *Escherichia coli*, whilst Cu NPs derived from *Mentha* species are efficient against a variety of bacterial and fungal pathogens, including *Candida albicans* [123, 124]. These plants' phytochemicals and essential oils not only aid in stabilising and lowering the nanoparticles but also work in concert to enhance their antimicrobial qualities, which makes them appropriate for use in topical treatments, wound dressings, and antimicrobial coatings on medical equipment [125]. Apart from their ability to fight diseases, these nanoparticles also have exceptional antioxidant properties. Oxidative stress, caused by an imbalance of free radicals and antioxidants, is a major contributor to the advancement of chronic ailments such as cardiovascular disease, diabetes, cancer, and neurological disorders such as Alzheimer's. Nanoparticles capped with aromatic plant-derived compounds—such as polyphenols, flavonoids, and terpenoids—have strong free radical scavenging properties. These characteristics are frequently verified by assays such as DPPH and ABTS [126]. For instance, extracts from *Lavandula angustifolia* and *Rosmarinus officinalis* were used to create gold nanoparticles, which demonstrated a potent antioxidant capability. These nanoparticles help shield cells from oxidative damage by directly neutralising ROS and increasing the production of important cellular antioxidant enzymes like catalase and superoxide dismutase. This implies that these nanoparticles may be used as dietary supplements or therapeutic agents to treat disorders connected to oxidative stress, such as neurodegenerative diseases [127, 128]. Another attractive approach is the possibility for drug delivery by metal nanoparticles made from aromatic plants. These nanoparticles can be modified into smart delivery systems by coating their surfaces with medicines, ligands, or polymers that respond to certain stimuli. In this regard, Au NPs and FeO_4 nanoparticles are particularly valued because of their positive interactions with biological systems and simplicity of surface modification [129]. When aromatic plant extracts are used to produce these nanoparticles, their biocompatibility, circulation time, and even intrinsic medicinal properties increase. One noteworthy example is the use of FeO_4 nanoparticles made from *eucalyptus* extracts, which have been coupled with anticancer drugs to provide targeted tumour therapy using hyperthermia and magnetic guidance. These methods reduce systemic toxicity, improve medication release management, and create promising opportunities for more precise treatment of infections, inflammation, and cancer [130]. Additionally, these biosynthesised nanoparticles are increasingly being used for anticancer purposes. Multidrug resistance and damage to healthy tissues are common concerns in cancer treatment. Aromatic plant-derived metal nanoparticles provide a novel remedy by exhibiting specific cytotoxicity to cancer cells [131]. For example, it has been demonstrated that silver nanoparticles made from *Ocimum sanctum* cause oxidative stress and death in breast cancer cells (MCF-7) whilst preserving healthy cells—a crucial characteristic for clinical use. By altering the development of the cell cycle, interfering with angiogenesis, and improving the effectiveness of conventional chemotherapies, additional nanoparticles, such as gold and platinum, which are made from different aromatic plants, have also demonstrated the capacity to stop the growth of tumours. The therapeutic potential is greatly increased when paired with naturally occurring anticancer compounds from the plants themselves [132].

5.2 *Environmental Applications*

Environmental pollution has become a major global issue, endangering ecosystems and human health whilst degrading the quality of the air, water, and soil. The search for environmentally acceptable and sustainable remediation techniques has focused a lot of interest on metal nanoparticles made from aromatic plants. These green-synthesised nanoparticles are perfect for fighting many types of pollution because they provide a special blend of environmental safety, catalytic efficiency, and biocompatibility [133]. One of the most important fields where aromatic plant-based nanoparticles have demonstrated great promise is water purification. Hazardous dyes, heavy metals, and microbiological infections are frequently present in domestic, industrial, and agricultural effluents. Plant extracts from species like *Mentha*, *Rosmarinus*, *Lavandula*, and *Eucalyptus* are used to create nanoparticles like ZnO and Ag NPs, which have shown strong properties to adsorb heavy metals and break down hazardous pigments [47]. For example, *Ocimum* species-derived ZnO nanoparticles have demonstrated excellent photocatalytic capabilities, achieving over 90% destruction of methylene blue in an hour under simulated sunlight. These nanoparticles produce ROS, such as hydroxyl radicals and superoxide ions, which degrade complex dye molecules into less toxic by-products [134]. Similar to this, silver nanoparticles' high surface-area-to-volume ratio and reactive surface sites allow them to degrade organic contaminants in addition to having potent antibacterial properties [135]. In the fight against heavy metal pollution, iron oxide and zero-valent iron nanoparticles biosynthesised from aromatic plants have demonstrated high affinity for binding and removing toxic metals like lead, mercury, and cadmium from water, effectively reducing their bioavailability and health risks. In addition to water treatment, these plant-derived nanomaterials have a potent use in the catalytic destruction of environmental contaminants. Green-synthesised nanoparticles accelerate degradation reactions in moderate environments and operate as environmentally beneficial catalysts [136]. Silver and gold nanoparticles made from *Ocimum* or *Lavandula* extracts have been demonstrated to convert nitroaromatic chemicals such as 4-nitrophenol, a major industrial contaminant, into less harmful aminophenol derivatives. This catalytic effectiveness is frequently boosted by phytochemical capping agents, which improve nanoparticle stability and electron transfer capacities. Furthermore, ZnO and TiO_2 nanoparticles generated from plants are efficient photocatalysts that use light energy to produce ROS, allowing the breakdown of persistent pollutants including pharmaceutical residues and chlorinated insecticides. These catalytic devices provide an environmentally friendly and reusable alternative to typical chemical treatments [137]. In the field of soil remediation, aromatic plant-derived nanoparticles have demonstrated significant promise for cleaning up contaminated areas. In addition to reducing agricultural output, heavy metals, herbicides, and hydrocarbons in the soil also endanger the safety of food. By immobilising hazardous elements through adsorption, precipitation, and redox transformations, iron-based nanoparticles made from species like *Mentha* and *Eucalyptus* can lessen their toxicity and mobility. Additionally, the phytochemical coatings on these nanoparticles can encourage enzymatic and microbial degradation processes in the soil, which would help organic contaminants break down naturally. These green nanomaterials reduce secondary contamination and support long-term soil health and ecosystem recovery, in contrast to traditional chemical techniques [138]. There are numerous benefits of the use of aromatic plant-derived nanoparticles in environmental applications. Their environmentally friendly synthesis eliminates the need for energy-intensive processes and dangerous chemicals. In environmental conditions, phytochemical capping improves stability, dispersibility, and biocompatibility. These nanoparticles are appropriate for challenging remediation activities because they exhibit multifunctional qualities, such as concurrent catalytic and antibacterial actions. Additionally, because aromatic plants are widely available, they are inexpensive, which opens the door for environmentally friendly technologies that are both scalable and profitable [139].

5.3 Agricultural Applications

Modern agriculture is facing a number of critical concerns, including decreased soil fertility and increasing pest resistance, as well as excessive fertiliser and pesticide use. Simultaneously, there is an increasing demand for increased crop productivity. In this regard, metal nanoparticles made from aromatic plants present a novel and environmentally friendly substitute for conventional agrochemicals [61]. These green-synthesised nanoparticles act as environmentally friendly pest and disease control agents in addition to improving nutrient efficiency and encouraging plant development. The creation of nano-fertilisers is one of the most exciting applications. In contrast to traditional fertilisers, which frequently result in nutrient losses due to runoff or leaching, nano-fertilisers allow for precise and regulated nutrient delivery [140]. Aromatic plants such as *Mentha*, *Ocimum*, and *Rosmarinus* have been successfully employed to produce metal nanoparticles with a large surface area and phytochemical coatings that improve nutrient uptake. Essential micronutrients including manganese, copper, iron, and zinc can be delivered more effectively by these nano-formulations. Over time, they maintain plant feeding because of their compact size, which improves root attachment and permits slow nutrient release [141]. For example, ZnO nanoparticles made from extracts of *Eucalyptus* and *Mentha* have demonstrated notable increases in yield and nutrient absorption in crops like wheat and maize with no harm to the environment [142]. Additionally, research has shown that these nano-fertilisers enhance photosynthetic activity, encourage root elongation, and stimulate the expression of nutrient transporter genes, making them an effective tool for environmentally friendly farming. Management of pests and diseases is another crucial issue [143]. Aromatic plants including *Mentha*, *Eucalyptus*, and *Ocimum* yield biogenic Ag NPs and Cu NPs, which have demonstrated potent antibacterial, insecticidal, and antifungal qualities. These nanoparticles function by rupturing cell membranes, causing oxidative stress through ROS, blocking essential enzymes, and affecting pests' ability to reproduce and develop [132]. For instance, silver nanoparticles made from *Mentha piperita* have shown promise in combating common agricultural pests such as the green peach aphid (*Myzus persicae*) and cotton bollworm (*Helicoverpa armigera*) [144]. Similarly, *Eucalyptus* oil-derived copper nanoparticles have demonstrated broad-spectrum antifungal efficacy against *Alternaria alternata* and *Fusarium oxysporum*. The synergistic action of the phytochemicals derived from plants that coat these nanoparticles frequently improves their ability to suppress pests whilst enabling lower dosages, limiting environmental damage and lowering the risk of pest resistance development [145]. In addition to delivering nutrients and controlling pests, these nanoparticles act as plant biostimulants, which encourage growth and improve stress tolerance. Studies show that biosynthesised nanoparticles can enhance photosynthetic activity, chlorophyll content, shoot and root development, and seed germination. Additionally, they activate antioxidant enzymes like peroxidase and catalase, which help plants better withstand abiotic challenges including heat, salinity, and drought [146]. For example, seeds treated with nano-ZnO made from extracts of *Ocimum* or *Mentha* show more strong seedling growth and faster germination. This is mostly due to increased mobilisation of stored nutrients, stimulation of growth-promoting genes, and improved micronutrient availability [147]. The priming effect is another way that nanoparticles can "prime" plants to react better to stresses in the future by eliciting subtle oxidative or hormonal signals. In certain instances, the nanoparticles' phytochemical coating functions as a stress-reduction or growth-promoting agent. The safety of green-synthesised metal nanoparticles for the environment is one of their main benefits. Because these nanoparticles are capped with plant-derived phytochemicals, they are generally more biocompatible and less harmful than their chemically produced counterparts. They decompose more readily and are less likely to release metal ions or bioaccumulate in the food chain. Dosage adjustment is crucial, though, because high doses can still be phytotoxic. However, these substances are regarded as safe, efficient, and ecologically friendly when used in the proper dosage range [148].

5.4 Industrial and Cosmetic Applications

Metal nanoparticles synthesised using aromatic plants have gained significant attention due to their unique attributes—eco-friendliness, biocompatibility, antimicrobial activity, and the presence of phytochemical-based surface functionalisation. These qualities have led to several practical uses in a variety of industries, most notably food packaging, cosmetics, and textiles. These plant-derived nanomaterials are progressively changing product innovation in consumer markets by providing a safer, more environmentally friendly substitute for traditional nanoparticles and synthetic additives. One of the main issues facing the food packaging sector is spoilage brought on by oxidation and microbial contamination, which not only results in financial loss but also presents health hazards [60]. Ag NPs and ZnO NPs, two metal nanoparticles generated from aromatic plants, have demonstrated significant promise when incorporated into biodegradable films composed of chitosan, cellulose, or starch. When it comes to common foodborne pathogens, including *Salmonella* species, *Listeria monocytogenes*, and *E. coli*, these nanoparticles have strong antibacterial activity [149]. For example, *Rosmarinus officinalis*-mediated Ag NPs have been successfully embedded in chitosan films, extending meat and fruit shelf life by suppressing microbial growth and oxidative degradation. In addition to their antibacterial properties, the natural aromatic chemicals found in these plants give the packaging odour-neutralising and antioxidant properties that enhance food preservation and the consumer experience. Furthermore, their biodegradability guarantees a low environmental impact after usage, making them a sustainable substitute for artificial antimicrobials and preservatives derived from plastic [150]. Nanotechnology is becoming more and more used in the cosmetics and personal care industry to improve the stability, potency, and bioavailability of products. Aromatic plant-based nanoparticles, particularly Ag and Au NPs derived from plant extracts such as *Lavandula angustifolia* and *Ocimum sanctum*, are being included in creams, lotions, shampoos, and deodorants. These nanoparticles help lessen body odour, irritation, and skin infections because of their antibacterial, anti-inflammatory, and skin-soothing qualities. These particles' nanoscale size allows them to enter the skin more deeply, improving absorption and extending the duration of active component retention [151]. The aromatic phytochemicals also serve as natural perfumes and antioxidants, which is ideal for the growing market for "green cosmetics" that do not include harsh chemicals. Examples include antiaging skin lotions that contain Au NPs, which are known for their ability to heal wounds and rejuvenate, and long-lasting deodorants that contain Ag NPs. These goods are safer and more appealing because they do not require artificial stabilisers and preservatives [152]. The use of metal nanoparticles obtained from aromatic plants is also revolutionising textile applications. Fabrics coated with silver or copper nanoparticles derived from the *Mentha* and *Eucalyptus* species have remarkable antibacterial properties, inhibiting the growth of odour-causing and fabric-degrading microbes. This is particularly helpful for bedding, sporting, and medical fabrics where hygienic conditions are crucial. These nanoparticles provide comfort and protection by contributing natural fragrance and antioxidant qualities in addition to microbial resistance [153]. Additionally, its capacity to reflect or absorb UV rays contributes to the UV-protective qualities of textiles, safeguarding the individual wearing them as well as the material. Research has shown that these functional qualities hold up even after several washings, indicating their longevity and economic feasibility [154]. Aromatic plant-derived nanoparticles have a number of long-term benefits over conventional industrial nanoparticles, which frequently depend on hazardous chemical reducing agents and stabilisers. Because they do not require the use of dangerous chemicals and are made from renewable plant resources, the procedure is safer for the environment. Additionally, the stability and functioning of nanoparticles are improved by the presence of naturally occurring phytochemicals on their surface, and long-term environmental issues are lessened by their biocompatibility and biodegradability. These characteristics make them particularly appropriate for uses where environmental impact and human safety are important factors [50].

6 Challenges and Future Prospects

Using aromatic plants to synthesise metal nanoparticles has become a sustainable and environmentally friendly substitute for traditional chemical processes. However, a number of challenges continue to impede the shift from lab-scale developments to large-scale, economically viable products. One of the most significant challenges is the lack of uniformity and reproducibility. The selection of plant parts, geographic origin, harvest timing, and phytochemical extraction techniques are some of the key reasons why various studies often yield contradictory results even when using the same plant species. These inconsistencies are exacerbated by variations in reaction parameters such as pH, temperature, metal salt concentration, and reaction duration, all of which have a significant impact on nanoparticle yield, shape, size, and surface properties [155]. In order to solve this, scientists are presently using sophisticated phytochemical profiling methods like LC-MS and NMR to link particular chemicals to the creation of nanoparticles and capping effectiveness. More consistency and easier protocol harmonisation throughout labs are also being fostered by the adoption of design of experiments (DoE) methodologies, standardised operating procedures (SOPs), and open-access databases [156]. Another significant challenge is the poor stability and shelf life of these nanoparticles, which are susceptible to oxidative deterioration, sedimentation, and aggregation, especially when stored in environments with varying conditions. Colloidal instability may result from the gradual degradation of natural capping agents made from plant metabolites. To address these limitations, researchers are investigating encapsulation technologies based on natural polymers such as chitosan and alginate, as well as freeze-drying techniques and optimal storage processes. Accelerated aging tests and other stability assessment techniques from pharmaceutical sciences are also being used to forecast the lifespan of nanoparticles in practical settings [157]. Toxicological knowledge gaps and regulatory ambiguities further hinder the path to clinical and industrial application. The majority of studies concentrate on in vitro cytotoxicity, with little in vivo study into biodistribution, metabolic breakdown, immunogenicity, and chronic toxicity. Furthermore, the complex organic corona generated by plant-derived chemicals on nanoparticle surfaces increases ambiguity about safety and immune system interactions [158]. It is also unclear how these particles would behave in the environment after being discharged into soil or water systems. Since chemically produced nanoparticles have traditionally been the focus of regulatory frameworks, green-synthesised equivalents do not yet readily fit into preexisting categories. Environmental risk assessments, imaging-based predictive modelling, and standardised toxicological testing procedures are being developed to help manage these uncertainties. To develop safety and effectiveness standards unique to nanoparticles, regulatory agencies and interdisciplinary consortia also engage in discussions regarding policy [159]. Scale-up continues to be a significant barrier to commercialisation. When applied to greater volumes, laboratory techniques that show promise frequently fall short because of difficulties in maintaining a steady supply of plant materials, reactor design, and cost-effectiveness. The complexity is increased by the availability of biomass that varies by season or location, particularly from uncultivated plant species. Furthermore, few engineered systems are presently available to facilitate continuous-flow or modular nanoparticle manufacturing from plant extracts. Innovative solutions, such as using agrobiotechnology to cultivate high-yield plant varieties, developing flow-based synthesis reactors, and incorporating circular economy models—where aromatic plants serve as a feedstock—are actively pursued to make the process economically and environmentally viable [160]. Future technological innovations in aromatic plant-based nanoparticle synthesis include multi-omics approaches like metabolomics and transcriptomics, which are enabling targeted plant selection and genetic engineering. Artificial intelligence and machine learning models are being trained to optimise synthesis parameters and predict nanoparticle properties and toxicity profiles. Hybrid nanoarchitectures are expanding their versatility across biomedical, environmental, and agricultural domains. However, the field must overcome barriers like reproducibility, stability, safety, scalability, and regulatory approval. A collaborative approach combining green chemistry, nanotechnology, bioinformatics, toxicology, and policy is essential to unlock the potential of aromatic plants in the next generation of sustainable nanomaterials [161].

7 Conclusion

Aromatic plants' role in the synthesis, characterisation, and application of nanoparticles represents a significant advance in sustainable nanotechnology. Aromatic plants, which contain a variety of bioactive phytochemicals such as phenolics, flavonoids, terpenoids, and essential oils, function as natural reducing, capping, and stabilising agents in the environmentally friendly green synthesis of metal and metal oxide nanoparticles. This biogenic approach not only offers an economical and safe substitute for conventional physical and chemical methods, but it also improves biocompatibility and gives the nanoparticles multifunctional biological characteristics like antimicrobial, antioxidant, anticancer, and anti-inflammatory effects. Characterisation techniques, such as spectroscopy, electron microscopy, and surface charge analysis, are critical for understanding the physicochemical properties, morphology, and stability of plant-mediated nanoparticles, ensuring their suitability for a variety of applications. Recent studies indicate that they hold great promise for use in industrial catalysis, environmental remediation, agriculture, and biomedical disciplines. Despite these benefits, standardising synthesis methods, scaling up for industrial production, and meeting strict regulatory requirements to ensure the safety and effectiveness of the resulting nanomaterials remain challenging tasks. Future directions suggest combining green synthesis with cutting-edge technology to maximise the production of nanoparticles and tailor their functions for sophisticated applications. Ultimately, the integration of modern nanotechnology with the diverse biochemical heritage of aromatic plants offers a new and sustainable path towards the development of safer, multifunctional nanoparticles with a broad range of applications, marking a substantial step towards more powerful and ecologically friendly nanomaterial solutions.

Competing Interests The authors declare that they have no competing interest.

Declaration of Generative AI and AI-Assisted Technologies in the Writing Process During the preparation of this work, the book chapter author(s) have not taken any assistance from AI, and the book chapter author(s) reviewed and edited the content as needed and take(s) full responsibility for the content of the publication.

References

1. Husen A (2024) Antimalarial medicinal plants. Taylor & Francis Group, New York. https://doi.org/10.1201/9781003378396
2. Husen A (2024) Medicinal spices and condiment crops. Taylor & Francis Group, New York. https://doi.org/10.1201/9781003387046
3. Husen A (2023) Antidiabetic medicinal plants and herbal treatments. CRC Press, Boca Raton. https://doi.org/10.1201/b23347
4. Husen A (2022) Herbs, shrubs and trees of potential medicinal benefits. CRC Press, Boca Raton. https://doi.org/10.1201/9781003205067
5. Husen A (2023) Exploring poisonous plants: medicinal values, toxicity response and other potential uses. CRC Press, Boca Raton. https://doi.org/10.1201/b23017
6. Salam U, Ullah S, Tang ZH, Elateeq AA, Khan Y, Khan J, Khan A, Ali S (2023) Plant metabolomics: an overview of the role of primary and secondary metabolites against different environmental stress factors. Life 13(3):706
7. Chen L, Liao P (2025) Current insights into plant volatile organic compound biosynthesis. Curr Opin Plant Biol 85:102708
8. Ramakrishna A, Ravishankar GA (2011) Influence of abiotic stress signals on secondary metabolites in plants. Plant Signal Behav 6(11):1720–1731
9. Ramos da Silva LR, Ferreira OO, Cruz JN, de Jesus Pereira Franco C, Oliveira Dos Anjos T, Cascaes MM, Almeida da Costa W, Helena de Aguiar Andrade E, Santana de Oliveira M (2021) Lamiaceae essential oils, phytochemical profile, antioxidant, and biological activities. Evid Based Complement Alternat Med 2021:6748052
10. Mueed A, Shibli SAD, Ashkan MF, Alharbi M, Alanazi H, Binothman N, Aljadani M, Majrashi KA, Huwaikem M, Abourehab MA, Korma SATM (2023) Extraction, characterization of polyphenols from certain medicinal plants and evaluation of their antioxidant, antitumor, antidiabetic, antimicrobial properties, and potential use in human nutrition. Front Nutr 10:1125106
11. Veeresham C (2012) Natural products derived from plants as a source of drugs. J Adv Pharm Technol Res 3(4):200
12. Guenther E (1952) Recent developments in essential oil production. Econ Bot 6:355–378
13. Joy PP, Thomas J, Mathew S, Jose G, Joseph J (2014) Aromatic plants. Trop Hortic 2:633–733

14. Pandey AK, Kumar P, Saxena M, Maurya P (2019) Distribution of aromatic plants in the world and their properties. In: Feed additives. Elsevier, Amsterdam, pp 89–114
15. Radulescu DM, Surdu VA, Ficai A, Ficai D, Grumezescu AM, Andronescu E (2023) Green Synthesis of metal and metal oxide nanoparticles: a review of the principles and biomedical applications. Int J Mol Sci 24(20):15397
16. Kumari P, Luqman S, Meena A (2019) Application of the combinatorial approaches of medicinal and aromatic plants with nanotechnology and its impacts on healthcare. Daru 27(1):475–489
17. Golthi V, Kommu J (2023) An eco-friendly and sustainable method for producing Fe3O4 nanoparticles using *Jatropha podagrica* leaf extract for efficient dye degradation and antibacterial uses. Hybrid Adv 4:100110
18. Thiyagarajan S, Kanchana S (2022) Green synthesis of silver nanoparticles using leaf extracts of *Mentha arvensis* Linn and demonstration of their in vitro antibacterial activities. Braz J Pharm Sci 58:e19898
19. Kumari P, Devi L, Kadian R, Waziri A, Alam MS (2024) Eco-friendly synthesis of *Azadirachta indica*-based metallic nanoparticles for biomedical application & future prospective. Pharm Nanotechnol. https://doi.org/10.2174/0122117385262947240206055107
20. Eltaweil AS, Fawzy M, Hosny M, Abd El-Monaem EM, Tamer TM, Omer AM (2022) Green synthesis of platinum nanoparticles using *Atriplex halimus* leaves for potential antimicrobial, antioxidant, and catalytic applications. Arab J Chem 15(1):103517
21. Seckin H, Tiri RNE, Meydan I, Aygun A, Gunduz MK, Sen F (2022) An environmental approach for the photodegradation of toxic pollutants from wastewater using Pt–Pd nanoparticles: antioxidant, antibacterial and lipid peroxidation inhibition applications. Environ Res 208:112708
22. Raghavendra VB, Shankar S, Govindappa M, Pugazhendhi A, Sharma M, Nayaka SC (2022) Green synthesis of zinc oxide nanoparticles (ZnO NPs) for effective degradation of dye, polyethylene and antibacterial performance in waste water treatment. J Inorg Organomet Polym Mater 32(2):614–630
23. Hosny M, Eltaweil AS, Mostafa M, El-Badry YA, Hussein EE, Omer AM, Fawzy M (2022) Facile synthesis of gold nanoparticles for anticancer, antioxidant applications, and photocatalytic degradation of toxic organic pollutants. ACS Omega 7(3):3121–3133
24. Hosny M, Fawzy M, El-Fakharany EM, Omer AM, El-Monaem EMA, Khalifa RE, Eltaweil AS (2022) Biogenic synthesis, characterization, antimicrobial, antioxidant, antidiabetic, and catalytic applications of platinum nanoparticles synthesized from Polygonum salicifolium leaves. J Environ Chem Eng 10(1):106806
25. Fouladi-Fard R, Aali R, Mohammadi-Aghdam S, Mortazavi-derazkola S (2022) The surface modification of spherical ZnO with Ag nanoparticles: a novel agent, biogenic synthesis, catalytic and antibacterial activities. Arab J Chem 15(3):103658
26. Vasantharaj S, Sathiyavimal S, Senthilkumar P, Kalpana VN, Rajalakshmi G, Alsehli M, Elfasakhany A, Pugazhendhi A (2021) Enhanced photocatalytic degradation of water pollutants using bio-green synthesis of zinc oxide nanoparticles (ZnO NPs). J Environ Chem Eng 9(4):105772
27. Vasantharaj S, Shivakumar P, Sathiyavimal S, Senthilkumar P, Vijayaram S, Shanmugavel M, Pugazhendhi A (2021) Antibacterial activity and photocatalytic dye degradation of copper oxide nanoparticles (CuONPs) using *Justicia gendarussa*. Appl Nanosci 13:2295–2302
28. Abdelfatah AM, Fawzy M, Eltaweil AS, El-Khouly ME (2021) Green synthesis of nano-zero-valent iron using *Ricinus communis* seeds extract: characterization and application in the treatment of methylene blue-polluted water. ACS Omega 6(39):25397–25411
29. Abdelfatah AM, Fawzy M, El-Khouly ME, Eltaweil AS (2021) Efficient adsorptive removal of tetracycline from aqueous solution using phytosynthesized nano-zero valent iron. J Saudi Chem Soc 25(12):101365

30. El-Borady OM, Fawzy M, Hosny M (2021) Antioxidant, anticancer and enhanced photocatalytic potentials of gold nanoparticles biosynthesized by common reed leaf extract. Appl Nanosci 13:3149–3160
31. Al-Senani GM, Al-Kadhi N (2020) The synthesis and effect of silver nanoparticles on the adsorption of Cu2+ from aqueous solutions. Appl Sci 10(14):4840
32. Vasantharaj S, Sathiyavimal S, Senthilkumar P, LewisOscar F, Pugazhendhi A (2019) Biosynthesis of iron oxide nanoparticles using leaf extract of *Ruellia tuberosa*: antimicrobial properties and their applications in photocatalytic degradation. J Photochem Photobiol B Biol 192:74–82
33. Oves M, Aslam M, Rauf MA, Qayyum S, Qari HA, Khan MS, Alam MZ, Tabrez S, Pugazhendhi A, Ismail IMI (2018) Antimicrobial and anticancer activities of silver nanoparticles synthesized from the root hair extract of *Phoenix dactylifera*. Mater Sci Eng C 89:429–443
34. Anitha J, Selvakumar R, Hema S, Murugan K, Premkumar T (2022) Facile green synthesis of nanosized ZnO using leaf extract of *Morinda tinctoria*: MCF-7 cell cycle arrest, antiproliferation, and apoptosis studies. J Ind Eng Chem 105:520–529
35. Thanh NC, Pugazhendhi A, Chinnathambi A, Alharbi SA, Subramani B, Brindhadevi K, Whangchai N, Pikulkaew S (2022) Silver nanoparticles (AgNPs) fabricating potential of aqueous shoot extract of *Aristolochia bracteolata* and assessed their antioxidant efficiency. Environ Res 208:112683
36. Kocak Y, Oto G, Meydan I, Seckin H, Gur T, Aygun A, Sen F (2022) Assessment of therapeutic potential of silver nanoparticles synthesized by ferula *Pseudalliacea Rech.* F Plant Inorg Chem Commun 140:109417
37. Hosny M, Fawzy M, Abdelfatah AM, Fawzy EE, Eltaweil AS (2021) Comparative study on the potentialities of two halophytic species in the green synthesis of gold nanoparticles and their anticancer, antioxidant and catalytic efficiencies. Adv Powder Technol 32(9):3220–3233
38. Hosny M, Fawzy M (2021) Instantaneous phytosynthesis of gold nanoparticles via *Persicaria salicifolia* leaf extract, and their medical applications. Adv Powder Technol 32(8):2891–2904
39. Naghizadeh A, Mizwari ZM, Ghoreishi SM, Lashgari S, Mortazavi-Derazkola S, Rezaie B (2021) Biogenic and eco-benign synthesis of silver nanoparticles using jujube core extract and its performance in catalytic and pharmaceutical applications: removal of industrial contaminants and in-vitro antibacterial and anticancer activities. Environ Technol Innov 23:101560
40. Kumari P, Meena A (2020) Green synthesis of gold nanoparticles from *Lawsonia inermis* and its catalytic activities following the Langmuir-Hinshelwood mechanism. Colloids Surf A Physicochem Eng Asp 606:125447
41. Ouassil M, Msairi S, Ainane T (2024) Medicinal and aromatic plants (MAP) based metal nanoparticles and their potential as nanopesticides in Morocco: a review study. Not Sci Biol 16(4):12043
42. Elmitwalli OSMMS, Kassim DAK, Algahiny AT, Henari FZ (2025) Green synthesis of metal nanoparticles using *Cinnamomum*-based extracts and their applications. Nanotechnol Sci Appl 18:93–114
43. Keskin C, Aslan S, Baran MF, Baran A, Eftekhari A, Adıcan MT, Ahmadian E, Arslan S, Mohamed AJ (2025) Green Synthesis and characterization of silver nanoparticles using *Anchusa Officinalis*: antimicrobial and cytotoxic potential. Int J Nanomed 20:4481–4502
44. Singh H, Desimone MF, Pandya S, Jasani S, George N, Adnan M, Aldarhami A, Bazaid AS, Alderhami SA (2023) Revisiting the green Synthesis of nanoparticles: uncovering influences of plant extracts as reducing agents for enhanced synthesis efficiency and its biomedical applications. Int J Nanomed 18:4727–4750
45. Sachan RS, Singh A, Karnwal A, Shidiki A, Kumar G (2024) Plant-mediated gold nanoparticles in cancer therapy: exploring anti-cancer mechanisms, drug delivery applications, and future prospects. Front Nanotechnol 6:1490980

46. Yang M, Yang L, Zheng J, Hondow N, Bourne RA, Bailey T, Irons G, Sutherland E, Lavric D, Wu K (2021) Mixing performance and continuous production of nanomaterials in an advanced-flow reactor. Chem Eng J 412:128565
47. Filho SA, Backx BP, Soran L, Opriş O, Lung I, Stegarescu A, Bououdina M (2023) Biosynthesis of nanoparticles using plant extracts and essential oils. Molecules 28(7):3060
48. Husen A (2023) Secondary metabolites based green synthesis of nanomaterials and their applications. Springer, Singapore. https://doi.org/10.1007/978-981-99-0927-8
49. Husen A (2023) Nanomaterials from agricultural and horticultural products. Springer, Singapore. https://link.springer.com/book/9789819934348
50. Khan F, Shariq M, Asif M, Siddiqui MA, Malan P, Ahmad F (2022) Green nanotechnology: plant-mediated nanoparticle synthesis and application. Nanomaterials 12(4):673
51. Sysak S, Czarczynska-Goslinska B, Szyk P, Koczorowski T, Mlynarczyk DT, Szczolko W, Lesyk R, Goslinski T (2023) Metal nanoparticle-flavonoid connections: synthesis, physicochemical and biological properties, as well as potential applications in medicine. Nanomaterials 13(9):1531
52. Chaachouay N, Azeroual A, Benkhnigue O, Zidane L (2023) Alkaloids: a suitable precursor for nanomaterials Synthesis, and their various applications. In: Husen A (ed) Secondary metabolites based green synthesis of nanomaterials and their applications. Springer, Singapore
53. Sun KY, Chueh CY, Wu MY, Wu T, Lin YW (2024) Facile preparation of tannic acid-gold nanoparticles for catalytic and selective detection of mercury(II) and iron(II) ions in the environmental water samples and commercial iron supplement. ACS Omega 9(40):41521–41531
54. Adnan M, Siddiqui AJ, Ashraf SA, Ashraf MS, Alomrani SO, Alreshidi M, Tepe B, Sachidanandan M, Danciu C, Patel M (2023) Saponin-derived silver nanoparticles from *Phoenix dactylifera* (Ajwa dates) exhibit broad-Spectrum bioactivities combating bacterial infections. Antibiotics 12(9):1415
55. Scroccarello A, Della Pelle F, Rojas D, Ferraro G, Fratini E, Gaggiotti S, Cichelli A, Compagnone D (2021) Metal nanoparticles based lab-on-paper for phenolic compounds evaluation with no sample pretreatment. Application to extra virgin olive oil samples. Anal Chim Acta 1183:338971
56. Ilavenil KK, Senthilkumar V, Kasthuri A (2025) Green synthesis of metal nanoparticles from three medicinal plants: a review of environmental and health applications. Discov Catal 2:3
57. Falsafi SR, Topuz F, Bajer D, Mohebi Z, Shafieiuon M, Heydari H, Rawal S, Sathiyaseelan A, Wang M, Khursheed R, Enayati M, Rostamabadi H (2023) Metal nanoparticles and carbohydrate polymers team up to improve biomedical outcomes. Biomed Pharmacother 168:115695
58. Theodosiou M, Chalmpes N, Gournis D, Sakellis E, Boukos N, Kostakis M, Thomaidis NS, Efthimiadou EK (2024) Amino acid driven synthesis of gold nanoparticles: a comparative study on their biocompatibility. Mater Chem Phys 319:129260
59. Marslin G, Siram K, Maqbool Q, Selvakesavan RK, Kruszka D, Kachlicki P, Franklin G (2018) Secondary metabolites in the green synthesis of metallic nanoparticles. Materials 11(6):940
60. Adeyemi JO, Oriola AO, Onwudiwe DC, Oyedeji AO (2022) Plant extracts mediated metal-based nanoparticles: synthesis and biological applications. Biomolecules 12(5):627
61. Makarov VV, Love AJ, Sinitsyna OV, Makarova SS, Yaminsky IV, Taliansky ME, Kalinina NO (2014) "Green" nanotechnologies: synthesis of metal nanoparticles using plants. Acta Nat 6(1):35–44
62. Chen Z, Liu J, Evans AJ, Alberch L, Wei A (2014) Calixarene-mediated synthesis of cobalt nanoparticles: an accretion model for separate control over nucleation and growth. Chem Mater 26(2):941–950
63. Sidhu AK, Verma N, Kaushal P (2022) Role of biogenic capping agents in the Synthesis of metallic nanoparticles and evaluation of their therapeutic potential. Front Nanotechnol 3:801620

64. Rocha FS, Gomes AJ, Lunardi CN, Kaliaguine S, Patience GS (2018) Experimental methods in chemical engineering: ultraviolet visible spectroscopy—UV-vis. Can J Chem Eng 96:2512–2517
65. Holder CF, Schaak RE (2019) Tutorial on powder X-ray diffraction for characterizing nanoscale materials. ACS Nano 13(7):7359–7365
66. Wattanawikkam C, Pecharapa W, Ishihara KN (2017) X-ray absorption spectroscopy analysis and magnetic properties of M-doped TiO2 nanoparticles (M=co, Mn, Ni and Zn) prepared by co-precipitation method. Ceram Int 43:S397–S402
67. Li T, Senesi AJ, Lee B (2016) Small angle X-ray scattering for nanoparticle research. Chem Rev 116:11128–11180
68. Greczynski G, Hultman L (2020) X-ray photoelectron spectroscopy: towards reliable binding energy referencing. Prog Mater Sci 107:100591
69. Pasieczna-Patkowska S, Cichy M, Flieger J (2025) Application of Fourier transform infrared (FTIR) spectroscopy in characterization of green synthesized nanoparticles. Molecules 30(3):684
70. Wolff N, Beuck C, Schaller T, Epple M (2024) Possibilities and limitations of solution-state NMR spectroscopy to analyze the ligand shell of ultrasmall metal nanoparticles. Nanoscale Adv 6(13):3285–3298
71. Loganathan S, Valapa RB, Mishra RK, Pugazhenthi G, Thomas S (2016) Thermogravimetric analysis for characterization of nanomaterials. In: Thomas S, Thomas R, Zachariah AK, Mishra RK (eds) Thermal and rheological measurement techniques for nanomaterials characterization. Elsevier, Amsterdam, pp 67–108
72. Rafati A, Ter Veen R, Castner DG (2013) Low energy ion scattering: determining overlayer thickness for functionalized gold nanoparticles. Surf Interface Anal 45(11-12). https://doi.org/10.1002/sia.5315
73. Zhang T, Lu G, Shen H, Shi K, Jiang Y, Xu D, Gong Q (2014) Photoluminescence of a single complex plasmonic nanoparticle. Sci Rep 4:3867
74. Rodriguez-Loya J, Lerma M, Gardea-Torresdey JL (2023) Dynamic light scattering and its application to control nanoparticle aggregation in colloidal systems: a review. Micromachines 15(1):24
75. Montoro Bustos AR (2024) ICP-MS-based characterization and quantification of nano- and microstructures. Nanomaterials 14(7):578
76. Shen Y, Howard L, Yu XY (2024) Secondary ion mass spectral imaging of metals and alloys. Materials 17(2):528
77. Yamamoto Y, Miura T, Teranishi T, Suzuki M, Kawamura N, Miyagawa H, Nakamura T, Kobayashi K, Hori H (2004) X-ray magnetic circular dichroism study of gold nanoparticles protected by polymer. J Magn Magn Mater 272:E1183–E1184
78. Zand F, Hangx S, Spiers CJ, Brink PJ, Burns JT, Boebinger MG, Poplawsky JD, Monai M, Weckhuysen BM (2023) Elucidating the structure and composition of individual bimetallic nanoparticles in supported catalysts by atom probe tomography. J Am Chem Soc 145(31):17299–17308
79. Sarkar A (2022) Biosensing, characterization of biosensors, and improved drug delivery approaches using atomic force microscopy: a review. Front Nanotechnol 3:798928
80. Brandt A, Kees K, Leopold K (2020) Characterization of various metal nanoparticles by graphite furnace atomic absorption spectrometry: possibilities and limitations with regard to size and shape. J Anal At Spectrom 35(11):2536–2544
81. Malatesta M (2021) Transmission electron microscopy as a powerful tool to investigate the interaction of nanoparticles with subcellular structures. Int J Mol Sci 22(23):12789
82. Ponce A, Mejía-Rosales S, José-Yacamán M (2012) Scanning transmission electron microscopy methods for the analysis of nanoparticles. Methods Mol Biol 906:453–471
83. Esparza R, Téllez Vázquez O, Rodríguez-Ortiz G, Ángeles-Pascual A, Velumani S, Pérez R (2014) Atomic structure characterization of Au–Pd bimetallic nanoparticles by aberration-corrected scanning transmission electron microscopy. J Phys Chem C 118(38):22383–22388

84. Jenkinson K, Liz-Marzán LM, Bals S (2022) Multimode electron tomography sheds light on synthesis, structure, and properties of complex metal-based nanoparticles. Adv Mater 34(36):2110394
85. Rades S, Hodoroaba V, Salge T, Wirth T, Lobera MP, Labrador RH, Natte K, Behnke T, Groß T, Unger WES (2014) High-resolution imaging with sem/t-sem, edx and sam as a combined methodical approach for morphological and elemental analyses of single engineered nanoparticles. RSC Adv 4(91):49577–49587
86. DeMott R, Haghdadi N, Kong C, Gandomkar Z, Kenney M, Collins P, Primig S (2021) 3D electron backscatter diffraction characterization of fine α titanium microstructures: collection, reconstruction, and analysis methods. Ultramicroscopy 230:113394
87. Altammar KA (2023) A review on nanoparticles: characteristics, synthesis, applications, and challenges. Front Microbiol 14:1155622
88. Arif M, Raza H, Akhter T (2024) UV-vis spectroscopy in the characterization and applications of smart microgels and metal nanoparticle-decorated smart microgels: a critical review. RSC Adv 14(51):38120–38134
89. Negri C, Colombo R, Bracconi M, Atzori C, Donazzi A, Lucotti A, Tommasini M, Maestri M (2024) Operando UV-vis spectroscopy for real-time monitoring of nanoparticle size in reaction conditions: a case study on rWGS over au nanoparticles. Cat Sci Technol 14(5):1318–1327
90. Dendisová M, Jeništová A, Parchaňská-Kokaislová A, Matějka P, Prokopec V, Švecová M (2018) The use of infrared spectroscopic techniques to characterize nanomaterials and nanostructures: a review. Anal Chim Acta 1031:1–14
91. Mishra RK, Zachariah AK, Thomas S (2017) Energy-dispersive X-ray spectroscopy techniques for nanomaterial. In: Thomas S, Thomas R, Zachariah AK, Mishra RK (eds) Microscopy methods in nanomaterials characterization. Elsevier, Amsterdam, pp 383–405
92. Slater TJ, Lewis EA, Haigh SJ (2016) Energy dispersive X-ray tomography for 3D elemental mapping of individual nanoparticles. J Vis Exp 113:52815
93. Dong C, Huang X, Ren J (2008) Characterization of water-soluble luminescent quantum dots by fluorescence correlation spectroscopy. Ann N Y Acad Sci 1130:253–261
94. Zhang B, Liu H, Huang X, Dong C, Ren J (2017) Size distribution of nanoparticles in solution characterized by combining resonance light scattering correlation spectroscopy with the maximum entropy method. Anal Chem 89(22):12609–12616
95. Yu L, Lei Y, Ma Y, Liu M, Zheng J, Dan D, Gao P (2021) A comprehensive review of fluorescence correlation spectroscopy. Front Phys 9:644450
96. Willets K, Mayer K (2018) Surface-enhanced Raman scattering (SERS) as a characterization method for metal-organic interactions. In: Ostroverkhova O (ed) Handbook of organic materials for electronic and photonic devices, 2nd edn. Woodhead Publishing, New Delhi, pp 529–549
97. Mehta M, Skinner WH, Gardner B, Mosca S, Palombo F, Matousek P, Stone N (2025) Gold nanoraspberries for surface-enhanced Raman scattering: synthesis, optimization, and characterization. ACS Omega 10(5):4588–4598
98. Cao Y, Sun M (2022) Tip-enhanced Raman spectroscopy. Rev Phys 8:100067
99. Bao Y, Zhu M, Zhao X, Chen H, Wang X, Ren B (2024) Nanoscale chemical characterization of materials and interfaces by tip-enhanced Raman spectroscopy. Chem Soc Rev 53(20):10044–10079
100. Costa-Fernández JM, Menéndez-Miranda M, Bouzas-Ramos D, Encinar JR, Sanz-Medel A (2016) Mass spectrometry for the characterization and quantification of engineered inorganic nanoparticles. Trends Anal Chem 84:139–148
101. Comby-Zerbino C, Dagany X, Chirot F, Dugourd P, Antoine R (2021) The emergence of mass spectrometry for characterizing nanomaterials. Atomically precise nanoclusters and beyond. Mater Adv 2(15):4896–4913
102. Pechyen C, Tangnorawich B, Toommee S, Marks R, Parcharoen Y (2023) Green synthesis of metal nanoparticles, characterization, and biosensing applications. Sens Int 5:100287

103. Spaeth P, Adhikari S, Le L, Jollans T, Pud S, Albrecht W, Bauer T, Caldarola M, Kuipers L, Orrit M (2019) Circular dichroism measurement of single metal nanoparticles using photothermal imaging. Nano Lett 19(12):8934–8940
104. Spaeth P, Adhikari S, Baaske MD, Pud S, Ton J, Orrit M (2021) Photothermal circular dichroism of single nanoparticles rejecting linear dichroism by dual modulation. ACS Nano 15(10):16277–16285
105. Härkönen VJ, Rossi T, Kuisma M, Rinke P, Chen X (2024) Enhancement of molecular circular dichroism with silver nanoparticles. Phys Rev B 110(11):115423
106. Lim J, Yeap SP, Che HX, Low SC (2013) Characterization of magnetic nanoparticle by dynamic light scattering. Nanoscale Res Lett 8(1):381
107. Cho EJ, Holback H, Liu KC, Abouelmagd SA, Park J, Yeo Y (2013) Nanoparticle characterization: state of the art, challenges, and emerging technologies. Mol Pharm 10(6):2093–2110
108. Sizochenko N, Mikolajczyk A, Syzochenko M, Puzyn T, Leszczynski J (2021) Zeta potentials (ζ) of metal oxide nanoparticles: a meta-analysis of experimental data and a predictive neural networks modeling. NanoImpact 22:100317
109. Németh Z, Csóka I, Semnani Jazani R, Sipos B, Haspel H, Kozma G, Kónya Z, Dobó DG (2022) Quality by design-driven zeta potential optimisation study of liposomes with charge imparting membrane additives. Pharmaceutics 14(9):1798
110. Bin Mobarak M, Hossain MS, Chowdhury F, Ahmed S (2022) Synthesis and characterization of CuO nanoparticles utilizing waste fish scale and exploitation of XRD peak profile analysis for approximating the structural parameters. Arab J Chem 15(10):104117
111. Burlec AF, Corciova A, Boev M, Batir-Marin D, Mircea C, Cioanca O, Danila G, Danila M, Bucur AF, Hancianu M (2023) Current overview of metal nanoparticles' synthesis, characterization, and biomedical applications, with a focus on silver and gold nanoparticles. Pharmaceuticals 16(10):1410
112. Plascencia-Villa G, Starr CR, Armstrong LS, Ponce A, José-Yacamán M (2012) Imaging interactions of metal oxide nanoparticles with macrophage cells by ultra-high resolution scanning electron microscopy techniques. Integr Biol 4(11):1358–1366
113. Hou Y, Kovács N, Xu H, Sun C, Erni R, Gálvez-Vázquez MDJ, Rieder A, Hu H, Kong Y, Liu M, Wiley BJ, Vesztergom S, Broekmann P (2021) Limitations of identical location SEM as a method of degradation studies on surfactant capped nanoparticle electrocatalysts. J Catal 394:58–66
114. Neděla V, Tihlaříková E, Cápal P, Doležel J (2024) Advanced environmental scanning electron microscopy reveals natural surface nano-morphology of condensed mitotic chromosomes in their native state. Sci Rep 14(1):12998
115. Xu C, Zou R, Peng Y, Liu Q, Ruan S, Hu J (2020) In situ transmission electron microscope studies on one-dimensional nanomaterials: manipulation, properties and applications. Prog Mater Sci 113:100674
116. Ii S (2024) Quantitative characterization by transmission electron microscopy and its application to interfacial phenomena in crystalline materials. Materials 17(3):578
117. Fujita D, Sagisaka K (2008) Active nanocharacterization of nanofunctional materials by scanning tunneling microscopy. Sci Technol Adv Mater 9(1):013003
118. Shnoudeh AJ, Hamad I, Abdo RW, Qadumii L, Jaber AY, Surchi HS, Alkelany SZ (2018) Synthesis, characterization, and applications of metal nanoparticles. In: Tekade RK (ed) Biomaterials and bionanotechnology. Elsevier, Amsterdam, pp 527–612
119. Witte AB, Leistra AN, Wong PT, Bharathi S, Refior K, Smith P, Kaso O, Sinniah K, Choi SK (2014) Atomic force microscopy probing of receptor-nanoparticle interactions for riboflavin receptor targeted gold-dendrimer nanocomposites. J Phys Chem B 118(11):2872–2882
120. Delvallée A, Feltin N, Ducourtieux S, Trabelsi M, Hochepied J (2015) Direct comparison of AFM and SEM measurements on the same set of nanoparticles. Measur Sci Technol 26:085601

121. Sharma D, Gautam S, Singh S, Srivastava N, Khan AM, Bisht D (2025) Unveiling the nano-world of antimicrobial resistance: integrating nature and nanotechnology. Front Microbiol 15:1391345
122. Aabed K, Mohammed AE (2021) Synergistic and antagonistic effects of biogenic silver nanoparticles in combination with antibiotics against some pathogenic microbes. Front Bioeng Biotechnol 9:652362
123. Gautam D, Dolma KG, Khandelwal B, Gupta M, Singh M, Mahboob T, Teotia A, Thota P, Bhattacharya J, Goyal R, Oliveira S, Pereira ML, Wiart C, Wilairatana P, Eawsakul K, Rahmatullah M, Saravanabhavan SS, Nissapatorn V (2023) Green synthesis of silver nanoparticles using *Ocimum sanctum* Linn and its antibacterial activity against multidrug resistant *Acinetobacter baumannii*. PeerJ 11:e15590
124. Jahan N, Hussain N, Touqeer SI, Khalil-Ur-Rahman SH, Abbas N (2024) Formulation of *Mentha piperita*-based nanobiopesticides and assessment of the pesticidal and antimicrobial potential. Life 14(1):144
125. Singaravelu S, Motsoene F, Abrahamse H, Dhilip Kumar SS (2025) Green-synthesized metal nanoparticles: a promising approach for accelerated wound healing. Front Bioeng Biotechnol 13:1637589
126. Balkrishna A, Kumar A, Arya V, Rohela A, Verma R, Nepovimova E, Krejcar O, Kumar D, Thakur N, Kuca K (2021) Phytoantioxidant functionalized nanoparticles: a green approach to combat nanoparticle-induced oxidative stress. Oxidative Med Cell Longev 2021:3155962
127. Alshahrani MY, Ibrahim EH (2022) Gold nanoparticles (AuNPs) and *Rosmarinus officinalis* extract and their potentials to prompt apoptosis and arrest cell cycle in HT-29 colon cancer cells. J King Saud Univ Sci 34(8):102304
128. Jadczak P, Kulpa D, Drozd R, Przewodowski W, Przewodowska A (2020) Effect of AuNPs and AgNPs on the antioxidant system and antioxidant activity of lavender (Lavandula angustifolia mill.) from in vitro cultures. Molecules 25(23):5511
129. Lv Y, Li W, Liao W, Jiang H, Liu Y, Cao J, Lu W, Feng Y (2024) Nano-drug delivery systems based on natural products. Int J Nanomed 19:541
130. Barathi S, Ramalingam S, Krishnasamy G, Lee J (2024) Exploring the biomedical Frontiers of plant-derived nanoparticles: synthesis and biological reactions. Pharmaceutics 16(7):923
131. Gaffar NA, Zahid M, Asghar A, Shafiq MF, Jelani S, Rehan F (2024) Biosynthesized metallic nanoparticles: a new era in cancer therapy. Arch Pharm 357(7):2300712
132. Ahmad N, Ansari MA, Al-Mahmeed A, Joji RM, Saeed NK, Shahid M (2024) Biogenic silver nanomaterials synthesized from *Ocimum sanctum* leaf extract exhibiting robust antimicrobial and anticancer activities: exploring the therapeutic potential. Heliyon 10(15):e35486
133. Mishra RK, Mentha SS, Misra Y, Dwivedi N (2023) Emerging pollutants of severe environmental concern in water and wastewater: a comprehensive review on current developments and future research. WENex 6:74–95
134. Hazrati Saadabadi R, Shariatmadar Tehrani F, Sabouri Z, Darroudi M (2024) Photocatalytic activity and anticancer properties of green synthesized ZnO-MgO-Mn_2O_3 nanocomposite via *Ocimum basilicum* L. seed extract. Sci Rep 14(1):29812
135. Eddy NO, Garg R, Garg R, Ukpe RA, Abugu H (2023) Adsorption and photodegradation of organic contaminants by silver nanoparticles: isotherms, kinetics, and computational analysis. Environ Monit Assess 196(1):65
136. Aragaw TA, Bogale FM, Aragaw BA (2021) Iron-based nanoparticles in wastewater treatment: a review on synthesis methods, applications, and removal mechanisms. J Saudi Chem Soc 25(8):101280
137. Mejía YR, Reddy Bogireddy NK (2022) Reduction of 4-nitrophenol using green-fabricated metal nanoparticles. RSC Adv 12(29):18661–18675
138. Puiatti GA, De Carvalho JP, De Matos AT, Lopes RP (2022) Green Synthesis of Fe0 nanoparticles using *Eucalyptus grandis* leaf extract: characterization and application for dye degradation by a (photo)Fenton-like process. J Environ Manag 311:114828

139. Zanbili F, Poursattar Marjani A (2025) Innovative green and bio-based approaches for photosensitive nanoparticle synthesis: a review on methodologies, characterization, and applications. Micro Nano Syst Lett 13(1):1–36
140. Garg D, Sridhar K, Stephen Inbaraj B, Chawla P, Tripathi M, Sharma M (2023) Nano-biofertilizer formulations for agriculture: a systematic review on recent advances and prospective applications. Bioengineering 10(9):1010
141. Sembada AA, Lenggoro IW (2024) Transport of nanoparticles into plants and their detection methods. Nano 14(2):131
142. Sharma P, Urfan M, Anand R, Sangral M, Hakla HR, Sharma S, Das R, Pal S, Bhagat M (2022) Green synthesis of zinc oxide nanoparticles using *Eucalyptus lanceolata* leaf litter: characterization, antimicrobial and agricultural efficacy in maize. Physiol Mol Biol Plants 28(2):363–381
143. Zafar S, Bilal M, Ali MF, Mahmood A, Kijsomporn J, Wong LS, Kumar MHV, Alotaibi SS (2024) Nano-biofertilizer an eco-friendly and sustainable approach for the improvement of crops under abiotic stresses. Environ Sustain Ind 24:100470
144. Khatoon A, Khan F, Ahmad N, Shaikh S, Rizvi SMD, Shakil S, Al-Qahtani MH, Abuzenadah AM, Tabrez S, Ahmed ABF, Alafnan A, Islam H, Iqbal D, Dutta R (2018) Silver nanoparticles from leaf extract of *Mentha piperita*: eco-friendly synthesis and effect on acetylcholinesterase activity. Life Sci 209:430–434
145. Ibarra-Laclette E, Blaz J, Pérez-Torres A, Villafán E, Lamelas A, Rosas-Saito G, Ibarra-Juárez LA, García-Ávila J, Martínez-Enriquez AI, Pariona N (2022) Antifungal effect of copper nanoparticles against *fusarium kuroshium*, an obligate symbiont of *Euwallacea kuroshio Ambrosia beetle*. J Fungi 8(4):347
146. Kumari S, Khanna RR, Nazir F, Albaqami M, Chhillar H, Wahid I, Khan MIR (2022) Biosynthesized nanoparticles in developing plant abiotic stress resilience: a new boon for sustainable approach. Int J Mol Sci 23(8):4452
147. Doğaroğlu ZG, Uysal Y, Çaylalı Z, Karakulak DS (2023) Green nanotechnology advances: green manufacturing of zinc nanoparticles, characterization, and foliar application on wheat and antibacterial characteristics using *Mentha spicata* (mint) and *Ocimum basilicum* (basil) leaf extracts. Environ Sci Pollut Res Int 30(21):60820–60837
148. Kumari A, Gupta AK, Sharma S, Jadon VS, Sharma V, Chun SC, Sivanesan I (2024) Nanoparticles as a tool for alleviating plant stress: mechanisms, implications, and challenges. Plants 13(11):1528
149. Couto C, Almeida A (2022) Metallic nanoparticles in the food sector: a mini-review. Foods 11(3):402
150. de Lima AF, Leite RHL, Pereira MWF, Silva MRL, de Araújo TLAC, de Lima Júnior DM, Gomes MNB, Lima PO (2024) Chitosan coating with rosemary extract increases shelf life and reduces water losses from beef. Foods 13(9):1353
151. Rathnasinghe NL, Kaushani KG, Rajapakshe PS, De Silva A, Jayasinghe RA, Liyanage RN, Tissera ND, Wijesena RN, Priyadarshana G (2024) Current trends on unique features and role of nanomaterials in personal care products. Cosmetics 11(5):152
152. Majerič P, Jović Z, Švarc T, Jelen Ž, Horvat A, Koruga D, Rudolf R (2022) Physicochemical properties of gold nanoparticles for skin care creams. Materials 16(8):3011
153. Xulu JH, Ndongwe T, Ezealisiji KM, Tembu VJ, Mncwangi NP, Witika BA, Siwe-Noundou X (2022) The use of medicinal plant-derived metallic nanoparticles in theranostics. Pharmaceutics 14(11):2437
154. Alhaji N, Sujatha S (2021) Synthesis of plant-mediated metal silver nanoparticles for fabric coating. Mater Today Proc 62:5684–5689
155. Thatyana M, Dube NP, Kemboi D, Manicum A-LE, Mokgalaka-Fleischmann NS, Tembu JV (2023) Advances in phytonanotechnology: a plant-mediated green synthesis of metal nanoparticles using Phyllanthus plant extracts and their antimicrobial and anticancer applications. Nano 13(19):2616

156. Acharya P, Jayaprakasha GK, Semper J, Patil BS (2020) ^{1}H nuclear magnetic resonance and liquid chromatography coupled with mass spectrometry-based metabolomics reveal enhancement of growth-promoting metabolites in onion seedlings treated with green-synthesized nanomaterials. J Agric Food Chem 68(46):13206–13220
157. Dikshit PK, Kumar J, Das AK, Sadhu S, Sharma S, Singh S, Gupta PK, Kim BS (2021) Green synthesis of metallic nanoparticles: applications and limitations. Catalysts 11(8):902
158. Karnwal A, Jassim AY, Mohammed AA, Sharma V, Al-Tawaha ARMS, Sivanesan I (2024) Nanotechnology for healthcare: plant-derived nanoparticles in disease treatment and regenerative medicine. Pharmaceuticals 17(12):1711
159. Martínez G, Merinero M, Pérez-Aranda M, Pérez-Soriano EM, Ortiz T, Begines B, Alcudia A (2020) Environmental impact of nanoparticles' application as an emerging technology: a review. Materials 14(1):166
160. Paliwal R, Babu RJ, Palakurthi S (2014) Nanomedicine scale-up technologies: feasibilities and challenges. AAPS PharmSciTech 15(6):1527–1534
161. Scott-Fordsmand JJ, Amorim MJB (2023) Using machine learning to make nanomaterials sustainable. Sci Total Environ 859(Pt 2):160303

Aromatic Plants and Their Role in Metal Oxide Nanoparticle Synthesis, Characterisation and Applications

Mohd Arsh Khan, Neda Tabassum, Abdul Rahman Khan, Fakhra Jabeen, and Qazi Inamur Rahman

Abstract The rapid advancement of nanotechnology has sparked growing interest in sustainable, green approach for fabrication of nanoparticles (NPs). Aromatic plants, which are rich in phytochemicals such as flavonoids, terpenoids, phenolics and alkaloids, offer an eco-friendly and cost-effective alternative to conventional chemical methods. This chapter highlights the significant role of aromatic plant extracts in the biosynthesis of metal oxide nanoparticles (MONPs), emphasising functions of phytochemical as reducing and stabilising agent and how various physicochemical parameters influence NP growth. Moreover, the chapter also explores the use of different analytical techniques such as UV-Vis spectroscopy, Fourier-transform infrared spectroscopy (FTIR), X-ray diffraction (XRD) and zeta potential to characterise the properties of synthesised NPs. In addition, the application of biosynthesised MONPs in catalysis, agriculture, environmental remediation and biomedicine has been overviewed, demonstrating their multifunctional potential.

Keywords Aromatic plants · Catalysis · Environmental remediation · Biomedicine · Green synthesis

1 Introduction

Nanoparticles (NPs) are typically defined as particles with dimensions ranging from 1 nm to 100 nm. However, in certain biotechnology and biomedical applications, their size range can be expanded up to 500 nm. Owing to exceptional physical, chemical and electrical properties, NPs have developed across various domains such as catalysis, electronics, targeted medication administration, water sensing and

M. A. Khan · N. Tabassum · A. R. Khan · Q. I. Rahman (✉)
Department of Chemistry, Integral University, Lucknow, Uttar Pradesh, India
e-mail: qirahman@iul.ac.in

F. Jabeen
Department of Physical Sciences, College of Science, Jazan University, Jazan, Saudi Arabia

A. Husen (ed.), *Sustainable Uses of Aromatic Plants in Nanotechnology*, Smart Nanomaterials Technology, https://doi.org/10.1007/978-981-95-7409-4_5

treatment, corrosion inhibition, oil recovery, etc. [1–3]. Typically, synthesised NPs are metallic and exhibit a phenomenon known as surface plasmon resonance (SPR), which contributes significantly to the quantum mechanical effects of light in UV-visible region, leading to unique optoelectronic and photonic properties, and changes in the size and shapes can significantly alter their light absorption behaviour and interaction with other particles [2, 4–9].

1.1 Nanotechnology and Green Synthesis

There is presently proof that the synergy of those substances is what gives plant-based synthesis its advantages. As a result, a sustainable method of creating nano-sized materials is to employ stabilisers and reducing agents with biological origin, such as microbiological entities and various other resources [10]. Their naturally complex biological machinery produces distinct NPs with varying chemical compositions, sizes and shapes which are used in microbe-mediated construction products [11]. However, scaling up can occasionally be difficult when it comes to microbiological preparations. Plant-based extracts provide a simple solution to this problem, and as a result production rates can be increased. Alkaloids, phenol and alcohol are just a few of the substances found in plants that have the ability to dissolve metallic ions into NPs that are quite stable [12, 13]. A very general approach for the "green synthesis of nano-sized material" is shown in Fig. 1.

1.2 MONs and Their Applications

With their many uses in optical and electronic devices, UV absorbents, adhesives, bio-imaging, catalytic sensing, coatings and other biomedical applications, metal oxide NPs (MONs) are an intriguing new technology. This has led to the need for their easy and repeatable preparation with improved switch ended their size, shape, crystal structure and surface chemistry. The creation of MON/polymer nanocomposites is among the most significant uses of MONs [14]. Because metal oxides provide special qualities in coatings, photocatalysis and UV absorbers MON/polymer nanocomposites have gathered a lot of interest in addition to carbon nanotube [15].

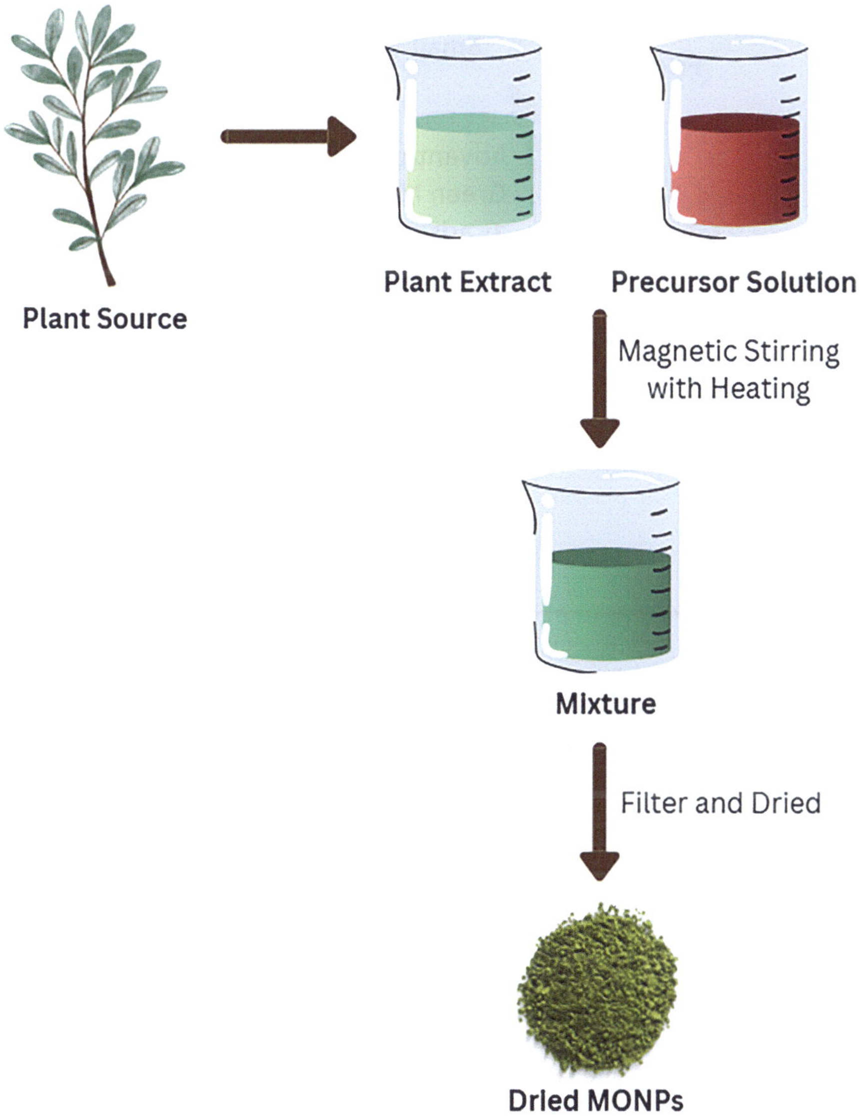

Fig. 1 Mechanistic overview of green synthesis of MONPs

Fig. 2 Various advantages of green MONPs as adsorbents

Insufficient MON density and dispersion can potentially impair the intrinsic properties of polymers in addition to making it impossible to achieve the intended attributes of nanocomposites. One of the better solutions to this problem is surface functionalisation [16–19]. MON surfaces can be altered by reacting with a coupling agent growing or linking polymer chains, and exchange existing capping agents. According to Lu et al. [20], they can be made in a lab setting from organic or inorganic precursors or by extracting plant leaves and flowers [21–26] (Fig. 2).

1.3 Sustainability in Nanoparticle Synthesis

The synthesis of NPs using traditional chemical processes has a number of drawbacks despite its enormous potential such as the usage of dangerous reagents and the production of hazardous waste [27]. Because of these challenges, sustainable methods for synthesising NPs have to be investigated [28]. Because of their distinct physicochemical characteristics that set them apart from their bulk complements, NPs defined as particles with sizes between 1 nm and 100 nm have attracted a lot of attention lately. According to Edo et al. [29], biosynthesis is the process of creating complex molecules using biological systems. Biosynthesis is relevant to the manufacture of NPs because it offers a more sustainable and environmentally friendly way than conventional chemical procedures [30, 31]. In addition to minimising the negative effects on the environment, using plant extracts in the production of NPs lowers the expenses related to the usage of hazardous chemicals and sophisticated equipment [32].

2 Aromatic Plants: A Biogenic Reservoir for Nanoparticle Synthesis

2.1 *Phytochemical Profile and Mechanistic Roles*

Biologically active naturally occurring chemical compounds found in plant phytochemicals offer additional health benefits to humans beyond those associated with macronutrients and micronutrients. Phytomedicines have included plant-based ingredients since the beginning of time. This can be derived from stem, bark, flowers, fruits and seeds [33–37]. Soxhlet, hydro-distillation solvent extraction, supercritical fluid extraction, microwave-assisted extraction (MAE), ultrasound-assisted extraction, pulsed electrical field extraction, enzyme-assisted extraction and high hydrostatic pressure extraction are some of the common extraction methods [38]. Because of their enormous surface area-to-volume ratio, which allows them to exhibit a variety of mechanical, physical, chemical NPs are completely different from bulk particles and can be used in a variety of contexts. Since metallic NPs have so many uses, an effective method for their synthesis is needed because the traditional chemical ways of creating them have drawbacks including expensive costs and detrimental effects on the environment. Green approaches to NP synthesis have become a viable and economical way to get around these problems. Unlike the chemical method for synthesising NPs, the green method for amalgamating NPs is both economical and environmentally beneficial. An alternative to the plant extract can be used to create NPs of different sizes. Furthermore, dangerous chemicals and outside stabilising agents are not needed [39]. The chemical general structure of flavonoids, which are found in many plants, is a 15-carbon skeleton made up of two benzene rings joined by a three-carbon bridge to form a heterocycle. By preventing the growth of new cancer cells, preventing carcinogens from reaching their activation sites and reducing the toxicity of certain substances by blocking their metabolism, a number of flavonoids have demonstrated chemopreventive benefits [40]. Because flavonoids can alter the signalling pathways that cause cancer cells to grow and proliferate, their potential as anticancer agents has been investigated. Furthermore, flavonoids control the start of angiogenesis, invasion and apoptosis. Nevertheless, flavonoids are hydrophobic and unstable in physiological fluids which restrict their biological activity and bioavailability [41]. Phenolics are highly reactive, sensitive substances that have immediate impacts when added to food systems. Thus, it is crucial to measure the rate at which phenolics are released from food samples. Over time, there was an increase in the release of phenolic chemicals, and the difference was statistically significant ($p < 0.05$) [42].

Alkaloids and other bioactive substances generated from plants mediate a range of metals and metal oxide-based NPs and have anticancer, antioxidant, anti-inflammatory and catalytic properties. Even if the effectiveness of these compounds has been proven, not enough research has been done to assess their safety and effectiveness in various scenarios, and not many studies have focused on the production of NPs employing alkaloid-mediated NP synthesis [43].

The most prevalent and structurally unique plant hydrocarbons that can convert metallic ions into their equivalent NPs are terpenoids. Generally speaking, terpenoids have strong antimicrobial qualities. Nevertheless, there has not been much in-depth research done on the synthesis mode of action and application. The importance of terpenoids, their role in the biological synthesis of metal NPs, their ability to reduce and stabilize metal ions, their biomedical uses such as anticancer and anti-fouling applications, and the possible future developments of terpenoid-mediated colloidal NPs has been also reported [44].

The pH of the surrounding medium determines the net surface charge of proteins, which are polypeptides with a certain shape. A number of factors, including van der Waals contacts, solvation forces and hydrogen bonds, facilitate the adsorption of proteins at the nanobio interface. A complex process, the overall NP-PC formation is influenced by the media, interacting proteins and the properties of the NP. The duration of each protein's interaction with the NP surface is determined by its unique rates of attachment and dissociation (Table 1).

Table 1 Mechanism of phytochemicals and their role in nanoparticle biosynthesis

Phytochemical class	Biochemical mechanism	Role in nanoparticle biosynthesis	Effect on nanoparticle properties
Flavonoids, phenolic acids, vitamins	Metal ion reduction	Convert metal ions into stable NPs; facilitate reduction	Enhance antioxidant activity, influence shape and size
Flavonoids, proteins, alkaloids, tannins	Capping and stabilisation	Prevent agglomeration and stabilise NPs	Influence stability, antimicrobial activity
Amino acids, terpenoids, polysaccharides	Shape and size regulation	Control nanoparticle morphology and size	Contribute to functionalisation, improve biocompatibility
Polyphenols, reducing sugars	Impact of redox potential	Govern synthesis rate and efficiency	Prevent oxidative degradation
Tannins, phenolics, amino acids	Chelation mechanism	Enhance ion stability and prevent unwanted reactions	Provide bioactive coatings
Lipids, glycoproteins	Biomolecular coating	Aid in biocompatibility and biomedical targeting	Improve surface functionalisation
Amino acids, sugars	Nucleation control	Initiate nanoparticle formation	Influence nucleation sites and particle distribution
Polyphenols, sugars	Electron transfer mechanisms	Facilitate nanoparticle growth via electron transfer	Enhance nanoparticle uniformity
Proteins, polysaccharides	Aggregation prevention	Enhance colloidal stability	Improve dispersion and suspension
Chlorophyll, carotenoids	Photoactivation mechanism	Light-induced synthesis via phytochemical excitation	Provide optical and photocatalytic functionality

3 Green Synthesis Using Aromatic Plants

Green synthesis refers to procedures that are ecologically approachable which generate NPs by using plant-based extracts as stabilising and reducing agents. This method promotes sustainability and green chemistry while avoiding the usage of hazardous chemicals. It has been noticed that pH, temperature, metal ion concentration and reaction time are some of the environmental variables that influence biogenesis of NPs [45–51]. Higher temperatures have the potential to improve the reduction reaction kinetics and hasten the formation of NPs. However, unintended side reactions may also be encouraged by extremely high temperatures leading to the creation of bigger aggregates as opposed to distinct NPs [52]. Increased nucleation rates could result from higher concentrations, but, if improperly managed, they could also encourage agglomeration. Although longer reaction periods may result in larger NPs, they may also make aggregation more likely [53]. Researchers can aid in the creation of ecologically friendly and supportable nanotechnology by utilising the special qualities of plant extracts and the inherent procedures that lead to the group of NPs.

Exploiting plant extracts for the manufacture of NPs is a complicated procedure that is impacted by a number of chemical and environmental factors. In order to optimise the synthesis procedure and produce NPs with the necessary size, shape, stability and usefulness, it is imperative to recognise these parameters [54]. This section examines important factors that affect the production of NPs such as pH, temperature and metal ion concentration.

3.1 pH

The synthesis of NPs can be greatly influenced by the pH of the reaction medium, which impacts not only the stability and solubility of metal ions but also the chemical reactivity of the phytochemicals utilised in the reduction process [55]. A slightly alkaline pH (approx. 7–9) can boost the reduction efficiency of flavonoids and phenolics, resulting in a higher production of NPs; severe acidic or basic conditions can denature particular biomolecules, decreasing their usefulness [56]. While lower pH values could trigger particle aggregation due to reduced surface charge, producing larger and less stable particles, higher pH levels may result in stronger electrostatic repulsion between NPs, minimising agglomeration and improving stability [57].

3.2 Temperature

In the biosynthesis of NPs, temperature is crucial since it affects the kinetics of the reaction as well as the characteristics of the final NPs. Faster reaction rates are typically the result of molecules' increased kinetic energy at higher temperatures [58].

In order to balance reaction kinetics with the stability of reducing agents' temperature, optimisation is therefore crucial [59].

3.3 Concentration of Metal Ions

There might not be enough metal ions present at lower concentrations to start nucleation which would result in smaller nanoparticle yields [60]. Due to restricted metal availability, lower concentrations frequently result in smaller NPs, whereas greater concentrations may produce larger NPs. NPs can be customised for certain uses by taking advantage of this interaction. Higher concentrations may encourage combination due to enhanced van der Waals forces, while lower concentrations can result in more stable and well-dispersed NPs.

3.4 Other Influential Factors

Thus, the selection of plant material can have a big impact on the production of NPs [61]. The homogeneity of the reaction is also affected by agitation and mixing. The reduction process is improved by stirring or shaking to improve interaction between the metal ions and plant extract [62].

4 Characterisations

The problem of characterising nanostructures was addressed by the employment of highly developed sophisticated instruments. FTIR, UV-Vis spectroscopy and zeta potential are used to verify the chemical composition, stability and purity [47].

4.1 UV-Vis Absorption Spectroscopy

The behaviour of light that is engrossed and detached by sample NPs is frequently studied using UV-visible spectroscopy. As the size of semiconductor NPs varies, so do the colours of their absorption spectra and the SPR band in metallic NPs. According to Jayachandran et al. [63], the UV-Vis absorption spectra revealed that the existence of secondary metabolites in *Cayratia pedata* extract was serving as a capping and reducing agent for the bio-fabrication of zinc oxide NPs. A peak at 320 nm was accredited to the creation of zinc oxide NPs.

4.2 Fourier Transform Infrared Spectroscopy

It is a crucial analytical instrument that is frequently employed to clarify the structure of both organic mixtures and individual molecules. Modulated mid-infrared energy is used in FTIR spectroscopy to examine a material. Absorption happens when the bond energy and the energy of undulating motion are equivalent.Organic compounds absorb infrared radiation and can be easily characterized by FTIR because they typically contain functional groups with atoms such as nitrogen, oxygen, and phosphorus. The use of plant extract from *Salvia officinalis* in the manufacture of ZnO NPs without the need of any other additives has recently been successfully investigated by Abomuti et al. [64]; the FTIR spectra of leaf extract revealed distinct peaks at 891/cm for 1- and 2-degrees N-H groups, and peaks at 1089 and 1026/cm represented stretching vibrations for -N-H, -C-O and 5C-H for various aliphatic amine, phenol and carboxylic acid groups.

By serving as both a capping and stabilising agent, these secondary metabolites may be essential to the biological synthesis of zinc oxide NPs. Unlikely, it has been noted that biological components of plant extract were still present on the surface of zinc oxide NPs after they were repeatedly cleaned with H_2O and ethanol and dried at 80 °C. These biological components vanished from the FTIR spectrum when the sample was calcined at 400 °C. However, the creation of ZnO NPs was confirmed by the appearance of peaks at 453 and 456/cm, which were ascribed to the Zn-O stretching vibration [65].

4.3 Other Important Characterisation Techniques

To ascertain the mechanical, chemical, microstructural and physical characteristics of materials, various analytical methods are employed such as XRD and zeta potential analysers. Since secondary metabolites which are in charge of the bio-decrease of metal ions constantly aid in the biogenic synthesis of NPs, zeta potential is very useful instrument for figuring out the stability and net surface charge change of the produced NPs.

5 Applications

ZnO and TiO_2 NPs derived from lemon fruit have shown anti-phytopathogenic action against the soft rot bacterium pathogen *Dickeyadadantii*. Another study on ZnO NPs generated using the extract of *Eucalyptus globulus* demonstrated the benefit of biogenic NPs in agriculture by finding that the NPs had fungicidal activity against apple orchard pathogens without harming the plants. These improvements contribute to sustainable farming practices by enhancing plant growth while limiting the excessive use of chemical fertilisers [66]. The use of *Punica granatum* (pomegranate) peel extracts in NP synthesis indicates an eco-friendly strategy that not only decreases environmental effect but also increases biological interactions in agricultural systems. These biologically synthesised NPs have demonstrated potential for enhancing soil health, stimulating plant growth and providing antimicrobial properties towards plant pathogens [67]. Owing to their small size, NPs are also utilised in agriculture for plant nutrition, plant protection, soil remediation, etc. The antimicrobial effect of NPs outlined in the preceding section can be used to protect crops against microbes as well as pests.

An important environmental hazard is the presence of non-biodegradable colours in water contamination from sewage waste and effluents from the paint, textile and dye industries. To combat this, scientists are always creating more eco-friendly and effective technology [68]. When organic and inorganic contaminants are absorbed on the outermost layer of ZnO, the OH radicals produced by the photocatalytic activity of the ZnO NPs react with them to produce harmless H_2O, CO_2 and inorganic compounds [69]. CuO NPs synthesised by *Rheum palmatum* root extract have been reported to break down the dyes rhodamine B and methylene blue [70] (Fig. 3).

Fig. 3 Application of NPs in various disciplines of science

6 Plant-Derived Metal Oxide NPs

Engineered metal oxide NPs have been shown to have an array of applications. In addition to increasing the rate of photosynthesis, TiO_2-based NPs have been shown to improve soybean nitrate reductase activity and fertiliser adsorption capacity [71]. Naturally occurring TiO_2 NPs have the ability to generate reactive oxygen species (ROS) when exposed to ultraviolet light [72]. ZnO NPs are suited for usage as semiconductors because of their significant excitation energy (around 60 meV) and band-gap energy of 3.37 eV [73].

6.1 Few Examples of Metal Oxide Synthesis

- *ZnO NPs*: ZnO NPs were shaped using an extract from *Pongamia pinnata* leaves. The characterisation showed that the NPs were highly pure and spherical with an average size of 100 nm [74]. Malachite green dye was shown to be degraded by *Rubus coreanus*, which was employed as a reducing agent for the fabrication of ZnO NPs [75].
- *CuO NPs*: TiO_2 NPs were synthesised from *Jatropha curcas*, while CuO NPs were obtained from *Aloe barbadensis* extracts [76]. CuO NPs have been synthesised using gum karaya. According to reports, CuO NPs from *Aloe vera* leaf extract have a monoclinic structure and a 20 nm particle size [77, 78]. Crystalline monoclinic black bean extract was used to create CuO NPs, which have an average size of 26.6 nm [79]. The reaction between $CuSO_4$ and OH_2, which is produced by the ionisation of the water molecule, is the suggested synthesis mechanism for CuO NPs. In an aqueous media, $CuSO_4$ stays ionised and can create $[Cu(H_2O)_6]$.
- *TiO_2 NPs*: The biosynthesis of TiO_2 NPs from several plants has been documented, and the TiO_2 NPs are readily replicable. TiO_2 NPs with a size range of 36–68 nm were created using the aqueous extract of *Eclipta prostrata* [80].
- *Comparative assessment*: The benefits of the plant-based green production of NPs have been repeatedly shown in recent research that have compared it with chemical approaches. A sustainable and eco-friendly method for producing nanomaterials with a wide range of applications, including metal and metal oxide NPs, is green synthesis [81]. Particularly for biomedical applications, plant-based green synthesis guarantees biocompatible NPs making it a secure and environmentally responsible substitute for chemical approaches [82, 83] (Fig. 4).

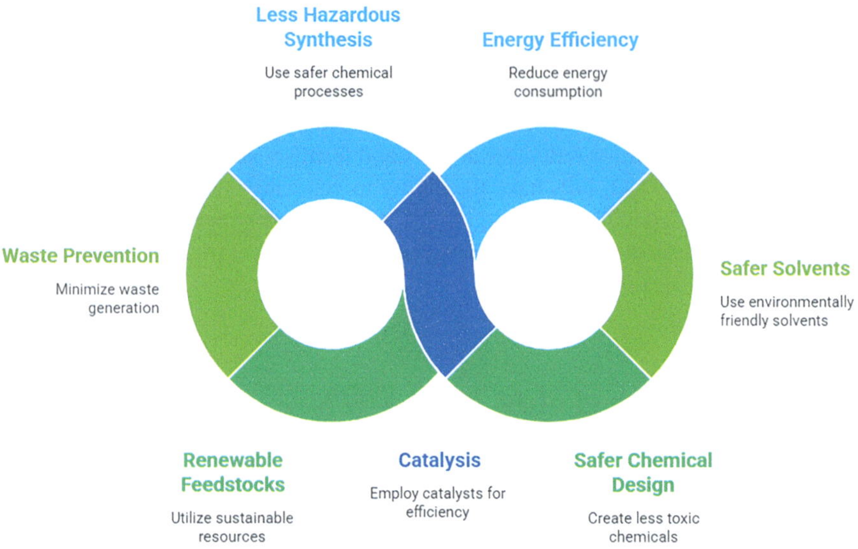

Fig. 4 Principles of green chemistry for a safe as well as sustainable environment

Accordingly, Aravind et al. [84] used both chemical and green synthesis techniques to create titanium dioxide NPs (TiO_2 NPs). Their existence in the visible spectrum was verified using UV-visible spectroscopy. Randomly organised spherical NPs were seen in SEM pictures. When compared to chemically synthesised NPs, the green-synthesised TiO_2 NPs showed better photocatalytic degradation efficiency for MB dye. These results imply that the green-synthesised TiO_2 NPs have the potential to be used in biological and environmental domains [85].

Using readily available plants instead of expensive chemicals or metal precursors makes the ecologically benign, plant-based synthesis of NPs cost-effective in addition to the previously listed advantages. Because it allows for a range of nanoparticle compositions and surface functionalities, it provides a high degree of tolerance and biocompatibility. Plant extracts not only work as natural reducing agents and stabilisers, but they also give precise control over the composition, size and form. Interestingly, a combination of plant extracts can enhance the properties of NPs. Plant-based green synthesis is therefore a controlled, adaptable and cost-effective technique for producing NPs with a range of biomedical applications [86, 87]. Green synthesis eliminates harmful solvents and requires little equipment which lowers costs. This approach is very sustainable and appropriate for use in agriculture and biomedicine due to the necessary toxicity and environmental effect (Table 2).

Table 2 Comparison of conventional chemical methods vs. emerging green/biosynthetic approaches

Parameter	Chemical synthesis techniques	Green/biosynthetic methods
Product yield	Typically high, but depending on the availability of reagents	Moderate, subject to biological fluctuations [88]
Material stability	Generally stable due to the use of synthetic stabilisers	Varies with biomolecular interactions [89]
Size uniformity	Consistently narrow size range	Broader distribution arising from natural diversity [90]
Energy usage	High consumption due to the heating and stirring process	Low energy usage, functions under mild, ambient conditions [91]
Scalability	Proven viability for industrial-scale operations	Requires further development for upscaling [92]
Toxicity risks	Possibility of toxic residual by-products	Nontoxic and environmentally benign [93]
Economic viability	Expensive due to chemical inputs and purification steps	Economical by leveraging natural, readily available material [94]
Purification requirements	Needs thorough and complex purification procedures	Minimal owing to the use of safe, nonhazardous reactants [95]
Eco-friendliness	Produces harmful by-products and waste	Sustainable and environmentally conscious [96]
Mechanistic clarity	Well-understood mechanisms with consistent reproducibility	Requires more research to establish a reliable mechanistic understanding [97]

7 Limitations and Future Directions

Additionally, evergreen plant sources are widely available and convenient which gives them a major edge over other traditional techniques. However, the feasibility of bulk synthesis of MONPs utilising plant sources is still threatened by significant issues with the method's application, energy requirements and possible harmful effects of synthesised compounds. Concerns have also been raised about MONP toxicity and bioaccumulative nature. A small number of previous studies have examined the harmful effects of MONPs on both single-celled and multicelled species, including fungi and bacteria.

Furthermore, the safety and harmless use of MONPs are seriously called into question due to the vast absence of sophisticated and thorough research on how MONPs interact with plants, animals and human health as well as the paucity of discussion regarding the ecotoxicological effects. Since even little variations in the quality of plant extract can result in notable morphological and qualitative differences in MONPs, emphasis must be placed on evaluating the kind and calibre of plant products being utilised for synthesis. The best results can be obtained without requiring significant changes to synthesis processes over time by addressing limitations related to the separation, purification, homogeneity and uniformity of synthesised compounds.

8 Conclusion

Green synthesis of metal oxide NPs with the use of aromatic plants exemplifies a radical change in nanoscience, thus keeping the balance between the extreme and the environment. Phytochemicals from aromatic plants not only provide the electron but also envelop the particle and allow carrying out of nanoparticle, on parameters like size, shape and surface functionality. Various studies of the method have confirmed a good structural and functional performance of these NPs, while the scope of their application is growing in the fields of agriculture, catalysis, environmental remediation and health. On the other hand, problems met in scaling-up procedures, reproducibility, variability of phytochemicals and ecotoxicological aspects require attention and adjacent effect: standardised protocols and multidisciplinary research. Through development and perseverance, the facilitated aromatic plant's synthesis approach will bring about the nanotechnologies of the future which are not only reliable but are easily integrated into industrial and environmental applications as well.

Competing Interests The authors declare that they have no competing interest.

Declaration of Generative AI and AI-Assisted Technologies in the Writing Process During the preparation of this work, the book chapter author(s) have not taken any assistance from AI, and the book chapter author(s) reviewed and edited the content as needed and take(s) full responsibility for the content of the publication.

References

1. Al-Anssari S, Ali M, Alajmi M, Akhondzadeh H, Khaksar Manshad A, Kalantariasl A et al (2021) Synergistic effect of NPs and polymers on the rheological properties of injection fluids: implications for enhanced oil recovery. Energy Fuel 35(7):6125–6135. https://doi.org/10.1021/acs.energyfuels.1c00105
2. Husen A, Siddiqi KS (2023) Advances in smart nanomaterials and their applications. Elsevier, Cambridge, MA. https://doi.org/10.1016/C2021-0-02202-1
3. Kaviya S (2020) Synthesis, self-assembly, sensing methods and mechanism of bio-source facilitated nanomaterials: a review with future outlook. Nano Struct Nano Objects 23:100498. https://doi.org/10.1016/j.nanoso.2020.100498
4. Jin-Chul K, Madhusudhan A, Husen A (2021) Smart nanomaterials in biomedical applications. Springer, Cham. https://doi.org/10.1007/978-3-030-84262-8
5. Lenders V, Koutsoumpou X, Sargsian A, Manshian BB (2020) Biomedical nanomaterials for immunological applications: ongoing research and clinical trials. Nanoscale Adv 2(11):5046–5089. https://doi.org/10.1039/D0NA00478B
6. Madhusudhan A, Purohit SD, Prasad R, Husen A (2024) Functional smart nanomaterials and their theranostics approaches. Springer, Singapore. https://doi.org/10.1007/978-981-99-6597-7
7. Das D, Chakraborty D, Barman S (2024) Emerging role of nanotechnology-based devices for detection of environmental contaminants. In: Mandal AK, Ghorai S, Husen A (eds) Functionalized smart nanomaterials for point-of-care testing. Smart nanomaterials technology. Springer, Singapore. https://doi.org/10.1007/978-981-99-5787-3_11

8. Mitarotonda R, Giorgi E, Eufrasio-da-Silva T, Dolatshahi-Pirouz A, Mishra YK, Khademhosseini A et al (2022) Immunotherapeutic NPs: from autoimmune disease control to the development of vaccines. Biomater Adv 135:212726. https://doi.org/10.1016/j.bioadv.2022.212726
9. Siddiqi KS, Husen A (2017) Recent advances in plant-mediated engineered gold nanoparticles and their application in biological system. J Trace Elem Med Biol 40:10–23. https://doi.org/10.1016/j.jtemb.2016.11.012
10. Potbhare AK, Chaudhary RG, Chouke PB, Yerpude S, Mondal A, Sonkusare VN et al (2019) Phytosynthesis of nearly monodisperse CuO nanospheres using Phyllanthus reticulatus/Conyza bonariensis and its antioxidant/antibacterial assays. Mater Sci Eng C 99:783–793. https://doi.org/10.1016/j.msec.2019.02.010
11. Antezana PE, Municoy S, Desimone MF (2022) Building nanomaterials with microbial factories. In: Biogenic sustainable nanotechnology. Elsevier, Amsterdam, pp 1–39. https://doi.org/10.1016/B978-0-323-88535-5.00012-3
12. Krestinin AV, Dremova NN, Knerel'Man EI, Blinova LN, Zhigalina VG, Kiselev NA (2015) Characterization of SWCNT products manufactured in Russia and the prospects for their industrial application. Nanotechnol Russia 10(7):537–548. https://doi.org/10.1134/S1995078015040096
13. Makarov VV, Love AJ, Sinitsyna OV, Makarova SS, Yaminsky IV, Taliansky ME, Kalinina NO (2014) "Green" nanotechnologies: synthesis of metal NPs using plants. Acta Naturae (англоязычнаяверсия) 6(1):35–44. https://pmc.ncbi.nlm.nih.gov/articles/PMC3999464/
14. Zadehnazari A, Khosropour A, Altaf AA, Amirjalayer S, Abbaspourrad A (2023) Pyrene-based polyimide covalent organic framework with temperature dependent fluorescence. Adv Opt Mater 11(14):2300412. https://doi.org/10.1002/adom.202370052
15. Mallakpour S, Zadehnazari A (2009) Fast synthesis, using microwave induction heating in ionic liquid and characterization of optically active aromatic polyamides. J Macromol Sci Part A Pure Appl Chem 46(8):783–789. https://doi.org/10.1080/10601320903004541
16. Berret JF, Graillot A (2022) Versatile coating platform for metal oxide NPs: applications to materials and biological science. Langmuir 38(18):5323–5338. https://doi.org/10.1021/acs.langmuir.2c00338
17. Kontham S, Mandava K, Dosa S, Mohd FU, Mohammed OA, Mohammad AU (2022) Review on facile synthesis of cerium oxide NPs and their biomedical applications. Inorg Nano Metal Chem 52(8):1183–1195. https://doi.org/10.1080/24701556.2021.1963284
18. Wang Z, Bockstaller MR, Matyjaszewski K (2021) Synthesis and applications of ZnO/polymer nanohybrids. ACS Mater Lett 3(5):599–621. https://doi.org/10.1021/acsmaterialslett.1c00145
19. Xu J, Yang W, Niu L, Dan X, Wang C, Wen C (2017) Study on the surface modification of Sb_2O_3 NPs by using mechanochemical method. Inorg Nano Metal Chem 47(5):697–702. https://doi.org/10.1080/15533174.2016.1212224
20. Lu W, Li Y, Yang M, Jiang X, Zhang Y, Xing Y (2020) Construction of hierarchical Mn_2O_3@MnO_2 core–shell nanofibers for enhanced performance supercapacitor electrodes. ACS Appl Energy Mater 3(9):8190–8197. https://doi.org/10.1021/acsaem.0c00392
21. Emmanuel SS, Adesibikan AA, Saliu OD (2023) Phytogenically bioengineered metal nanoarchitecture for degradation of refractory dye water pollutants: a pragmatic minireview. Appl Organomet Chem 37(2):e6946. https://doi.org/10.1002/aoc.6946
22. Faisal S, Jan H, Shah SA, Shah S, Khan A, Akbar MT et al (2021) Green synthesis of zinc oxide (ZnO) NPs using aqueous fruit extracts of Myristica fragrans: their characterizations and biological and environmental applications. ACS Omega 6(14):9709–9722. https://doi.org/10.1021/acsomega.1c00310
23. Gharpure S, Yadwade R, Ankamwar B (2022) Non-antimicrobial and non-anticancer properties of ZnO NPs biosynthesized using different plant parts of Bixa orellana. ACS Omega 7(2):1914–1933. https://doi.org/10.1021/acsomega.1c05324

24. Joudeh N, Linke D (2022) Nanoparticle classification, physicochemical properties, characterization, and applications: a comprehensive review for biologists. J Nanobiotechnol 20(1):262. https://doi.org/10.1186/s12951-022-01477-8
25. Katoch V, Singh J, Sharma NR, Singh RP (2024) Synthesis and characterization of mesoporous zinc oxide NPs. Inorg Nano Metal Chem 54(1):58–66. https://doi.org/10.1080/24701556.2021.1998121
26. Matussin SN, Tan AL, Harunsani MH, Mohammad A, Cho MH, Khan MM (2020) Effect of Ni-doping on properties of the SnO_2 synthesized using Tradescantia spathacea for photoantioxidant studies. Mater Chem Phys 252:123293. https://doi.org/10.1016/j.matchemphys.2020.123293
27. Osman AI, Zhang Y, Farghali M, Rashwan AK, Eltaweil AS, Abd El-Monaem EM et al (2024) Synthesis of green NPs for energy, biomedical, environmental, agricultural, and food applications: a review. Environ Chem Lett 22(2):841–887. https://ui.adsabs.harvard.edu/link_gateway/2024EnvCL..22..841O/doi:10.1007/s10311-023-01682-3
28. Ying S, Guan Z, Ofoegbu PC, Clubb P, Rico C, He F, Hong J (2022) Green synthesis of NPs: current developments and limitations. Environ Technol Innov 26:102336. https://doi.org/10.1016/j.eti.2022.102336
29. Edo GI, Samuel PO, Oloni GO, Ezekiel GO, Ikpekoro VO, Obasohan P et al (2024) Environmental persistence, bioaccumulation, and ecotoxicology of heavy metals. Chem Ecol 40(3):322–349. https://doi.org/10.1080/02757540.2024.2306839
30. Banjara RA, Kumar A, Aneshwari RK, Satnami ML, Sinha SK (2024) A comparative analysis of chemical vs green synthesis of NPs and their various applications. Environ Nanotechnol Monit Manag 22:100988. https://doi.org/10.1016/j.enmm.2024.100988
31. Zuhrotun A, Oktaviani DJ, Hasanah AN (2023) Biosynthesis of gold and silver NPs using phytochemical compounds. Molecules 28(7):3240. https://doi.org/10.3390/molecules28073240
32. Haider FU, Zulfiqar U, ul Ain N, Hussain S, Maqsood MF, Ejaz M et al (2024) Harnessing plant extracts for eco-friendly synthesis of iron nanoparticle (Fe-NPs): characterization and their potential applications for ameliorating environmental pollutants. Ecotoxicol Environ Saf 281:116620. https://doi.org/10.1016/j.ecoenv.2024.116620
33. Husen A (2022) Herbs, shrubs and trees of potential medicinal benefits. CRC, Boca Raton, FL. https://doi.org/10.1201/9781003205067
34. Husen A (2023a) Antidiabetic medicinal plants and herbal treatments. CRC, Boca Raton, FL, pp 33487–32742. https://doi.org/10.1201/b23347
35. Husen A (2023b) Exploring poisonous plants: medicinal values, toxicity response and other potential uses. CRC, Boca Raton, FL, pp 33487–32742. https://doi.org/10.1201/b23017
36. Husen A (2024a) Antimalarial medicinal plants. Taylor & Francis Group, LLC, Boca Raton, FL. https://doi.org/10.1201/9781003378396
37. Husen A (2024b) Medicinal spices and condiment crops. Taylor & Francis Group, LLC, Boca Raton, FL. https://doi.org/10.1201/9781003387046
38. Said M, Ibrahim MM (2024) Antioxidants and phytochemical profiles in medicinal plants: unraveling molecular mechanisms. Sch Int J Biochem 7(7):106–117. https://doi.org/10.36348/sijb.2024.v07i07.003
39. Ritu, Verma KK, Das A, Chandra P (2023) Phytochemical-based synthesis of silver nanoparticle: mechanism and potential applications. BioNanoScience 13(3):1359–1380. https://doi.org/10.1007/s12668-023-01125-x
40. Khan H, Ullah H, Martorell M, Valdes SE, Belwal T, Tejada S et al (2021) Flavonoids NPs in cancer: treatment, prevention and clinical prospects. Semin Cancer Biol 69:200–211. https://doi.org/10.1016/j.semcancer.2019.07.023
41. Sharma G, Nim S, Alle M, Husen A, Kim JC (2022) Nanoparticle-mediated delivery of flavonoids for cancer therapy: prevention and treatment. In: Smart nanomaterials in biomedical applications. Springer, Cham, pp 61–100. https://doi.org/10.1007/978-3-030-84262-8_3

42. Esmaeilzadeh Kenari R, Razavi R (2022) Phenolic profile and antioxidant activity of free/bound phenolic compounds of sesame and properties of encapsulated NPs in different wall materials. Food Sci Nutr 10(2):525–535. https://doi.org/10.1002/fsn3.2712
43. Agarwal S, Chauhan A, Anand J, Abate L, Rai N (2023) Medicinal plant-based alkaloids: a suitable precursor for nanoparticle synthesis and their various applications. In: Secondary metabolites from medicinal plants, pp 111–132. https://doi.org/10.1201/9781003213727-5
44. Venkata ALK, Sivaram S, Srilakshman G, Muthuraman MS (2022) Review on terpenoid mediated NPs: significance, mechanism, and biomedical applications. Adv Nat Sci Nanosci Nanotechnol 13(3):033003
45. Ansari M, Ahmed S, Abbasi A, Khan MT, Subhan M, Bukhari NA et al (2023a) Plant mediated fabrication of silver NPs, process optimization, and impact on tomato plant. Sci Rep 13(1):18048. https://doi.org/10.1038/s41598-023-45038-x
46. Husen A (2019) Natural product-based fabrication of zinc oxide nanoparticles and their application. In: Husen A, Iqbal M (eds) Nanomaterials and plant potential. Springer, Cham, pp 193–291. https://doi.org/10.1007/978-3-030-05569-1_7
47. Husen A (2020) Introduction and techniques in nanomaterials formulation. In: Husen A, Jawaid M (eds) Nanomaterials for agriculture and forestry applications. Elsevier, Cambridge, MA, pp 1–14. https://doi.org/10.1016/B978-0-12-817852-2.00001-9
48. Husen A, Rahman QI, Iqbal M, Yassin MO (2019) Plant-mediated fabrication of gold nanoparticles and their applications. In: Husen A, Iqbal M (eds) Nanomaterials and plant potential. Springer, Cham, pp 71–110. https://doi.org/10.1007/978-3-030-05569-1_3
49. Khan T, Rahman QI, Raza S, Zehra S, Ahmad N, Husen A (2023) Nano-dimensional materials: an approach towards the biogenic synthesis and their applications. In: Husen A, Siddiqi KS (eds) Advances in smart nanomaterials and their applications. Elsevier, Cambridge, MA, pp 523–568. https://doi.org/10.1016/B978-0-323-99546-7.00009-4
50. Mishra VK, Husen A, Rahman QI, Iqbal M, Sohrab SS, Yassin MO, Bachheti RK (2019) Plant-based fabrication of silver nanoparticles and their application. In: Husen A, Iqbal M (eds) Nanomaterials and plant potential. Springer, Cham, pp 135–175. https://doi.org/10.1007/978-3-030-05569-1_5
51. Morgan RN, Aboshanab KM (2024) Green biologically synthesized metal NPs: biological applications, optimizations and future prospects. Future Sci OA 10(1):FSO935. https://doi.org/10.2144/fsoa-2023-0196
52. Amor LB, Toumi H, Ameur H, Hamida RB, Bennia H (2024) Examining the drivers of environmental sustainability in OPEC countries: a symbiotic influence index-STIRPAT-Kaya-EKC approach. Emiss Control Sci Technol 10(2):223–243. https://doi.org/10.1007/s40825-024-00247-2
53. Szczyglewska P, Feliczak-Guzik A, Nowak I (2023) Nanotechnology–general aspects: a chemical reduction approach to the synthesis of NPs. Molecules 28(13):4932. https://doi.org/10.3390/molecules28134932
54. Antunes Filho S, Dos Santos MS, Dos Santos OAL, Backx BP, Soran ML, Opriş O et al (2023) Biosynthesis of NPs using plant extracts and essential oils. Molecules 28(7):3060. https://doi.org/10.3390/molecules28073060
55. Perumalsamy J, Punniakodi SB, Selvam C, Senthil R (2025) Enhancing phase change characteristics of hybrid nanocomposites for latent heat thermal energy storage. J Compos Sci 9(3):120. https://doi.org/10.3390/jcs9030120
56. Ketemepi HK, Awang MAB, Seelan JSS, Noor NQIM (2024) Extraction process and applications of mushroom-derived protein hydrolysate: a comprehensive review. Future Foods 9:100359. https://doi.org/10.1016/j.fufo.2024.100359
57. Khoramian R, Salaudeen I, Pourafshary P, Riazi M, Kharrat R (2025) Impact of relative permeability hysteresis on CO storage in saline aquifers. Greenhouse Gases Sci Technol 15(1):3–12. https://doi.org/10.1002/ghg.2319

58. Miu BA, Dinischiotu A (2022) New green approaches in NPs synthesis: an overview. Molecules 27(19):6472. https://doi.org/10.3390/molecules27196472
59. Mungwari CP, Obadele BA, King'ondu CK (2025) Phytochemicals as green and sustainable corrosion inhibitors for mild steel and aluminium. Results Surf Interfaces 18:100374. https://doi.org/10.1016/j.rsurfi.2024.100374
60. Silva JWJ, Bustamante AG, Codaro EN, Nakazato RZ, Hein LRO (2004) Morphological analysis of pits formed on Al 2024-T3 in chloride aqueous solution. Appl Surf Sci 236(1–4):356–365. https://doi.org/10.1016/j.apsusc.2004.05.007
61. Uthra C, Muralitharan G, Nagaraj K, Thajjudin N, Kaliyaperumal R, Badgujar NP et al (2024) Exploring biogenic microbial synthesis and biomedical applications of silver NPs: a comprehensive review. Results Chem 8:101555. https://doi.org/10.1016/j.rechem.2024.101555
62. Ansari N, Rahman QI, Khan T, Khan A, Khan R, Wagay JA, Husen A (2023b) Mycogenic-assisted synthesis of nanoparticles and their efficient applications. In: Husen A, Siddiqi KS (eds) Advances in smart nanomaterials and their applications. Elsevier, Cambridge, MA, pp 569–595. https://doi.org/10.1016/B978-0-323-99546-7.00020-3
63. Jayachandran A, Aswathy TR, Nair AS (2021) Green synthesis and characterization of zinc oxide NPs using Cayratia pedata leaf extract. Biochem Biophys Rep 26:100995. https://doi.org/10.1016/j.bbrep.2021.100995
64. Abomuti MA, Danish EY, Firoz A, Hasan N, Malik MA (2021) Green synthesis of zinc oxide NPs using Salvia officinalis leaf extract and their photocatalytic and antifungal activities. Biology 10(11):1075. https://doi.org/10.3390/biology10111075
65. Hosny M, Fawzy M, Eltaweil AS (2022) Green synthesis of bimetallic Ag/ZnO@ Biohar nanocomposite for photocatalytic degradation of tetracycline, antibacterial and antioxidant activities. Sci Rep 12(1):7316. https://doi.org/10.1038/s41598-022-11014-0
66. Zafar S, Bilal M, Ali MF, Mahmood A, Kijsomporn J, Wong LS et al (2024) Nano-biofertilizer an eco-friendly and sustainable approach for the improvement of crops under abiotic stresses. Environ Sustain Indic 24:100470. https://doi.org/10.1016/j.indic.2024.100470
67. Atanda SA, Shaibu RO, Agunbiade FO (2025) NPs in agriculture: balancing food security and environmental sustainability. Discov Agric 3(1):26. https://doi.org/10.1007/s44279-025-00159-x
68. Fu L, Fu Z (2015) Plectranthus amboinicus leaf extract–assisted biosynthesis of ZnO NPs and their photocatalytic activity. Ceram Int 41(2):2492–2496. https://doi.org/10.1016/j.ceramint.2014.10.069
69. Rauf MA, Ashraf SS (2009) Fundamental principles and application of heterogeneous photocatalytic degradation of dyes in solution. Chem Eng J 151(1–3):10–18. https://doi.org/10.1016/j.cej.2009.02.026
70. Bordbar M, Sharifi-Zarchi Z, Khodadadi B (2017) Green synthesis of copper oxide NPs/clinoptilolite using Rheum palmatum L. root extract: high catalytic activity for reduction of 4-nitro phenol, rhodamine B, and methylene blue. J Sol Gel Sci Technol 81(3):724–733. https://doi.org/10.1007/s10971-016-4239-1
71. Hong F, Zhou J, Liu C, Yang F, Wu C, Zheng L, Yang P (2005) Effect of nano-TiO on photochemical reaction of chloroplasts of spinach. Biol Trace Elem Res 105(1):269–279. https://doi.org/10.1385/bter:105:1-3:269
72. Armelao L, Barreca D, Bottaro G, Gasparotto A, Tondello E, Ferroni M, Polizzi S (2004) Au/TiO nanosystems: a combined RF-sputtering/sol–gel approach. Chem Mater 16(17):3331–3338. https://doi.org/10.1021/cm0353308
73. Matinise N, Fuku XG, Kaviyarasu K, Mayedwa N, Maaza MJASS (2017) ZnO NPs via Moringa oleifera green synthesis: physical properties & mechanism of formation. Appl Surf Sci 406:339–347. https://doi.org/10.1016/j.apsusc.2017.01.219
74. Rolim WR, Pelegrino MT, de Araújo Lima B, Ferraz LS, Costa FN, Bernardes JS et al (2019) Green tea extract mediated biogenic synthesis of silver NPs: characterization, cytotoxicity evaluation and antibacterial activity. Appl Surf Sci 463:66–74. https://doi.org/10.1016/j.apsusc.2018.08.203

75. Ali K, Ahmed B, Dwivedi S, Saquib Q, Al-Khedhairy AA, Musarrat J (2015) Microwave accelerated green synthesis of stable silver NPs with Eucalyptus globulus leaf extract and their antibacterial and antibiofilm activity on clinical isolates. PLoS One 10(7):e0131178. https://doi.org/10.1371/journal.pone.0131178
76. Jeevanandam J, Chan YS, Danquah MK (2016) Biosynthesis of metal and metal oxide NPs. ChemBioEng Rev 3(2):55–67. https://doi.org/10.1002/cben.201500018
77. Kumar PV, Shameem U, Kollu P, Kalyani RL, Pammi SVN (2015) Green synthesis of copper oxide NPs using Aloe vera leaf extract and its antibacterial activity against fish bacterial pathogens. BioNanoScience 5(3):135–139. https://doi.org/10.1007/s12668-015-0171-z
78. Vishveshvar K, Aravind Krishnan MV, Haribabu K, Vishnuprasad S (2018) Green synthesis of copper oxide NPs using Ixora coccinea plant leaves and its characterization. BioNanoScience 8(2):554–558. https://doi.org/10.1007/s12668-018-0508-5
79. Nagajyothi PC, Muthuraman P, Sreekanth TVM, Kim DH, Shim J (2017) Green synthesis: in-vitro anticancer activity of copper oxide NPs against human cervical carcinoma cells. Arab J Chem 10(2):215–225. https://doi.org/10.1016/j.arabjc.2016.01.011
80. Rajakumar G, Rahuman AA, Priyamvada B, Khanna VG, Kumar DK, Sujin PJ (2012) Eclipta prostrata leaf aqueous extract mediated synthesis of titanium dioxide NPs. Mater Lett 68:115–117. https://doi.org/10.1016/j.matlet.2011.10.038
81. Singh J, Dutta T, Kim KH, Rawat M, Samddar P, Kumar P (2018) 'Green' synthesis of metals and their oxide NPs: applications for environmental remediation. J Nanobiotechnol 16(1):84. https://doi.org/10.1186/s12951-018-0408-4
82. Mustapha T, Misni N, Ithnin NR, Daskum AM, Unyah NZ (2022) A review on plants and microorganisms mediated synthesis of silver NPs, role of plants metabolites and applications. Int J Environ Res Public Health 19(2):674. https://doi.org/10.3390/ijerph19020674
83. Ramazanli VN, Ahmadov IS (2022) Synthesis of silver NPs by using extract of olive leaves. Adv Biol Earth Sci 7(3):238–244. https://jomardpublishing.com/UploadFiles/Files/journals/ABES/V7N3/Ramazanli_Ahmadov.pdf
84. Aravind M, Amalanathan M, Mary MSM (2021) Synthesis of TiO NPs by chemical and green synthesis methods and their multifaceted properties. SN Appl Sci 3(4):409. https://doi.org/10.1007/s42452-021-04281-5
85. Hano C, Abbasi BH (2021) Plant-based green synthesis of NPs: production, characterization and applications. Biomolecules 12(1):31. https://doi.org/10.3390/biom12010031
86. Helmy A, El-Shazly M, Seleem A, Abdelmohsen U, Salem MA, Samir A et al (2020) The synergistic effect of biosynthesized silver NPs from a combined extract of parsley, corn silk, and gum arabic: in vivo antioxidant, anti-inflammatory and antimicrobial activities. Mater Res Express 7(2):025002. https://doi.org/10.1088/2053-1591/ab6e2d
87. Teli S, Sethiya A, Agarwal S (2024) Pioneering synthetic strategies of 2-substituted benzothiazoles using 2-aminothiophenol. Chemistry 6(1):165–206. https://doi.org/10.3390/chemistry6010009
88. Ramanathan A, Robb GB, Chan SH (2016) mRNA capping: biological functions and applications. Nucleic Acids Res 44(16):7511–7526. https://doi.org/10.1093/nar/gkw551
89. Pulingam T, Foroozandeh P, Chuah JA, Sudesh K (2022) Exploring various techniques for the chemical and biological synthesis of polymeric NPs. Nanomaterials 12(3):576. https://doi.org/10.3390/nano12030576
90. de Castro M, Baptista J, Matos C, Valente A, Briga-Sá A (2024) Energy efficiency in winemaking industry: challenges and opportunities. Sci Total Environ 930:172383. https://doi.org/10.1016/j.scitotenv.2024.172383
91. Galúcio JM, de Souza SGB, Vasconcelos AA, Lima AKO, da Costa KS, de Campos Braga H, Taube PS (2022) Synthesis, characterization, applications, and toxicity of green synthesized NPs. Curr Pharm Biotechnol 23(3):420–443. https://doi.org/10.2174/1389201022666210521102307
92. Yuan H, Ma Q, Ye L, Piao G (2016) The traditional medicine and modern medicine from natural products. Molecules 21(5):559. https://doi.org/10.3390/molecules21050559

93. ElFessi R, Khamessi O, Srairi-Abid N, Sabatier JM, Tytgat J, Peigneur S, Kharrat R (2022) Purification and characterization of Bot33: a non-toxic peptide from the venom of Buthus occitanus tunetanus scorpion. Molecules 27(21):7278. https://doi.org/10.3390/molecules27217278
94. Zhang Z, Malik MZ, Khan A, Ali N, Malik S, Bilal M (2022) Environmental impacts of hazardous waste, and management strategies to reconcile circular economy and eco-sustainability. Sci Total Environ 807:150856. https://doi.org/10.1016/j.scitotenv.2021.150856
95. Jalab J, Abdelwahed W, Kitaz A, Al-Kayali R (2021) Green synthesis of silver NPs using aqueous extract of Acacia cyanophylla and its antibacterial activity. Heliyon 7(9):e08033. https://doi.org/10.1016/j.heliyon.2021.e08033
96. Kodagoda GK, Hong HT, O'Hare TJ, Topp B, Sultanbawa Y, Netzel ME (2024) Storage effects on the physicochemical properties, phytochemical composition, and sugars in red-fleshed cultivars, 'Rubycot' plumcot, and 'queen garnet' plum. Molecules 29(19):4641. https://doi.org/10.3390/molecules29194641
97. Mafe AN, Iruoghene Edo G, Akpoghelie PO, Gaaz TS, Yousif E, Zainulabdeen K et al (2025) Probiotics and food bioactives: unraveling their impact on gut microbiome, inflammation, and metabolic health. Probiotics Antimicrob Proteins 17:1851–1892. https://doi.org/10.1007/s12602-025-10452-2

Role of Aromatic Plant-Based Bioactive Compounds in Synthesis of Various Nanoparticles, Nanocomposites, Characterization, and Their Applications

Irtiza Binti Farooq, Danish Ahmad Shergujri, Murtaza Ahmad Khanday, Mahrukh Shafiq, Andleeb Altaf, Roohi Mohi-ud-din, Sameer Ullah Khan, and Reyaz Hassan Mir

Abstract Aromatic plants are a rich source of bioactive compounds, particularly essential oils, phenolics, terpenes, and alkaloids, which possess unique physicochemical and biological properties. In recent years, these natural compounds have gained increasing attention as sustainable, eco-friendly agents in the green synthesis of nanoparticles and nanocomposites. The integration of aromatic plant-derived bioactive compounds into nanotechnology represents a paradigm shift toward eco-friendly and sustainable material synthesis. This chapter explores the cutting-edge role of phytochemicals such as essential oils, polyphenols, terpenoids, and alkaloids from plants like *Ocimum sanctum*, *Eucalyptus globulus*, and *Mentha piperita* in the green fabrication of metallic (Ag, Au, Cu), metal oxide (ZnO, TiO_2), and hybrid nanoparticles. Unlike conventional chemical methods, these bioactives act as dual reducing and stabilizing agents, enabling precise control over nanoparticle morphology (spherical, anisotropic, core–shell) while eliminating toxic byproducts.

I. B. Farooq
Food Science and Nutrition, Institute of Home Science, University of Kashmir, Hazratbal, Srinagar, Jammu and Kashmir, India

D. A. Shergujri · M. A. Khanday
Pharmaceutics Division, Department of Pharmaceutical Sciences, University of Kashmir, Hazratbal, Srinagar, Jammu and Kashmir, India

M. Shafiq · A. Altaf · R. H. Mir (✉)
Pharmaceutical Chemistry Division, Department of Pharmaceutical Sciences, University of Kashmir, Hazratbal, Srinagar, Jammu and Kashmir, India
e-mail: mirreyaz.phscct@uok.edu.in

R. Mohi-ud-din
Department of General Medicine, Sher-I-Kashmir Institute of Medical Sciences (SKIMS), Srinagar, Jammu and Kashmir, India

S. U. Khan
Pediatric Oncology, Oregon Health and Science University, Portland, OR, USA

A. Husen (ed.), *Sustainable Uses of Aromatic Plants in Nanotechnology*, Smart Nanomaterials Technology, https://doi.org/10.1007/978-981-95-7409-4_6

Keywords Aromatic plants · Functional foods · Phytopharmaceuticals · Essential oils · Pharmacological effects

1 Introduction

Nanotechnology represents one of the most dynamic areas of scientific fields that involve the design and application of materials within the nanoscale range of 1–100 nm [1]. At this scale, materials display unique physiochemical properties, such as high surface-to-volume ratios, quantum effects, and altered electronic structures, which enhance catalytic activity, optical behavior, and biological interactions. These features have driven widespread applications in medicine, agriculture, drug delivery, biosensing, food preservation, and environmental remediation [2–6].

The functionality of nanomaterials is largely determined during synthesis. For control over nucleation, growth and surface chemistry dictates particle size, morphology, and stability, which in turn influence performance and safety [7, 8]. Conventional approaches, including chemical vapor deposition, sol–gel techniques, and chemical reduction, offer precision but often require high energy input and hazardous reagents, raising concerns about sustainability and toxicity [9–11].

Green synthesis has emerged as a pragmatic response to these limitations. Green approaches use biological systems, including plants, algae, fungi, and bacteria, as natural factories for nanoparticle production to mediate reduction and stabilization under mild, often aqueous conditions [12–20]. Plant-mediated synthesis has become particularly attractive due to its simplicity, low cost, rapidity, and eco-friendliness. Plant extracts contain a complex mixture of phytochemicals that act as reducing, capping, and stabilizing agents, eliminating the need for external chemical additives [21, 22]. Nanoparticle synthesis requires a precursor source, which may be organic, inorganic, or metallic, along with biomolecules that act as reducing and stabilizing agents. The first step of green synthesis is bioreduction, which involves plant-derived compounds such as flavonoids, phenols, or polysaccharides to convert precursor ions into atoms. This step is critical, as ion reduction influences the biological activity and cytotoxicity of the nanomaterial. The subsequent stage, nucleation, involves atom aggregation assisted by bioactive molecules, forming nanoparticles that stabilize within a nanosystem. During this process, physiochemical factors such as light, temperature, electrical energy, and pH strongly affect nanoparticle stability and the functional properties of the final nanomaterial [23]. This natural route provides nanoparticles with enhanced biocompatibility, making suitable for applications in biomedicine and food industries. Within this context, aromatic plants occupy a significant niche due to their long history of medicinal and culinary use, combined with their rich reservoir of bioactive compounds ([9, 22, 24, 25]).

Aromatic plants are well known for their fragrance and essential oils, but they are also a rich source of secondary metabolites with important pharmacological and technological potential. Their bioactive compounds contribute to antimicrobial, antioxidant, anti-inflammatory, and anticancer activities, which are central to both traditional and modern medicine [26, 27]. Importantly, these same metabolites enable aromatic plants to serve as efficient reducing and stabilizing agents in nanoparticle synthesis. Among the many aromatic species studied, basil (*Ocimum* spp.), clove (*Syzygium aromaticum*), cinnamon (*Cinnamon verum*), rosemary (*Salvia rosmarinus*), and mint (*Mentha* spp.) stand out as model plants due to their diverse phytochemical profiles and strong bioactivity [25, 28].

2 Aromatic Plants as Sources of Bioactive Compounds

Aromatic plants are distinctive group of medicinal and culinary herbs that produce a wide range of secondary metabolites with important biological activities. Their aroma arises mainly from volatile compounds, yet their chemical profiles extend far beyond fragrance [29, 30]. They contain phenolics, flavonoids, terpenoids, alkaloids, and essential oils, all of which contribute to their pharmacological, preservative, and ecological function. These metabolites not only support the plants in defense against pathogens and herbivores but also offer therapeutics, nutritional, and technological applications of humans. Recent research highlights their central role in green nanotechnology, where they serve as natural sources of reducing and stabilizing agents for nanoparticle synthesis [31].

Phenolic compounds are the most abundant secondary metabolite in aromatic plants [32]. They are characterized by one or more hydroxyl group attached to aromatic rings, and they are broadly classified into flavonoids, phenolic acids, tannins, lignans, stilbenes, and coumarins. This chemical diversity underlies their wide range of biological activities with strong redox potential. This class includes simple phenolic acids (such as caffeic, gallic, and ferulic acids) and complex esters like rosmarinic acid, which is especially common in rosemary and basil [33–35]. These compounds act as powerful antioxidants by donating electrons or hydrogen atoms to neutralize free radicals. In green nanoparticle synthesis, the phenolic hydroxyl and carboxyl groups reduce metal ions (e.g., Ag^+ or Au^{3+}) into stable nanoparticles and bind to particle surfaces, thereby limiting aggregation. In clove, eugenol, a phenolic compound, shows strong reducing and stabilizing effects in silver nanoparticle formation [33, 36, 37]. In spectroscopic studies, rosmarinic acid and caffeic acid were identified as active agents in nanoparticle reduction and surface capping [34, 38–40]. They serve structural, protective, and signaling roles in plants. In humans, phenolic acids exhibit important antioxidant, antiaging and anti-inflammatory activities, and are being explored for their role in preventing chronic diseases [41, 42].

Flavonoids are a major class of phytochemicals. They are polyphenolic compounds with a C_6-C_3-C_6 structure, widely distributed across aromatic plants [43, 44]. They include flavones (luteolin, apigenin), flavonols (quercetin, kaempferol), and others compounds like anthocyanins, chalcones, and aurones, present mostly in plants and fruits like rosemary, basil, mint, cinnamon leaves, naringenin, and hesperetin. Their hydroxyl and carbonyl groups contribute to antioxidant, anti-inflammatory, and antimicrobial activities [45–47]. In nanoparticle synthesis, flavonoids undergo enol-to-keto tautomeric shifts, releasing hydrogen atoms that reduce metal ions. Their carbonyl and hydroxyl sites further chelate and stabilize the growing nanoparticles [21, 39]. In mint (*Mentha* spp.), quercetin and glycosides contribute to its antioxidant and anti-inflammatory properties [48]. In rosemary and basil, flavonoids complement phenolic acids in enhancing radical scavenging capacity and protecting lipids and proteins from oxidative stress [49]. These compounds are also known for their vasodilatory, anticarcinogenic, and anti-inflammatory property. In plants, they prevent pigmentation and give resilience against environmental stressors [50, 51].

Terpenoids, also known as isoprenoids, are a significant class of secondary plant metabolites, with over 55,000 identified structures [52]. The fundamental structure of terpenes is derived from five-carbon isoprene units. They are classified based on the number of isoprene units, hemiterpenoids C_5, monoterpenoids C_{10}, sesquiterpenoids C_{15}, diterpenoids C_{20}, and large polyterpenes (e.g., triterpenes C_{30}, tetraterpenes C_{40} [53]. Terpenes are hydrocarbons, while their oxygen-containing derivatives are known as terpenoids [54]. Monoterpenes such as linalool, menthol, thymol, and carvacrol are found in basil, mint, and thyme, while diterpenes like carnosic acid and carnosol are characteristic of rosemary [55–58]. Terpenoids often contain hydroxyl or carbonyl groups, enabling them to act as both reducing and capping agents. Thymol and carvacrol have the capacity to disrupt microbial membranes, while their hydroxyl groups donate electrons in nanoparticle formation [59–61]. In rosemary, diterpenes such as carnosic acid contribute to nanoparticle stability by adsorbing to particle surfaces, which enhances both colloidal stability and retained bioactivity [62]. Moreover, these compounds exhibit a wide range of biological activities and have gained considerable attention for their beneficial effects on human health, with studies exploring their roles in inflammatory diseases, tumorigenesis, and neurodegeneration [63–66].

Alkaloids are nitrogen-containing secondary metabolites derived mainly from amino acids such as tryptophan, tyrosine, and phenylalanine. They are structurally diverse, including pyridine, indole, isoquinoline, tropane, and purine alkaloids [67–69]. They are best known for pharmacological agents like morphine, caffeine, and vincristine but also occur in small amounts in aromatic plants [68]. Basil (*Ocimum basilicum*) contain indolic and pyridine-type alkaloids, while mint (*Mentha* spp.), rosemary (*Salvia rosmarinus*), clove (*Syzygium aromaticum*), and cinnamon (*Cinnamomum verum*) also yield minor alkaloid fractions alongside their phenolics and terpenoids [70–73].

In nanoparticle synthesis, alkaloids play a complementary role. Their heterocyclic nitrogen and amine groups coordinate with metal ions and nanoparticle surfaces, acting as capping and stabilizing agents [74–76]. Spectroscopic evidence and extract studies indicate that some alkaloids also assist in reduction through conjugated nitrogen systems, while hydroxyl and carbonyl substituents enhance chelation. Spectroscopic studies also show shifts in N-H and C-N bands, confirming their adsorption [36, 77–80]. Though less abundant than phenolics, alkaloids support nanoparticle stability and functionalization in aromatic plant extracts.

Essential oils are volatile, hydrophobic mixtures composed mainly of terpenes, terpenoids, and phenylpropanoid derivatives, and they represent a significant class of bioactive compounds in aromatic plants that can mediate green synthesis of nanoparticles [81, 82]. Their diverse functional groups include hydroxyl, carbonyl, aldehyde, and ether moieties, allowing them to act simultaneously as reducing and stabilizing agents. Phenolic constituents such as eugenol in clove and cinnamaldehyde in cinnamon provide electron-donating hydroxyl or aldehyde groups facilitate reduction of metal ions, while oxygenated terpenes like linalool and 1.8-cineole contribute to capping and stabilization through coordination with nanoparticle surfaces [83]. Studies have demonstrated that *Mentha spicata* essential oil can reduce and stabilize gold and silver nanoparticles, with FTIR spectra confirming involvement of N-H and C-O groups in surface adsorption [84, 85]. Essential oils from rosemary, basil, and mint commonly contain compounds such as thymol, carvacrol, and linalool which adsorb on nanoparticle surfaces and FTIR band shift for O—H C=ON—H and C—H are consistent with capping and coordination. While isolated mechanistic proof that these single components alone drive nucleation is scarce [82, 86–88]. Thus, essential oils provide a complex but effective phytochemical reservoir of nanoparticle biosynthesis, with their structural diversity enabling multifunctional roles in reduction, capping, and stabilization, making them indispensable mediators in plant-based nanomaterial production (Table 1).

Table 1 Summary of major secondary metabolites from aromatic plants with examples, structural features, and functional contributions to green nanoparticle synthesis

Category	Example molecules	Structural features	Functional role
Polyphenols	Gallic acid, tannic acid, catechins	Aromatic rings with multiple hydroxyl groups (phenol as the core unit)	Act as strong reducers that convert metal ions into nanoparticles
Flavonoids	Quercetin, rutin, kaempferol	C_6-C_3-C_6 diphenylpropane backbone with ketone and hydroxyl substitutions	Chelate and reduce metal ions improving nanoparticle stability
Terpenes	Limonene, linalool, huperzine A	Chains or cyclic forms built from isoprene units	Serve as reducing and capping agents influencing nanoparticle shape
Alkaloids	Indole alkaloids (*Ocimum*), pyridine alkaloids (mint), morphine	Nitrogen-containing heterocyclic structures (pyridine indole, isoquinoline, tropane, purine)	Act as capping and stabilizing agents via N-H and C-N coordination; sometimes assist in reduction
Essential oils	Eugenol (clove), cinnamaldehyde (cinnamon), linalool (mint), thymol (rosemary)	Volatile mixtures of terpenes, terpenoids, and phenylpropanoids, with hydroxyl, aldehyde, and ether groups	Reduce metal ions (-OH cap nanoparticles via O-H, C=O, and C-O groups) and provide multifunctional stabilization

3 Green Synthesis of Nanoparticles

Green synthesis using aromatic plants refers to the environmentally friendly method of producing nanoparticles, bioactive compounds, or chemicals by utilizing natural extracts from aromatic plants as reducing, capping, and stabilizing agents—without using toxic chemicals or high-energy processes [89–91] (Fig. 1) [92]. In recent years, the convergence of green chemistry and nanotechnology has given rise to sustainable methods of nanoparticle synthesis. Among the myriad of natural resources explored, aromatic plants have emerged as a potent and eco-friendly

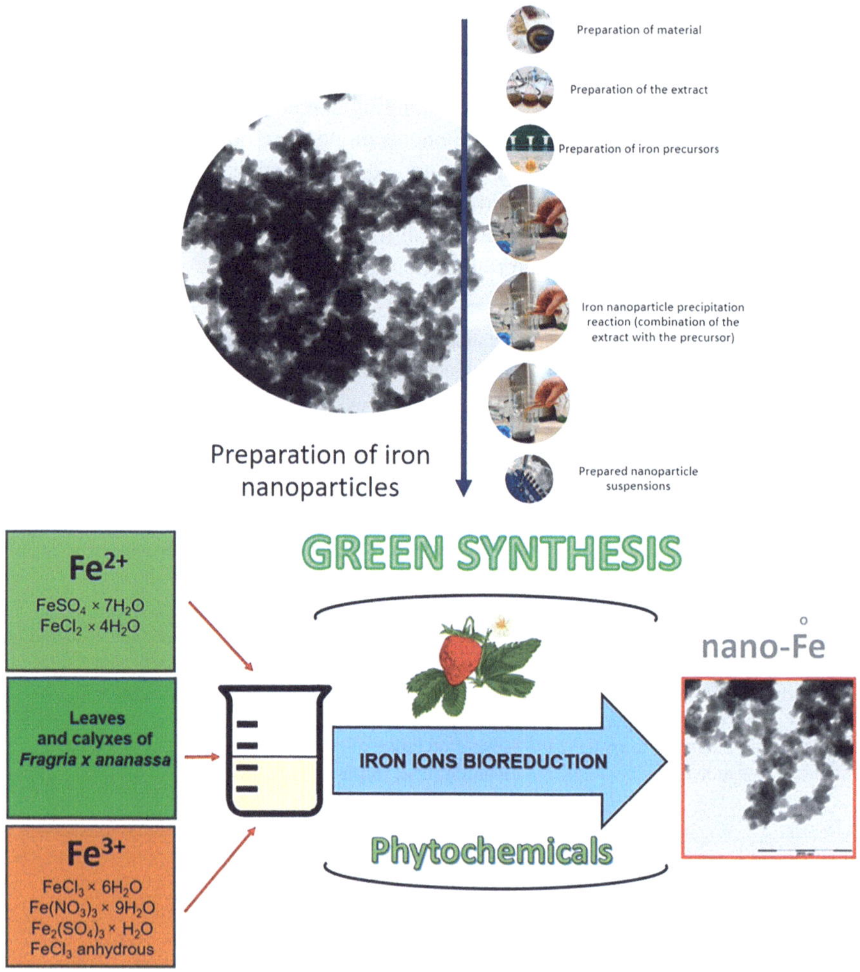

Fig. 1 General process of green synthesis of nanoparticles (iron nanoparticles) [92]

source for the green synthesis of nanoparticles. Traditionally valued for their fragrance and medicinal properties, aromatic plants are now being harnessed in cutting-edge nanotechnological applications due to their rich phytochemical profiles [23, 91, 93]. Aromatic plants such as *Ocimum sanctum* (holy basil), *Mentha* spp. (mint), *Lavandula* spp. (lavender), and *Cymbopogon* spp. (lemongrass) are rich in secondary metabolites, including essential oils (e.g., linalool, eugenol, citronellal), flavonoids, terpenoids, phenolic acids, alkaloids, and tannins. These bioactive compounds possess strong antioxidant, reducing, and metal-chelating properties, making them ideal for the green synthesis of metallic and metal oxide nanoparticles. The phytochemicals act as natural reducing agents, converting metal ions into nanoparticles, while also capping and stabilizing them to prevent aggregation [23, 93–95]. The development of green synthesis methodologies has been significantly advanced by the incorporation of aromatic plant extracts, which are rich in secondary metabolites such as phenolics, flavonoids, terpenoids, and essential oils. These bioactive compounds not only facilitate the reduction of metal ions to metallic nanoparticles but also contribute to nucleation and subsequent stabilization of the nanoparticles. Understanding the mechanistic aspects of these stages is essential for controlling nanoparticle characteristics such as size, shape, dispersity, and functionality [91, 96, 97].

3.1 Mechanism of Reduction

The initial and most crucial step in the green synthesis process is the reduction of metal ions to their corresponding zero-valent metallic form. Aromatic plants, such as *Ocimum sanctum* (basil), *Syzygium aromaticum* (clove), and *Rosmarinus officinalis* (rosemary), contain a wide array of phytochemicals capable of donating electrons to metal ions. Compounds such as phenolic acids (e.g., rosmarinic acid), flavonoids (e.g., quercetin), terpenoids, essential oils (e.g., eugenol in clove, citral in lemongrass), and ascorbic acid act as natural reducing agents. These molecules contain functional groups such as hydroxyl (-OH), carbonyl (-C=O), and aldehyde (-CHO) that can donate electrons to metal ions [98].

3.2 Mechanism of Nucleation

Following the reduction of metal ions, the resulting metal atoms undergo nucleation, a process by which they aggregate into small clusters or "nuclei." This stage determines the initial size and morphology of the nanoparticles [99].

3.2.1 Types of Nucleation

Homogeneous nucleation: Occurs uniformly throughout the solution when a critical concentration of reduced metal atoms is reached.

Burst nucleation: A rapid process where a high number of nuclei are formed almost simultaneously, resulting in smaller and more uniform particles.

Various factors affecting nucleation are concentration of metal ions, rate of reduction, temperature and pH of the medium, and nature and concentration of phytochemicals. A faster rate of reduction often leads to a higher number of nucleation sites, thereby producing smaller nanoparticles.

3.3 Mechanism of Stabilization (Capping)

Newly formed nanoparticles tend to be thermodynamically unstable and have a high surface energy, leading to agglomeration. To prevent this, phytochemicals from the plant extract also serve as capping agents, binding to the surface of nanoparticles and providing steric or electrostatic stabilization. Phytochemicals that assist in stabilization include proteins and amino acids, polysaccharides, polyphenols, terpenes, and essential oil components. These molecules possess functional groups such as –OH, –COOH, $-NH_2$, and aromatic rings that can adsorb onto the nanoparticle surface, thus forming a protective layer [100].

3.3.1 Types of Stabilization

Steric stabilization: Physical hindrance by large molecules prevents particles from approaching each other closely.

Electrostatic stabilization: Charged phytochemicals generate a surface charge on the nanoparticles, creating repulsion between particles [101].

4 Types of Nanoparticles Synthesized

Nanoparticles are particles with at least one dimension in the range of 1–100 nm [101]. When synthesized via green routes using aromatic plants, the phytochemicals act as both reducing and stabilizing agents, yielding nanoparticles with distinct morphologies, surface properties, and biological functionalities. The specific plant used and its phytochemical composition significantly influence the type and quality of nanoparticles synthesized.

4.1 Metal Nanoparticles

4.1.1 Silver Nanoparticles (AgNPs)

Most widely synthesized metal nanoparticle via plant-based green synthesis. It shows strong antibacterial, antifungal, and anticancer properties. Synthesized using plant extracts from: *Ocimum sanctum* (basil), *Syzygium aromaticum* (clove), *Mentha piperita* (peppermint), *Lavandula angustifolia* (lavender).

Applications: Biomedical (wound dressings, drug delivery), textiles, and water purification [102–105].

4.1.2 Gold Nanoparticles (AuNPs)

Biocompatible, nontoxic, and widely used in diagnostics, imaging, and cancer therapy. It is synthesized using *Eucalyptus globulus*, *Rosmarinus officinalis*, and *Cymbopogon citratus* (lemongrass) [106–109].

4.1.3 Copper Nanoparticles (CuNPs)

It exhibits antimicrobial, antioxidant, and catalytic activities. Less expensive than Ag or Au but prone to oxidation. Synthesized using *Ocimum basilicum* (basil) and *Cinnamomum zeylanicum* (cinnamon) and finds its application in agriculture (nanopesticides), antimicrobial coatings, and catalysis [110, 111].

4.2 Metal Oxide Nanoparticles

4.2.1 Zinc Oxide Nanoparticles (ZnO NPs)

It is well known for UV absorption, antibacterial activity, and photocatalysis. Synthesized using aromatic plants like *Mentha arvensis* (wild mint), *Cymbopogon flexuosus* (lemongrass), and *Lavandula officinalis* (lavender).

Applications: Sunscreens, paints, antimicrobial formulations, and biosensors [112, 113].

4.2.2 Iron Oxide Nanoparticles (Fe_3O_4 NPs)

Magnetic nanoparticles used in MRI contrast agents, drug delivery, and environmental remediation. Synthesized using extracts of *Rosmarinus officinalis* (rosemary) and *Cymbopogon citratus* (lemongrass) [114].

4.2.3 Titanium Dioxide Nanoparticles (TiO_2 NPs)

It is known for photocatalytic and self-cleaning properties, synthesized using aromatic herbs such as: *Thymus vulgaris* (thyme), *Ocimum sanctum* (holy basil) [115].

5 Factors Influencing Nanoparticle Synthesis

5.1 Concentration of Plant Extract

The quantity and type of phytochemicals present in the plant extract play a pivotal role in reducing metal ions and stabilizing the formed nanoparticles. Higher extract concentration generally leads to faster reduction of metal ions, smaller particle sizes due to rapid nucleation, and increased capping (resulting in better stability), and lower extract concentration may result in incomplete reduction and formation of larger, polydisperse particles. For example, in the synthesis of silver nanoparticles using *Ocimum sanctum*, a higher volume of extract resulted in smaller, spherical nanoparticles with enhanced antimicrobial activity [116].

5.2 pH of the Reaction Mixture

pH affects the ionization of phytochemicals, such as phenols and flavonoids; the rate of reduction and nucleation; and shape and size of the resulting nanoparticles. For example, at alkaline pH, eug enol from clove extract more effectively reduces silver ions, leading to smaller and well-dispersed AgNPs. Table 2 summarizes the effect of pH [117].

Table 2 Effect of pH on synthesis of nanoparticles

pH level	Effect
Acidic	Slower reduction; larger and less stable nanoparticles
Neutral	Balanced reaction rate; moderate particle size
Alkaline	Faster reduction; smaller and more uniform nanoparticles

5.3 Temperature

Temperature directly influences the kinetic energy of molecules and the rate of reaction. Higher temperatures accelerate reduction and nucleation, promote formation of smaller particles due to rapid supersaturation, and can sometimes degrade sensitive phytochemicals. Lower temperatures lead to slower reactions and may result in larger, irregularly shaped nanoparticles. For example, in the green synthesis of ZnO nanoparticles using *Mentha* extract, optimal particle formation was observed at 70–80 °C [118].

5.4 Reaction Time

The duration of the synthesis process impacts the growth and stabilization of nanoparticles. Shorter reaction time may lead to incomplete reduction and unstable particles. Longer reaction time allows for complete reduction and growth of larger nanoparticles. Excessive time may cause aggregation or degradation of capping agents [119].

Many of the aforementioned parameters are interdependent, and altering one can affect others. For example, increasing temperature accelerates reduction but may degrade thermolabile capping agents. Alkaline pH enhances reduction but may cause precipitation of metal hydroxides if not carefully controlled. Thus, optimization using a one-variable-at-a-time (OVAT) or design of experiments (DoE) approach is often recommended for reproducible synthesis [120].

6 Mechanistic Role of Aromatic Plants in Nanocomposite Formation

Nanocomposites are materials composed of a nanoscale phase dispersed within a matrix, where the combination imparts superior mechanical, electrical, thermal, or biological properties compared to the individual components. In green nanotechnology, aromatic plants serve as sustainable and bioactive mediators in the synthesis of such nanocomposites, offering both a green route to nanoparticle formation and functionalization capabilities for complex material architectures. Aromatic plants such as basil (*Ocimum basilicum*), clove (*Syzygium aromaticum*), lemongrass (*Cymbopogon citratus*), and rosemary (*Rosmarinus officinalis*) are widely utilized due to their high content of flavonoids, terpenoids, phenolic acids, essential oils, and antioxidants.

6.1 *Types of Green-Synthesized Nanocomposites Using Aromatic Plants*

6.1.1 Metal–Polymer Nanocomposites

Aromatic plant-mediated nanoparticles (e.g., AgNPs, AuNPs) can be incorporated into biopolymers such as chitosan, gelatin, starch, or cellulose to form metal–polymer nanocomposites. Example includes silver nanoparticles synthesized using clove extract incorporated into chitosan films that exhibit strong antimicrobial activity, suitable for wound dressings or food packaging.

6.1.2 Clay-Based Nanocomposites

Aromatic plant-derived nanoparticles can be embedded into clay matrices (e.g., montmorillonite, bentonite), forming eco-friendly nanocomposites for water filtration, controlled drug release, and soil remediation.

6.1.3 Polymer–Metal Oxide Hybrid Nanocomposites

Nanocomposites such as ZnO–chitosan, where ZnO nanoparticles synthesized via lemongrass extract are incorporated into a biopolymer, show synergistic antibacterial, UV-blocking, and mechanical reinforcement properties.

6.2 *Mechanisms of Synergistic Interaction*

Nanotechnology has revolutionized the delivery and performance of bioactive compounds. However, many plant-derived compounds, though rich in therapeutic properties, suffer from limitations such as poor water solubility, low bioavailability, and rapid degradation in physiological environments.

By incorporating these phytochemicals into or onto nanomaterials, a synergistic effect often emerges, where the combined action is greater than the sum of individual effects. These effects arise from the physicochemical interactions between plant molecules and nanostructures, enhancing the efficacy, stability, and target specificity of both. The synergy between plant compounds and nanomaterials is governed by several mechanisms:

6.2.1 Enhanced Bioavailability

Nanocarriers increase the solubility and systemic absorption of poorly soluble plant compounds. For example, curcumin-loaded nanoparticles show enhanced anticancer activity compared to free curcumin due to improved absorption and retention.

6.2.2 Controlled and Targeted Release

Nanomaterials enable sustained release of plant-derived compounds and can be tailored for target-specific delivery, improving therapeutic efficiency.

6.2.3 Stabilization of Phytochemicals

Plant compounds that are sensitive to light, heat, or oxidation are protected within the nanoparticle matrix or on the nanoparticle surface.

6.2.4 Multimodal Activity

Plant compounds and nanomaterials can target different biological pathways, leading to broad-spectrum activity or multi-targeted therapeutic action. For example, silver nanoparticles have membrane-disrupting effects, while polyphenols interfere with intracellular signaling—together, they exhibit potent antibacterial synergy [121].

Examples

Syzygium aromaticum (Clove) + Silver Nanoparticles

- Phytochemicals: Eugenol, flavonoids.
- Effect: Enhanced antibacterial and antifungal activity due to synergistic ROS production and membrane disruption.

Ocimum sanctum (Holy Basil) + Iron Oxide Nanoparticles

- Phytochemicals: Eugenol, ursolic acid.
- Effect: Improved antioxidant and anti-inflammatory effects with magnetic targeting capabilities.

Curcumin + Gold Nanoparticles

- Phytochemical: Curcumin (from *Curcuma longa*).
- Effect: Enhanced bioavailability, photothermal therapy potential, and anticancer activity.

Advantages of Synergistic Systems [121]

Advantage	Description
Eco-friendly and biocompatible	Derived from natural sources; reduced toxicity
Multi-target activity	Combines different mechanisms of action for broader therapeutic use
Reduced drug resistance	Less likely to induce resistance compared to synthetic drugs alone
Stability and shelf life	Enhanced by nanoparticle encapsulation or surface adsorption
Dose reduction	Potentiation allows for lower doses, reducing side effects

7 Characterization Techniques

Medicinal plant analysis has a lengthy history, particularly when it comes to determining a plant's quality. Utilizing the physical sensations of taste, smell, and appearance, the initial methods were organoleptic. These eventually gave way to increasingly complex instrumental methods [122]. Flavonoids, phenolics, and essential oils are only a few of the many secondary metabolites found in aromatic plants that have strong biological effects. Understanding these compounds' structural, chemical, and stability properties is crucial to maximizing their potential. Advanced analytical approaches must be used since conventional methods frequently fail to capture nanoscale structures or real-time dynamics. Key instruments for characterizing aromatic plants are examined in this study, with a focus on their applicability to both fundamental and applied research [123].

Advanced techniques such as SEM, TEM, XRD, UV–Vis spectroscopy, FTIR, and dynamic light scattering are frequently used to assess particle size, shape, surface charge, and aggregation, which are influenced by the presence and interaction of bioactive compounds during the synthesis process. For instance, spherical, rod-like, and polygonal morphologies are confirmed by SEM analysis, and FTIR can be used to authenticate the chemical fingerprints of functional plant metabolites used as capping agents. Variability in plant extract composition and extraction techniques has a major impact on repeatability and stability [25].

7.1 Spectroscopic Methods

7.1.1 In Situ UV–Vis Kinetics

A popular analytical method in many scientific fields, such as chemistry, biochemistry, and materials science, is UV–visible spectroscopy. It involves measuring how much ultraviolet (UV) and visible light a sample absorbs or transmits, giving important details on its electrical structure and chromophores [124]. The ease of use and accessibility of UV–visible spectroscopy are two of its main benefits. A UV–visible spectrophotometer that emits a wide spectrum of UV and visible light wavelengths makes up the comparatively simple apparatus. To enable quantitative analysis, the sample is usually put in a transparent cuvette, and the absorbance or transmittance of light is measured. It works especially well for studying molecules having conjugated pi-electron systems, like transition metal complexes and organic compounds. Because of electronic transitions between energy levels, these compounds display distinctive absorption bands in the visible and ultraviolet spectrums. The electronic characteristics, concentration, and purity of the sample can all be learned by examining the absorption spectra [125]. Compounds at low quantities can be detected, thanks to the technique's exceptional sensitivity. For quantitative analysis, such as figuring out a compound's concentration in a solution or tracking chemical reactions in real time, UV–visible spectroscopy is therefore helpful [126]. Additionally, UV–visible spectroscopy is used for component identification and qualitative analysis. Every substance has a different absorption spectrum, which is frequently distinguished by different intensities and peak positions. An unknown substance can be identified or categorized by comparing its absorption spectrum to reference spectra or databases [127]. Furthermore, there is little sample preparation needed for UV–visible spectroscopy because it is a nondestructive method. Liquids, solids, and gasses are just a few of the many sample types that can be analyzed with it. Because of its adaptability, it is a useful instrument in many scientific domains, including quality control, environmental monitoring, and pharmaceutical analysis [128].

By examining how materials absorb or transmit light at various wavelengths, UV–Vis spectroscopy aids in the evaluation of molecular and nanoscale materials' size, shape, stability, and structure.

Structure and Morphology Evaluation

The existence and kind of functional groups as well as conjugation are linked to electronic transitions in molecules, which are revealed by UV–Vis spectra. Chromophores and their electronic surroundings can be identified by their position (wavelength) and intensity (absorbance or molar extinction coefficient).

The shape and structure of metal nanoparticles, like gold and silver, are directly reflected by their distinctive surface plasmon resonance (SPR) peaks, which are detected by UV–Vis spectroscopy. While rod-shaped gold nanorods display two peaks (e.g., 514 nm for transverse and 689 nm for longitudinal SPR), suggesting rod-like morphology, spherical gold nanoparticles usually display a significant SPR peak around 520 nm. In the ultraviolet spectrum, uniformly formed ZnO nanoparticles have a pronounced absorption peak at 376 nm, which is indicative of their inherent band gap and production quality [129].

Size Determination

Because of surface plasmon resonance phenomena and quantum size effects, the absorption spectra of nanoparticles drastically alter with size. The absorption peaks may move to longer wavelengths as nanoparticle size grows, and the spectra's width and form may reveal the size distribution.

The size of the nanoparticles affects the UV–Vis spectrum's peak position and width. The SPR peak broadens and moves to longer wavelengths (redshift) as particle diameter rises. For example, size-dependent variations are confirmed by microscopy for 16 nm AuNPs, which absorb at 518 nm, 30 nm at 526 nm, and 55 nm at 532 nm. These spectra are frequently fitted and interpreted using the Mie theory model, which is also frequently used to predict nanoparticle size distributions from UV–Vis data for spheres [130].

Stability Analysis

UV–Vis spectroscopy works quite well for tracking the stability of nanoparticles. Sharp, recognizable absorption peaks are maintained by stable nanoparticles. Peak broadening, intensity changes, or peak shifts are caused by aggregation or destabilization and can occasionally be seen as a change in the suspension's hue. Absorbance spectra measure colloidal stability both numerically and visually because the distortion or removal of absorption peaks over time indicates the disintegration or aggregation of nanoparticles [131].

The fundamental method for analyzing phytochemicals is still UV–Vis spectroscopy. Reaction pathways and chemical transformations can be monitored in real time using the in situ use of UV–Vis spectroscopy, especially in kinetic mode. When evaluating oxidation, light-induced alterations, or enzymatic degradation in aromatic extracts, this is especially helpful. Spectral variations over time can be tracked to provide information about stability profiles and reaction processes [132]. Transmission electron microscopy and X-ray diffraction can be used to examine the structural properties of the produced nanostructures, including their morphological aspects and crystal structure. X-ray photoelectron spectroscopy can be used to analyze the elemental composition and oxidation states, while UV–Vis absorption spectroscopy can be used to determine the optical properties of the biosynthesized nanostructures [133].

7.1.2 FTIR Spectroscopy

In the realm of nanotechnology, Fourier transform infrared (FTIR) spectroscopy has become a vital characterization method, especially for verifying the surface functionalization of nanoparticles with bioactive substances. Because of their therapeutic potential and chemical adaptability, aromatic plant-based bioactives—rich in phenolics, flavonoids, and essential oils—are being used more and more for green synthesis and functionalization of nanoparticles. Bioactive compounds derived from aromatic plants, such as flavonoids, terpenoids, and polyphenols, contain a variety of functional groups, such as -OH, -COOH, C=O, and aromatic rings. By interacting with nanoparticles, these groups can offer stability and biological activity. FTIR is a very sensitive and nondestructive technique for confirming that these bioactives are successfully integrated and interacting on the surface of nanoparticles [134, 135].

Fourier transform infrared (FTIR) spectroscopy is a ubiquitous and extensively used spectral technology for the identification of internal molecular structures in a wide range of disciplines. Its basic idea is based on the vibration and rotation of atoms. Functional groups with a permanent dipole moment and chemical bonds can be thoroughly studied using FTIR spectroscopy. Because of its great versatility, FTIR spectroscopy finds use in a wide range of fields, such as biology, materials science, and environmental studies. It is especially useful for researching nanoparticles and their interactions with biological systems since it can examine complicated combinations without requiring a lot of sample preparation. A key method in the study of nanoparticles, FTIR spectroscopy offers comprehensive information about the chemical makeup and interactions of materials at the nanoscale. Many important aspects of the synthesis, stability, and use of green-generated nanoparticles (NPs) motivate the use of FTIR in their study. Certain absorption peaks that correspond to different functional groups can be identified by FTIR analysis, indicating their role in the reduction of metal ions or the creation of metal oxide nanoparticles and their subsequent stability. By demonstrating the existence of biomolecules that aid in nanoparticle formation, distinctive peaks in the FTIR spectrum help to validate the effective synthesis of nanoparticles. Additionally, FTIR can be used to investigate the interactions between the capping agents and the produced nanoparticles, offering insights into how these interactions affect particle stability. This knowledge is crucial to guaranteeing that the NPs retain their intended characteristics throughout time, which qualifies them for a range of uses, including medication delivery and bioremediation [136].

A strong method for locating functional groups and tracking chemical changes is Fourier transform infrared (FTIR) spectroscopy. FTIR is frequently employed in the context of aromatic plants to verify surface functionalization, particularly in formulation or nanoencapsulation procedures. The presence and interaction of bioactive substances with carriers or excipients can be determined by detecting distinctive bands that correspond to hydroxyl, carbonyl, and ether groups [137]. Investigations into potential functional groups that work as capping and reducing agents can be conducted using Fourier transform infrared spectroscopy [133].

Evaluation of Structural and Chemical Composition

By measuring distinctive vibrational modes, FTIR assists in identifying functional groups involved in capping or nanoparticle production. In order to confirm successful synthesis or functionalization without causing structural damage to the compounds involved, peaks corresponding to O–H, C=O, N–H, and C–H stretching and bending vibrations, for instance, show the presence of organic molecules such as proteins, amino acids, or plant metabolites bound to the nanoparticle surface. The structure and chemical makeup of the drug-loaded nanoparticle system were confirmed by FTIR, which showed amide and sulfate peaks that indicated drug entrapment and liposome coating [138].

Size and Morphology

FTIR offers indirect proof even though it cannot assess particle size or shape directly. Because smaller nanoparticles have a larger surface-to-volume ratio that affects the molecular vibrations of surface-bound ligands, changes in peak intensities, shifts in wavenumbers, or broadening of bands may indicate size-dependent surface chemistry alterations or morphological abnormalities. Changes in surface chemistry in nanoparticles of various sizes or aggregated states, for example, have been linked to shifts in infrared absorption bands, offering hints that supplement microscopy (SEM/TEM) studies [139].

Stability Assessment

By tracking alterations in the surface functional groups over time or under various circumstances, FTIR spectroscopy assesses the stability of nanoparticles. The permanence or changes of distinctive peaks are frequently indicative of stability. Consistent spectral peaks, for instance, signify ligands or stable capping layers that stop aggregation or disintegration. According to a study on silver nanoparticles, the particles were stabilized and capped by amine and carboxyl groups from biomolecules, which demonstrated chemical stability through enduring FTIR signals [140].

7.2 Structural and Morphological Analysis

7.2.1 Synchrotron-Assisted X-Ray Diffraction (XRD)

A basic, nondestructive analytical method that is widely used to describe the crystalline structure, phase composition, crystallite size, and strain in nanomaterials is X-ray diffraction (XRD). XRD is still a crucial tool in nanoscience and nanotechnology research for comprehending material properties at the atomic scale because of the distinct physicochemical characteristics of nanoparticles, which are primarily controlled by their crystallographic structure [139]. XRD operates on the principle of elastic scattering of monochromatic X-rays by the atomic planes in a crystal lattice. According to Bragg's law,

$$n\lambda = 2d\sin\theta,$$

where λ is the wavelength of incident X-rays, d is the interplanar spacing between crystal planes, θ is the diffraction angle, and n is the order of reflection, constructive interference occurs when reflected X-rays from different planes are in phase, producing characteristic diffraction peaks [141].

A monochromatic X-ray beam is focused onto a crystalline sample in XRD, and as the sample or detector rotates, the diffracted X-rays are detected. Constructive interference creates a distinctive diffraction pattern with peaks that correspond to particular crystal planes when the diffraction criteria are satisfied. By calculating these peaks' angles and intensities, data like crystal structure and lattice parameters, phase identification and composition, crystallite size and strain in the material, and thin-film thickness and multilayer composition is obtainable [139]. The following are some ways that X-ray diffraction (XRD) aids in the assessment of crystalline materials' size, shape, stability, and structure.

Structure

By examining the positions and intensities of diffraction peaks, which represent the arrangement and spacing of atoms or molecules in the crystal lattice, XRD is able to ascertain the crystal structure and phase composition.

By comparing with reference databases, the diffraction pattern serves as a "fingerprint" of the periodic atomic arrangement, enabling the identification of crystalline phases and lattice characteristics.

By resolving strain, crystal flaws, and lattice distortions, high-resolution XRD can reveal information about the internal structure of the material [142].

Size

The broadening of diffraction peaks is influenced by the size of the crystallite. Larger crystals provide sharper peaks, while smaller crystallites give broader peaks.

If the particles are single crystals, peak width analysis using the Scherrer equation can determine the average size of coherent diffracting domains (crystallite size), which may correspond with particle size. XRD is helpful for determining the size of nanomaterials because very small crystallites (less than 10 nm) result in substantial peak widening and decreased signal intensity [143].

Morphology

By examining the relative strengths of diffraction peaks, XRD can provide details on the preferred crystal orientation (texture) in polycrystalline or thin-film samples.

It has the ability to distinguish between crystalline and amorphous phases; amorphous materials exhibit diffuse, wide patterns.

Detailed morphology typically necessitates supplementary procedures, even if XRD offers indirect information about morphology through crystallite forms and orientation.

Stability

By tracking changes in diffraction patterns over time or under different settings, XRD can identify phase transformations and polymorphism alterations that impact material stability.

It tracks variations in lattice strain and crystallinity, which may signal the start of mechanical stress or degeneration.

XRD evaluates structural stability in response to mechanical, chemical, or thermal stressors by detecting new phases or breakdown products [142].

7.2.2 High-Resolution and Cryogenic Transmission Electron Microscopy (HR-TEM and Cryo-TEM)

TEM can be used to analyze the size and shape of peptide nanostructures with excellent resolution at a scale of a few nanometers or even smaller. By passing an electron beam through the object, this approach gathers data for creating images by interacting with interior structures. TEM permits imaging with greater magnification and resolution than light microscopy and for the examination of internal material features because the electron beam's wavelength is lower than that of light. TEM is used to determine the specifics of structures at the nanoscale and even to carry out atomic resolution characterization of materials, which includes determining the elemental composition and chemical bonding in addition to structural characterization [144]. Advanced imaging methods such as high-resolution transmission electron microscopy (HR-TEM) and cryogenic transmission electron microscopy (Cryo-TEM) offer comprehensive information about the shape, crystallinity, and behavior of nanoparticles in their natural state. Atomic arrangements within a nanoparticle can be seen, thanks to HR-TEM's high spatial resolution, which is achieved by using electron beams. This method produces fine-grained pictures of the atomic-scale defects, lattice fringes, and crystallographic structure [145].

Size Measurement

With nanometer to sub-nanometer accuracy, TEM allows direct imaging of nanoparticles, clusters, and thin films. Particle thicknesses and size distributions can be quantified precisely, which is essential for managing functional performance, particularly in nanoscale devices and materials [146].

Morphology Evaluation

Because of its great resolution, TEM can distinguish between various morphologies (such as spherical, rod-like, and sheet-like particles) and surface coatings. It can also visualize forms, surface textures, interior porosity, and aggregation states. This aids in the design and optimization of materials according to their qualities that rely on their shape [146].

Stability Assessment

When exposed to in situ thermal, electrical, or chemical stimuli, TEM can track morphological and structural changes. Degradation, coalescence, or other stability problems are revealed by imaging materials before and after stress or over time. Additionally, TEM-integrated analytical modes such as energy-dispersive X-ray spectroscopy (EDS) and electron energy loss spectroscopy (EELS) can monitor compositional or stability-related oxidation state changes [146].

7.3 *Stability and Bioactivity Assessment*

7.3.1 Zeta Potential and Dynamic Light Scattering (DLS)

Because of their potential biomedical uses and environmentally friendly nature, the use of bioactive compounds derived from aromatic plants for the synthesis of nanoparticles has attracted a lot of attention. Understanding these nanoparticles' physicochemical characteristics which affect their stability, bioavailability, and therapeutic efficacy requires characterization. Two essential methods used to evaluate the size distribution and surface charge of nanoparticles are zeta potential and dynamic light scattering (DLS), respectively. An environmentally friendly substitute for conventional chemical synthesis techniques is nanoparticles made from aromatic plant-based bioactive substances like flavonoids, polyphenols, and essential oils. These biogenic nanoparticles have special qualities, such as natural antioxidant and antimicrobial activity, biocompatibility, and biodegradability. It is essential to fully characterize these nanoparticles in order to realize their full potential. In this effort, zeta potential and DLS have become essential instruments, offering information on the size distribution and surface charge of nanoparticles, which are crucial for their stability and performance [147].

A noninvasive method for determining the Brownian motion of suspended nanoparticles is DLS. DLS calculates the hydrodynamic diameter and size distribution of nanoparticles by examining variations in light scattering brought on by particle motion. When evaluating the colloidal stability and aggregation behavior of nanoparticles in diverse media, this method is especially helpful [148].

The electrokinetic potential at a suspended particle's slipping plane is known as the zeta potential. It sheds light on the colloidal stability and surface charge of nanoparticles. Stable colloidal dispersions result from strong electrostatic repulsion between particles, which is indicated by a high absolute value of zeta potential (usually $> \pm 30$ mV). On the other hand, van der Waals forces may cause aggregation at low zeta potential values [139, 149, 150]. By revealing the degree of electrostatic repulsion between particles, zeta potential measurements can forecast the long-term stability of nanoparticle suspensions. Variations in zeta potential can show whether surface alterations or nanoparticle functionalization were successful. Zeta potential affects how nanoparticles interact with biological membranes, which in turn affects how well they are absorbed by cells and distributed throughout the body. When characterizing bioactive nanoparticles derived from aromatic plants, DLS and zeta potential are essential instruments. They facilitate the design and optimization of nanomaterials for a range of applications by offering comprehensive insights into the stability, surface charge, and size distribution of nanoparticles. Zeta potential is a crucial measure of colloidal stability, particularly in aromatic plant-based emulsions or nano-formulations. In order to assess formulation uniformity and stability over time, dynamic light scattering provides size distribution and polydispersity index (PDI) data [151].

8 Applications of Aromatic Plant-Derived Nanomaterials

Many natural phytochemicals, such as polysaccharides, terpenoids, alkaloids, and flavonoids, have been utilized to treat and prevent diseases. In addition to their high nutritional content, they exhibit extraordinary bioactivity, strong biodegradability as well as biocompatibility, and structural variety [152]. Specifically, a thorough examination of medications manufactured between 1981 and 2019 showed that natural phytochemicals or their derivatives accounted for 32% of recently authorized small-molecule medications [153] (Fig. 2). When compared to the already employed nanocarriers, natural phytochemicals provide the following benefits: (a) certain natural phytochemicals have the ability to self-assemble, which facilitates drug loading; (b) natural phytochemicals decrease adverse effects and immunogenicity because of their superior biocompatibility and controlled degradability [154].

8.1 Biomedical Applications

8.1.1 Antibacterial Activities

Since the number of cases of antibiotic resistance is steadily increasing, researchers are always looking for new antibacterial solutions that are more effective and have fewer negative effects. It has been observed that, in recent years, plant-derived nanoparticles (PDNPs) have become more useful as antibacterial agents [155]. The accumulation of PDNPs on bacterial surfaces can prevent beneficial signal inductions together with cellular transport. Meanwhile, PDNP may enter the cytoplasm of the microorganism, which would then cause the release of the nanoparticles and alter the activity of the bacterium. Positively charged green-produced silver nanoparticles (AgNPs) were more efficient against negatively charged bacterial cells, according to a recent study by Banasiuk et al. [156]. According to Buarki et al. [157], iron nanoparticles made from a 1:1 ratio of *Hibiscus rosa-sinensis* flower extract and $FeCl_36H_2O$ solution demonstrated antibacterial activity against a number of bacterial strains, including *Salmonella typhi, Pseudomonas aeruginosa, Staphylococcus aureus*, and *Klebsiella pneumoniae*. Natural flavonoids like apigenin triphosphate, quercetin pentaphosphide, and quercetin sulfonic acid have been effectively utilized to create AgNPs with

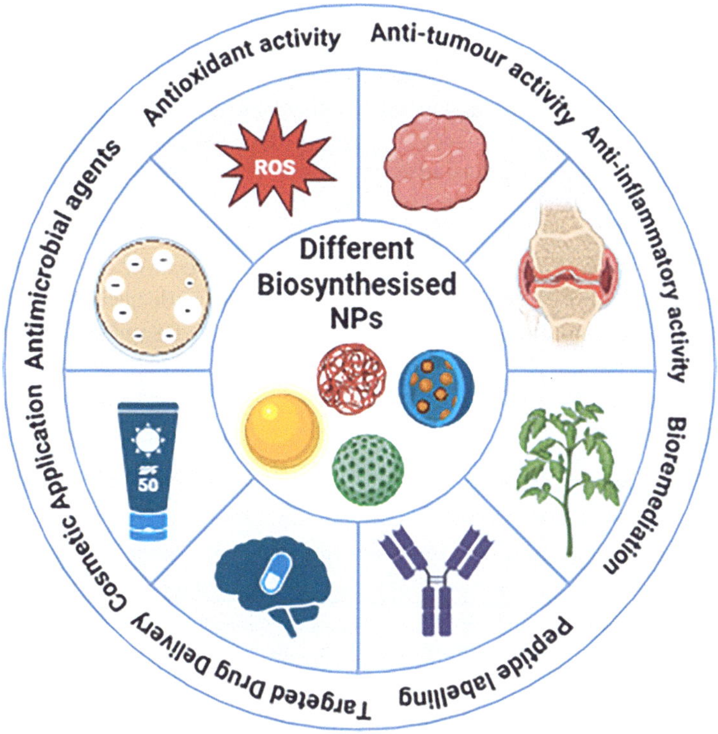

Fig. 2 Schematic representation of various medical applications. (Created with Biorender)

potent antibacterial properties against *Citrobacter freundii*, *E. coli*, and *Staphylococcus epidermidis* [158].

8.1.2 Antiviral Activities

The antiviral effects of ZnNP and AgNPs made from natural honey and olive leaves were recently described by Zeedan et al. [159]. At 25 μg/mL, these NPs significantly reduced cytopathic effect (CPE) in MDBK cells infected with bovine herpesvirus-1 [159]. Catechin, a plant-purified metabolite, has been used in an earlier work to create gold nanoparticles with an average size of 50 nm. The study demonstrated the formulation's exceptional antiviral qualities by preventing 100% of CPE brought on by the Japanese encephalitis virus, even at extremely low concentrations of 0.04 to 5.85 μg/mL [160].

8.1.3 Antitumor Activity

According to a recent study, AgNPs generated from plants had fewer cytotoxic and genotoxic effects than those manufactured chemically. AgNPs made using plant extract from *Nepeta deflersiana* recently demonstrated anticancer properties against human cervical cancer cells, causing concentration-dependent cytotoxicity and lowering glutathione levels in the cells [161]. According to a recent study, AuNPs made from *A. leptopus* had strong anticancer efficacy against MCF-7 breast cancer cells at a concentration of 257.8 μg/mL [162]. When applied to Dalton's lymphoma cells, silver–selenium (Ag-Se) bimetallic nanoparticles made with quercetin and gallic acid showed possible anticancer activity [163]. Nanoparticles made from the roots and leaves of *Jurinea dolomiaea* have shown antitumor efficacy against HeLa and MCF-7 cell lines [164].

8.1.4 Antioxidant Activities

The antioxidant and anti-inflammatory qualities of *Uncaria rhynchophylla*, *Sanguisorba officinalis*, *Salvia miltiorrhiza*, and *Ligustrum lucidum* are correlated with their significant concentrations of phenolics, flavonoids, and trace elements like selenium, zinc, and magnesium [165]. Because of its high polyphenolic content, the antioxidant activity of the nanoparticle generated from *Terminalia chebula* extract was assessed, demonstrating a strong reduction capacity and scavenging of radicals caused by hydrogen peroxide and nitric oxide [166].

8.1.5 Antidiabetic Activity

A nanoparticle made from *Dioscorea bulbifera* tuber extract had a strong antioxidant effect and inhibits α-amylase and α-glucosidase, hence showing promising antidiabetic activity [167]. The use of *Solanum nigrum*-synthesized nanoparticles to treat rats with alloxan-induced diabetes mellitus improved their dyslipidemic state and decreased their blood glucose levels [168]. When compared to metformin, a conventional antidiabetic drug, glycyrrhizin encapsulated in NPs demonstrated notable antidiabetic and antihyperlipidemic effects in type 2 diabetic mice [169].

8.1.6 Wound Healing

A diverse array of plant extracts has been utilized to produce green-synthesized AgNPs and formulate appropriate treatments for wounds. Several well-known examples of these plants include *Azadirachta indica* [170], *Cassia roxburghii* [171], *Carica papaya L.* [172], *Bryonia laciniosa L.* [173], *Indigofera aspalathoides DC* [174], etc. The majority of these extracts are suitable for wound care since they naturally have antibacterial and antimicrobial properties. The use of antimicrobial ingredients in wound dressings is crucial because they play a significant part in preventing the colonization as well as subsequent growth of microorganisms, which speeds up the healing process [175–177]. Compared to conventional gauze dressings or silver sulfadiazine, AgNPs in wound dressings reduce the healing period of injuries by an average of 3.35 days while preventing bacterial development without causing any adverse effects on patients [178]. Yao et al. [179] fabricated nanofiber dressing material for wound healing using extract of *Centella asiatica*. AgNPs produced using *Aspergillus niger* were able to enhance wound healing through the inhibition of harmful microorganisms and the modification of cytokines linked to wound healing [180].

8.2 Agricultural Applications

8.2.1 Bioherbicidal Applications

The primary objectives for the nanoscale synthesis of bioherbicides are great selectivity and enhanced solubility [181]. The existence of phytotoxic allelochemicals, which are naturally occurring substances emitted by plants that prevent the development of nearby plants, gives certain plant-derived components their bioherbicidal qualities [182–184]. Through a variety of approaches, these bioherbicides can prevent the growth and development of weeds. They can interfere with vital biological functions like respiration and photosynthesis (like camphor), affect the germination of seeds (like eugenol), and affect the division of cells (like thymol by changing the order of chromosomes) [29]. Taban et al. [185] investigated the specific herbicidal benefits of *S. hortensis* essential oil nanoencapsulated into organic polymer-based nanocarriers using tomato and *A. retroflexus*. While having a less severe effect on tomato plants, the nanocapsules, which were made using gum arabic, gum tragacanth, and a gum tragacanth–gelatin combination, had a notable herbicidal effectiveness against *A. retroflexus*. Eldarier et al. [186] examined the herbicidal effects of *H. tuberculatum* raw aqueous extract and a similar extract enhanced with AgNPs on *T. aestivum* and *Phalaris minor L.* According to the authors, at every tested dose, *P. minor* seed germination was totally suppressed by *H. tuberculatum* aqueous extract enhanced with AgNPs.

8.2.2 Soil Remediation

Soil contamination is a global environmental issue that is mostly caused by inappropriate waste disposal, industrial operations, and agricultural practices. In addition to posing serious health hazards to humans, the buildup of heavy metals and organic pollutants, along with other toxic compounds in soil, can negatively impact soil quality and impede plant development [187]. It has been demonstrated that phyto-nanoremediation, a well-known fusion of nanotechnology and

Table 3 Nanoparticles and plant species employed for soil remediation

Nanoparticle	Plant species	Effective concentration	References
AgNPs	*Boswellia ovalifoliolata*	10–30 g mL^{-1}	He et al. [191]
MnNPs	*Vigna radiata*	50–1000 mg L^{-1}	Pradhan et al. [192]
Iron oxide NPs	*Arachis hypogaea*	2–1000 mg kg^{-1}	Awasthi et al. [193]; Kumar et al. [194]; Singh et al. [195]
Zinc oxide NPs	*Arachis hypogaea*	1000 ppm	Ma et al. [196]
Copper oxide NPs	Common wheat	500 mg kg^{-1}	Yanık et al. [197]
Nano zero-valent iron	*Ocimum sanctum*	–	Dimkpa et al. [198]
Salicylic acid NPs	*Isatis cappadocica*	–	Cañas et al. [199]; Khodakovskaya et al. [200]

phytoremediation, is effective in breaking down and eliminating environmental contaminants [188]. In order to assess the remediation potential for textile dyes like methylene blue from water resources, Ahmed Rather's study used *Bergenia ciliata* extract as a reducing and capping agent for the green production of ZnNPs. Of MB dye, 82% could be broken down by zinc nanoparticles in 52 min when exposed to sunshine for photocatalytic degradation [189]. In contrast, 80% of MB was broken down by green-generated Fe-NP from *Citrus paradisi* extract in 6 h at room temperature [190]. Table 3 lists some plant species and NPs used for phytoremediation.

8.3 Environmental Applications

8.3.1 Water Purification

Metal oxide nanoparticles have been widely employed to remediate various wastewater types. AgNPs are utilized to effectively remove heavy metals (iron, copper, lead, and magnesium) from simulated wastewater. They were created using *P. thonningii* leaf extract [201]. Banerjee et al. [202] reported the green production of a silver nanocomposite for the treatment of textile dye using *Ocimum tenuiflorum* (Black Tulsi) leaf extract. Both residential and commercial applications can effectively utilize this treated wastewater. According to Ehrampoush et al. [203], tangerine peel extract was used to create green synthetic iron oxide nanoparticles that effectively removed cadmium from wastewater. Up to 90% of the cadmium ions were removed in this research. A green chemistry technique for creating gold nanoparticles (AuNPs) utilizing

Lagerstroemia speciosa leaf extract was described by Chaudhary et al. [204]. When exposed to visible light and $NaBH_4$, this approach exhibited high photocatalytic capability in lowering the levels of dyes, including methylene blue, methyl orange, bromophenol blue, and bromocresol green, as well as 4-nitrophenol [204].

8.3.2 Pollutant Degradation

Plant extract-derived nanoparticles are a potential method for degrading a variety of contaminants [205]. In order to increase their efficacy, phytonanoparticles use a variety of methods to break down contaminants, frequently in tandem [205]. These mechanisms include adsorption [206], photocatalytic degradation [207], and reductive dechlorination. Silver nanoparticles are created using *Azadirachta indica* extract for the photocatalytic degradation of heavy metals along with other environmental pollutants [25]. A variety of metal NPs, including iron, gold, and silver, are synthesized by using different species of the genus *Phyllanthus*. These have antibacterial and anticancer qualities, as well as uses in pollution degradation [80]. AuNPs were created using fruit extract from *Capsicum baccatum* L., and its photocatalytic efficiency was examined. These AuNPs showed significant photocatalytic capability in the 50% degradation of methylene blue after 300 min of sun irradiation [208]. Gold NPs fabricated from leaf extract of *Limnophila rugosa* exhibit significant catalytic activity in the reduction of different nitrophenols [209]. Iron NPs synthesized from leaf extract of *Syzygium jambos (L.) Alston* are utilized in removing chromium metals from the environment [210].

8.3.3 Biosensing

Both industries and researchers are very interested in the potential of nanoparticles for biosensing. The ability of biomolecules to attach to metal nanoparticle surfaces and then alter the nanoparticles' plasmonic characteristics is necessary for biomolecule sensing [211]. Because of their special qualities, particularly in the identification of environmental pollutants and biomolecules, plant-extracted silver NPs are becoming increasingly attractive for biosensing applications. Silver NPs are perfect for creating sophisticated biosensors because of their special properties, which include high electrical conductivity and a huge surface area [212]. Silver NPs extracted from *Citrus sinensis* [213], *Piper longum* [214], and *Azadirachta indica* [215] are used for the identification of heavy metals, bacterial infections, and pathogens, respectively. Carbon dots made from plant parts have also been used as biosensors to identify bioactive substances like proteins, amino acids, thiols, vitamins, enzymes, microorganisms, biological colors, and nucleic acids [216].

8.4 Industrial Applications

8.4.1 Cosmetics

The cosmetics sector frequently uses nanoparticles to enhance the qualities of various formulations and products. Additionally, nanomaterials can give cosmetic goods better stability and sensory qualities [217]. Phytosynthesized AgNPs can be used to create antimicrobial cream products. Synthetic silver nanoparticles created from *Withania somnifera* extract were added to a cream base and evaluated for their ability to inhibit human infections such *S. aureus*, *P. aeruginosa*, and *E. coli*. The results demonstrated the greater inhibitory potential of creams loaded with AgNPs against these microorganisms [218]. Likewise, AgNPs made from extract of *Solanum trilobatum* Linn, which has antibacterial qualities against a variety of germs and fungus, can be utilized as an active component in shampoos [219]. ZnO nanoparticles synthesized from the leaves of *Azadirachta indica* and flowers of *Vitex negundo* are widely used in sunscreen creams [220].

8.4.2 Food Preservation

Nanoparticles are an extremely important and successful food preservation intervention among all other techniques. As preservatives, nanoparticles can be added to food packaging or combined with the food [221]. Numerous nanoparticles are being studied extensively for food preservation, including metal (alloys and oxides), nanocomposite, liposomes, lysosomes, etc. Copper, gold, and silver are examples of metals, alloys, and oxide nanoparticles that are gaining popularity because of their biological qualities, which include antioxidant along with antibacterial effects [222]. The plant-mediated synthesis of zinc oxide NPs from plants like *Azadirachta indica*, *Aloe vera*, and *Catharanthus roseus* is used in food preservatives. Extracts from *C. citratus* successfully extended the shelf life and enhanced the overall quality of chicken sausage in meat preservation [223].

8.4.3 Food Packaging

The manufacturing of packaging materials with antibacterial and antifouling qualities heavily relies on nanoparticles. *Protium serratum* leaf extract was used to create silver nanoparticles, which were recommended for use in food packing and preservation because of their capacity to fight off a variety of foodborne pathogenic bacteria, including *E. coli* and *Bacillus subtilis* [224]. In another investigation, starch-based films containing AgNPs made from active lemon juice were used in active food packaging [225].

9 Challenges and Future Prospects

Despite the evident advantage of green synthesis, several challenges hinder its wider applications in the production of nanoparticles from aromatic plants. A major limitation lies in the intrinsic variability of plant extracts as their phytochemical composition is strongly influenced by seasonal, geographical, and physiological factors [23, 226–228]. This variability often results in poor reproducibility of nanoparticle size, morphology, and yield, thereby restricting the establishment of standardized protocols [229]. Scaling up synthesis remains another critical obstacle. Most reports describe laboratory-scale experiments, yet the translation of these methods to pilot or industry level is complicated by issues of reaction control, process uniformity, and downstream separation [230].

In addition, the stability of nanoparticles synthesized from plant extracts continues to present difficulties. Many formulations are prone to agglomerations, oxidations, or chains in surface chemistry during storage which in turn compromised their functionality and shelf life [231]. Mechanistic understanding is also limited as the precise biomolecules responsible for reduction and stabilization are rarely identified. Without such knowledge, rational control over nucleation and growth remains elusive making it difficult to design particles with predictable physiochemical properties [232, 233]. Furthermore, although plant-mediated synthesis avoids the use of toxic reagents, the nanoparticles themselves may still induce cytotoxicity, oxidative stress, and genotoxic effects. Their long-term environmental fate and ecological consequences remain insufficiently studied underscoring the need for comprehensive toxicological evaluation [234–237].

Future research should prioritize the development of standardized methodologies that ensure reproducibility and comparability across studies. The adoption of consistent extraction protocols, coupled with metabolic and physiochemical profiling, would enable the identification of active biomolecules and facilitate more precise control of the synthesis process. Scaling up production requires the integration of continuous flow reactors, real-time monitoring, and optimized purification methods to bridge the gap between laboratory-scale synthesis and industrial applications [21, 228, 238–243].

Advances in process optimization could also be achieved by correlating specific phytochemical constraints with reaction parameter such as pH temperature and extract concentration, thereby allowing deliberate manipulation of particle nucleation and growth [80, 244–247]. In parallel, safety assessments must be advanced through systematic in vitro and in vivo studies focusing on biodistribution, biodegradation, and ecotoxicological profile [248–250]. Integrating life cycle assessment into the development pipeline will ensure that the technology progresses responsibly. Moreover, valorization of agricultural and industrial waste from aromatic plant processing represents an underexplored but highly sustainable feedstock for nanoparticle synthesis, aligning with the principle of circular bioeconomy [251, 252].

10 Conclusion

Aromatic plants have emerged as valuable resources in green technology offering phytochemicals that act simultaneously as reducing and stabilizing agents during nanoparticle synthesis. This natural route not only ensures particle stability but also preserves intrinsic bioactivities such as antioxidant, antimicrobial, and anti-inflammatory effects. Compared with conventional methods, plant-based synthesis provides sustainable, biocompatible, and cost-effective alternatives with wide-ranging applications in biomedicine, food preservation agriculture, and environmental remediation. While the diversity of plant metabolites presents challenge in controlling nanoparticle uniformity, it also opens opportunity for multifunctional materials. Future research should therefore emphasize mechanistic insights, standardized protocols, and scale-up strategies positioning aromatic plants as a bridge between traditional phytochemistry and modern nanoscience.

Competing Interests The authors declare that they have no competing interest.

Declaration of Generative AI and AI-Assisted Technologies in the Writing Process During the preparation of this work, the book chapter author(s) have not taken any assistance from AI, and the book chapter author(s) reviewed and edited the content as needed and take(s) full responsibility for the content of the publication.

References

1. Bayda S, Adeel M, Tuccinardi T, Cordani M, Rizzolio F. The history of nanoscience and nanotechnology: from chemical–physical applications to nanomedicine. Molecules. 2019 Dec 27;25(1):112
2. Harish V, Tewari D, Gaur M, Yadav AB, Swaroop S, Bechelany M, Barhoum A. Review on nanoparticles and nanostructured materials: Bioimaging, biosensing, drug delivery, tissue engineering, antimicrobial, and agro-food applications. Nanomaterials. 2022 Jan 28;12(3):457
3. Husen A, Siddiqi KS (2023) Advances in smart nanomaterials and their applications. Elsevier, Cambridge, MA. https://doi.org/10.1016/C2021-0-02202-1
4. Jin-Chul K, Madhusudhan A, Husen A (2021) Smart nanomaterials in biomedical applications. Springer Nature Switzerland AG, Cham. https://doi.org/10.1007/978-3-030-84262-8
5. Madhusudhan A, Purohit SD, Prasad R, Husen A (2024) Functional smart nanomaterials and their theranostics approaches. Springer Nature, Singapore
6. Su H, Wang Y, Gu Y, Bowman L, Zhao J, Ding M (2018) Potential applications and human biosafety of nanomaterials used in nanomedicine 38(1):3–24. https://doi.org/10.1002/jat.3476
7. Mozaffari S, Li W, Thompson C, Ivanov S, Seifert S, Lee B et al (2017) Colloidal nanoparticle size control: experimental and kinetic modeling investigation of the ligand–metal binding role in controlling the nucleation and growth kinetics 9(36):13772–13785
8. Wu K-J, Edmund C, Shang C, Guo Z (2022) Nucleation and growth in solution synthesis of nanostructures–from fundamentals to advanced applications 123:100821

9. Ahmed SF, Mofijur M, Rafa N, Chowdhury AT, Chowdhury S, Nahrin M et al (2022) Green approaches in synthesising nanomaterials for environmental nanobioremediation 204:111967
10. Gupta D, Boora A, Thakur A, Gupta T (2023) Green and sustainable synthesis of nanomaterials: recent advancements and limitations 231:116316
11. Singh YR, Selvam A, Lokhande P, Chakrabarti S (2023b) Sustainability: an emerging design criterion in nanoparticles synthesis and applications, pp 65–113
12. Husen A (2019) Natural product-based fabrication of zinc oxide nanoparticles and their application. In: Husen A, Iqbal M (eds) Nanomaterials and plant potential. Springer International Publishing AG, Cham, pp 193–291. https://doi.org/10.1007/978-3-030-05569-1_7
13. Husen A (2020) Introduction and techniques in nanomaterials formulation. In: Husen A, Jawaid M (eds) Nanomaterials for agriculture and forestry applications. Elsevier, Cambridge, MA, pp 1–14. https://doi.org/10.1016/B978-0-12-817852-2.00001-9
14. Husen A (2023) Secondary metabolites based green synthesis of nanomaterials and their applications. Springer Nature, Singapore. https://doi.org/10.1007/978-981-99-0927-8
15. Siddiqi KS, Husen A (2016a) Fabrication of metal nanoparticles from fungi and metal salts: scope and application. Nanoscale Res Lett 11(98):1–15
16. Siddiqi KS, Husen A (2016b) Fabrication of metal and metal oxide nanoparticles by algae and their toxic effects. Nanoscale Res Lett 11(363):1–11
17. Husen A, Siddiqi KS (2014) Phytosynthesis of nanoparticles: concept, controversy and application. Nanoscale Res Lett 9(229):1–24
18. Khafaga DS, Radwan MG, Muteeb G, Aatif M, Farhan M (2023) Green synthesis of biocatalysts based on nanocarriers promises an effective role in pharmaceutical and biomedical fields 13(11):1448
19. Singh J, Dutta T, Kim K-H, Rawat M, Samddar P, Kumar P (2018) 'Green' synthesis of metals and their oxide nanoparticles: applications for environmental remediation 16(1):84
20. Zhang D, Ma X, Gu Y, Huang H, Zhang G (2023) Green synthesis of metallic nanoparticles and their potential applications to treat cancer. Front Chem 8:799
21. Lakhani KG, Hamid R, Motamedi E, Marviya G (2025) A review on plant metabolite-mediated nanoparticle synthesis: sustainable applications in horticultural crops 7:1545413
22. Radulescu D-M, Surdu V-A, Ficai A, Ficai D, Grumezescu A-M, Andronescu E (2023) Green synthesis of metal and metal oxide nanoparticles: a review of the principles and biomedical applications 24(20):15397
23. Antunes Filho S, Dos Santos MS, Dos Santos OAL, Backx BP, Soran M-L, Opriş O et al (2023) Biosynthesis of nanoparticles using plant extracts and essential oils. Molecules 28(7):3060
24. Selvakesavan RK, Franklin G (2021) Prospective application of nanoparticles green synthesized using medicinal plant extracts as novel nanomedicines, pp 179–195
25. Shahzadi S, Fatima S, Shafiq Z, Janjua MRSA (2025) A review on green synthesis of silver nanoparticles (SNPs) using plant extracts: a multifaceted approach in photocatalysis, environmental remediation, and biomedicine 15(5):3858–3903
26. Elansary HO, Abdelgaleil SA, Mahmoud EA, Yessoufou K, Elhindi K, El-Hendawy S (2018) Effective antioxidant, antimicrobial and anticancer activities of essential oils of horticultural aromatic crops in northern Egypt 18(1):214
27. Reichling J, Schnitzler P, Suschke U, Saller R (2009) Essential oils of aromatic plants with antibacterial, antifungal, antiviral, and cytotoxic properties—an overview 16(2):79–90
28. Mohammed HA, Sulaiman GM, Khan RA, Al-Saffar AZ, Mohsin MH, Albukhaty S, Ismail A (2024) Essential oils pharmacological activity: chemical markers, biogenesis, plant sources, and commercial products 144:112–132
29. Hernández-Bolaños E, Sánchez-Retuerta V, Matías-Hernández L, Cuyas L (2025) Promising applications on the use of medicinal and aromatic plants in agriculture 3(1):36
30. Pinto T, Aires A, Cosme F, Bacelar E, Morais MC, Oliveira I et al (2021) Bioactive (poly) phenols, volatile compounds from vegetables, medicinal and aromatic plants 10(1):106

31. Grigore-Gurgu L, Dumitrașcu L, Aprodu I (2025) Aromatic herbs as a source of bioactive compounds: an overview of their antioxidant capacity, antimicrobial activity, and major applications 30(6):1304
32. Salam U, Ullah S, Tang Z-H, Elateeq AA, Khan Y, Khan J et al (2023) Plant metabolomics: an overview of the role of primary and secondary metabolites against different environmental stress factors 13(3):706
33. Chen Z, Świsłocka R, Choińska R, Marszałek K, Dąbrowska A, Lewandowski W, Lewandowska H (2024) Exploring the correlation between the molecular structure and biological activities of metal–phenolic compound complexes: research and description of the role of metal ions in improving the antioxidant activities of phenolic compounds 25(21):11775
34. Guo D, Dou D, Ge L, Huang Z, Wang L, Gu N (2015) A caffeic acid mediated facile synthesis of silver nanoparticles with powerful anti-cancer activity 134:229–234
35. Swallah MS, Sun H, Affoh R, Fu H, Yu H (2020) Antioxidant potential overviews of secondary metabolites (polyphenols) in fruits 2020(1):9081686
36. Sharma NK, Vishwakarma J, Rai S, Alomar TS, AlMasoud N, Bhattarai A (2022b) Green route synthesis and characterization techniques of silver nanoparticles and their biological adeptness 7(31):27004–27020
37. Zuhrotun A, Oktaviani DJ, Hasanah AN (2023) Biosynthesis of gold and silver nanoparticles using phytochemical compounds 28(7):3240
38. Amini SM, Akbari A (2019) Metal nanoparticles synthesis through natural phenolic acids. IET Nanobiotechnol 13(8):771–777
39. El-Seedi HR, El-Shabasy RM, Khalifa SA, Saeed A, Shah A, Shah R et al (2019) Metal nanoparticles fabricated by green chemistry using natural extracts: biosynthesis, mechanisms, and applications 9(42):24539–24559
40. Netala VR, Hou T, Sana SS, Li H, Zhang ZJM (2024) Rosmarinic acid-rich perilla frutescens extract-derived silver nanoparticles: a green synthesis approach for multifunctional biomedical applications including antibacterial, antioxidant, and anticancer activities. Molecules 29(6):1250
41. Mohagheghian B, Saeidi G, Arzani A (2025) Phenolic compounds, antioxidant enzymes, and oxidative stress in barley (Hordeum vulgare L) genotypes under field drought-stress conditions 25(1):709
42. Shin S-A, Moon SY, Kim W-Y, Paek S-M, Park HH, Lee CS (2018) Structure-based classification and anti-cancer effects of plant metabolites 19(9):2651
43. Santos-Buelga C, Feliciano AS (2017) Flavonoids: from structure to health issues 22:477
44. Yonekura-Sakakibara K, Higashi Y, Nakabayashi R (2019) The origin and evolution of plant flavonoid metabolism 10:943
45. Alam F, Mohammadin K, Shafique Z, Amjad ST, Asad MHH (2022) Citrus flavonoids as potential therapeutic agents: a review. Phytother Res 36(4):1417–1441
46. Ciriminna R, Petri GL, Angellotti G, Luque R, Fabiano Tixier AS, Meneguzzo F (2025) Citrus flavonoids as antimicrobials 22(6):e202403210
47. Shehata MG, Awad TS, Asker D, El Sohaimy SA, Abd El-Aziz NM, Youssef MM (2021) Antioxidant and antimicrobial activities and UPLC-ESI-MS/MS polyphenolic profile of sweet orange peel extracts 4:326–335
48. Pavlešić T, Poljak S, Mišetić Ostojić D, Lučin I, Reynolds CA, Kalafatovic D (2022) Mint (Mentha spp) honey: analysis of the phenolic profile and antioxidant activity 60(4):509–519
49. Peixoto JAB, Álvarez-Rivera G, Alves RC, Costa AS, Machado S, Cifuentes A et al (2021) Comprehensive phenolic and free amino acid analysis of rosemary infusions: Influence on the antioxidant potential 10(3):500
50. Hanafy DM, Burrows GE, Prenzler PD, Hill R (2020) Potential role of phenolic extracts of mentha in managing oxidative stress and Alzheimer's disease 9(7):631
51. Park JB (2011) Identification and quantification of a major anti-oxidant and anti-inflammatory phenolic compound found in basil, lemon thyme, mint, oregano, rosemary, sage, and thyme 62(6):577–584

52. Wang Q, Zhao X, Jiang Y, Jin B, Wang L (2023) Functions of representative terpenoids and their biosynthesis mechanisms in medicinal plants 13(12):1725
53. terpenes (2025) IUPAC Compendium of Chemical Terminology. Online version 5.0.0, 5th edn. International Union of Pure and Applied Chemistry, Research Triangle Park, NC
54. Adefegha SA, Oboh G, Oluokun OOJ (2022) Food bioactives: the food image behind the curtain of health promotion and prevention against several degenerative diseases. Stud Nat Prod Chem 72:391–421
55. Habtemariam S (2023) Anti-inflammatory therapeutic mechanisms of natural products: insight from rosemary diterpenes, carnosic acid and carnosol 11(2):545
56. Habtemariam S (2016) The therapeutic potential of rosemary (Rosmarinus officinalis) diterpenes for Alzheimer's disease 2016(1):2680409
57. Iftikhar T, Majeed H, Zahra SS, Waheed M, Niaz M, Bano N (2023) Thyme. In: Essentials of medicinal and aromatic crops. Springer, Cham, pp 399–429
58. Ramadan KM, El-Beltagi HS, Bendary ES, Ali HM (2022) Experimental evaluation of the antioxidant and antitumor activities of thyme and basil essential oils and their phenolic constituents. Theor Antioxidant Eval 9(1):102
59. Mishra P, Rafi Z, Faruqui T, Mansoor S, Ahmad I, Ahmad I et al (2024) Exploring the significance and cutting-edge applications of terpenes and terpenoid-derived inorganic nanoparticles 16(6):665–681
60. Parlinska-Wojtan M, Kus-Liskiewicz M, Depciuch J, Sadik O (2016) Green synthesis and antibacterial effects of aqueous colloidal solutions of silver nanoparticles using camomile terpenoids as a combined reducing and capping agent 39(8):1213–1223
61. Romero-Montero A, Melgoza-Ramírez LJ, Ruíz-Aguirre JA, Chávez-Santoscoy A, Magaña JJ, Cortés H et al (2023) Essential-oils-loaded biopolymeric nanoparticles as strategies for microbial and biofilm control: a current status 25(1):82
62. Hcini K, Lozano-Pérez AA, Luis Cenis J, Quílez M, Jose Jordan M (2021) Extraction and encapsulation of phenolic compounds of tunisian rosemary (Rosmarinus officinalis L) extracts in silk fibroin nanoparticles 10(11):2312
63. Kayashima T, Matsubara K (2012) Antiangiogenic effect of carnosic acid and carnosol, neuroprotective compounds in rosemary leaves 76(1):115–119
64. Mirza FJ, Zahid S, Holsinger RD (2023) Neuroprotective effects of carnosic acid: insight into its mechanisms of action 28(5):2306
65. Satoh T, Trudler D, Oh C-K, Lipton SA (2022) Potential therapeutic use of the rosemary diterpene carnosic acid for Alzheimer's disease, Parkinson's disease, and long-COVID through NRF2 activation to counteract the NLRP3 inflammasome 11(1):124
66. Yana NTY, Shipon MNH, Bristy AH, Safa FA, Hossain MA, Al Hasan MS (2025) Carnosic acid as a promising anticancer agent: mechanisms of action and therapeutic potential across multiple cancer types 1(01):1–5
67. Kukula-Koch W, Widelski J (2017) Alkaloids. In: Pharmacognosy. Elsevier, Amsterdam, pp 163–198
68. Singh S, Pathak N, Fatima E, Negi AS (2021) Plant isoquinoline alkaloids: advances in the chemistry and biology of berberine 226:113839
69. Yu Q, Fan L (2021) Improving the bioactive ingredients and functions of asparagus from efficient to emerging processing technologies: a review 358:129903
70. Adarsh A, Chettiyar B, Kanthesh B, Raghu N (2020) Phytochemical screening and antimicrobial activity of "Cinnamon zeylanicum". Int J Pharmaceut Res Innov 13:22–33
71. Hudz N, Kobylinska L, Pokajewicz K, Horčinová Sedláčková V, Fedin R, Voloshyn M et al (2023) Mentha piperita: Essential oil and extracts, their biological activities, and perspectives on the development of new medicinal and cosmetic products 28(21):7444
72. Putri SA, Julaeha E, Kagawa N, Kurnia D (2025) Exploring Ocimum basilicum-derived indole alkaloid: s dual approach of antibacterial effects and computational study into enzymatic inhibition of Streptococcus sanguinis 16:102467

73. Sharma Y, Velamuri R, Fagan J, Schaefer J (2020) Full-spectrum analysis of bioactive compounds in rosemary (Rosmarinus officinalis L) as influenced by different extraction methods 25(20):4599
74. Bouttier-Figueroa DC, Cortez-Valadez M, Flores-Acosta M, Robles-Zepeda RE (2023) Synthesis of metallic nanoparticles using plant's natural extracts: synthesis mechanisms and applications 25(3):125–139
75. Chopra H, Bibi S, Singh I, Hasan MM, Khan MS, Yousafi Q (2022) Green metallic nanoparticles: biosynthesis to applications 10:874742
76. El-Seedi HR, Omara MS, Omar AH, Elakshar MM, Shoukhba YM, Duman H et al (2024) Updated review of metal nanoparticles fabricated by green chemistry using natural extracts: biosynthesis, mechanisms, and applications 11(11):1095
77. Fahmy SA, Fawzy IM, Saleh BM, Issa MY, Bakowsky U, Azzazy HME-S (2021a) Green synthesis of platinum and palladium nanoparticles using Peganum harmala L. seed alkaloids: biological and computational studies 11(4):965
78. Fahmy SA, Issa MY, Saleh BM, Meselhy MR, Azzazy HM (2021b) Peganum harmala alkaloids self-assembled supramolecular nanocapsules with enhanced antioxidant and cytotoxic activities 6(18):11954–11963
79. Singaravelu S, Motsoene F, Abrahamse H, Dhilip Kumar SS (2025) Green-synthesized metal nanoparticles: a promising approach for accelerated wound healing 13:1637589
80. Thatyana M, Dube NP, Kemboi D, Manicum A-LE, Mokgalaka-Fleischmann NS, Tembu JV (2023) Advances in phytonanotechnology: a plant-mediated green synthesis of metal nanoparticles using phyllanthus plant extracts and their antimicrobial and anticancer applications 13(19):2616
81. Albuquerque PM, Azevedo SG, de Andrade CP, D'Ambros NC, Pérez MTM, Manzato L (2022) Biotechnological applications of nanoencapsulated essential oils: a review. Polymers 14(24):5495
82. Mohamed AA, Alotaibi BM (2023) Essential oils of some medicinal plants and their biological activities: a mini review 9(1):40–49
83. Vilas V, Philip D, Mathew J (2016) Essential oil mediated synthesis of silver nanocrystals for environmental, anti-microbial and antioxidant applications 61:429–436
84. Khan MA, Khan T, Nadhman A (2016) Applications of plant terpenoids in the synthesis of colloidal silver nanoparticles 234:132–141
85. Moosavy M-H, de la Guardia M, Mokhtarzadeh A, Khatibi SA, Hosseinzadeh N, Hajipour N (2023) Green synthesis, characterization, and biological evaluation of gold and silver nanoparticles using Mentha spicata essential oil 13(1):7230
86. Dzimitrowicz A, Berent S, Motyka A, Jamroz P, Kurcbach K, Sledz W, Pohl P (2019) Comparison of the characteristics of gold nanoparticles synthesized using aqueous plant extracts and natural plant essential oils of Eucalyptus globulus and Rosmarinus officinalis 12(8):4795–4805
87. Jabir M, Sahib UI, Taqi Z, Taha A, Sulaiman G, Albukhaty S et al (2020) Linalool-loaded glutathione-modified gold nanoparticles conjugated with CALNN peptide as apoptosis inducer and NF-κB translocation inhibitor in SKOV-3 cell line. Int J Nanomedicine 15:9025–9047
88. Malathy B, Ajitha SP, Sangeetha K, Thampy S, Kamala G (2021) Antimicrobial activity of commercial essential oils on human pathogens 14(8):4440–4444
89. Hano C, Abbasi BH (2021) Plant-based green synthesis of nanoparticles: production, characterization and applications 12:31
90. Huston M, DeBella M, DiBella M, Gupta A (2021) Green synthesis of nanomaterials. Nanomaterials 11(8):2130
91. Kumar H, Bhardwaj K, Kuča K, Kalia A, Nepovimova E, Verma R, Kumar D (2020b) Flower-based green synthesis of metallic nanoparticles: applications beyond fragrance. Nanomaterials 10(4):766

92. Góral-Kowalczyk M, Grządka E, Orzeł J, Góral D, Skrzypek T, Kobus Z, Nawrocka A (2024) Green synthesis of iron nanoparticles using an aqueous extract of strawberry (Fragaria× ananassa Duchesne) leaf waste. Materials 17(11):2515
93. Süntar I (2020) Importance of ethnopharmacological studies in drug discovery: role of medicinal plants. Phytochem Rev 19(5):1199–1209
94. Eddin LB, Jha NK, Meeran MN, Kesari KK, Beiram R, Ojha S (2021) Neuroprotective potential of limonene and limonene containing natural products. Molecules 26(15):4535
95. Mahmoudi R, Aghaei S, Salehpour Z, Mousavizadeh A, Khoramrooz SS, Taheripour Sisakht M et al (2020) Antibacterial and antioxidant properties of phyto-synthesized silver nanoparticles using Lavandula stoechas extract. Appl Organomet Chem 34(2):e5394
96. Balasooriya ER, Jayasinghe CD, Jayawardena UA, Ruwanthika RWD, Mendis de Silva R, Udagama PV (2017) Honey mediated green synthesis of nanoparticles: new era of safe nanotechnology. J Nanomater 2017(1):5919836
97. Khandel P, Shahi S (2018) Mycogenic nanoparticles and their bio-prospective applications: current status and future challenges. J Nanostruct Chem 8:369–391
98. Deka K, Nongbet RD, Das K, Saikia P, Kaur S, Talukder A, Thakuria B (2025) Understanding the mechanism underlying the green synthesis of metallic nanoparticles using plant extract (s) with special reference to silver, gold, copper and zinc oxide nanoparticles. Hybrid Adv 9:100399
99. Azad A, Zafar H, Raza F, Sulaiman M (2023) Factors influencing the green synthesis of metallic nanoparticles using plant extracts: a comprehensive review. Pharmaceut Fronts 5(03):e117–e131
100. Pirsaheb M, Gholami T, Seifi H, Dawi EA, Said EA, Hamoody A-HM et al (2024) Green synthesis of nanomaterials by using plant extracts as reducing and capping agents. Environ Sci Pollut Res 31(17):24768–24787
101. Szczyglewska P, Feliczak-Guzik A, Nowak I (2023) Nanotechnology–general aspects: a chemical reduction approach to the synthesis of nanoparticles. Molecules 28(13):4932
102. Aftab R, Akbar F, Afroz A, Asif A, Khan MR, Rehman N, Zeeshan N (2024) Mentha piperita silver nanoparticle-loaded hydrocolloid film for enhanced diabetic wound healing in rats. J Wound Care 33(Sup3a):xlviii–lx
103. Mačák L, Velgosova O, Dolinská S (2023) Impact of two lavender extracts on silver nanoparticle synthesis, and the study of nanoparticles' antibiofilm properties and their ability to transfer them into a nontoxic polymer. Paper presented at the Micro
104. Mustapha T, Misni N, Ithnin NR, Daskum AM, Unyah NZ (2022) A review on plants and microorganisms mediated synthesis of silver nanoparticles, role of plants metabolites and applications. Int J Environ Res Public Health 19(2):674
105. Sharma NK, Vishwakarma J, Rai S, Alomar TS, AlMasoud N, Bhattarai A (2022a) Green route synthesis and characterization techniques of silver nanoparticles and their biological adeptness. ACS Omega 7(31):27004–27020
106. Hameed Hameed MK, Nematollahi F, Rouhani M (2024) Green synthesis of gold nanoparticles using salvia rosmarinus essential oil. J Med Plants Byprod 13(3):582–589
107. Katyal P (2018) Review on gold nanoparticles and their applications. Int J Compos Const Mater 4(2):32–39
108. Kiran M, Betageri VS, Vinay S, Latha M (2020) Synthesis of gold nanoparticles using novel Eucalyptus tereticornis and their in-vitro antibacterial, antioxidant and anticancer studies. Adv Nat Sci Nanosci Nanotechnol 11(4):045012
109. Pathania D, Sharma M, Thakur P, Chaudhary V, Kaushik A, Furukawa H, Khosla A (2022) Exploring phytochemical composition, photocatalytic, antibacterial, and antifungal efficacies of au NPs supported by Cymbopogon flexuosus essential oil. Sci Rep 12(1):14249
110. Ashokkumar M, Palanisamy K, Ganesh Kumar A, Muthusamy C, Senthil Kumar K (2024) Green synthesis of silver and copper nanoparticles and their composites using Ocimum sanctum leaf extract displayed enhanced antibacterial, antioxidant and anticancer potentials. Artif Cells Nanomed Biotechnol 52(1):438–448

111. Sarwar N, Humayoun UB, Kumar M, Zaidi SFA, Yoo JH, Ali N et al (2021) Citric acid mediated green synthesis of copper nanoparticles using cinnamon bark extract and its multifaceted applications. J Clean Prod 292:125974
112. Mehdipour A, Reza MAS, Rasouli A, Baravati MHJ, Jafari GA, Heidari F (2024) Green synthesis of zinc nanoparticles by hydroalcoholic extract of lavender (Lavandula stoechas L.), characterization, and cytotoxic effects on human breast and colon cancer. Sci Rep 14(1):29543
113. Stoyanova D, Stambolova I, Blaskov V, Georgieva P, Shipochka M, Zaharieva K et al (2022) Modified approach using Mentha arvensis in the synthesis of ZnO nanoparticles—textural, structural, and photocatalytic properties. Appl Sci 12(3):1096
114. Koudehi MF, Zibaseresht R (2023) Green synthesis and antibacterial effect of Fe3O4/cu nanocomposite using Rosmarinus officinalis L. (rosemary) aqueous extracts. Proc Natl Acad Sci Sect B Biol Sci 93(1):71–78
115. Ahmad MZ, Alasiri AS, Ahmad J, Alqahtani AA, Abdullah MM, Abdel-Wahab BA et al (2022) Green synthesis of titanium dioxide nanoparticles using Ocimum sanctum leaf extract: in vitro characterization and its healing efficacy in diabetic wounds. Molecules 27(22):7712
116. Dikshit PK, Kumar J, Das AK, Sadhu S, Sharma S, Singh S et al (2021) Green synthesis of metallic nanoparticles: applications and limitations. Catalysts 11(8):902
117. Długosz O, Chwastowski J, Banach M (2020) Hawthorn berries extract for the green synthesis of copper and silver nanoparticles. Chem Pap 74(1):239–252
118. Dejen KD, Kibret DY, Mengesha TH, Bekele ET, Tedla A, Bafa TA, Derib FT (2023) Green synthesis and characterisation of silver nanoparticles from leaf and bark extract of Croton macrostachyus for antibacterial activity. Mater Technol 38(1):2164647
119. Behravan M, Panahi AH, Naghizadeh A, Ziaee M, Mahdavi R, Mirzapour A (2019) Facile green synthesis of silver nanoparticles using Berberis vulgaris leaf and root aqueous extract and its antibacterial activity. Int J Biol Macromol 124:148–154
120. Malik K, Kazmi A, Sultana T, Raja NI, Bibi Y, Abbas M et al (2025) A mechanistic overview on green assisted formulation of nanocomposites and their multifunctional role in biomedical applications. Heliyon 11(3):e41654
121. Rivero-Montejo S d J, Vargas-Hernandez M, Torres-Pacheco I (2021) Nanoparticles as novel elicitors to improve bioactive compounds in plants. Agriculture 11(2):134
122. Fitzgerald M, Heinrich M, Booker A (2020) Medicinal plant analysis: a historical and regional discussion of emergent complex techniques 10:1480
123. Krüger H, Schulz H (2007) Analytical techniques for medicinal and aromatic plants 3(4):1–12
124. Rawat B, Garg AP (2021) Characterization of phytochemicals isolated from Cucurbita pepo seeds using UV-VIS and FTIR spectroscopy 21(1)
125. Mir-Cerdà A, Nunez O, Granados M, Sentellas S, Saurina J (2023) An overview of the extraction and characterization of bioactive phenolic compounds from agri-food waste within the framework of circular bioeconomy 161:116994
126. Liu M, Wang J, Yang K, Zhao Z, Zhou Z, Ma Y et al (2021) Highly sensitive, broad-band organic photomultiplication-type photodetectors covering UV-Vis-NIR 9(19):6357–6364
127. Savitharani M, Kumar V, Manjunatha T, Ruksana F, Vishala E, Prathap H (2024) Biopesticide potential of Anisomeles indica Kuntze leaf essential oil and its nanoemulsion against bacterial blight-Xanthomonas strains 8:100075
128. Akhtar MS, Swamy MK, Sinniah U (2019) Natural bio-active compounds, vol 1, pp 1–608
129. Arif M, Raza H, Akhter T (2024) UV-vis spectroscopy in the characterization and applications of smart microgels and metal nanoparticle-decorated smart microgels: a critical review. RSC Adv 14(51):38120–38134. https://doi.org/10.1039/d4ra07643e
130. Amendola V, Meneghetti M (2009) Size evaluation of gold nanoparticles by UV−Vis spectroscopy. J Phys Chem C 113(11):4277–4285. https://doi.org/10.1021/jp8082425
131. Saira F, Saleemi S, Razzaq H, Qureshi R (2021) Spectrophotometric analysis of stability of gold nanoparticles during catalytic reduction of 4-nitrophenol. Turk J Chem 45(1):82–91. https://doi.org/10.3906/kim-2004-52

132. Murali S, Ramaiah B, Golakoti NR (2016) Liquid chromatography coupled to mass spectrometry based identification of elite Chemotypes of Adhatoda vasica Nees for profitable agronomy—a farmer centric approach. Indian J Sci Technol 9(26). https://doi.org/10.17485/ijst/2016/v9i26/89180
133. Mallikarjuna K, Nasif O, Ali Alharbi S, Chinni SV, Reddy LV, Reddy MRV, Sreeramanan SJB (2021) Phytogenic synthesis of Pd-Ag/rGO nanostructures using stevia leaf extract for photocatalytic H2 production and antibacterial studies 11(2):190
134. Iravani S (2011) Green synthesis of metal nanoparticles using plants 13(10):2638–2650
135. Smith BC (2011) Fundamentals of Fourier transform infrared spectroscopy. CRC, Boca Raton, FL
136. Pasieczna-Patkowska S, Cichy M, Flieger J (2025) Application of Fourier transform infrared (FTIR) spectroscopy in characterization of green synthesized nanoparticles 30(3):684
137. Kadam J, Madiwale S, Bashte B, Dindorkar S, Dhawal P, More P (2020) Green mediated synthesis of palladium nanoparticles using aqueous leaf extract of Gymnema sylvestre for catalytic reduction of Cr (VI). SN Appl Sci 2:1854. https://doi.org/10.1007/s42452-020-03663-5
138. Jayawardena HSN, Liyanage SH, Rathnayake K, Patel U, Yan M (2021) Analytical methods for characterization of nanomaterial surfaces. Anal Chem 93(4):1889–1911. https://doi.org/10.1021/acs.analchem.0c05208
139. Mourdikoudis S, Pallares RM, Thanh NT (2018) Characterization techniques for nanoparticles: comparison and complementarity upon studying nanoparticle properties 10(27):12871–12934
140. Kamaraj N, Rajaguru PY, Issac PK, Sundaresan S (2017) Fabrication, characterization, in vitro drug release and glucose uptake activity of 14-deoxy, 11, 12-didehydroandrographolide loaded polycaprolactone nanoparticles. Asian J Pharm Sci 12(4):353–362. https://doi.org/10.1016/j.ajps.2017.02.003
141. Majithia M, Barretto DA (2023) Biocompatible green-synthesized nanomaterials for therapeutic applications. In: Advances in nano and biochemistry. Elsevier, Amsterdam, pp 285–367
142. Lee J, Oba J, Ohba N, Kajita S (2023) Creation of crystal structure reproducing X-ray diffraction pattern without using database. NPJ Comput Mater 9(1):135. https://doi.org/10.1038/s41524-023-01096-3
143. Holder CF, Schaak RE (2019) Tutorial on powder X-ray diffraction for characterizing nanoscale materials. ACS Nano 13(7):7359–7365. https://doi.org/10.1021/acsnano.9b05157
144. Mammadov R, Tekinay AB, Dana A, Guler MO (2012) Microscopic characterization of peptide nanostructures 43(2–3):69–84
145. Kuntsche J, Horst JC, Bunjes H (2011) Cryogenic transmission electron microscopy (cryo-TEM) for studying the morphology of colloidal drug delivery systems. Int J Pharm 417(1–2):120–137
146. Gauvin R (1997) Review of transmission electron microscopy for the characterization of materials. SPIE 10291:200
147. Carvalho PM, Felício MR, Santos NC, Gonçalves S, Domingues MM (2018) Application of light scattering techniques to nanoparticle characterization and development. Front Chem 6:237. https://doi.org/10.3389/fchem.2018.00237
148. Rodriguez-Loya J, Lerma M, Gardea-Torresdey JL (2024) Dynamic light scattering and its application to control nanoparticle aggregation in colloidal systems: a review 15(1):24
149. Akhter MS, Rahman MA, Ripon RK, Mubarak M, Akter M, Mahbub S et al (2024) A systematic review on green synthesis of silver nanoparticles using plants extract and their biomedical applications. Heliyon 10(11):e29766. https://doi.org/10.1016/j.heliyon.2024.e29766
150. Sánchez-Pérez DM, Flores-Loyola E, Márquez-Guerrero SY, Galindo-Guzman M, Marszalek JE (2023) Green synthesis and characterization of zinc oxide nanoparticles using Larrea tridentata extract and their impact on the in-vitro germination and seedling growth of Capsicum annuum 15(4):3080

151. Mansoori A, Singh N, Dubey S, Thakur T, Alkan N, Das S, Kumar A (2020) Phytochemical characterization and assessment of crude extracts from Lantana camara L. for antioxidant and antimicrobial activity. Front Agron 2:582268
152. Zhu J, Zhang Z, Wang R, Zhong K, Zhang K, Zhang N et al (2022) Review of natural phytochemical-based self-assembled nanostructures for applications in medicine 5(3):3146–3169
153. Newman DJ, Cragg GM (2020) Natural products as sources of new drugs over the nearly four decades from 01/1981 to 09/2019 83(3):770–803
154. Liu Y, Yang S, Wang K, Lu J, Bao X, Wang R et al (2020b) Cellular senescence and cancer: focusing on traditional Chinese medicine and natural products. Cell Prolif 53(10):e12894
155. Okeke E, Nweze E, Anaduaka E, Okoye C, Anosike C, Joshua P, Ezeorba T (2023) Plant-derived nanomaterials (PDNM): a review on pharmacological potentials against pathogenic microbes, antimicrobial resistance (AMR) and some metabolic diseases. 3 Biotech 13(9):291
156. Banasiuk R, Krychowiak M, Swigon D, Tomaszewicz W, Michalak A, Chylewska A et al (2020) Carnivorous plants used for green synthesis of silver nanoparticles with broad-spectrum antimicrobial activity 13(1):1415–1428
157. Can M (2020) Green gold nanoparticles from plant-derived materials: an overview of the reaction synthesis types, conditions, and applications 36(7):859–877
158. Özen T, Çöllü Z, Korkmaz H (2010) Antioxidant properties of Urtica pilulifera root, seed, flower, and leaf extract 13(5):1224–1231
159. Zeedan G, El-Razik K, Allam A, Abdalhamed A, Zeina H (2020) Evaluations of potential antiviral effects of green zinc oxide and silver nanoparticles against bovine herpesvirus-1 8(4):433–443
160. Chowdhury P, Bora T, Khan S, Chakraborty B, Senapati K, Sengupta M et al (2016) Inhibition of Japanese encephalitis virus infection by biogenic catechin silver nanoparticles: an in-vitro study 45(Suppl 1):276
161. Al-Sheddi ES, Farshori NN, Al-Oqail MM, Al-Massarani SM, Saquib Q, Wahab R (2018) Anticancer potential of green synthesized silver nanoparticles using extract of Nepeta deflersiana against human cervical cancer cells (HeLA). Bioinorg Chem Appl 2018(1):9390784
162. Balasubramani G, Ramkumar R, Krishnaveni N, Pazhanimuthu A, Natarajan T, Sowmiya R (2015) Structural characterization, antioxidant and anticancer properties of gold nanoparticles synthesized from leaf extract (decoction) of Antigonon leptopus Hook. & Arn. J Trace Elem Med Biol 30:83–89
163. Mittal AK, Kumar S, Banerjee UC (2014) Quercetin and gallic acid mediated synthesis of bimetallic (silver and selenium) nanoparticles and their antitumor and antimicrobial potential. J Colloid Interface Sci 431:194–199
164. Ahmed MJ, Murtaza G, Rashid F, Iqbal J (2019) Eco-friendly green synthesis of silver nanoparticles and their potential applications as antioxidant and anticancer agents. Drug Dev Ind Pharm 45(10):1682–1694
165. Ravipati AS, Zhang L, Koyyalamudi SR, Jeong SC, Reddy N, Bartlett J (2012) Antioxidant and anti-inflammatory activities of selected Chinese medicinal plants and their relation with antioxidant content 12(1):173
166. Saha S, Verma R (2016) Antioxidant activity of polyphenolic extract of Terminalia chebula Retzius fruits 10(6):805–812
167. Ghosh S, Jagtap S, More P, Shete UJ, Maheshwari NO, Rao S et al (2015) Dioscorea bulbifera mediated synthesis of novel AucoreAgshell nanoparticles with potent antibiofilm and antileishmanial activity 2015(1):562938
168. Sengottaiyan A, Aravinthan A, Sudhakar C, Selvam K, Srinivasan P, Govarthanan M et al (2016) Synthesis and characterization of Solanum nigrum-mediated silver nanoparticles and its protective effect on alloxan-induced diabetic rats 6(1):41–48

169. Rani R, Dahiya S, Dhingra D, Dilbaghi N, Kim K-H, Kumar S (2017) Evaluation of anti-diabetic activity of glycyrrhizin-loaded nanoparticles in nicotinamide-streptozotocin-induced diabetic rats 106:220–230
170. Bansod SD, Bawaskar MS, Gade AK, Rai MK (2015) Development of shampoo, soap and ointment formulated by green synthesised silver nanoparticles functionalised with antimicrobial plants oils in veterinary dermatology: treatment and prevention strategies 9(4):165–171
171. Pannerselvam B, Jothinathan MKD, Rajenderan M, Perumal P, Thangavelu KP, Kim HJ et al (2017) An in vitro study on the burn wound healing activity of cotton fabrics incorporated with phytosynthesized silver nanoparticles in male Wistar albino rats. Eur J Pharm Sci 100:187–196
172. Jain D, Daima HK, Kachhwaha S, Kothari SL (2009) Synthesis of plant-mediated silver nanoparticles using papaya fruit extract and evaluation of their anti microbial activities. Digest J Nanomater Biostruct 4(3):557–563
173. Dhapte V, Kadam S, Moghe A, Pokharkar V (2014) Probing the wound healing potential of biogenic silver nanoparticles 23(9):431–441
174. Arunachalam KD, Annamalai SK, Arunachalam AM, Kennedy S (2013) Green synthesis of crystalline silver nanoparticles using indigofera aspalathoides-medicinal plant extract for wound healing applications. Asian J Chem 25:S311
175. Ali A, Haq IU, Akhtar J, Sher M, Ahmed N, Zia MJ (2017) Synthesis of Ag-NPs impregnated cellulose composite material: its possible role in wound healing and photocatalysis. IET Nanobiotechnol 11(4):477–484
176. Bhide B, Ashok B, Acharya R, Ravishankar B (2011) Anti-microbial and wound healing activities of Cordia macleodii Hook. f & Thoms. leaves. Indian J Nat Prod Resour 1(2):198–203
177. Krychowiak M, Grinholc M, Banasiuk R, Krauze-Baranowska M, Głód D, Kawiak A, Królicka A (2014) Combination of silver nanoparticles and Drosera binata extract as a possible alternative for antibiotic treatment of burn wound infections caused by resistant Staphylococcus aureus 9(12):e115727
178. Imchen L, Manisekaran R, Jamir I, Rathore HS, Senthilvelan T (2025) A review on plant-mediated synthesis of AgNPs and their formulations for wound healing application 52(1):419
179. Yao CH, Yeh JY, Chen YS, Li MH, Huang CH (2017) Wound-healing effect of electrospun gelatin nanofibres containing Centella asiatica extract in a rat model 11(3):905–915
180. Kalaiselvan V, Rajasekaran A (2009) Biosynthesis of silver nanoparticles from Aspergillus niger and evaluation of its wound healing activity in experimental rat model 4:1523–1529
181. La Iacona M, Scavo A, Lombardo S, Mauromicale G (2025) The exploitation of nanotechnology in herbicides and bioherbicides: a novel approach for sustainable weed management 15(1):228
182. Abd-ElGawad AM, El Gendy AE-NG, Assaeed AM, Al-Rowaily SL, Alharthi AS, Mohamed TA et al (2020) Phytotoxic effects of plant essential oils: a systematic review and structure-activity relationship based on chemometric analyses 10(1):36
183. Khamare Y, Chen J, Marble S (2022) Allelopathy and its application as a weed management tool: a review 13:1034649
184. Kisiriko M, Anastasiadi M, Terry LA, Yasri A, Beale MH, Ward J (2021) Phenolics from medicinal and aromatic plants: characterisation and potential as biostimulants and bioprotectants 26(21):6343
185. Taban A, Saharkhiz MJ, Khorram M (2020) Formulation and assessment of nano-encapsulated bioherbicides based on biopolymers and essential oil 149:112348
186. Eldarier SM, Abou-Zeid HM, Marzouk RI, Abo Hatab A (2020) Biosynthesis of silver nanoparticles via Haplophyllum tuberculatum (Forssk.) A. Juss (Rutaceae) and its use as bioherbicide 60(1):25–40
187. Avni A. Point-of-care (POC) Assay to Detect Dengue Exposure. Biosensors and Nanotheranostics. 2022 ;1(1):1–6
188. Srivastav A, Yadav KK, Yadav S, Gupta N, Singh JK, Katiyar R, Kumar V (2019) Nano-phytoremediation of pollutants from contaminated soil environment: current scenario and

future prospects. In: Phytoremediation: management of Environmental Contaminants, vol 6. Springer, Cham, pp 383–401
189. Rather GA, Nanda A, Pandit MA, Yahya S, Barabadi H, Saravanan M (2021) Biosynthesis of zinc oxide nanoparticles using Bergenia ciliate aqueous extract and evaluation of their photocatalytic and antioxidant potential 134:109020
190. Kumar B, Smita K, Galeas S, Sharma V, Guerrero VH, Debut A, Cumbal L (2020a) Characterization and application of biosynthesized iron oxide nanoparticles using Citrus paradisi peel: a sustainable approach 119:108116
191. He S, Feng Y, Ni J, Sun Y, Xue L, Feng Y et al (2016) Different responses of soil microbial metabolic activity to silver and iron oxide nanoparticles 147:195–202
192. Pradhan S, Patra P, Das S, Chandra S, Mitra S, Dey KK (2013) Photochemical modulation of biosafe manganese nanoparticles on Vigna radiata: a detailed molecular, biochemical, and biophysical study 47(22):13122–13131
193. Awasthi G, Nagar V, Mandzhieva S, Minkina T, Sankhla MS, Pandit PP et al (2022) Sustainable amelioration of heavy metals in soil ecosystem: existing developments to emerging trends 12(1):85
194. Kumar V, Guleria P, Kumar V, Yadav SK (2013) Gold nanoparticle exposure induces growth and yield enhancement in Arabidopsis thaliana 461:462–468
195. Singh A, Sharma A, Verma RK, Chopade RL, Pandit PP, Nagar V et al (2022) Heavy metal contamination of water and their toxic effect on living organisms. In: The toxicity of environmental pollutants. IntechOpen, London
196. Ma X, Geiser-Lee J, Deng Y, Kolmakov A (2010) Interactions between engineered nanoparticles (ENPs) and plants: phytotoxicity, uptake and accumulation 408(16):3053–3061
197. Yanık F, Vardar F (2015) Toxic effects of aluminum oxide (Al2O3) nanoparticles on root growth and development in Triticum aestivum 226(9):296
198. Dimkpa CO, McLean JE, Latta DE, Manangón E, Britt DW, Johnson WP et al (2012) CuO and ZnO nanoparticles: phytotoxicity, metal speciation, and induction of oxidative stress in sand-grown wheat. J Nanoparticle Res 14(9):1125
199. Caiñas JE, Long M, Nations S, Vadan R, Dai L, Luo M (2008) Effects of functionalized and nonfunctionalized single-walled carbon nanotubes on root elongation of select crop species 27(9):1922–1931
200. Khodakovskaya M, Dervishi E, Mahmood M, Xu Y, Li Z, Watanabe F, Biris AS (2009) Carbon nanotubes are able to penetrate plant seed coat and dramatically affect seed germination and plant growth 3(10):3221–3227
201. Shittu K, Ihebunna O (2017) Purification of simulated waste water using green synthesized silver nanoparticles of Piliostigma thonningii aqueous leave extract 8(4):045003
202. Banerjee P, Sau S, Das P, Mukhopadhyay A (2014) Green synthesis of silver-nanocomposite for treatment of textile dye 1(2):1–6
203. Ehrampoush MH, Miria M, Salmani MH, Mahvi AH (2015) Cadmium removal from aqueous solution by green synthesis iron oxide nanoparticles with tangerine peel extract 13(1):84
204. Choudhary BC, Paul D, Gupta T, Tetgure SR, Garole VJ, Borse AU, Garole D (2017) Photocatalytic reduction of organic pollutant under visible light by green route synthesized gold nanoparticles 55:236–246
205. Kristanti RA, Hadibarata T, Niculescu A-G, Mihaiescu DE, Grumezescu AM (2025) Nanomaterials for persistent organic pollutants decontamination in water: mechanisms, challenges, and future perspectives 15(14):1133
206. Thakur S, Ojha A, Kansal SK, Gupta NK, Swart HC, Cho J et al (2024) Advances in powder nano-photocatalysts as pollutant removal and as emerging contaminants in water: analysis of pros and cons on health and environment 3(6):100233
207. Asghar N, Hussain A, Nguyen DA, Ali S, Hussain I, Junejo A, Ali A (2024) Advancement in nanomaterials for environmental pollutants remediation: a systematic review on bibliometrics analysis, material types, synthesis pathways, and related mechanisms. J Nanobiotechnol 22(1):26

208. Kumar B, Smita K, Cumbal LJCC (2015) Phytosynthesis of gold nanoparticles using Andean Ajı′ (Capsicum baccatum L) 1(1):1120982
209. El-Saber Batiha G, Alkazmi LM, Wasef LG, Beshbishy AM, Nadwa EH, Rashwan EK (2020) Syzygium aromaticum L. (Myrtaceae): traditional uses, bioactive chemical constituents, pharmacological and toxicological activities 10(2):202
210. Khalil AT, Ovais M, Ullah I, Ali M, Shinwari ZK, Maaza M (2020) Physical properties, biological applications and biocompatibility studies on biosynthesized single phase cobalt oxide (Co3O4) nanoparticles via Sageretia thea (Osbeck) 13(1):606–619
211. Augustine R, Abraham AR, Kalarikkal N, Thomas S (2016) Monitoring and separation of food-borne pathogens using magnetic nanoparticles. In: Novel approaches of nanotechnology in food. Elsevier, Amsterdam, pp 271–312
212. Trang NLN, Nga DTN, Hoang V-T, Ngo X-D, Nhung PT, Le A-T (2022) Bio-AgNPs-based electrochemical nanosensors for the sensitive determination of 4-nitrophenol in tomato samples: the roles of natural plant extracts in physicochemical parameters and sensing performance 12(10):6007–6017
213. Jayaprakash M, Kannappan S (2022) An overview of a sustainable approach to the biosynthesis of AgNPs for electrochemical sensors 15(12):104324
214. Zhang X-F, Liu Z-G, Shen W, Gurunathan S (2016) Silver nanoparticles: synthesis, characterization, properties, applications, and therapeutic approaches 17(9):1534
215. Beck F, Loessl M, Baeumner A (2023) Signaling strategies of silver nanoparticles in optical and electrochemical biosensors: considering their potential for the point-of-care 190(3):91
216. Zulfajri M, Abdelhamid HN, Sudewi S, Dayalan S, Rasool A, Habib A, Huang GG (2020) Plant part-derived carbon dots for biosensing 10(6):68
217. Augustine R, Hasan A (2020) Multimodal applications of phytonanoparticles. In: Phytonanotechnology. Elsevier, Amsterdam, pp 195–219
218. Marslin G, Selvakesavan RK, Franklin G, Sarmento B, Dias AC (2015) Antimicrobial activity of cream incorporated with silver nanoparticles biosynthesized from Withania somnifera, pp 5955–5963
219. Pant G, Nayak N, Gyana Prasuna R (2013) Enhancement of antidandruff activity of shampoo by biosynthesized silver nanoparticles from Solanum trilobatum plant leaf 3(5):431–439
220. Andal V, Kannan K, Selvaraj V, Suba K (2022) Plant-derived nanoparticles for heavy metal remediation. In: Phytonanotechnology. Springer, Cham, pp 59–76
221. Jadhav R, Pawar P, Choudhari V, Topare N, Raut-Jadhav S, Bokil S, Khan A (2023) An overview of antimicrobial nanoparticles for food preservation 72:204–216
222. Ghosh C, Bera D, Roy L (2019) Role of nanomaterials in food preservation. In: Microbial nanobionics: basic research and applications. Springer, Cham, pp 181–211
223. Boeira CP, Piovesan N, Flores DCB, Soquetta MB, Lucas BN, Heck RT et al (2020) Phytochemical characterization and antimicrobial activity of Cymbopogon citratus extract for application as natural antioxidant in fresh sausage 319:126553
224. Mohanta YK, Panda SK, Bastia AK, Mohanta TK (2017) Biosynthesis of silver nanoparticles from Protium serratum and investigation of their potential impacts on food safety and control 8:626
225. Ortega F, Arce VB, Garcia, M. A. (2021) Nanocomposite starch-based films containing silver nanoparticles synthesized with lemon juice as reducing and stabilizing agent 252:117208
226. Martins-Gomes C, Nunes FM, Silva AM (2024) Linking variability in phytochemical composition with safety profile of Thymus carnosus Boiss. Extracts: effect of major compounds and evaluation of markers of oxidative stress and cell death. Int J Mol Sci 25(10):5343
227. Pereira CG, Rodrigues MJ, Nawrot-Hadzik I, Matkowski A, Custódio LJM (2024) Seasonal and geographic dynamics in bioproperties and phytochemical profile of Limonium algarvense Erben 29(2):481
228. Singh H, Desimone MF, Pandya S, Jasani S, George N, Adnan M et al (2023a) Revisiting the green synthesis of nanoparticles: uncovering influences of plant extracts as reducing

agents for enhanced synthesis efficiency and its biomedical applications. Int J Nanomedicine 18:4727–4750

229. Islam MS, Sams SU, Malitha SB, Alam M (2025) Trend analysis of reaction parameters and properties of plant-mediated green synthesized selenium nanoparticles using non-parametric statistical methods (Mann–Kendall trend test, Sen's slope estimator, ANOVA) and their applications in the modern world: a critical perspective 15(34):28155–28180
230. Simeonidis K, Liébana-Viñas S, Wiedwald U, Ma Z, Li Z-A, Spasova M et al (2016) A versatile large-scale and green process for synthesizing magnetic nanoparticles with tunable magnetic hyperthermia features 6(58):53107–53117
231. Osman AI, Zhang Y, Farghali M, Rashwan AK, Eltaweil AS, Abd El-Monaem EM et al (2024) Synthesis of green nanoparticles for energy, biomedical, environmental, agricultural, and food applications: a review 22(2):841–887
232. Adeyemi JO, Oriola AO, Onwudiwe DC, Oyedeji AO (2022) Plant extracts mediated metal-based nanoparticles: synthesis and biological applications. Biomolecules 12(5):627
233. Herlekar M, Barve S, Kumar R (2014) Plant-mediated green synthesis of iron nanoparticles 2014(1):140614
234. Bin-Jumah MN, Al-Abdan M, Al-Basher G, Alarifi S (2020) Molecular mechanism of cytotoxicity, genotoxicity, and anticancer potential of green gold nanoparticles on human liver normal and cancerous cells 18(2):1559325820912154
235. Ibrahim ATA, Banaee M, Sureda A (2021) Genotoxicity, oxidative stress, and biochemical biomarkers of exposure to green synthesized cadmium nanoparticles in Oreochromis niloticus (L.). Comp Biochem Physiol C Toxicol Pharmacol 242:108942
236. Kumari P, Panda PK, Jha E, Kumari K, Nisha K, Mallick MA, Verma SK (2017) Mechanistic insight to ROS and apoptosis regulated cytotoxicity inferred by green synthesized CuO nanoparticles from Calotropis gigantea to embryonic zebrafish 7(1):16284
237. Tran T-K, Nguyen M-K, Lin C, Hoang T-D, Nguyen T-C, Lone AM et al (2024) Review on fate, transport, toxicity and health risk of nanoparticles in natural ecosystems: emerging challenges in the modern age and solutions toward a sustainable environment 912:169331
238. Abegunde SM, Afolayan BO, Ilesanmi TM (2024) Ensuring sustainable plant-assisted nanoparticles synthesis through process standardization and reproducibility: challenges and future directions—a review. Sustain Chem One World 3:100014
239. Huanbutta K, Suwanpitak K, Weeranoppanant N, Sriamornsak P, Garg K, Sharma S (2024) Continuous flow synthesis: a promising platform for the future of nanoparticle-based drug delivery 91:105265
240. Khan T, Ullah I, Ibrahim AA, Khan A, Jafari MS, Vidar WS, LaJeunesse D (2025b) Mechanistic insights and metabolic profiling of sodium hydroxide-treated green synthesized silver nanoparticles: optimization of structural properties and assessment of antibacterial activity 11(10):e43427
241. Liu Z, Wang C, Lu J, Yin Y, Zhang Z, Gao W et al (2024) Continuous flowing technology for efficient and stable scale-up production of Pt nanoparticles catalysts in PEMFC 50:1584–1591
242. Pukkella AK, Nadimpalli NRV, Runkana V, Subramanian S (2022) A novel spiral infinity reactor for continuous hydrothermal synthesis of nanoparticles 12(1):8616
243. Silva S, Dias MC, Pinto DC, Silva AM (2023) Metabolomics as a tool to understand nano-plant interactions: the case study of metal-based nanoparticles 12(3):491
244. Akpobolokemi T, Chung E, Martinez-Nunez RT, Ren G, Abraham BTR, Griffiths A (2025) Influence of Spinacia oleracea leaf extract concentration on silver nanoparticle formation and evaluation of antimicrobial properties. RSC Pharmaceut 2(2):353–368
245. Gwada CA, Ndivhuwo PS, Matshetshe K, Aradi E, Mdluli P, Moloto N et al (2025) Phytochemical-assisted synthesis, optimization, and characterization of silver nanoparticles for antimicrobial activity 15(18):14170–14181
246. Liu H, Zhang H, Wang J, Wei J (2020a) Effect of temperature on the size of biosynthesized silver nanoparticle: deep insight into microscopic kinetics analysis 13(1):1011–1019

247. Mutukwa D, Taziwa RT, Khotseng L (2024) A review of plant-mediated ZnO nanoparticles for photodegradation and antibacterial applications 14(14):1182
248. Jacinto C, Javed Y, Lavorato G, Tarraga WA, Conde BIC, Orozco JM et al (2025) Biotransformation and biological fate of magnetic iron oxide nanoparticles for biomedical research and clinical applications. Nanoscale Adv 7:2818
249. Khan IA, Yu T, Yang M, Liu J, Chen Z (2025a) A systematic review of toxicity, biodistribution, and biosafety in upconversion nanomaterials: critical insights into toxicity mitigation strategies and future directions for safe applications 6:0120
250. Khlebtsov N, Dykman L (2011) Biodistribution and toxicity of engineered gold nanoparticles: a review of in vitro and in vivo studies 40(3):1647–1671
251. Flores-Contreras EA, González-González RB, Pablo Pizaña-Aranda JJ, Parra-Arroyo L, Rodríguez-Aguayo AA, Iñiguez-Moreno M et al (2024) Agricultural waste as a sustainable source for nanoparticle synthesis and their antimicrobial properties for food preservation 6:1346069
252. Gunasekaran VV, Ranjani S, Hemalatha S. Sustainable and eco-friendly silver nanoparticles synthesized from Arachis hypogaea nutshell extract to control phytopathogens. Environmental Sciences Europe. 2025 ;37(1):142

Aromatic Plant-Based Nanoparticles for Antibacterial and Antiviral Performance

Saumya Pandey and Amrendra Kumar

Abstract Aromatic medicinal plants, rich in essential oils and phytochemicals, have demonstrated excellent antimicrobial properties; however, low bioavailability, instability, and high volatility limit their practical application. Nanotechnology facilitates the synthesis of plant-based metallic nanoparticles (NPs) (Ag, Au, Cu, Zn) using plant extracts as reducing, capping, and stabilizing agents. These NPs have improved antibacterial, antiviral, and antioxidant activities through mechanisms including ROS generation, disruption of microbial membranes, enzyme inhibition, and interference with viral replication. Additionally, synergistic combinations of NPs with antibiotics or essential oils further improve antimicrobial efficacy and reduce dosage and resistance development. The integration of phytochemical constituents with nanotechnology provides a versatile and eco-friendly platform to combat infections, improve therapeutic outcomes, and address the growing challenge of antimicrobial resistance.

Keywords Toxicity · Synergistic effect · Nanoformulations · Encapsulation · Essential oils · Multiple drug resistance

1 Introduction

The global rise in antibiotic resistance represents a significant threat, diminishing the efficacy of conventional antibiotics against a broad spectrum of bacterial pathogens. According to a report published by the WHO, antimicrobial resistance is responsible for over 4.95 million deaths globally in 2019. Of which, the death of 1.27 million individuals was mainly due to antimicrobial resistance [1]. Thus, a

S. Pandey (✉)
Department of Biotechnology, IILM University, Greater Noida, Uttar Pradesh, India
e-mail: saumya.pandey@iilm.edu

A. Kumar
College of Medicine, The Pennsylvania State University, Hershey, PA, USA

A. Husen (ed.), *Sustainable Uses of Aromatic Plants in Nanotechnology*, Smart Nanomaterials Technology, https://doi.org/10.1007/978-981-95-7409-4_7

growing demand for novel antimicrobials as a potential alternative to conventional antibiotics has driven research and development of new therapeutic strategies.

Medicinal and aromatic plants are a rich reservoir of abundant phytochemicals and essential oils (EOs). Thus, it has been fundamental to herbal medicine for the treatment of various diseases and disorders [2–8]. Essential oils comprise a complex blend of volatile bioactive, primarily terpenes, aldehydes, alcohols, esters, phenolics, ethers, and ketones [9, 10]. They are widely used in aromatherapy. Besides, they have significant therapeutic applications in cardiovascular, metabolic, neurodegenerative, and cancer-related disorders [11]. Furthermore, EOs derived from aromatic medicinal plants exhibit a wide-spectrum antimicrobial activity, including viruses, bacteria, yeasts, and filamentous fungi [12, 13].

The antimicrobial potential of EO is mainly due to its capacity to disrupt cellular homeostasis, causing growth inhibition and apoptosis [14, 15]. Among secondary metabolites present in EOs, phenolic compounds containing hydroxyl groups bound to a phenyl ring, such as the monoterpenes carvacrol and thymol and the phenylpropene eugenol, have been recognized as the most efficient antimicrobial agents [16, 17]. Non-phenolic metabolites exhibit variable antimicrobial potential. For example, p-cymene and γ-terpinene showed limited activity compared to phenolic monoterpenes, whereas cinnamaldehyde, a non-phenolic phenylpropene, demonstrates strong antimicrobial efficacy [18]. Furthermore, their chemical profile and antimicrobial activity vary greatly depending on factors such as plant origin, species, and subspecies, genetic and environmental factors, geographical location, seasonal variations, plant part, and extraction method. Despite efficient biological activity, several medicinal and aromatic plant-derived drugs offer various limitations. The nanotechnological approach has been employed to overcome limitations and enhance the properties and behaviour of phytochemicals in the biological environment.

Advancements in nanoscience and nanotechnology have significantly impacted the development and application of nanomaterials and expanded their rapid integration in diverse industrial sectors [19–21]. The drawbacks of conventional methods can be addressed by nanomaterials, such as nanofibres and nanoparticles (NPs), which offer unique physical and chemical characteristics that improve antibacterial activity [22, 23]. These characteristics can be further modified to improve the mechanical, biological, and physicochemical performance of NPs for targeted applications.

Recently, the increase in microbial resistance to metal ions and conventional antibiotics and the rise of multidrug-resistant (MDR) strains have led to the development of a novel approach involving the integration of the therapeutic potential of plant-based phytochemicals with the efficacy of nanostructures. Drugs encapsulated in nanostructures can enhance the permeability and solubility of the drug and improve their potential as an efficient drug delivery system [24]. Furthermore, NPs can also improve the biopharmaceutical properties of therapeutic drugs, specifically for compounds with low bioavailability [25, 26]. Among others, silver NPs (AgNPs), due to their remarkable chemical stability, conductivity, catalytic capacity, and broad-spectrum biological activities (antibacterial, antiviral, antifungal, and anti-inflammatory), have been extensively used in pharmaceutical industries [27, 28].

Synthetic drugs often fail to provide adequate inhibition against many pathogenic microorganisms, and their use poses further risks such as carcinogenicity, acute toxicity, and environmental hazards [29, 30]. For example, methanol isolated from peppermint oil exhibits significant bactericidal and fungicidal activity. However, it is highly volatile, insoluble, and unstable and presents high crystallization in aqueous medium, which is a major threat to its application and shelf life [31]. These limitations highlight the need to explore safer and more effective alternatives. In this context, EOs extracted from aromatic medicinal plants have been reported as a potent source of novel antimicrobial agents. The integration of nanotechnology has profound effects on aromatic plant-based herbal medicine regarding enhanced stability, targeted delivery, and improved therapeutic efficiency. The present chapter, therefore, highlights the antibacterial and antiviral efficacy of such NPs, elucidates their therapeutic relevance, and discusses the underlying mechanisms by which they suppress microbial growth (Fig. 1). Furthermore, it emphasizes future research directions aimed at advancing aromatic plant-based nanomaterials as innovative tools to combat antibiotic resistance and infectious diseases.

2 Synthesis and Characterization of Aromatic Plant-Based NPs

Using plant-based routes for synthesizing NPs is recognized as a more sustainable and viable alternative to conventional chemical methods. In addition to higher abundance and diversity, aromatic plants are a rich source of phytochemicals. These phytochemicals are responsible for the reduction, stability, and synthesis of NPs [32]. The binding of bioactive compounds to the surface of NPs influences their interaction with cells and cellular components, thereby improving biological activity [28, 33]. Also, the reducing properties of plant extract and EOs can effectively address the toxicity issues commonly associated with chemical synthesis.

Several studies have demonstrated the significant biomedical potential of aromatic plant-derived NPs, more specifically the antimicrobial properties and efficient drug delivery systems to combat various infectious diseases [34, 35]. Besides offering an eco-friendly, biodegradable, biocompatible, and cost-effective approach for NP synthesis, plant extracts often impart synergistic antibacterial and antioxidant effects. Various plant parts, including seeds, roots, stems, leaves, flowers, and fruits, have been utilized as natural reducing agents for the green fabrication of metallic NPs.

AgNPs were developed from the *Satureja hortensis* leaf extract treated with varying concentrations of NaCl. Advanced techniques such as UV-Vis spectroscopy (SPR at 450 nm), FESEM, XRD, and FTIR have been used for the characterization of NPs. FESEM analysis demonstrated the size of spherical AgNPs ranging from 2.9 to 3.4 nm. FTIR analysis indicated that functional groups, particularly aromatic

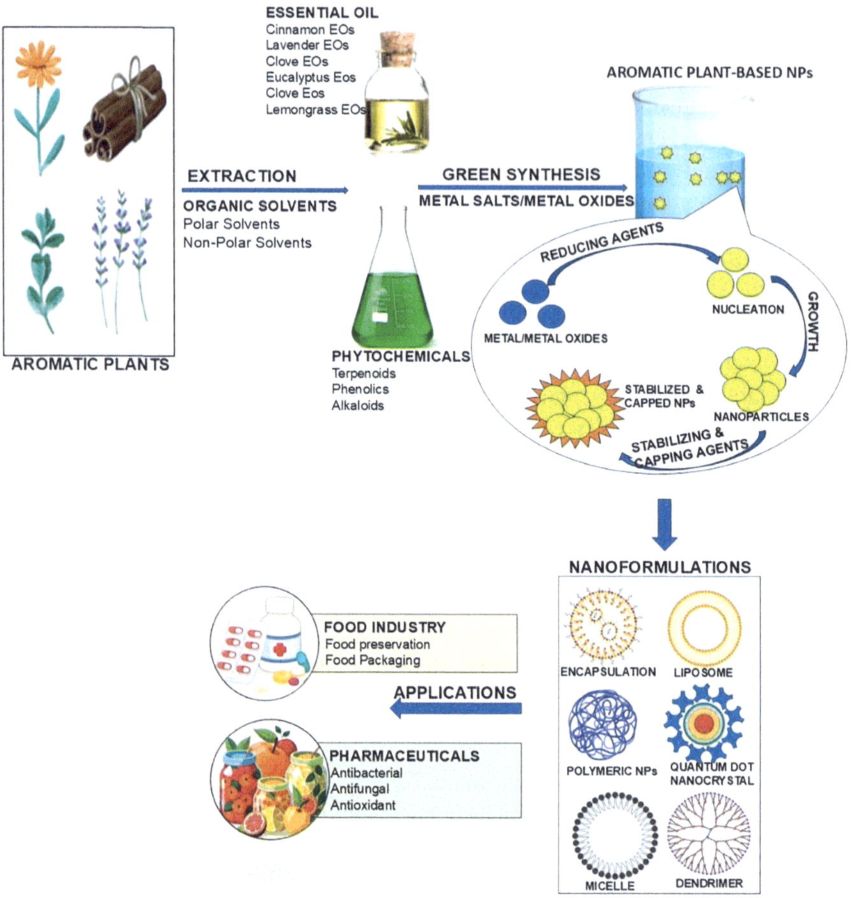

Fig. 1 Aromatic plant-based NP synthesis and applications

bicyclic monoterpenes, played a critical role in silver ion reduction and NP stabilization. The biosynthesized AgNPs exhibited significant antibacterial activity, with the strongest effect against *Bacillus subtilis* [36].

Similarly, AgNPs (size 44 nm) developed using the aromatic plant *Coleus aromaticus* leaf extract have shown high antibacterial activity against *Bacillus subtilis* and *Klebsiella planticola*. The characterization was performed by various analytical methods, including UV-Vis spectroscopy, XRD, SEM, energy-dispersive X-ray spectroscopy, and FTIR [37].

Ahmad et al. [34] developed spherical-shaped stable NPs of an average size of ~10 nm from the room-dried stem and root of *Ocimum sanctum*. They have exhibited significant antibacterial properties. Furthermore, AgNPs have been successfully fabricated using *Taraxacum officinale* hydroalcoholic extract. They demonstrate remarkable antibacterial, antidiabetic, antioxidant, and photocatalytic activities.

The precursor salt silver nitrate (AgNO3) reduces to AgNPs, evident by a visible change in colour from colourless to brown. Phenolics, carotenoids, and flavonoids act as capping and stabilizing agents [38]. Barabadi et al. [39] fabricated *Zataria multiflora*-derived AgNPs and demonstrated that higher pH improves the reducing properties of phytochemicals [39]. These results can be explained based on the effect of pH on the interfacial free energy, the dissociation, and the total electric charge of the complexing agents.

The CuO NPs have been synthesized using a green synthesis technique from aromatic plants such as *Stachys lavandulifolia* [40], *Ephedra alata* [41], and *Rosmarinus officinalis* [35]. The aqueous plant extract acts as a reducing and capping agent. The biosynthesized Cu/CuO NPs showed impressive antibacterial and antifungal activity against *Staphylococcus aureus*, *Bacillus subtilis*, *B. cereus*, *C. albicans*, *S. cerevisiae*, *Pseudomonas aeruginosa*, and *E. coli*.

Similar to Cu/CuO NPs, several studies have demonstrated the green synthesis of gold NPs (AuNPs) from aromatic plant extract. Spherical AuNPs with uniform distribution were synthesized using aqueous leaf extracts of *Eucalyptus globulus* and *Rosmarinus officinalis*. However, NPs from *E. globulus* EO showed variable morphologies with partial aggregation [42]. The effect of different reaction parameters on the formation and size of NPs was demonstrated by comparing AuNP synthesis in aqueous extracts of *Mentha piperita*, *Melissa officinalis*, and *Salvia officinalis*. The reaction parameters include temperature, extract type, and precursor concentration. Among the tested extracts, the aqueous leaf extract of *M. piperita* produced AuNPs with the smallest volume-weighted diameter. The ATR-FTIR and Folin–Ciocalteu assay confirmed the critical role of phenolic compounds in NP synthesis [43]. Lee et al. [44] synthesized AuNPs using *Ocimum sanctum* extracts in solvents with different polarities. The finding showed that the shape and size of AuNPs varied according to the extraction solvents. The NP shapes varied from spherical in hexane, circular disc with rough edges in chloroform, large and anisotropic platelets with defined edges in aqueous solvent, and aggregates in n-butanol solvent. Other plant-derived NPs, such as nickel (Ni) [45], zinc (Zn) [46], and functionalized magnetic NPs [47], are gaining appeal due to their wide range of applications in medicine.

Compared to their bulk counterparts, plant-derived nanomaterials offer benefits including controlled release of active compounds, reduced dosage requirements to achieve comparable efficacy, and lower toxicity to non-target organisms, overall leading to reduced environmental pollution and costs [48, 49].

3 Antibacterial Properties of Aromatic Plant-Based NPs

Metallic NPs show efficient antibacterial properties through several interconnected mechanisms, including the generation of ROS, excessive metal ion accumulation in cell membranes, disruption of microbial plasma membrane and respiratory chain enzymes, electrostatic interactions that impair metabolic processes, and inhibition of microbial proteins or enzymes via enhanced H_2O_2 production [50, 51]. Among NPs, AgNPs have demonstrated great potential in treating microbial infections and reducing microbial pathogens. AgNPs prepared from aqueous extract of *Artemisia scoparia* demonstrated effectiveness in inhibiting the growth of *E. faecalis* and *P. aeruginosa*, indicating potential in treating MDR bacterial infections. Rai et al. [52] suggest that efficient antimicrobial activity of AgNPs is strongly associated with a large surface area that maximizes bacterial contact. Based on particle size, NPs can either penetrate the cell or remain attached to the bacterial cell surface.

Agnihotri et al. [53] suggested that AgNPs exhibit strong antibacterial activity, which increases with a decrease in particle size. AgNPs can bind to sulfur-containing proteins in bacterial membranes and intracellular components and phosphorus-containing biomolecules like DNA. These interactions can disrupt vital cellular functions, leading to bacterial cell death [54]. AgNPs obtained from plant leaf extract showed potent bacterial inhibition against *Bacillus subtilis* and *Klebsiella planticola* [37].

Many drugs derived from aromatic plants have shown efficient antimicrobial activity. But limitations such as low absorption, high toxicity, and other side effects affect bioavailability and efficacy. However, the synergistic effect between plant-based extract nanostructures can play a crucial role in activating and enhancing antibacterial potential. Also, studies revealed the significance of aromatic plant-based NPs in combating MDR pathogens and thus facilitating the development of new therapeutic approaches [55]. As an example, SeS2 NPs made from cinnamon, curcumin, and pepper extract (average crystallite size = 28–44 nm) have shown increased bioactivity towards microorganisms such as *S. aureus*, *E. coli*, *P. aeruginosa*, and *C. albicans* strains [56]. Similarly, circular AuNPs (40–60 nm) were synthesized from the leaves and fruit extracts of *Pistacia atlantica*. They have demonstrated significant potential for treating infections caused by both Gram-negative bacteria (*P. aeruginosa*, *E. coli*) and Gram-positive bacteria (*B. subtilis*, *S. aureus*) [57]. These bacteria showed multiple drug susceptibility and are commonly found in hospital environments. Moreover, CuO NPs produced from the extract of *Ephedra alata*, *Catha edulis*, and *Rosmarinus officinalis* showed spherical morphologies, high crystallinity, and strong structural stability. They show low MIC and an increased inhibition zone. Thus, inhibiting strains of *Staphylococcus aureus*, *Bacillus subtilis*, *E. coli*, *Klebsiella pneumoniae*, and *Candida albicans*, compared to their chemically synthesized counterpart [35, 41, 58]. Zinc oxide NPs (ZnO NPs) formed from the leaf extracts of the *Vitex negundo* plant have demonstrated a 12 mm zone of inhibition against *Bacillus subtilis*. However, standard streptomycin formed a 3 mm zone of inhibition [46].

MubarakAli et al. [59] found that AuNPs obtained from *M. piperita* leaf extract have potent antibacterial activity against *Escherichia coli* and *Staphylococcus aureus*. Similar studies on *Stevia rebaudiana*-derived Ni NPs showed remarkable antioxidant and antimicrobial activities against Gram-positive strains of *Bacillus* and *Streptococcus*, as well as Gram-negative *E. coli* [60].

4 Antiviral Performance of Aromatic Plant-Based NPs

The NPs primarily target key stages of the viral life cycle, including attachment, penetration, replication, and budding. The possible mode of action may involve either direct or indirect viral inactivation, inhibition of viral attachment to host cells, or disruption of replication processes. However, efficacy is largely dependent on the type and specific form of NPs [61]. Lysenko et al. (62) demonstrated that NPs can interact directly with viral surface receptors via local-field effects, altering membrane potential. They can interact indirectly via blocking viral penetration by modifying host cell membranes. Overall, NPs exert antiviral effects through multiple pathways, including interaction with viral proteins like gp120, competitive binding with host receptors, inhibition of attachment and entry, plasma membrane binding, viral particle interaction, and viral DNA interference. Collectively, these actions disrupt viral replication and prevent infection [63, 64].

Khandelwal et al. [65] synthesized AgNPs (particle size 5–30 nm) from the leaf extract of *Argemone mexicana* with the ability to control peste des petits ruminants virus infections. The TEM analysis showed that AgNPs interact at both the virion surface and the virion core. This interaction, however, blocks the entry of the virus into the target cells rather than directly killing the virus.

MnO_2 NPs synthesized using Arabic gum have shown strong antiviral activity against influenza A/H1N. The study indicated the preventive and protective effect of MnO2 NPs on the cell viability against viral cytopathic effects [66]. Similarly, SeS2 NPs made from the extract of cinnamon, curcumin, and pepper with an average crystallite size of 28–44 nm have shown increased antiviral bioactivity against human influenza virus A/H1N1 and Betacoronavirus 1 (human coronavirus HCoV-OC43), exhibiting a reduction of over 99% [56].

Furthermore, *P. guajava* AgNPs demonstrated the anti-chikungunya activities in the Vero cell. Compared to virus control, cell viability increases by 40% and 60% on treatment with a minimum non-toxic dosage of *P. guajava* NPs and extracts, respectively. The mechanism of antiviral activity was evaluated by docking the replication-essential cysteine protease (nsP2) protein of chikungunya with phytochemicals in *P. guajava*. This study revealed significant anti-chikungunya properties of phytochemicals longifolene and quercetin [67].

AuNPs and other photosynthesized NPs can inhibit a variety of viruses, including foot-and-mouth disease virus, human immunodeficiency virus (HIV), influenza virus, chikungunya virus, herpes simplex virus (HSV), hepatitis B virus, and bovine viral diarrhoea virus [68]. Functionalization as multivalent-, gallic acid-, tannic acid-, or curcumin has been used for AuNPs [69, 70]. Notable examples include the following: tannic acid-modified AgNPs from *Astragalus membranaceus*, effective against HSV [71]; AgNPs synthesized from the extract of *Tinospora cordifolia*, *Phyllanthus niruri*, and *Andrographis paniculata*, which inhibit chikungunya virus [72]; AuNPs based on *Malephora lutea* and *Lampranthus coccineus*, which inhibit hepatitis B virus, influenza virus, and HSV; and *Plumbago indica*-derived ZnO NPs with anti-HSV potential [73]. Although studies have shown phyto-NPs' ability to inhibit viral replication in infected cells, the exact mechanisms of inhibition are still not well known.

5 Synergistic Effects of Antibiotics or Antivirals and NPs

Functionalizing aromatic plant-based NPs with antibiotics is promising for treating resistant bacterial infections because of targeted drug delivery directly to infection sites with reduced toxicity and low drug doses [74]. Moreover, considering their ability to delay the onset of resistance in bacteria, these NPs have been used in combination with antibiotics for treating multiple drug resistance [75]. For example, rind extract from *Garcinia mangostana* L. was used to synthesize AgNPs, AuNPs, and PtNPs. AgNPs showed the strongest antibacterial activity, whereas AuNPs and PtNPs showed moderate antibacterial activity. However, the combination of commercial antibiotics with these NPs exhibits synergistic effects, significantly enhancing antibacterial activity. Furthermore, the study revealed that streptomycin-resistant *Bacillus* sp. became highly susceptible when treated with streptomycin in combination with AuNPs. This observation unveils possibilities for employing NP–antibiotic conjugates to overcome antibiotic resistance in clinical settings. Another example of conjugates of antibiotics (cefepime, ceftazidime, ceftriaxone, and cefotaxime) and *Mentha longifolia*-based AgNPs that showed significantly higher activity against bacterial pathogens, including *S. aureus*, *P. aeruginosa*, *E. faecalis*, and *K. pneumoniae*. Except Cpm-AgNPs, the inhibitory zones in each kind of AgNP demonstrated that antibiotic conjugates (AgNPs–antibiotics) showed enhanced antibacterial efficacy than antibiotics and AgNPs alone. The cellular toxicity of antibiotic conjugates and AgNPs alone in RBCs was found to be reduced at lower doses. Therefore, the produced antibiotic-supplemented AgNPs would open the door to efficient therapy and might be used as a substitute medication to combat infections resistant to cephalosporins.

The quantitative evaluation of the synergistic effect of NPs in combination with antibiotics can be evaluated by measuring the fold increase in the inhibition zone diameter after combination with NPs [76]:

$$\text{The fold increase} = \left(b^2 - a^2\right) / a^2$$

where (a) is the inhibition zone of the antibiotic alone and (b) is the inhibition zone of the antibiotic plus NPs.

Several studies have shown that combining NPs with antibiotics inhibits the growth of MDR microorganisms. AgNPs, for instance, are used in the development of antibiotic conjugates containing antibiotics such as clotrimazole [77], vancomycin [78], and ciprofloxacin [79]. These conjugates inhibit vancomycin-resistant enterococci and methicillin-resistant *Staphylococcus aureus* (MRSA). Similarly, conjugates of AuNPs and antibiotics (vancomycin, ampicillin) [80,

81] and ZnO NPs with ciprofloxacin [82] have demonstrated significant antibacterial activity against MRSA and MDR *A. baumannii*. These conjugates have been shown to prevent the growth of both Gram-positive and Gram-negative bacteria. Thus, it provides an advantage over the limitations faced while using only conventional antibiotics to treat MDR organisms [74]. Therefore, further investigations into nanomaterial–antibiotic synergies are needed. This may lead to a significant breakthrough in the treatment of diseases caused by pathogenic MDR organisms.

AgNPs synthesized using *Curcuma aromatica* rhizome extract (CAAgNPs) demonstrated potent antibacterial, antibiofilm, and synergistic activities with conventional antibiotics against ESKAPE pathogens (*Enterococcus faecium*, *Staphylococcus aureus*, *Klebsiella pneumoniae*, *Acinetobacter baumannii*, *Pseudomonas aeruginosa*, and *Enterobacter* species). The phytochemicals not only stabilize NPs but also enhance biological activity, reduce drug dosage, and lower toxicity [83].

Nano-enabled delivery systems maximize the therapeutic potential of antiviral agents. This can be achieved through overcoming heterogeneous biological and physicochemical barriers, including poor stability, matrix interactions, low solubility and availability, untargeted release, toxicity, and development of antiviral drug resistance. A nanocarrier-based combinatorial delivery system overcomes these restrictions. This allows better drug stability, synergistic therapeutic efficacy, and reduced emergence of viral resistance. For this purpose, various delivery systems, such as microemulsions, micelles, nanoliposomes, solid lipid NPs, nanoemulsions, biopolymer NPs, and nanogels, are available.

For example, Lakshmi et al. [84] developed a dual-drug formulation to prevent HIV infection. The formulation was developed through the co-encapsulation of the dual drugs efavirenz (a synthetic antiretroviral) and curcumin (a phytoconstituent) in lactoferrin NPs. The result showed improved vaginal drug delivery and entrapment efficiency. Further, in vivo studies revealed significantly higher drug retention at the target site, alongside reduced inflammatory responses, demonstrating the potential of plant-based compounds when integrated into nanoscale drug delivery systems. Further, triple-drug nanoformulations (AZT, EFV, and 3TC) and hybrid systems combining herbal and metallic nanocomponents have been developed that showed improved bioavailability, half-life, and therapeutic profiles, while minimizing systemic toxicity [85].

Moreover, the integration of phytochemicals such as curcumin with metallic or polymeric nanocarriers offers not only antiviral efficacy but also reduced toxicity and additional therapeutic benefits. For example, the synergistic use of AgNPs with a curcumin–copper–β-cyclodextrin complex demonstrated both contraceptive and antimicrobial effects [86]. Thus, the combination of plant-based antiviral agents with nanocarrier technologies can be further explored for the development of novel HIV therapeutics.

6 Toxicological Considerations and Biocompatibility of Aromatic Plant-Derived NPs

Despite their promising potential, challenges and concerns surrounding toxicity and harm to beneficial bacteria in humans are also associated with NPs [87, 88]. The mechanism of NPs remains unclear; however, some reports suggest that NPs and their toxic transformation products may induce destruction of RBC and interrupt blood coagulation pathways [89].

The NPs' toxicity depends upon their size, concentration, and treatment duration [90]. For instance, Ag–ZnO–graphene nanocomposites are not toxic at low concentrations but are toxic at higher concentrations [91]. The oxidative damage and genotoxic effects of CuO, ZnO, and TiO_2 NPs limit their use [74].

AgNPs derived from aromatic plants have widespread applications as antiviral and antibacterial agents. However, several reports showed increased drug resistance in bacterial pathogens due to genetic modification [92]. AgNPs build up in critical organs, including the spleen, liver, and lungs. The accumulation causes toxicity, decreasing their efficiency and eventually impairing organ function [92]. Additionally, prolonged accumulation of NPs in tissue results in nephrotoxicity and hepatotoxicity [93, 94].

Several innovative strategies involving combination therapies and surface functionalization with biocompatible polymers have been developed. These approaches reduce NPs' toxicity, while preserving their antimicrobial efficacy [95]. For instance, it has been proven that encapsulating AgNPs in polymers like chitosan, polyethylene glycol, or polyvinyl alcohol improves biocompatibility and antimicrobial activity and decreases toxicity caused by NPs [96, 97]. Abdallah et al. [98] found that AgNPs capped with chitosan and polyvinyl alcohol demonstrated improved antibacterial activity against clinical isolates and successfully prevented the formation of biofilms. The clinical isolates include *Escherichia coli*, *Klebsiella pneumoniae*, *Staphylococcus aureus*, and *Staphylococcus epidermidis*.

7 Applications of Aromatic Plant-Based NPs

7.1 Role of Aromatic Plant-Based NPs in Food Preservation and Packaging

The combination of EOs and NPs has been widely used in inhibiting food spoilage pathogens. These conjugates enhance antimicrobial efficacy and reduce resistance development. For example, Begum et al. [99] developed conjugates of AgNPs and EOs that showed bactericidal and fungicidal activities against pathogenic bacteria (*Salmonella typhimurium* and *E. coli* O157:H7) and spoilage fungi (*P. chrysogenum*, *A. niger*, and *M. circinelloides*). Compared to EO alone, conjugates with spherical AgNPs exhibited strong synergistic effects, significantly lowering MIC values. Growth modelling revealed that these combinations reduced fungal growth rate and colony size, while prolonging the lag phase. Essential oil–AgNP formulations suppressed microbial growth for up to 28 days in rice.

Essential oils are a rich source of bioactive compounds with excellent antioxidant and antimicrobial properties. Owing to this, their combination with nanopolymers has been widely used in the food packaging industry. However, poor water solubility and loss in bioactivity under unfavourable conditions limit its use in food packaging [100]. These limitations can be overcome by employing encapsulation techniques or blending with polymeric matrices. Encapsulation not only ensures controlled and targeted release of active compounds but also protects them from adverse environmental conditions [100, 101].

The encapsulated EOs in biopolymer-based structures enhance their antimicrobial and antioxidant activity. In addition, they also improve other critical packaging parameters such as tensile strength, hydrophobicity, barrier properties, and optical characteristics [101]. Some of the most common EOs used in packaging systems include thyme, clove, oregano, cinnamon, and lemon. For example, encapsulation of clove oil in PLA-PBAT and PLA films inhibits the growth of *E. coli* and *S. aureus* at concentrations of 10 wt% and 15–30%, respectively [102, 103].

7.2 Role of Aromatic Plant-Based NPs in Pharmaceuticals and Medicine

Aromatic plant extracts act as nanocarriers or stabilizers in the green synthesis of metallic NPs (e.g. Ag, Au, Cu, and Zn) by reducing their metal salts or oxides through their high antioxidant activity, producing biocompatible NPs with promising antibacterial and pharmaceutical applications [104, 105].

Several studies demonstrate the pharmaceutical potential of aromatic plant-derived NPs owing to their biocompatibility, stability, and enhanced delivery capabilities. For instance, Wintachai et al. [106] developed metallic AgNPs from eucalyptus leaf extract. The result showed that AgNPs inhibited *Acinetobacter baumannii* (MDR) growth. Additionally, these AgNPs have shown effective antibacterial activity against *Bacillus subtilis*, *Escherichia coli*, *Staphylococcus aureus*, and *Pseudomonas aeruginosa* [107]. Another example is nanocurcumin (size 2–40 nm) developed using the wet milling process. They cause higher inhibition of *Bacillus subtilis*, *Escherichia coli*, *Staphylococcus aureus*, and *Pseudomonas aeruginosa* than free curcumin [108, 109].

8 Conclusion

Aromatic plant-derived NPs have emerged as an efficient and multifunctional approach to address microbial infections and MDR pathogens. These NPs have stronger antibacterial, antiviral, and antioxidant properties than their bulk counterparts because they use bioactive phytochemicals and EOs as reducing and stabilizing agents. However, a thorough probing of toxicity and biocompatibility assessment of NPs are required to guarantee their safety for biomedical purposes. Future studies should concentrate on developing standardized protocols for aromatic plant-derived NP synthesis. Long-term safety assessments of NPs, including in vitro and in vivo studies, are crucial for a better understanding of their toxicity. Additionally, the development of regulatory frameworks and standardized protocols is crucial to apply NPs safely and effectively in clinical settings.

Competing Interests The authors declare that they have no competing interest.

Declaration of Generative AI and AI-Assisted Technologies in the Writing Process During the preparation of this work, the book chapter author(s) have not taken any assistance from AI, and the book chapter author(s) reviewed and edited the content as needed and take(s) full responsibility for the content of the publication.

References

1. Díaz-Puertas R, Álvarez-Martínez FJ, Falco A, Barrajón-Catalán E, Mallavia R (2023) Phytochemical-based nanomaterials against antibiotic-resistant bacteria: an updated review. Polymers 15:1392
2. Husen A (2024a) Antimalarial medicinal plants. Taylor & Francis Group, LLC, Boca Raton, FL. https://doi.org/10.1201/9781003378396
3. Husen A (2024b) Medicinal spices and condiment crops. Taylor & Francis Group, LLC, Boca Raton, FL. https://doi.org/10.1201/9781003387046
4. Husen A (2023a) Antidiabetic medicinal plants and herbal treatments. CRC, Boca Raton, FL. https://doi.org/10.1201/b23347
5. Husen A (2022) Herbs, shrubs and trees of potential medicinal benefits. CRC, Boca Raton, FL. https://doi.org/10.1201/9781003205067
6. Husen A (2023b) Exploring poisonous plants: medicinal values, toxicity response and other potential uses. CRC, Boca Raton, FL. https://doi.org/10.1201/b23017
7. Salvi P, Kumar G, Gandass N, Kajal Verma A, Rajarammohan S, Rai N, Gautam V (2022) Antimicrobial potential of essential oils from aromatic plant Ocimum sp.; a comparative biochemical profiling and in-silico analysis. Agronomy 12:627
8. Vasinauskiene M, Radusiene J, Zitikaite I, Surviliene E (2006) Antibacterial activities of essential oils from aromatic and medicinal plants against growth of phytopathogenic bacteria. Agron Res 4:437–440
9. Kaplan A (2022) The nanocomposites designs of phytomolecules from medicinal and aromatic plants: promising anticancer-antiviral applications. Beni Suef Univ J Basic Appl Sci 11:17

10. Singh S, Panwar N, Kumar SS, Singh R, Anand G, Kumar A, Malyan SK (2023) Aromatic oils from medicinal plants and their role in nanoparticles synthesis, characterization, and applications. In: Secondary metabolites from medicinal plants. CRC Press, Boca Raton, FL, pp 305–314
11. Ali B, Al-Wabel NA, Shams S, Ahamad A, Khan SA, Anwar F (2015) Essential oils used in aromatherapy: a systemic review. Asian Pac J Trop Biomed 5:601–611
12. Akthar MS, Degaga B, Azam T (2014) Antimicrobial activity of essential oils extracted from medicinal plants against the pathogenic microorganisms: a review. Iss Biol Sci Pharmaceut Res 2:001–007
13. Reichling J, Schnitzler P, Suschke U, Saller R (2009) Essential oils of aromatic plants with antibacterial, antifungal, antiviral, and cytotoxic properties—an overview. Forsch Komplementmed 16:79–90
14. Devi KP, Nisha SA, Sakthivel R, Pandian SK (2010) Eugenol (an essential oil of clove) acts as an antibacterial agent against salmonella typhi by disrupting the cellular membrane. J Ethnopharmacol 130:107–115
15. Lorenzi V, Muselli A, Bernardini AF, Berti L, Pages JM, Amaral L, Bolla JM (2009) Geraniol restores antibiotic activities against multidrug-resistant isolates from gram-negative species. Antimicrob Agents Chemother 53:2209–2211
16. Dorman HD, Deans SG (2000) Antimicrobial agents from plants: antibacterial activity of plant volatile oils. J Appl Microbiol 88:308–316
17. Lambert RJW, Skandamis PN, Coote PJ, Nychas GJ (2001) A study of the minimum inhibitory concentration and mode of action of oregano essential oil, thymol and carvacrol. J Appl Microbiol 91:453–462
18. Christaki E, Bonos E, Giannenas I, Florou-Paneri P (2012) Aromatic plants as a source of bioactive compounds. Agriculture 2:228–243
19. Husen A, Siddiqi KS (2023) Advances in smart nanomaterials and their applications. Elsevier, Cambridge, MA. https://doi.org/10.1016/C2021-0-02202-1
20. Jagessar RC (2020) Plant extracts based nanoparticles, a good perspective in the development of drugs in nanomedicine. Mod Approach Drug Des 3:10–31031
21. Kaviya S, Santhanalakshmi J, Viswanathan B (2011) Green synthesis of silver nanoparticles using Polyalthia longifolia leaf extract along with D-sorbitol: study of antibacterial activity. J Nanotech 1:152970
22. Barani M, Zeeshan M, Kalantar-Neyestanaki D, Farooq MA, Rahdar A, Jha NK, Sargazi S, Gupta PK, Thakur VK (2021) Nanomaterials in the management of gram-negative bacterial infections. Nanomaterials 11:2535
23. Hamdan N, Yamin A, Hamid SA, Khodir WKWA, Guarino V (2021) Functionalized antimicrobial nanofibers: design criteria and recent advances. J Funct Biomater 12:59
24. Zielińska A, Carreiró F, Oliveira AM, Neves A, Pires B, Venkatesh DN, Durazzo A, Lucarini M, Eder P, Silva AM, Santini A (2020) Polymeric nanoparticles: production, characterization, toxicology and ecotoxicology. Molecules 25:3731
25. Yetisgin AA, Cetinel S, Zuvin M, Kosar A, Kutlu O (2020) Therapeutic nanoparticles and their targeted delivery applications. Molecules 25:2193
26. Zhuo Y, Zhao YG, Zhang Y (2024) Enhancing drug solubility, bioavailability, and targeted therapeutic applications through magnetic nanoparticles. Molecules 29:4854
27. Ahmad A, Mukherjee P, Senapati S, Mandal D, Khan MI, Kumar R, Sastry M (2003) Extracellular biosynthesis of silver nanoparticles using the fungus fusarium oxysporum. Colloids Surf B Biointerfaces 28:313–318
28. Kumari P, Luqman S, Meena A (2019) Application of the combinatorial approaches of medicinal and aromatic plants with nanotechnology and its impacts on healthcare. DARU J Pharma Sci 27:475–489

29. Mulyaningsih S, Sporer F, Zimmermann S, Reichling J, Wink M (2010) Synergistic properties of the terpenoids aromadendrene and 1,8-cineole from the essential oil of Eucalyptus globulus against antibiotic-susceptible and antibiotic-resistant pathogens. Phytomedicine 17:1061–1066
30. Swamy MK, Akhtar MS, Sinniah UR (2016) Antimicrobial properties of plant essential oils against human pathogens and their mode of action: an updated review. Evid Based Complement Alternat Med 2016:3012462
31. Piran P, Kafil HS, Ghanbarzadeh S, Safdari R, Hamishehkar H (2017) Formulation of menthol-loaded nanostructured lipid carriers to enhance its antimicrobial activity for food preservation. Adv Pharm Bull 7:261
32. Cardoso B, Nobrega G, Afonso IS, Ribeiro JE, Lima RA (2025) Sustainable green synthesis of metallic nanoparticle using plants and microorganisms: a review of biosynthesis methods, mechanisms, toxicity, and applications. J Environ Chem Eng 13:116921
33. Thiruvengadam R, Kondapavuluri BK, Thangavelu L, Thiruvengadam M, Hatami M, Kim JH (2025) Nanoparticle-based strategies with bioactive compounds for targeting oxidative stress in therapeutic interventions: a comprehensive review. Ind Crop Prod 227:120804
34. Ahmad N, Sharma S, Alam MK, Singh VN, Shamsi SF, Mehta BR, Fatma A (2010) Rapid synthesis of silver nanoparticles using dried medicinal plant of basil. Colloids Surf B Biointerfaces 81:81–86
35. Moroda MD, Deressa TL, Tiwikrama AH, Chala TF (2025) Green synthesis of copper oxide nanoparticles using Rosmarinus officinalis leaf extract and evaluation of its antimicrobial activity. Next Mater 7:100337
36. Rasaee I, Ghannadnia M, Baghshahi S (2018) Biosynthesis of silver nanoparticles using leaf extract of Satureja hortensis treated with NaCl and its antibacterial properties. Micropor Mesopor Mater 264:240–247
37. Vanaja M, Annadurai G (2013) Coleus aromaticus leaf extract mediated synthesis of silver nanoparticles and its bactericidal activity. Appl Nanosci 3:217–223
38. Yousefzadeh-Valendeh S, Fattahi M, Asghari B, Alizadeh Z (2023) Dandelion flower-fabricated ag nanoparticles versus synthetic ones with characterization and determination of photocatalytic, antioxidant, antibacterial, and α-glucosidase inhibitory activities. Sci Rep 13:15444
39. Barabadi H, Mojab F, Vahidi H, Marashi B, Talank N, Hosseini O, Saravanan M (2021) Green synthesis, characterization, antibacterial and biofilm inhibitory activity of silver nanoparticles compared to commercial silver nanoparticles. Inorg Chem Commun 129:108647
40. Khatami M, Heli H, Mohammadzadeh Jahani P, Azizi H, Lima Nobre MA (2017) Copper/copper oxide nanoparticles synthesis using Stachys lavandulifolia and its antibacterial activity. IET Nanobiotechnol 11:709–713
41. Atri A, Echabaane M, Bouzidi A, Harabi I, Soucase BM, Chaâbane RB (2023) Green synthesis of copper oxide nanoparticles using ephedra Alata plant extract and a study of their antifungal, antibacterial activity and photocatalytic performance under sunlight. Heliyon 9
42. Dzimitrowicz A, Berent S, Motyka A, Jamroz P, Kurcbach K, Sledz W, Pohl P (2019a) Comparison of the characteristics of gold nanoparticles synthesized using aqueous plant extracts and natural plant essential oils of Eucalyptus globulus and Rosmarinus officinalis. Arab J Chem 12:4795–4805
43. Dzimitrowicz A, Jamróz P, Sergiel I, Kozlecki T, Pohl P (2019b) Preparation and characterization of gold nanoparticles prepared with aqueous extracts of Lamiaceae plants and the effect of follow-up treatment with atmospheric pressure glow microdischarge. Arab J Chem 12:4118–4130
44. Lee SY, Krishnamurthy S, Cho CW, Yun YS (2016) Biosynthesis of gold nanoparticles using Ocimum sanctum extracts by solvents with different polarity. ACS Sustain Chem Eng 4:2651–2659
45. Ezhilarasi AA, Vijaya JJ, Kaviyarasu K, Kennedy LJ, Ramalingam RJ, Al-Lohedan HA (2018) Green synthesis of NiO nanoparticles using Aegle marmelos leaf extract for the

evaluation of in-vitro cytotoxicity, antibacterial and photocatalytic properties. J Photochem Photobiol B Biol 180:39–50
46. Mohammad MS, Perugu S (2025) An aromatic plant bioactive compound corymbosin from Vitex negundo-mediated synthesis of zinc oxide nanoparticles, characterization, and their bioactivity against selected cancer cell lines and microbial pathogens. Biomass Convers Biorefin 15:2097–2110
47. Hashemi-Moghaddam H, Mohammadhosseini M, Azizi Z (2018) Impact of amine- and phenyl-functionalized magnetic nanoparticles impacts on microwave-assisted extraction of essential oils from root of Berberis integerrima Bunge. J Appl Res Med Aromat Plants 10:1–8
48. Antunes Filho S, Dos Santos MS, Dos Santos OA, Backx BP, Soran ML, Opriş O, Lung I, Stegarescu A, Bououdina M (2023) Biosynthesis of nanoparticles using plant extracts and essential oils. Molecules 28:3060
49. Jampilek JO, Králová K (2018) Application of nanobioformulations for controlled release and targeted biodistribution of drugs. In: Nanobiomaterials. Apple Academic Press, Palm Bay, FL, pp 131–208
50. Nisar P, Ali N, Rahman L, Ali M, Shinwari ZK (2019) Antimicrobial activities of biologically synthesized metal nanoparticles: an insight into the mechanism of action. J Biol Inorg Chem 24:929–941
51. Velusamy P, Kumar GV, Jeyanthi V, Das J, Pachaiappan R (2016) Bio-inspired green nanoparticles: synthesis, mechanism, and antibacterial application. Toxicol Res 32:95–102
52. Rai M, Yadav A, Gade A (2009) Silver nanoparticles as a new generation of antimicrobials. Biotechnol Adv 27:76–83
53. Agnihotri S, Mukherji S, Mukherji S (2014) Size-controlled silver nanoparticles synthesized over the range 5–100 nm using the same protocol and their antibacterial efficacy. RSC Adv 4:3974–3983
54. Park Y (2014) A new paradigm shift for the green synthesis of antibacterial silver nanoparticles utilizing plant extracts. Toxicol Res 30:169–178
55. Bhati P, Tiwari R, Kothari SL, Ray K, Lamba NP, Kumar Y, Binsuwaidan R, Saeed M, Obaidur S, Srivastava SC, Sambhav K (2025) Green synthesis of silver NPs using aqueous extract of Artemisia scoparia for hydrogenation of aromatic nitro compounds and their biological activity. Front Microbiol 16:1584066
56. Długosz O, Ochnik M, Sochocka M, Franz D, Orzechowska B, Anna CK, Agata D, Banach M (2022) Antimicrobial and antiviral activity of selenium sulphide nanoparticles synthesised in extracts from spices in natural deep eutectic solvents (NDES). Sustain Mater Technol 32:e00433
57. Hamelian M, Hemmati S, Varmira K, Veisi H (2018) Green synthesis, antibacterial, antioxidant and cytotoxic effect of gold nanoparticles using Pistacia Atlantica extract. J Taiwan Inst Chem Eng 93:21–30
58. Andualem WW, Sabir FK, Mohammed ET, Belay HH, Gonfa BA (2020) Synthesis of copper oxide nanoparticles using plant leaf extract of Catha edulis and its antibacterial activity. J Nanotech 2020:2932434
59. MubarakAli D, Thajuddin N, Jeganathan K, Gunasekaran M (2011) Plant extract mediated synthesis of silver and gold nanoparticles and its antibacterial activity against clinically isolated pathogens. Colloids Surf B Biointerfaces 85:360–365
60. Srihasam S, Thyagarajan K, Korivi M, Lebaka VR, Mallem SP (2020) Phytogenic generation of NiO nanoparticles using stevia leaf extract and evaluation of their in-vitro antioxidant and antimicrobial properties. Biomolecules 10:89
61. Chen L, Liang J (2020) An overview of functional nanoparticles as novel emerging antiviral therapeutic agents. Mater Sci Eng C 112:110924
62. Lysenko V, Lozovski V, Lokshyn M, Gomeniuk YV, Dorovskih A, Rusinchuk N, Pankivska Y, Povnitsa O, Zagorodnya S, Tertykh V, and Bolbukh Y, (2018). Nanoparticles as antiviral agents against adenoviruses. Advances in Natural Sciences: *Nanoscience and Nanotechnology,* 9(2), p.025021.

63. Speshock JL, Murdock RC, Braydich-Stolle LK, Schrand AM, Hussain SM (2010) Interaction of silver nanoparticles with Tacaribe virus. J Nanobiotechnol 8(1):19
64. Sun RW, Chen R, Chung NP, Ho CM, Lin CL, Che CM (2005) Silver nanoparticles fabricated in Hepes buffer exhibit cytoprotective activities toward HIV-1 infected cells. Chem Commun 40:5059–5061
65. Khandelwal N, Kaur G, Chaubey KK, Singh P, Sharma S, Tiwari A, Singh SV, Kumar N (2014) Silver nanoparticles impair Peste des Petits ruminants virus replication. Virus Res 190:1–7
66. Baghban N, Momeni S, Behboudi E, Dianat-Moghadam H, Darabi A, Targhi HS, Keshavarz M (2024) Green synthesis of MnO2 NPs using Arabic gum: assessing its potential antiviral activity against influenza a/H1N1. Virol J 21:48
67. Sharma Y, Kawatra A, Sharma V, Dhull D, Kaushik S, Yadav JP, Kaushik S (2021) In-vitro and in-silico evaluation of the anti-chikungunya potential of Psidium Guajava leaf extract and their synthesized silver nanoparticles. Virus Dis 32:260–265
68. Yang XX, Li CM, Huang CZ (2016) Curcumin modified silver nanoparticles for highly efficient inhibition of respiratory syncytial virus infection. Nanoscale 8:3040–3048
69. Papp I, Sieben C, Ludwig K, Roskamp M, Böttcher C, Schlecht S, Herrmann A, Haag R (2014) Inhibition of influenza virus infection by multivalent sialic-acid-functionalized gold nanoparticles. Small 6:2900–2906
70. Wei J, Zheng L, Lv X, Bi Y, Chen W, Zhang W, Shi Y, Zhao L, Sun X, Wang F, Cheng S (2014) Analysis of influenza virus receptor specificity using glycan-functionalized gold nanoparticles. ACS Nano 8:4600–4607
71. Fu J, Wang Z, Huang L, Zheng S, Wang D, Chen S, Zhang H, Yang S (2014) Review of the botanical characteristics, phytochemistry, and pharmacology of Astragalus membranaceus (Huangqi). Phytother Res 28:1275–1283
72. Sharma V, Kaushik S, Pandit P, Dhull D, Yadav JP, Kaushik S (2019) Green synthesis of silver nanoparticles from medicinal plants and evaluation of their antiviral potential against chikungunya virus. Appl Microbiol Biotechnol 103:881–891
73. Parvez MK, Ahmed S (2023) Phytosynthesized nanoparticles against viral diseases: current updates. Future Integr Med 2:32–35
74. Hemeg HA (2017) Nanomaterials for alternative antibacterial therapy. Int J Nanomedicine 12:8211–8225
75. Vicente A, Sohm B, Flayac J, Rousselle P, Bauda P, Pagnout C (2019) Toxicity mechanisms of ZnO UV-filters used in sunscreens toward the model cyanobacteria Synechococcus elongatus PCC 7942. Environ Sci Pollut Res 26:22450–22463
76. Sindhu PD, Mukherjee A, Chandrasekaran N (2013) Synergistic effect of biogenic silver nanocolloid in combination with antibiotics: a potent therapeutic agent. Int J Pharm Pharm Sci 51:292–295
77. Kalhapure RS, Sonawane SJ, Sikwal DR, Jadhav M, Rambharose S, Mocktar C, Govender T (2015) Solid lipid nanoparticles of clotrimazole silver complex: an efficient nano antibacterial against Staphylococcus aureus and MRSA. Colloids Surf Biointerfaces 136:651–658
78. Esmaeillou M, Zarrini G, Rezaee MA, Mojarrad JS, Bahadori A (2017) Vancomycin capped with silver nanoparticles as an antibacterial agent against multi-drug resistance bacteria. Adv Pharm Bull 7:479–483
79. Naqvi SZH, Kiran U, Ali MI, Jamal A, Hameed A, Ahmed S, Ali N (2013) Combined efficacy of biologically synthesized silver nanoparticles and different antibiotics against multidrug-resistant bacteria. Int J Nanomedicine 8:3187–3195
80. Brown AN, Smith K, Samuels TA, Lu J, Obare SO, Scott ME (2012) Nanoparticles functionalized with ampicillin destroy multiple-antibiotic-resistant isolates of Pseudomonas aeruginosa and Enterobacter aerogenes and methicillin-resistant Staphylococcus aureus. Appl Environ Microbiol 78:2768–2774

81. Hur YE, Park Y (2016) Vancomycin-functionalized gold and silver nanoparticles as an antibacterial nanoplatform against methicillin-resistant Staphylococcus aureus. J Nanosci Nanotechnol 16:6393–6399
82. Ghasemi F, Jalal R (2016) Antimicrobial action of zinc oxide nanoparticles in combination with ciprofloxacin and ceftazidime against multidrug-resistant Acinetobacter baumannii. J Glob Antimicrob Resist 6:118–122
83. Tawre MS, Shiledar A, Satpute SK, Ahire K, Ghosh S, Pardesi K (2022) Synergistic and antibiofilm potential of Curcuma aromatica derived silver nanoparticles in combination with antibiotics against multidrug-resistant pathogens. Front Chem 10:1029056
84. Lakshmi YS, Kumar P, Kishore G, Bhaskar C, Kondapi AK (2016) Triple combination MPT vaginal microbicide using curcumin and efavirenz loaded lactoferrin nanoparticles. Sci Rep 6:25479
85. Kumar P, Lakshmi YS, Kondapi AK (2017) Triple drug combination of zidovudine, efavirenz and lamivudine loaded lactoferrin nanoparticles: an effective nano first-line regimen for HIV therapy. Pharm Res 34:257–268
86. Gaurav C, Goutam R, Rohan KN, Sweta KT, Abhay CS, Amit GK (2015) In situ stabilized AgNPs and (cu-cur)CD dispersed gel, a topical contraceptive antiretroviral (ARV) microbicide. RSC Adv 101:1–31
87. Khan MF, Ansari AH, Hameedullah M, Ahmad E, Husain FM, Zia Q, Baig U, Zaheer MR, Alam MM, Khan AM, AlOthman ZA (2016) Sol-gel synthesis of thorn-like ZnO nanoparticles endorsing mechanical stirring effect and their antimicrobial activities: potential role as nano-antibiotics. Sci Rep 6:27689
88. Zhang L, Pornpattananangkul D, Hu CM, Huang CM (2010) Development of nanoparticles for antimicrobial drug delivery. Curr Med Chem 17:585–594
89. Kandi V, Kandi S (2013) Antimicrobial properties of nanomolecules: potential candidates as antibiotics in the era of multi-drug resistance. Epidemiol Health 37:e2015020
90. Dobrucka R, Szymanski M, Przekop R (2019) The study of toxicity effects of biosynthesized silver nanoparticles using Veronica officinalis extract. Int J Environ Sci Technol 16:8517–8526
91. Naskar A, Bera S, Bhattacharya R, Saha P, Roy SS, Sen T, Jana S (2016) Synthesis, characterization and antibacterial activity of ag incorporated ZnO–graphene nanocomposites. RSC Adv 6:88751–88761
92. Graves JL, Tajkarimi M, Cunningham Q, Campbell A, Nonga H, Harrison SH, Barrick JE (2015) Rapid evolution of silver nanoparticle resistance in Escherichia coli. Front Genet 6:42
93. Wei L, Lu J, Xu H, Patel A, Chen ZS, Chen G (2015) Silver nanoparticles: synthesis, properties, and therapeutic applications. Drug Discov Today Technol 20:595–601
94. Zazo H, Colino CI, Lanao JM (2016) Current applications of nanoparticles in infectious diseases. J Control Release 224:86–102
95. Adeniji OO, Nontongana N, Okoh JC, Okoh AI (2022) The potential of antibiotics and nanomaterial combinations as therapeutic strategies in the management of multidrug-resistant infections: a review. Int J Mol Sci 23:15038
96. Cinteza LO, Scomoroscenco C, Voicu SN, Nistor CL, Nitu SG, Trica B, Jecu ML, Petcu C (2018) Chitosan-stabilized ag nanoparticles with superior biocompatibility and their synergistic antibacterial effect in mixtures with essential oils. Nanomaterials 8:826
97. Meshram JV, Koli VB, Kumbhar SG, Borde LC, Phadatare MR, Pawar SH (2018) Structural, spectroscopic and anti-microbial inspection of PEG capped ZnO nanoparticles for biomedical applications. Mater Res Express 5:045016
98. Abdallah OM, El-Baghdady KZ, Khalil MMH, El Borhamy MI, Meligi GA (2020) Antibacterial, antibiofilm and cytotoxic activities of biogenic polyvinyl alcohol-silver and chitosan-silver nanocomposites. J Polym Res 27:74
99. Begum T, Follett PA, Mahmud J, Moskovchenko L, Salmieri S, Allahdad Z, Lacroix M (2022) Silver nanoparticles-essential oils combined treatments to enhance the antibacterial

and antifungal properties against foodborne pathogens and spoilage microorganisms. Microb Pathog 164:105411
100. Devi LS, Das B, Dutta D, Kumar S (2024) Essential oils as functional agents in biopolymer-based sustainable food packaging system: a review. Sustain Chem Pharm 39:101563
101. Zubair M, Shahzad S, Hussain A, Pradhan RA, Arshad M, Ullah A (2022) Current trends in the utilization of essential oils for polysaccharide- and protein-derived food packaging materials. Polymers 14:1146
102. Arfat YA, Ahmed J, Ejaz M, Mullah M (2018) Polylactide/graphene oxide nanosheets/clove essential oil composite films for potential food packaging applications. Int J Biol Macromol 107:194–203
103. Sharma S, Barkauskaite S, Duffy B, Jaiswal AK, Jaiswal S (2020) Characterization and antimicrobial activity of biodegradable active packaging enriched with clove and thyme essential oil for food packaging application. Foods 9:1117
104. Jayachandran A, Aswathy TR, Nair AS (2021) Green synthesis and characterization of zinc oxide nanoparticles using Cayratia pedata leaf extract. Biochem Biophys Rep 26:100995
105. Mali SC, Dhaka A, Githala CK, Trivedi R (2020) Green synthesis of copper nanoparticles using Celastrus paniculatus Willd. Leaf extract and their photocatalytic and antifungal properties. Biotechnol Rep (Amst) 27:e00518
106. Wintachai P, Paosen S, Yupanqui CT, Voravuthikunchai SP (2019) Silver nanoparticles synthesized with eucalyptus critriodora ethanol leaf extract stimulate antibacterial activity against clinically multidrug-resistant Acinetobacter baumannii isolated from pneumonia patients. Microb Pathog 126:245–257
107. Sulaiman GM, Mohammed WH, Marzoog TR, Al-Amiery AA, Kadhum AA, Mohamad AB (2013) Green synthesis, antimicrobial and cytotoxic effects of silver nanoparticles using Eucalyptus chapmaniana leaves extract. Asian Pac J Trop Biomed 3:58–63
108. Bhawana BRK, Buttar HS, Jain VK, Jain N (2011) Curcumin nanoparticles: preparation, characterization, and antimicrobial study. J Agric Food Chem 59(5):2056–2061
109. Sun Y, Du L, Liu Y, Li X, Li M, Jin Y, Qian X (2014) Transdermal delivery of the in situ hydrogels of curcumin and its inclusion complexes of hydroxypropyl-β-cyclodextrin for melanoma treatment. Int J Pharm 469:31–39

Aromatic Plant-Based Nanoparticles for Antifungal Performance

Weria Weisany

Abstract Fungal pathogens pose persistent threats to global agriculture and human health, exacerbated by rising resistance to conventional antifungal agents and environmental concerns associated with synthetic fungicides. This chapter explores the emerging role of nanotechnology, particularly nanoparticles synthesized via aromatic plant extracts as a sustainable and potent alternative for fungal control. Aromatic plants, rich in phytochemicals such as terpenoids, flavonoids, and essential oils, serve as natural biofactories for green nanoparticle synthesis, offering both reducing and capping capabilities. The resulting nanoparticles exhibit multifaceted antifungal mechanisms, including membrane disruption, enzymatic inhibition, oxidative stress induction, and biofilm suppression. Their integration into agricultural sprays, edible coatings, and postharvest treatments demonstrates significant potential in crop protection and food preservation. Additionally, the chapter discusses cytotoxicity, environmental safety, and regulatory considerations essential for real-world application. It concludes with perspectives on multifunctional hybrid nanostructures, smart delivery systems, and the challenges of scale-up and commercialization. Collectively, this work highlights the transformative potential of plant-based nanotechnology as an eco-friendly and effective solution to global antifungal challenges.

Keywords Aromatic plants · Nanotechnology · Antifungal nanoparticles · Essential oils · Postharvest preservation

W. Weisany (✉)
Department of Agriculture Science and Engineering, SRC, Islamic Azad University, Tehran, Iran
e-mail: w.weisany@iau.ac.ir

A. Husen (ed.), *Sustainable Uses of Aromatic Plants in Nanotechnology*, Smart Nanomaterials Technology, https://doi.org/10.1007/978-981-95-7409-4_8

1 Introduction

1.1 Overview of Fungal Pathogens in Agriculture and Medicine

Fungal pathogens present a persistent challenge across both agriculture and medicine. In crop systems, fungal infections significantly reduce yield, compromise food quality, and contribute to large-scale economic losses. Common phytopathogenic fungi such as *Fusarium*, *Botrytis*, *Alternaria*, and *Aspergillus* cause devastating diseases that not only impact productivity but also lead to contamination with mycotoxins, rendering food and feed unsafe for consumption [1]. The growing occurrence of these infections is exacerbated by shifting climatic patterns, monoculture practices, and intensive pesticide use.

In the medical field, fungal pathogens are increasingly recognized as major contributors to opportunistic infections, especially in immunocompromised individuals. Infections caused by species like *Candida albicans*, *Aspergillus fumigatus*, and *Cryptococcus neoformans* are associated with high morbidity and mortality rates [2, 3]. The rise in antifungal-resistant strains and the formation of biofilms complicate treatment outcomes. These concerns are magnified by the limited arsenal of effective antifungal drugs and the slow pace of new drug discovery [4]. Addressing fungal threats in both agriculture and human health requires integrated, innovative, and sustainable approaches.

1.2 Limitations of Conventional Antifungal Agents

While conventional antifungal agents have long served as the primary means of fungal control, they are increasingly limited in effectiveness. In agriculture, chemical fungicides often face challenges related to environmental persistence, non-target toxicity, and the evolution of resistant fungal strains [5]. Continuous application of these agents has led to a decline in soil health and a disruption of beneficial microbial communities. Moreover, regulatory restrictions on chemical residues in food products necessitate the development of cleaner alternatives [6].

In clinical applications, antifungal therapies suffer from toxicity, limited drug selectivity, and interaction with host metabolism. Many standard treatments are less effective against fungal biofilms, which provide a protective environment for pathogens [7]. Additionally, the emergence of multidrug-resistant fungal strains is rendering some of the most commonly used antifungal drugs ineffective [8]. These limitations have created an urgent need for alternative antifungal strategies that are both effective and sustainable, capable of minimizing environmental and health risks while maintaining broad-spectrum activity.

1.3 Emergence of Nanotechnology in Fungal Control

Nanotechnology offers a paradigm shift in the way fungal pathogens are managed. Nanoparticles possess unique physicochemical properties, including high surface area-to-volume ratio, quantum effects, and customizable surface functionalities [9]. These features enable nanoparticles to interact with fungal cells at the molecular level, often leading to enhanced antifungal performance compared to bulk materials [10]. Mechanistically, nanoparticles can induce oxidative stress, disrupt membrane integrity, interfere with metabolic pathways, and penetrate biofilms [11].

In agricultural settings, nanomaterials are being integrated into seed coatings, foliar sprays, and postharvest treatments to improve crop protection [12]. In medicine, they are being developed into targeted delivery systems, antifungal dressings, and topical formulations [13]. The nanoscale delivery of antifungal compounds can reduce effective doses, lower toxicity, and enhance treatment precision [14]. Despite these advances, concerns about the environmental impact and long-term safety of chemically synthesized nanoparticles remain. This has sparked interest in green nanotechnology approaches that leverage plant-based resources for nanoparticle synthesis.

1.4 Role of Aromatic Plants in Nanoparticle Synthesis

Aromatic plants serve as natural bioreactors for the synthesis of biofunctional nanoparticles. These plants are rich in secondary metabolites such as flavonoids, terpenoids, alkaloids, and essential oils, which can act as both reducing and capping agents in the green synthesis of nanoparticles [15]. This method eliminates the need for hazardous chemicals and high-energy processes, making it an environmentally friendly alternative to conventional synthesis.

The use of aromatic plant extracts or essential oils enables the production of nanoparticles with inherent antimicrobial properties. These phytochemically coated nanoparticles often exhibit synergistic antifungal effects, combining the antimicrobial activity of the nanoparticle core with the bioactivity of the plant-derived surface compounds [13]. This synergy enhances efficacy against a range of fungal targets, including resistant strains and biofilm-forming pathogens. The green synthesis of nanoparticles using aromatic plants thus represents a promising intersection of nanotechnology, phytochemistry, and antifungal therapy [16]. It aligns with the goals of sustainable development by offering biodegradable, non-toxic, and efficient antifungal solutions for both agricultural and clinical applications.

2 Aromatic Plants as Biofactories

2.1 Definition and Classification of Aromatic Plants

Aromatic plants are a diverse group of botanical species characterized by their ability to produce and store volatile aromatic compounds, primarily within specialized glandular structures such as trichomes or oil ducts [17]. These compounds, known collectively as essential oils, contribute to the distinctive fragrance and therapeutic properties of aromatic plants. Beyond their sensory appeal, these plants play a vital role in traditional medicine, culinary practices, cosmetics, and, more recently, nanotechnology [18].

Aromatic plants are taxonomically diverse and span multiple families, including Lamiaceae (e.g., *Thymus*, *Lavandula*, *Ocimum*), Myrtaceae (e.g., *Eucalyptus*), Apiaceae (e.g., *Foeniculum*, *Coriandrum*), and Rutaceae (e.g., *Citrus*). Classification of aromatic plants can be based on the dominant class of volatile compounds they produce—such as terpenes, phenylpropanoids, or sulfur-containing compounds—or according to their therapeutic or ethnobotanical use. This phytochemical diversity makes aromatic plants valuable resources for bioactive compound extraction and green synthesis of antimicrobial nanomaterials.

2.2 Phytochemical Profile Relevant to Antifungal Activity

The antifungal efficacy of aromatic plants is primarily attributed to their rich phytochemical profile, especially the volatile and semi-volatile constituents present in essential oils and extracts [13, 19]. These bioactives include monoterpenes (e.g., thymol, carvacrol, menthol), sesquiterpenes (e.g., caryophyllene, farnesene), aldehydes (e.g., cinnamaldehyde, citral), phenolic acids, flavonoids, and tannins. The composition and concentration of these compounds vary with species, plant part, geographic origin, and extraction method.

Many of these phytochemicals have demonstrated strong antifungal activity by targeting multiple cellular structures and processes. Terpenes and phenolic derivatives can disrupt fungal cell membranes, interfere with mitochondrial function, inhibit spore germination, and reduce virulence factor production [20]. Moreover, non-volatile compounds such as flavonoids and tannins contribute to the antioxidant and synergistic properties of plant extracts, enhancing the overall antifungal spectrum [21].

The multifunctionality of these bioactives makes aromatic plants particularly attractive for use in nanoparticle synthesis. When utilized as reducing and stabilizing agents, they confer bioactivity onto nanoparticles, resulting in hybrid systems with superior antifungal performance.

2.3 Mechanisms of Action of Essential Oils and Extracts

Essential oils and plant extracts exert their antifungal effects through diverse and often synergistic mechanisms. One of the primary actions involves the disruption of fungal cell membrane integrity [22]. Lipophilic compounds such as thymol and eugenol integrate into lipid bilayers, increasing membrane permeability and causing leakage of essential intracellular components [23].

Additionally, these compounds can interfere with the function of membrane-bound enzymes, alter ion gradients, and collapse membrane potential. At the intracellular level, essential oil components may induce oxidative stress by generating reactive oxygen species (ROS), leading to the damage of nucleic acids, proteins, and organelles such as mitochondria [24, 25]. Some phytochemicals also inhibit key biosynthetic pathways, including ergosterol synthesis and chitin formation, impairing fungal growth and reproduction [26] (Fig. 1).

Another critical aspect of essential oil action is their ability to prevent fungal adhesion and biofilm formation [27]. This property is particularly important in both agricultural and medical contexts, where biofilms are associated with increased resistance and persistence of fungal infections. The broad-spectrum and multitargeted action of aromatic plant-derived compounds reduces the likelihood of resistance development and enhances their effectiveness when used in combination with nanoparticles or conventional antifungal agents.

2.4 Examples of Key Aromatic Species

Several aromatic plant species have been extensively studied for their antifungal potential and utility in green nanoparticle synthesis:

- ***Thymus vulgaris*** (thyme): Known for its high thymol and carvacrol content, thyme essential oil exhibits strong fungicidal activity against a range of phytopathogens and human pathogens. Its extract has also been used to synthesize silver and zinc oxide nanoparticles with enhanced antifungal efficiency [28].
- ***Lavandula angustifolia*** (lavender): Rich in linalool and linalyl acetate, lavender oil has shown antifungal properties against dermatophytes and *Candida* species. Lavender extract-mediated nanoparticles demonstrate potential for topical antifungal formulations [29].
- ***Ocimum sanctum*** (holy basil): Containing eugenol, methyl chavicol, and rosmarinic acid, this species is valued for both antimicrobial and antioxidant activities. Its phytochemicals facilitate the green synthesis of metal nanoparticles with broad-spectrum antifungal action [30].
- ***Eucalyptus globulus*** (eucalyptus): Its essential oil, dominated by 1,8-cineole, exhibits antimicrobial activity, especially in respiratory pathogens. Eucalyptus extracts have been successfully employed in the biosynthesis of antifungal metal and metal oxide nanoparticles [31].

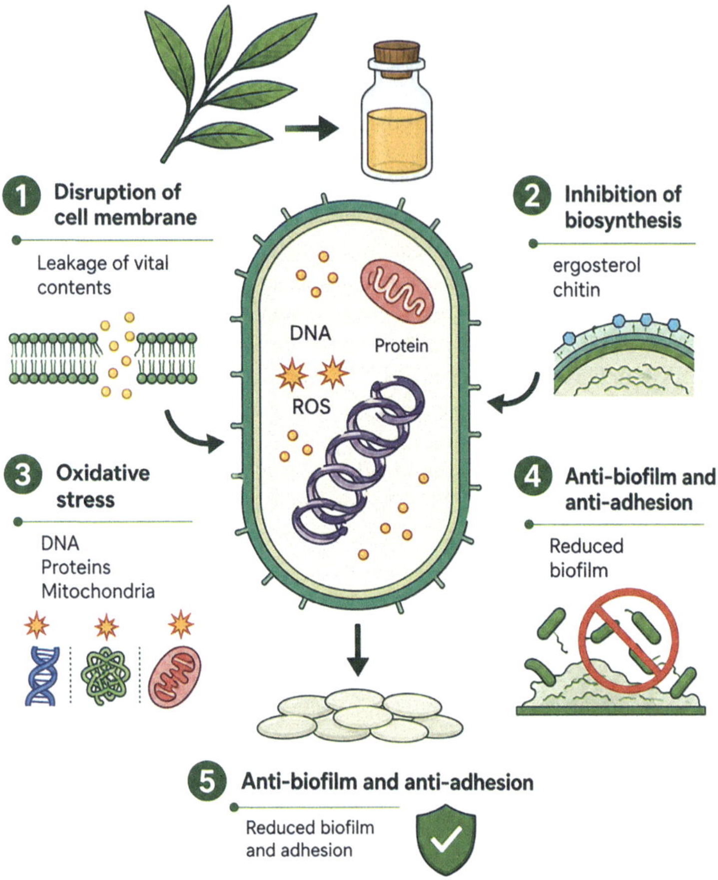

Fig. 1 Enhanced visualization of the antifungal mechanisms of essential oils and plant extracts. Essential oil components exert multifaceted effects on fungal cells by (1) disrupting cell membrane integrity, leading to leakage of vital contents; (2) inhibiting key enzymatic pathways involved in the biosynthesis of structural components like ergosterol and chitin; (3) inducing oxidative stress through the generation of ROS, damaging DNA, proteins, and mitochondria; and (4) suppressing biofilm formation and fungal adhesion, which are critical for persistent infections. The synergy of these actions supports their potent and broad-spectrum antifungal efficacy

These species exemplify the potential of aromatic plants as bioresources for developing multifunctional nanoparticle systems. Their wide distribution, renewable nature, and inherent antimicrobial potency make them ideal candidates for incorporation into antifungal nanotechnological applications.

3 Green Synthesis of Nanoparticles Using Aromatic Plants

3.1 *Principles of Green Nanotechnology*

Green nanotechnology refers to the development of nanomaterials and nanoscale processes that are environmentally benign, energy-efficient, and compatible with sustainable development goals. Unlike conventional nanoparticle synthesis, which often relies on toxic solvents, harsh reducing agents, and high-temperature conditions, green nanotechnology emphasizes the use of natural, renewable resources—such as plant extracts—as functional agents in nanoparticle production.

The core principles of green nanotechnology are rooted in green chemistry, aiming to minimize the generation of hazardous substances and promote the use of safe materials and energy-efficient methods [32]. In this context, plant-based synthesis methods provide a dual advantage: they eliminate the need for hazardous chemicals and simultaneously introduce phytochemicals that impart biological functionality to the nanoparticles [33]. This eco-friendly approach aligns well with the goals of circular bioeconomy and sustainable agricultural and biomedical innovation.

3.2 *Role of Plant Extracts in Reducing and Capping Nanoparticles*

Plant extracts, especially those derived from aromatic species, play a critical role in the biosynthesis of nanoparticles through two main functions: reduction and capping. The reducing activity is facilitated by bioactive phytochemicals such as phenolic acids, flavonoids, terpenoids, alkaloids, and sugars, which donate electrons to metal ions (e.g., Ag^+, Au^{3+}, Zn^{2+}), leading to their reduction into zero-valent metal atoms that nucleate to form nanoparticles (Fig. 2).

Simultaneously, these phytochemicals also act as capping or stabilizing agents. They bind to the surface of the forming nanoparticles, preventing agglomeration and controlling their growth, size, and morphology. The capping layer is not only important for stabilization but also enhances biocompatibility and endows the particles with additional biological activities, such as antifungal and antioxidant properties [34]. The composition of the plant extract, including pH, temperature, and concentration of phytochemicals, directly influences nanoparticle characteristics. Extracts from aromatic plants, due to their abundance of reactive secondary metabolites, are particularly efficient at both reducing and capping functions, making them ideal candidates for green nanofabrication [35].

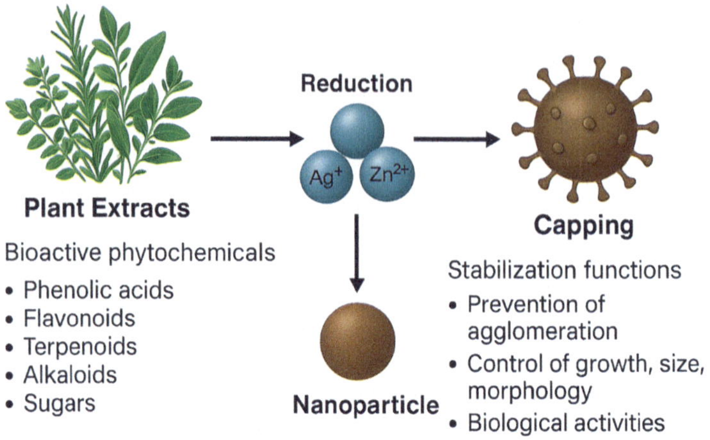

Fig. 2 Mechanism of plant extract-mediated nanoparticle synthesis. Aromatic plant extracts rich in phytochemicals such as phenolic acids, flavonoids, terpenoids, alkaloids, and sugars serve dual functions in green nanofabrication. During the reduction phase, these compounds donate electrons to metal ions (Ag^+, Zn^{2+}), converting them into zero-valent atoms that nucleate into nanoparticles. In the capping phase, the same phytochemicals bind to nanoparticle surfaces, preventing agglomeration and controlling growth, size, and morphology. The capping layer also imparts additional biological activities, enhancing the nanoparticles' stability and functional efficacy, particularly in antifungal applications

3.3 Methods for Biosynthesis of Metal and Metal Oxide Nanoparticles

The biosynthesis of metal and metal oxide nanoparticles using plant extracts can be achieved through several straightforward and scalable techniques. The most common method involves mixing an aqueous solution of a metal salt (such as silver nitrate, zinc acetate, or copper sulfate) with the plant extract under controlled conditions of temperature, pH, and stirring [36]. The reaction leads to a visible color change, often within minutes to hours, indicating the formation of nanoparticles. Parameters such as reaction time, metal ion concentration, and extract-to-metal ratio can be optimized to tailor the particle size, shape, and dispersion.

Metal nanoparticles (e.g., silver, gold, copper) are typically synthesized for their strong antimicrobial properties [11, 14], whereas metal oxide nanoparticles (e.g., ZnO, TiO_2, Fe_3O_4) offer advantages in photocatalytic activity, stability, and biocompatibility [37]. These biosynthesized nanostructures can be further integrated into formulations for antifungal applications in agriculture, medicine, and food packaging. Solid-state biosynthesis, microwave-assisted synthesis, and ultrasonic-assisted green synthesis are emerging variations that enhance reaction efficiency and control over nanoparticle properties. These methods maintain the eco-friendly ethos, while improving reproducibility and scalability for industrial applications.

3.4 Characterization Techniques (UV-Vis, XRD, TEM, FTIR, DLS, SEM)

Accurate characterization of green-synthesized nanoparticles is essential to confirm their formation, understand their properties, and predict their biological performance. Several analytical techniques are employed for this purpose (Fig. 3):

- *UV-Visible Spectroscopy (UV-Vis)*: Used for initial confirmation of nanoparticle formation based on surface plasmon resonance. Each type of nanoparticle exhibits a specific absorbance peak depending on size, shape, and composition [38].
- *X-Ray Diffraction (XRD)*: Determines the crystalline structure and phase purity of nanoparticles. The XRD patterns help identify the formation of metallic or metal oxide phases and allow estimation of particle size using the Scherrer equation [39].
- *Transmission Electron Microscopy (TEM)*: Provides high-resolution images that reveal the shape, size distribution, and internal morphology of nanoparticles at the nanoscale level.

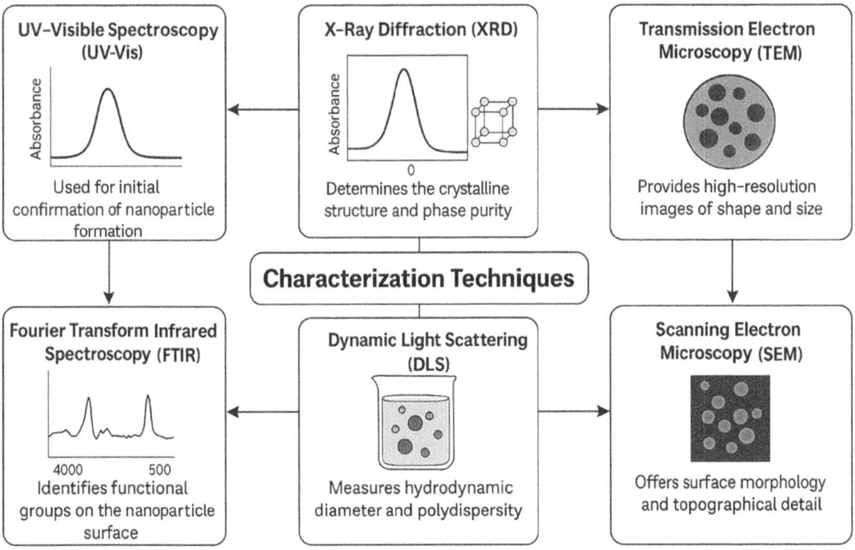

Fig. 3 Characterization techniques for green-synthesized nanoparticles. This diagram illustrates six core analytical methods: UV-Vis spectroscopy for confirming nanoparticle formation via surface plasmon resonance, XRD for determining crystalline structure and phase purity, TEM for visualizing nanoparticle morphology at the nanoscale, FTIR for identifying surface functional groups and plant-derived capping agents, DLS for measuring hydrodynamic size and polydispersity, and SEM for examining surface morphology and particle distribution. These complementary techniques provide a comprehensive profile essential for evaluating nanoparticle structure, stability, and antifungal performance

- *Fourier Transform Infrared Spectroscopy (FTIR)*: Identifies functional groups present on the nanoparticle surface, revealing the phytochemical compounds responsible for reduction and capping. The FTIR spectra offer insight into the chemical interactions between plant biomolecules and the nanoparticle surface [40].
- *Dynamic Light Scattering (DLS)*: Measures the hydrodynamic diameter and polydispersity index of nanoparticles in colloidal suspensions. This is critical for assessing stability and dispersion quality in aqueous media.
- *Scanning Electron Microscopy (SEM)*: Offers surface morphology and topographical detail of nanoparticles, often used in combination with energy-dispersive X-ray analysis to determine elemental composition.

Together, these techniques provide a comprehensive understanding of nanoparticle characteristics and ensure that the synthesized materials meet the desired specifications for antifungal performance and safety.

4 Mechanisms of Antifungal Activity

4.1 Disruption of Cell Membrane Integrity

One of the primary antifungal mechanisms of both aromatic plant-based nanoparticles and essential oil components is the disruption of fungal cell membrane integrity [41]. The fungal plasma membrane, composed predominantly of ergosterol and phospholipids, is vital for maintaining cellular homeostasis, ion gradients, and metabolic flux. Lipophilic compounds present in essential oils—such as thymol, carvacrol, and eugenol—intercalate into the lipid bilayer, altering membrane fluidity and permeability [23].

When these compounds are incorporated into or capped onto nanoparticles, their bioavailability and stability are enhanced, leading to stronger interactions with the fungal membrane [11]. Nanoparticles themselves, particularly metal and metal oxide types, can adhere to the fungal cell wall and exert localized physical or chemical stress [14]. This can result in pore formation, leakage of intracellular contents, and irreversible membrane damage. The combined action of the nanocarrier and the bioactive phytochemicals intensifies the disruption process, leading to rapid fungal cell death.

The degree of membrane disruption depends on nanoparticle characteristics such as size, shape, charge, and surface functionalization. Smaller, spherical nanoparticles with positive surface charges tend to exhibit higher affinity toward the negatively charged fungal membrane, thereby facilitating more effective antifungal action [42] (Fig. 4).

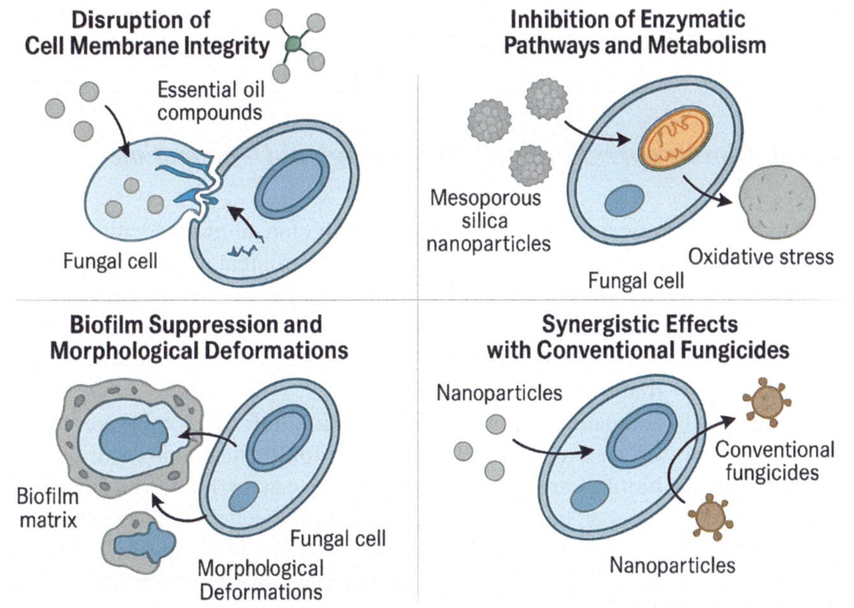

Fig. 4 Mechanisms of antifungal action of aromatic plant-based nanoparticles and essential oils. The four main modes of action include the following: (1) disruption of cell membrane integrity by lipophilic essential oil compounds, leading to leakage of intracellular contents; (2) inhibition of enzymatic pathways and metabolism through ROS generation and interference with mitochondrial function by bioactive-loaded nanoparticles; (3) biofilm suppression and morphological deformations that impair fungal adhesion and cause hyphal damage; and (4) synergistic effects with conventional fungicides, enhancing antifungal efficacy and reducing resistance development

4.2 Inhibition of Enzymatic Pathways and Metabolism

Beyond physical damage, aromatic plant-derived nanoparticles interfere with critical enzymatic and metabolic pathways in fungal cells. Key targets include enzymes involved in ergosterol biosynthesis, respiratory electron transport, and antioxidant defense [43]. For instance, nanoparticles functionalized with phenolic compounds can inhibit cytochrome P450 enzymes that are essential for ergosterol production, compromising membrane stability from within [43].

Additionally, nanoparticles can penetrate the fungal cytoplasm and generate ROS, either intrinsically or via photoactivation (in the case of ZnO or TiO_2). The accumulation of ROS induces oxidative stress, leading to the oxidation of proteins, nucleic acids, and lipids. This oxidative damage disrupts mitochondrial respiration, impairs ATP production, and ultimately compromises energy-dependent cellular processes [44].

Essential oils, known for their redox-active components, synergize with the ROS-generating capacity of nanoparticles. Together, they amplify metabolic disruption by inhibiting fungal detoxification pathways, depleting glutathione levels, and blocking oxidative stress response enzymes. This multitargeted mode of action reduces the likelihood of fungal adaptation or resistance development [45].

4.3 Biofilm Suppression and Morphological Deformations

Fungal biofilms represent a major challenge in both clinical and agricultural contexts due to their inherent resistance to conventional antifungal agents. These structured microbial communities are composed of fungal cells embedded in an extracellular polymeric matrix, which acts as a barrier to drug penetration [46]. Aromatic plant-based nanoparticles exhibit strong antibiofilm properties, which is a key advantage over traditional fungicides.

The antifungal nanomaterials can inhibit early biofilm formation by preventing fungal adhesion to surfaces, a process that is critical for the initial stages of biofilm development. Once biofilms are formed, nanoparticles can disrupt the matrix structure, penetrate into deeper biofilm layers, and kill embedded fungal cells [47]. Essential oil-loaded nanoparticles enhance this effect by delivering hydrophobic bioactives that can diffuse through the matrix and target metabolically active cells within the biofilm. Moreover, the interaction of nanoparticles with fungal cells often results in visible morphological deformations, including swelling, shrinking, irregular budding, and rupture of hyphae [10]. These structural changes are indicative of cellular stress and loss of homeostatic control, marking the effectiveness of the treatment.

4.4 Synergistic Effects with Conventional Fungicides

One of the most promising aspects of aromatic plant-based nanoparticles is their ability to synergize with existing antifungal agents. This synergy enables a reduction in the minimum inhibitory concentration of conventional fungicides, minimizing cytotoxicity and delaying the onset of resistance. Nanoparticles can act as carriers that enhance the solubility, stability, and targeted delivery of synthetic fungicides, while simultaneously contributing their own antifungal properties [48]. In agricultural applications, the co-application of plant-derived nanoparticles and chemical fungicides has shown improved efficacy against resistant strains, better coverage, and longer residual activity [49].

Mechanistically, this synergism results from multiple simultaneous attacks on fungal cells—membrane disruption by phytochemicals, metabolic inhibition by metal ions, and intracellular interference by synthetic agents. By targeting different cellular components and pathways concurrently, these combined treatments reduce the fungal adaptive response and prevent the development of tolerance. This multipronged strategy aligns with the principles of integrated disease management and rational drug design, offering a new avenue for effective and sustainable antifungal therapy across diverse sectors.

5 Application in Agriculture and Postharvest Preservation

5.1 Foliar Sprays and Soil Amendments

One of the most direct and effective applications of aromatic plant-based nanoparticles in agriculture is their use as foliar sprays and soil amendments for the management of fungal phytopathogens. Foliar applications enable nanoparticles to interact directly with fungal spores or hyphae on the leaf surface, where rapid action is crucial for disease suppression. The lipophilic properties of many plant-derived bioactive compounds allow for effective penetration into the fungal cell membranes, while the nano-size enhances surface coverage and retention [50].

As soil amendments, these nanoparticles improve the root zone environment by suppressing soilborne fungal pathogens such as *Fusarium* and *Rhizoctonia*, while potentially promoting beneficial microbial communities [51]. The gradual release of bioactive molecules from capping agents in the soil matrix allows for sustained antifungal activity. Additionally, these treatments can be tailored to specific crops and pathogens by selecting appropriate aromatic plant sources, such as *Thymus vulgaris* or *Ocimum sanctum*, known for their antifungal potency [52].

5.2 Integration into Edible and Biodegradable Coatings

Aromatic plant-based nanoparticles have shown substantial promise in the formulation of edible and biodegradable coatings for perishable commodities such as fruits, vegetables, and grains [53, 54]. These coatings act as physical and biochemical barriers, protecting produce from fungal infections during storage and transportation [55].

The incorporation of nanoparticles synthesized using essential oils or plant extracts provides a dual mode of protection: the physical barrier created by the coating and the biochemical inhibition of fungal growth by the embedded nanomaterials. Metal and metal oxide nanoparticles (e.g., AgNPs, ZnO NPs) synthesized with plant-derived capping agents can be uniformly distributed in polysaccharide- or protein-based matrices to enhance antimicrobial activity, extend shelf-life, and maintain nutritional and sensory quality [56].

Such coatings are not only biodegradable and non-toxic but can also be consumed with the produce, eliminating waste and ensuring food safety [55]. Their controlled release behavior and high stability make them particularly suitable for maintaining postharvest quality in high-value fruits and vegetables [15].

5.3 *Postharvest Disease Management in Fruits and Vegetables*

Postharvest fungal diseases are a major cause of food loss worldwide, particularly in tropical and subtropical regions. Pathogens such as *Botrytis cinerea*, *Penicillium expansum*, and *Aspergillus niger* are responsible for significant spoilage of fresh produce [10, 41]. Aromatic plant-based nanoparticles offer an innovative solution to this challenge through their antifungal, antioxidant, and surface-adhesive properties [57].

Applications can include dipping treatments, vapor-phase exposure, or spray-coating of harvested produce. Nanoparticles functionalized with essential oils or phenolic compounds from aromatic plants can effectively reduce pathogen load, delay the onset of spoilage, and maintain fruit firmness and color. These nanoparticles also exhibit a strong capacity to penetrate biofilms and fungal cuticles, ensuring deeper action and prolonged protection [58]. Moreover, the biodegradable nature of the nanoparticle matrix and the absence of synthetic chemical residues meet the rising consumer and regulatory demands for clean-label and eco-friendly preservation strategies (Fig. 5).

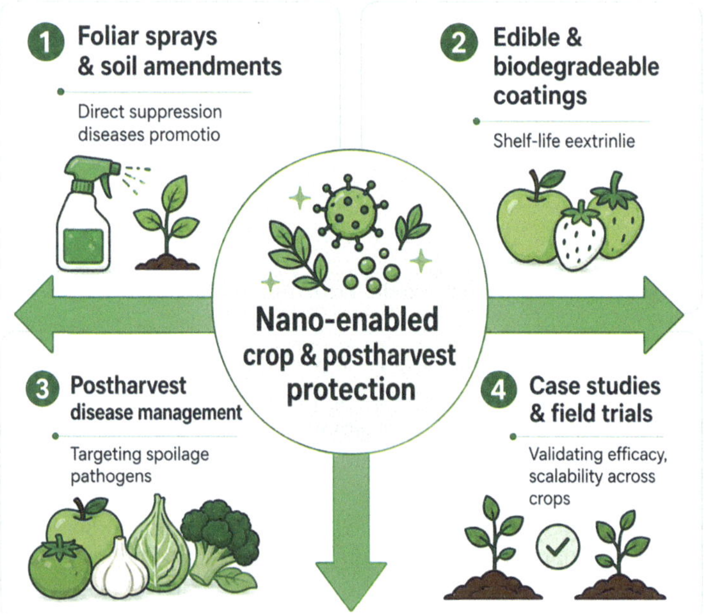

Fig. 5 Visual summary of agricultural and food preservation applications of aromatic plant-based nanoparticles. Central to the diagram is the concept of nano-enabled crop and postharvest protection, supported by four key applications: (1) foliar sprays and soil amendments for direct disease suppression and growth promotion, (2) integration into edible and biodegradable coatings for shelf-life extension and fungal inhibition, (3) postharvest disease management targeting spoilage pathogens, and (4) case studies and field trials validating efficacy and scalability across crop systems. Each application is linked through the multifunctional role of bioactive, eco-friendly nanoparticles derived from aromatic plants

6 Toxicological and Environmental Considerations

6.1 Cytotoxicity and Biocompatibility Studies

Before clinical or agricultural deployment, the cytotoxicity and biocompatibility of aromatic plant-based nanoparticles must be rigorously evaluated. Although these nanoparticles are synthesized using biogenic routes and plant-derived agents, their interaction with mammalian and plant cells must be carefully assessed to ensure safe application. Furthermore, hemocompatibility, genotoxicity, and immunogenicity studies are important for pharmaceutical applications. In agricultural systems, phytotoxicity assays evaluate whether the nanoparticles impact seed germination, root elongation, or photosynthetic efficiency. A favorable cytocompatibility profile across biological systems reinforces their suitability for broad-spectrum antifungal use.

6.2 Biodegradability and Eco-Safety Assessment

Aromatically derived nanoparticles are generally assumed to be biodegradable and eco-safe, particularly when compared to their chemically synthesized counterparts. However, detailed assessments are necessary to confirm their environmental fate, degradation kinetics, and ecotoxicological impact across soil, aquatic, and microbial ecosystems [59]. Biodegradability is influenced by nanoparticle composition (metal vs. metal oxide), surface chemistry, size, and environmental conditions [60]. Plant-based capping agents typically enhance biodegradability by promoting enzymatic and microbial breakdown. Long-term eco-safety evaluations are crucial, particularly for field applications where repeated use may lead to accumulation. Evaluating nanoparticle leaching, bioaccumulation in plants or food chains, and their influence on beneficial soil microbiota ensures responsible use in agroecosystems and aligns with sustainable agricultural practices [61].

6.3 Regulatory Aspects and Risk Management

Despite their potential, regulatory frameworks for plant-based nanoparticles are still in development and often lag behind technological advancements. The dual nature of these materials—as both nanostructures and botanical products—complicates their classification and oversight, particularly in terms of labeling, usage limits, and safety standards [62]. In most jurisdictions, regulatory approval requires comprehensive data on nanoparticle synthesis, composition, stability, toxicity, environmental behavior, and efficacy. However, the absence of standardized protocols and international consensus creates uncertainty for producers and users. Harmonization of guidelines by regulatory agencies—such as EFSA, EPA, or FDA—is urgently needed to accelerate responsible commercialization [63]. Risk management strategies include thorough pre-market evaluation, lifecycle assessments, and implementation of safe-by-design principles. Encouraging transparency in nanoparticle formulation and usage practices will help bridge the gap between innovation and regulation.

6.4 Public Perception and Acceptance

The successful adoption of aromatic plant-based nanoparticles in antifungal applications also depends on public perception and consumer trust. While "green" or "plant-based" labels may suggest safety and sustainability, the term "nanoparticle" often raises concerns related to toxicity, environmental impact, and long-term health effects. Consumer education and clear communication of benefits, safety data, and regulatory compliance are essential for public acceptance. This is particularly relevant in food-related applications, such as edible coatings and postharvest treatments, where transparency is critical. Engaging stakeholders—including farmers, healthcare providers, and consumers—in participatory decision-making can improve acceptance and address misconceptions. Furthermore, public trust can be strengthened through third-party certifications, adherence to eco-labeling standards, and open-access safety testing data. Balancing innovation with ethical responsibility and inclusive dialogue will be key to establishing a positive perception of plant-based nanotechnologies.

7 Future Trends and Research Perspectives

7.1 Multifunctional Hybrid Nanoparticles

The next generation of antifungal nanomaterials is increasingly moving toward multifunctional hybrid nanoparticles. These systems integrate two or more functional components—such as metals, metal oxides, biopolymers, or bioactive plant compounds—within a single nanostructure to achieve synergistic effects. In the context of aromatic plant-based nanoparticles, hybridization may involve combining silver or zinc oxide cores with essential oil-functionalized surfaces and biodegradable polymeric shells. Such multifunctional systems can simultaneously offer antimicrobial activity, antioxidant protection, and targeted release properties. Additionally, hybrid nanoparticles can be engineered to respond to multiple biological or environmental triggers, expanding their versatility for agricultural, biomedical, and packaging applications. These advanced materials not only enhance antifungal efficacy but also reduce required dosages and extend functional longevity under field or clinical conditions.

7.2 Smart Delivery Systems and Stimuli-Responsive Release

A major trend in nanoparticle research is the development of smart delivery systems capable of responding to specific stimuli—such as pH, temperature, moisture, enzymatic activity, or fungal metabolites. Aromatic plant-derived nanoparticles can be incorporated into stimuli-responsive matrices or surface-modified to react under targeted conditions, ensuring precise release of antifungal agents where and when needed [64]. For example, in postharvest disease control, a nanoparticle embedded in a biodegradable coating could remain inert under dry storage conditions but activate upon detecting moisture or fungal enzymes on the fruit surface. In topical medical formulations, pH-sensitive nanoparticles may selectively release their cargo in infected or inflamed tissues. Such approaches minimize off-target effects, improve bioavailability, and extend the duration of antifungal protection. By leveraging plant-derived compounds with natural affinity for biological membranes, smart delivery platforms can be tailored for high specificity, reduced resistance development, and enhanced therapeutic outcomes.

7.3 Challenges in Scale-Up and Commercialization

Despite promising laboratory and pilot-scale results, the scale-up and commercialization of aromatic plant-based nanoparticles face several challenges. One of the most pressing issues is maintaining consistency in nanoparticle quality during large-scale production. Variability in plant extract composition—due to geographic, seasonal, or processing differences—can lead to batch-to-batch inconsistencies, affecting safety, efficacy, and regulatory compliance.

Moreover, industrial-scale green synthesis must meet cost efficiency, scalability, and environmental sustainability criteria without compromising product stability. Formulation challenges include ensuring long shelf-life, user-friendly delivery formats, and compatibility with existing agricultural or pharmaceutical infrastructure. Regulatory hurdles, intellectual property barriers, and limited public awareness further slow market adoption. Addressing these obstacles requires robust interdisciplinary collaboration between researchers, engineers, industrial partners, and policy-makers. Investment in process standardization, automation, and validation studies is critical to transition from proof-of-concept to real-world application.

8 Conclusion

8.1 Current Gaps and Research Needs

Despite considerable progress, several critical gaps remain that must be addressed to enable broader implementation. The complexity and variability of plant extracts pose challenges in achieving reproducible nanoparticle characteristics, especially for regulatory approval and commercial scale-up. Additionally, the long-term environmental fate, potential accumulation in ecosystems, and interactions with non-target organisms require more comprehensive and standardized eco-toxicological assessments.

On the clinical side, robust pharmacokinetic and toxicological data, as well as validation in advanced in vivo models, are essential to ensure safe translation into pharmaceutical use. Regulatory uncertainty and lack of harmonized global guidelines further hinder product development pipelines.

8.2 Roadmap for Next-Generation MOF–EO Platforms

The future of plant-based antifungal nanotechnology lies in the integration of green nanoscience with digital technologies, including AI-based design, precision delivery, and smart-responsive systems. Multifunctional hybrid nanoparticles and stimuli-triggered release platforms will enable tailored, efficient, and targeted antifungal therapy with minimal environmental footprint.

To bridge laboratory innovation and real-world application, interdisciplinary collaboration must be strengthened across materials science, microbiology, agronomy, pharmacology, and regulatory science. Public engagement, transparent risk communication, and education will also be vital in shaping acceptance and guiding responsible deployment.

In conclusion, aromatic plant-based nanoparticles offer a transformative solution at the intersection of biotechnology, nanoscience, and sustainable development. With focused research and coordinated innovation, they hold immense potential to redefine antifungal strategies for agriculture, medicine, and beyond.

Competing Interests The authors declare that they have no competing interest.

Declaration of Generative AI and AI-Assisted Technologies in the Writing Process During the preparation of this work, the book chapter author(s) have not taken any assistance from AI, and the book chapter author(s) reviewed and edited the content as needed and take(s) full responsibility for the content of the publication.

References

1. Pouris J, Kolyva F, Bratakou S, Vogiatzi CA, Chaniotis D, Beloukas A (2024) The role of fungi in food production and processing. Appl Sci 14:5046
2. Karkowska-Kuleta J, Rapala-Kozik M, Kozik A (2009) Fungi pathogenic to humans: molecular bases of virulence of Candida albicans, Cryptococcus neoformans and aspergillus fumigatus. Acta Biochim Pol 56:211
3. Zhao Y, Ye L, Zhao F, Zhang L, Lu Z, Chu T, Wang S, Liu Z, Sun Y, Chen M (2023) Cryptococcus neoformans, a global threat to human health. Infect Dis Poverty 12:1–18
4. Vitiello A, Ferrara F, Boccellino M, Ponzo A, Cimmino C, Comberiati E, Zovi A, Clemente S, Sabbatucci M (2023) Antifungal drug resistance: an emergent health threat. Biomedicine 11:1063
5. Islam T, Danishuddin, Tamanna NT, Matin MN, Barai HR, Haque MA (2024) Resistance mechanisms of plant pathogenic fungi to fungicide, environmental impacts of fungicides, and sustainable solutions. Plants 13:2737
6. Okon OG, Antia UE (2023) Plants and soil microbiota health implications of agrochemicals: potential alternatives for the safe propagation of food crops. In: One health implications of agrochemicals and their sustainable alternatives. Springer, Cham, pp 441–460
7. Wu S, Song R, Liu T, Li C (2023) Antifungal therapy: novel drug delivery strategies driven by new targets. Adv Drug Deliv Rev 199:114967
8. Belay WY, Getachew M, Tegegne BA, Teffera ZH, Dagne A, Zeleke TK, Wondm SA, Abebe RB, Gedif AA, Fenta A (2025) Antimicrobial resistance with a focus on antibacterial, antifungal, antimalarial, and antiviral drugs resistance, its threat, global priority pathogens, prevention, and control strategies: a review. Ther Adv Infect Dis 12:20499361251340144
9. Khan S, Hossain MK (2022) Classification and properties of nanoparticles. In: Nanoparticle-based polymer composites. Elsevier, Amsterdam, pp 15–54
10. Weisany W, Soufiania SP, Razmi J, Eshaghadadi AH (2024a) Nano-encapsulation of fenugreek and coriander essential oils using copper oxide NPs: novel approach for augmenting their effectiveness against Colletotrichum nymphaeae. Ind Crop Prod 219:119051
11. Weisany W, Amini J, Samadi S, Hossaini S, Yousefi S, Struik PC (2019a) Nano silver-encapsulation of thymus daenensis and Anethum graveolens essential oils enhances antifungal potential against strawberry anthracnose. Ind Crop Prod 141:111808
12. Weisany W, Yousefi S, Soufiani SP, Pashang D, McClements DJ, Ghasemlou M (2024b) Mesoporous silica nanoparticles: a versatile platform for encapsulation and delivery of essential oils for food applications. Adv Colloid Interface Sci 325:103116
13. Yousefi S, Weisany W, Hosseini SE, Ghasemlou M (2024) Mechanisms of nanoencapsulation to boost the antimicrobial efficacy of essential oils: a review. Food Hydrocoll 150:109655
14. Weisany W, Samadi S, Amini J, Hossaini S, Yousefi S, Maggi F (2019b) Enhancement of the antifungal activity of thyme and dill essential oils against Colletotrichum nymphaeae by nano-encapsulation with copper NPs. Ind Crop Prod 132:213–225
15. Weisany W, Yousefi S, Tahir NA-r, Golestanehzadeh N, McClements DJ, Adhikari B, Ghasemlou M (2022b) Targeted delivery and controlled released of essential oils using nano-encapsulation: a review. Adv Colloid Interface Sci 303:102655
16. Singh A, Pal V, Aggarwal S, Pant M (2023) Aromatic oil from plants, and their role in nanoparticle synthesis, characterization and applications. In: Secondary metabolites based green synthesis of nanomaterials and their applications. Springer, Berlin, pp 173–190
17. Butnariu M (2021) Plants as source of essential oils and perfumery applications. In: Bioprospecting of plant biodiversity for industrial molecules, pp 261–292
18. Zhang J, Zhang M, Bhandari B, Wang M (2024) Basic sensory properties of essential oils from aromatic plants and their applications: a critical review. Crit Rev Food Sci Nutr 64:6990–7003
19. Hegazy MM, Mekky RH, Ibrahim AE, Abouelela ME, Kedra TA, Al-Harrasi A (2025) Essential oils: the science of extraction and its implications for composition and biological activity—a review. Food Anal Methods 18:1483–1513

20. Raj N, Parveen, Khatoon S, Manzoor N (2024) Antifungal efficacy of terpenes and mechanism of action against human pathogenic fungi. In: Advances in antifungal drug development: natural products with antifungal potential. Springer, Berlin, pp 315–341
21. Huang J, Zaynab M, Sharif Y, Khan J, Al-Yahyai R, Sadder M, Ali M, Alarab SR, Li S (2024) Tannins as antimicrobial agents: understanding toxic effects on pathogens. Toxicon 247:107812
22. Yap PSX, Yusoff K, Lim S-HE, Chong C-M, Lai K-S (2021) Membrane disruption properties of essential oils—a double-edged sword? Processes 9:595
23. Nasr G, Greige-Gerges H, Elaissari A, Khreich N (2021) Liposome permeability to essential oil components: a focus on cholesterol content. J Membr Biol 254:381–395
24. Chen X, Shang S, Yan F, Jiang H, Zhao G, Tian S, Chen R, Chen D, Dang Y (2023) Antioxidant activities of essential oils and their major components in scavenging free radicals, inhibiting lipid oxidation and reducing cellular oxidative stress. Molecules 28:4559
25. Spisni E, Valerii MC, Massimino ML (2023) Essential oil molecules can break the loop of oxidative stress in neurodegenerative diseases. Biology 12:1504
26. Tessa S, Mia MR, Dadan S, Dikdik K (2023) Phytochemical profiling of Piper crocatum and its antifungal mechanism action as Lanosterol 14 alpha demethylase CYP51 inhibitor: a review. F1000Res 11:1115
27. El-Tarabily KA, El-Saadony MT, Alagawany M, Arif M, Batiha GE, Khafaga AF, Elwan HA, Elnesr SS, Abd El-Hack ME (2021) Using essential oils to overcome bacterial biofilm formation and their antimicrobial resistance. Saudi J Biol Sci 28:5145–5156
28. Aldosary S, El-Rahman S, Al-Jameel S, Alromihi N (2021) Antioxidant and antimicrobial activities of Thymus vulgaris essential oil contained and synthesis thymus (vulgaris) silver nanoparticles. Braz J Biol 83:e244675
29. D'Auria FD, Tecca M, Strippoli V, Salvatore G, Battinelli L, Mazzanti G (2005) Antifungal activity of Lavandula angustifolia essential oil against Candida albicans yeast and mycelial form. Med Mycol 43:391–396
30. Sharifi-Rad J, Adetunji CO, Olaniyan OT, Ojo SK, Samuel MO, Temitayo BT, Roli OI, Nimota OO, Oluwabunmi BT, Adetunji JB (2023) Antimicrobial, antioxidant and other pharmacological activities of Ocimum species: potential to be used as food preservatives and functional ingredients. Food Rev Intl 39:1547–1577
31. Čmiková N, Galovičová L, Schwarzová M, Vukic MD, Vukovic NL, Kowalczewski PŁ, Bakay L, Kluz MI, Puchalski C, Kačániová M (2023) Chemical composition and biological activities of Eucalyptus globulus essential oil. Plants 12:1076
32. Aithal S, Aithal P (2021) Green and eco-friendly nanotechnology—concepts and industrial prospects. Int J Manage Technol Soc Sci 6:1–31
33. Mohammadzadeh V, Barani M, Amiri MS, Yazdi MET, Hassanisaadi M, Rahdar A, Varma RS (2022) Applications of plant-based nanoparticles in nanomedicine: a review. Sustain Chem Pharm 25:100606
34. Ahmad U, Khan MA, Zahid MU, Alarjani KM, Gawwad MRA, Hussain SJ, Bokhari SAI (2024) Effect of surface capping on the biological properties of calcium oxide nanoparticles (CaO-NPs). Chem Pap 78:4971–4990
35. Shahzadi S, Fatima S, Shafiq Z, Janjua MRSA (2025) A review on green synthesis of silver nanoparticles (SNPs) using plant extracts: a multifaceted approach in photocatalysis, environmental remediation, and biomedicine. RSC Adv 15:3858–3903
36. Jeevanandam J, Kiew SF, Boakye-Ansah S, Lau SY, Barhoum A, Danquah MK, Rodrigues J (2022) Green approaches for the synthesis of metal and metal oxide nanoparticles using microbial and plant extracts. Nanoscale 14:2534–2571
37. Dabodabo DJG, Solidor CJH (2025) Metal oxide nanoparticles for various applications. In: Carrageenan-mediated green synthesis of metal oxide nanoparticles for various applications. CRC, Boca Raton, FL, pp 71–88

38. Gaur R, Manikandan P, Manikandan D, Umapathy S, Padhy HM, Maaza M, Elayaperumal M (2021) Noble metal ion embedded nanocomposite glass materials for optical functionality of UV–visible surface plasmon resonance (SPR) surface-enhanced Raman scattering (SERS) X-ray and electron microscopic studies: an overview. Plasmonics 16:1461–1493
39. Li H, Naeem U, Li P-J, Wu L-J, Yue H-W, Alshoaibi A, Wang D, Deghfel B, Kheawhom S, Mohamad AA (2025) Structural analysis by X-ray diffraction technique of transition metal doped zinc oxide and its applications in energy storage systems: a critical review. Int J Energy Res 2025:2587682
40. Pasieczna-Patkowska S, Cichy M, Flieger J (2025) Application of Fourier transform infrared (FTIR) spectroscopy in characterization of green synthesized nanoparticles. Molecules 30:684
41. Weisany W, Samadi S, Tahir NA-r, Amini J, Hossaini S (2022a) Nano-encapsulated with mesoporous silica enhanced the antifungal activity of essential oil against Botrytis cinerea (Helotiales; Sclerotiniaceae) and Colletotrichum nymphaeae (Glomerellales; Glomerellaceae). Physiol Mol Plant Pathol 122:101902
42. Ebrahimi H, Siavoshi F, Jazayeri MH, Sarrafnejad A, Saniee P, Mobini M (2023) Physicochemical properties of intact fungal cell wall determine vesicles release and nanoparticles internalization. Heliyon 9:e13834
43. Jaiswal N, Kumar A (2024) Harnessing the antifungal potential of natural products. In: Advances in antifungal drug development: natural products with antifungal potential. Springer, Berlin, pp 233–250
44. Slavin YN, Bach H (2022) Mechanisms of antifungal properties of metal nanoparticles. Nanomaterials 12:4470
45. Hou T, Sana SS, Li H, Xing Y, Nanda A, Netala VR, Zhang Z (2022) Essential oils and its antibacterial, antifungal and anti-oxidant activity applications: a review. Food Biosci 47:101716
46. Touati A, Mairi A, Ibrahim NA, Idres T (2025) Essential oils for biofilm control: mechanisms, synergies, and translational challenges in the era of antimicrobial resistance. Antibiotics 14:503
47. Fayed B (2025) Nanoparticles in the battle against Candida auris biofilms: current advances and future prospects. Drug Deliv Transl Res 15:1496–1512
48. Khosropour E, Hakimi L, Weisany W (2023) Role of engineered nanomaterials in biotic stress managements. In: Engineered nanomaterials for sustainable agricultural production, soil improvement and stress management. Elsevier, Amsterdam, pp 257–272
49. Kapeleka JA, Mwema MF (2024) State of nano pesticides application in smallholder agriculture production systems: human and environmental exposure risk perspectives. Heliyon 10:e39225
50. Aziz ZAA, Setapar SHM (2022) Current status and future prospect of nanotechnology incorporated plant-based extracts in cosmeceuticals. In: Nanotechnology for the preparation of cosmetics using plant-based extracts. Elsevier, Amsterdam, pp 235–261
51. Li X, Liu J (2024) Soil amendments with nanoparticles control cucumber wilt caused by fusarium oxysporium. S Afr J Bot 168:384–393
52. Santra HK, Dutta R, Banerjee D (2024) Antifungal activity of bio-active cell-free culture extracts and volatile organic compounds (VOCs) synthesised by endophytic fungal isolates of garden nasturtium. Sci Rep 14:11228
53. Arabpoor B, Yousefi S, Weisany W, Ghasemlou M (2021) Multifunctional coating composed of Eryngium campestre L. essential oil encapsulated in nano-chitosan to prolong the shelf-life of fresh cherry fruits. Food Hydrocoll 111:106394
54. Daeialiakbar M, Yousefi S, Weisany W (2025) Enhanced properties of chitosan-PVA nanocomposite films with lemongrass oil microcapsules. Carbohydr Polym Technol Appl 9:100668
55. Soufiani SP, Yousefi S, Honarvar M, Weisany W, Asadi G (2025) Polyvinyl alcohol films incorporating cellulose nanocrystals, chitosan, and mesoporous silica nanoparticles for enhanced stability and controlled release of beet extracts. Carbohydr Polym Technol Appl 10:100740
56. Qu B, Xiao Z, Luo Y (2025) Sustainable nanotechnology for food preservation: synthesis, mechanisms, and applications of zinc oxide nanoparticles. J Agric Food Res 19:101743

57. Karimi K, Ahari AB, Weisany W, Pertot I, Vrhovsek U, Arzanlou M (2016) Funneliformis mosseae root colonization affects Anethum graveolens essential oil composition and its efficacy against Colletotrichum nymphaeae. Ind Crop Prod 90:126–134
58. Sahli C, Moya SE, Lomas JS, Gravier-Pelletier C, Briandet R, Hémadi M (2022) Recent advances in nanotechnology for eradicating bacterial biofilm. Theranostics 12:2383
59. Pinaeva LG, Noskov AS (2024) Biodegradable biopolymers: real impact to environment pollution. Sci Total Environ 947:174445
60. Remya R, Julius A, Suman T, Mohanavel V, Karthick A, Pazhanimuthu C, Samrot AV, Muhibbullah M (2022) Role of nanoparticles in biodegradation and their importance in environmental and biomedical applications. J Nanomater 2022:6090846
61. Atanda SA, Shaibu RO, Agunbiade FO (2025) Nanoparticles in agriculture: balancing food security and environmental sustainability. Discov Agric 3:26
62. Karnwal A, Jassim AY, Mohammed AA, Sharma V, Al-Tawaha ARMS, Sivanesan I (2024) Nanotechnology for healthcare: plant-derived nanoparticles in disease treatment and regenerative medicine. Pharmaceuticals 17:1711
63. Hristozov D, Badetti E, Bigini P, Brunelli A, Dekkers S, Diomede L, Doak SH, Fransman W, Gajewicz-Skretna A, Giubilato E, Gómez-Cuadrado L, Grafström R, Gutleb AC, Halappanavar S, Hischier R, Hunt N, Katsumiti A, Kermanizadeh A, Marcomini A, Moschini E, Oomen A, Pizzol L, Rumbo C, Schmid O, Shandilya N, Stone V, Stoycheva S, Stoeger T, Merino BS, Tran L, Tsiliki G, Vogel UB, Wohlleben W, Zabeo A (2024) Next generation risk assessment approaches for advanced nanomaterials: current status and future perspectives. NanoImpact 35:100523
64. Murugan S, Senthilvelan T, Govindasamy M, Thangavel K (2025) A comprehensive review on exploring the potential of phytochemicals and biogenic nanoparticles for the treatment of antimicrobial-resistant pathogenic bacteria. Curr Microbiol 82:1–18

Aromatic Plants-Based Nanoparticles for Anticancer Activities

Sourya Jyoti Dash, Rajinder Kaur, Navami Pai, Nimisha Patel, Ahmad Ali, and Hina Alim

Abstract Cancer remains a major global health challenge, with conventional therapies often limited by systemic toxicity, multidrug resistance, and poor patient outcomes. In recent years, the convergence of nanotechnology and plant-based medicine has opened new avenues for safer and more effective interventions. Aromatic plants, rich in essential oils, flavonoids, terpenoids, and alkaloids, have gained attention not only for their long history in traditional medicine but also as biofactories for the green synthesis of nanoparticles. These nanoparticles, typically silver, gold, or zinc oxide, are stabilized and functionalized by plant phytochemicals, endowing them with unique physicochemical and biological properties. Extensive in vitro studies across diverse cancer cell lines, including breast (MCF-7), lung (A549), colon (HT-29), liver (HepG2), and cervical (HeLa), demonstrate that aromatic plant-based nanoparticles induce apoptosis, generate reactive oxygen species, disrupt the cell cycle, inhibit angiogenesis, and suppress metastatic pathways. Compared to conventional therapies, they offer greater biocompatibility, selective targeting, reduced toxicity, and the ability to overcome drug resistance. Moreover, their eco-friendly synthesis is cost-effective and sustainable. Collectively, these findings highlight the immense potential of aromatic plant-derived nanoparticles as next-generation anticancer agents, bridging traditional herbal wisdom with cutting-edge nanomedicine for more precise and holistic cancer therapy.

S. J. Dash · A. Ali · H. Alim (✉)
Department of Life Sciences, University of Mumbai, Vidya Nagari, Mumbai, Maharashtra, India
e-mail: hina.alim@lifesciences.mu.ac.in

R. Kaur
Department of Plant Sciences, University of Idaho, Moscow, ID, USA

N. Pai
School of Biotechnology and Bioinformatics, D.Y. Patil Deemed to be University, Navi Mumbai, Maharashtra, India

N. Patel
Department of Life Sciences, J.C. Bose University of Science and Technology, YMCA, Faridabad, Haryana, India

A. Husen (ed.), *Sustainable Uses of Aromatic Plants in Nanotechnology*, Smart Nanomaterials Technology, https://doi.org/10.1007/978-981-95-7409-4_9

Keywords Aromatic plants · Nanoparticles · Anticancer activity · Eco-friendly synthesis · Sustainable uses

Abbreviations

Ag^+	Silver ion (oxidation state +1)
Ag^0	Zerovalent silver
$AgNO_3$	Silver nitrate
AgNPs	Silver nanoparticles
Au^0	Zerovalent gold
Au^{3+}	Gold ion (oxidation state +3)
AuNPs	Gold nanoparticles
Bax	Pro-apoptotic protein Bax
Bcl-2	B-cell lymphoma 2
Bcl-xL	B-cell lymphoma-extra large
Caov-4	Human ovarian carcinoma cell line
CDK1/CDKs	Cyclin-dependent kinase 1/cyclin-dependent kinases
COLO-205	Human colon adenocarcinoma cell line
COX-2	Cyclooxygenase-2
DAPI	4′,6-Diamidino-2-phenylindole
DISC	Death-inducing signaling complex
DNA	Deoxyribonucleic acid
EMT	Epithelial-mesenchymal transition
EPR	Enhanced permeability and retention
FTIR	Fourier-transform infrared spectroscopy
G0/G1, G1, S, G2/M	Cell-cycle phases
$HAuCl_4$	Gold (III) chloride
HCT116	Human colorectal carcinoma cell line
HeLa	Henrietta Lacks (human cervical cancer cell line)
HepG2	Human hepatocellular carcinoma cell line HepG2
HT-29 (also *HT 29*)	Human colon adenocarcinoma cell line HT-29
IC_{50}	Half-maximal inhibitory concentration
IL-6	Interleukin-6
JC-1	5,5′,6,6′-Tetrachloro-1,1′,3,3′-tetraethylbenzimidazolylcarbocyanine iodide (mitochondrial membrane potential dye)
KDa/kDa	*Not found* (no kDa molecular-weight abbreviations detected)
MAPK	Mitogen-activated protein kinase (MAPK)
MCF-7	Michigan Cancer Foundation-7 (human breast adenocarcinoma cell line)
MDR	Multidrug resistance
MMPs/MMP-2/MMP-9	Matrix metalloproteinases
NK	Natural killer cells
PANC-1	Human pancreatic cancer cell line

PLGA	Poly(lactic-*co*-glycolic acid)
RAW 264.7	Murine macrophage cell line
ROS	Reactive oxygen species
SCC-40	Squamous cell carcinoma cell line
SEM	Scanning electron microscopy
TEM	Transmission electron microscopy
TNF-α	Tumor necrosis factor-alpha
TRAIL	TNF-related apoptosis-inducing ligand
TUNEL	Terminal deoxynucleotidyl transferase dUTP nick-end labelling
UV-vis/UV-vis	Ultraviolet-visible spectrophotometry
VEGF	Vascular endothelial growth factor
XRD	X-ray diffraction
$Zn(CH_3COO)_2$	Zinc acetate
Zn^0	Zerovalent zinc
Zn^{2+}	Zinc ion (oxidation state +2)
ZnO NPs	Zinc oxide nanoparticles

1 Introduction

Cancer remains one of the major health problems of our times, ranking among the leading causes of death worldwide. Millions of people are diagnosed with differing sorts of disease every year [1]. The reason why cancer has been so difficult to defeat lies in its complex biology, which includes uncontrolled cell division, ability of cancer cells to avoid programmed cell death, metastasis of foreign organs, and drug resistance [2]. Although significant progress has been made with conventional treatments such as chemotherapy, radiotherapy, immunotherapy, and surgery, these therapies also have serious limitations. They are usually extremely cytotoxic for normal cells, disabling and leading to side effects, are expensive in most cases, and, most severely, lose effectiveness with time due to the formation of resistant cancer cells [3]. This long-standing war has forced researchers to seek safer, more specific, and affordable ones able to complement or ideally substitute conventional methods of therapy [4].

Nanotechnology has introduced viable prospects pertaining to oncology. The ability of engineering materials at the nanolevel has transformed our mindset of drug delivery, diagnosis, and treatment monitoring [5]. Due to their higher surface area, capability of tuning up the size, and possibilities of surface customization, nanoparticles (NPs) have the potential of delivering therapeutic entities directly into tumors, thereby reducing systemic toxicity and increasing efficacy of treatment. It has evoked a new discipline of nanomedicine, which has been enthusiastically explored as an active tool for cancer diagnosis and therapy [6]. Among different methods of synthesizing the nanoparticles, chemical, physical, and biological, biological or so-called greener synthesis route has attracted certain interest [7–12].

Compared with state-of-the-art chemical methods that often require unwanted reagents and environments, greener synthesis emerges as economical and environment friendly and often provides biocompatible nanoparticles for application in biomedicine [13, 14]. One of the defining features of green synthesis lies in its employment of living entities or biological extracts as reducing and stabilizing agents [7]. Microorganisms such as fungi, algae, and bacteria have been employed for the same, but plants have very attractive advantages. Plant extracts are rich sources of phytochemicals, which are natural compounds such as flavonoids, terpenoids, alkaloids, phenolic acids, and essential oils, that not only help in the formation and stabilization of the nanoparticle but also have inherent therapeutic effects [15]. Aromatic medicinal plants are of specific interest, however, due to their very high concentration of bioactive volatile oils and their long history of use as folk drugs. Their employment as both nanoparticle synthesizers and sources of potent anticancer agents makes them very useful for the construction of new therapeutic platforms. These herbs are *Ocimum sanctum* (holy basil), *Curcuma longa* (turmeric), *Zingiber officinale* (ginger), *Cymbopogon citratus* (lemongrass), and *Rosmarinus officinalis* (rosemary). These are aromatic herbs that have been cherished for centuries for their health benefits and have gone beyond their use for cooking and for fragrances, which are extensively utilized in Ayurveda, traditional Chinese medicine, and Unani medicine [16]. These essential oils and phytoconstituents derived from herbs are found to have antioxidant, anti-inflammatory, antimicrobial, and anticancer activities [17, 18]. Curcumin, eugenol, gingerol, citral, and rosmarinic acid are some compounds that have been extensively studied and shown to exert their effects by inducing programmed cell death, preventing the development of the new vascular system of tumors, and regulating key signaling pathways of cancer development [19, 20].

Such plants have a twofold advantage when employed for nanoparticle biosynthesis. First, they provide an environment-friendly and sustainable method of nanoparticle synthesis without the employment of harmful reagents. Then, they yield particles with imbedded bioactivity of the plant constituents themselves, which give better therapeutic efficacy compared to crude plant extracts [21]. The small size of nanoparticles increases cellular internalization, enhances bioavailability, and ensures more controlled delivery of drug entities [22]. Moreover, aromatic plant-sourced nanoparticles are amenable for functionalization for target delivery, which ensures delivery of drug entities will especially target cancer end tissues sparing normal cells [23]. In itself, the synthesis method is simple. A metal salt solution, for example, silver nitrate, chloroauric acid, or zinc acetate, and an extract from plants are mixed under specific temperature and pH conditions. Phytochemicals from the extract reduce the metal ions from their ion state down to their elemental nanoparticulate state and simultaneously cap and stabilize the particles so that they do not clump together [24]. Shaped and size particles varying from spherical to triangular or rodlike are achievable based on the botanical species and reaction parameters [7].

These biogenic nanoparticles have demonstrated disparate anticancer mechanisms. Some of them are able to generate reactive oxygen species (ROS), causing cancer cell-specific damage due to the induction of oxidative stress, and the rest interfere with cell cycle progression and arrest improper cell division. Others control molecular signaling pathways PI3K/Akt, MAPK/ERK, and NF-κB, which are crucial for the development of tumors and metastasis [25]. Remarkably, several investigations have illustrated that nanoparticles from natural sources possess preferential cancer cell cytotoxicity with regard to normal cells and are hence of significant interest for drug candidates [21]. Evidence from animal and laboratory studies further strengthens this view. For instance, silver nanoparticles and *Ocimum sanctum* extracts have shown notable cytotoxicity on colon and breast cancer cell lines. Gold particles and *Curcuma longa* extracts have demonstrated intensified anticancer activities on cervical and colorectal cancer models [26, 27]. In the same line of evidence, *Cymbopogon citratus*-sourced zinc oxide particles have demonstrated intense antitumor activities in liver and lung cancer models [10, 11, 28]. These findings imply that such particles not only maintain their parent plants' therapeutic capabilities but also have benefits from the greater efficacy of nanotechnology.

Despite such encouraging findings, there are several challenges ahead. For reproducibility, standardization of synthesis protocols is needed. Extensive toxicological analyses must be carried out before clinical translation [29]. In addition, regulatory guidelines for plant-originated nanomedicines are still under development. At present, the majority of reports are limited to in vitro or small animal in vivo models, and broad-spectrum preclinical and clinical studies will be required for safety and efficacy validation for humans. These challenges will have to be overcome with interdisciplinary collaboration involving researchers from fields of ethnobotany, pharmacology, nanoscience, and oncology. Merging of classical botanical intelligence and advanced nanotechnology offers a viable model for developing outstanding anticancer therapy. Aromatic medicinal herbs constitute an environmentally safe and natural foundation for cancer therapy with nanoparticles. Plant-sourced nanoparticles could potentially play a central role for the introduction of safer, more effective, and customized cancer therapies with further innovation, exploration, and interdisciplinarity in the near future.

2 Green Synthesis of Nanoparticles Using Aromatic Plants

2.1 Green Nanotechnology

The onset of nanotechnology has caused a dramatic paradigm shift in a number of scientific fields, particularly in those allied sciences pertaining to biomedicine. Its most revolutionary application is in generating nanoparticles (NPs), which are materials whose minimum size is in between 1 and 100 nm [22]. Because of their size at the nanoscale level, such materials hold specific physical, chemical, and biological characteristics not observed in their bulk state such as increased surface area, quantum effects, and increased reactivity [6]. Nanoparticles surfaced in recent decades for their capability in target drug delivery, diagnostics, imaging, and notably cancer therapy [5, 23].

Conventional modes for nanoparticle synthesis largely consist of physical and chemical routes such as laser ablation, ball milling, and chemical reduction [14]. Despite their efficiency, these routes involve toxic chemicals and high energy input and form hazardous by-products, thereby creating severe risks for environmental safety as well as for medical compatibility in a biome [10, 11, 28]. As a sustainable alternative, green routes for nanoparticle synthesis were formulated based on biological entities such as microorganisms, enzymes, and plant extracts for nanoparticle production [15, 21]. Of all the biological entities, plants, especially aromatic medicinal plants, are the brightest horizons for greener nanoparticle production [24]. Plant entities have several merits: Plants are economical, environment friendly, and easy to scale up and do not require cell culture maintenance [14]. Plants are a repository for a plethora of bioactive molecules serving as in-built reducing as well as capping agents for obtaining thermostable therapeutically active nanoparticles [7]. Synthesis with aromatic plant extracts is an example illustrating a holistic approach where ancient knowledge about ethnobotany is integrated with cutting-edge nanotechnology.

2.2 *Need for Aromatic Plants*

Aromatic plants are plants whose leaves or other parts possess volatile aromatic compounds in the form chiefly of essential oils that confer upon them distinctive odors and pharmacological effects [30]. These plants are widely used in folk medicine for antimicrobial, anti-inflammatory, antioxidant, and anticancer activity [18, 25]. Their medicinal uses are due to an immense variety of secondary metabolites including terpenoids, flavonoids, polyphenols, tannins, alkaloids, and glycosides [16].

2.3 *Green Synthesis of Nanoparticles Using Aromatic Medicinal Plants*

When medicinal aromatic plants are used in the green synthesis of nanoparticles, their phytochemicals play two major roles that are crucial for nanoparticle formation and stability. These roles are as follows:

(a) Reduction of Metal Ions

Phytochemicals in aromatic medicinal plants act as effective reducing agents. When introduced into a medium containing metal salts such as silver nitrate ($AgNO_3$), gold chloride ($HAuCl_4$), or zinc acetate [$Zn(CH_3COO)_2$], they donate electrons to the cationic metal ions, reducing them from their ionic forms (e.g., Ag^{+}, Au^{3+}, Zn^{2+}) to their zerovalent states (Ag^{0}, Au^{0}, Zn^{0}) [14, 29]. This reduction is primarily caused by flavonoids, polyphenols, terpenoids, and other antioxidant-like phytochemicals, which are abundantly present in aromatic medicinal plants [7, 21]. These bioactive molecules act as green, nontoxic reducing agents, without recourse to the toxic chemicals typical of conventional chemical preparations [15].

(b) Stabilization of Nanoparticles

After downsizing nanoparticles, it is imperative for them to be stabilized in such a manner not to clump and become dysfunctional. Natural capping around nanoparticles is provided by secondary metabolites in aromatic plant extracts such that a barrier is created that retains particle dispersion in position and clumping is averted [24]. Such stabilization is equally essential in defining morphology, size, and charge on the surface of nanoparticles [15]. These are molecules like tannins, saponins, alkaloids, and glycosides that come into contact with nanoparticles' surfaces for steric as well as charge electrostatic stabilization [7]. Such dual functionality, reducing and stabilizing, is a key advantage in using aromatic plants in making nanoparticles as it does not allow for the use of synthetic capping agents or surfactants while guaranteeing increased biocompatibility and drug synergy [7, 21].

2.4 Common Aromatic Medicinal Plants in Green Synthesis

Most aromatic medicinal plants are used in biosynthesis of nanoparticles as they are abundant in pharmacologically active phytochemicals and essential oils (Fig. 1). The following are just a few examples:

(a) *Ocimum sanctum* (holy basil): Rich in eugenol, ursolic acid, and flavonoids, which are effective reducing agents for stabilizing metal ions [18, 31].
(b) *Curcuma longa* (turmeric): Contains curcumin, a potent antioxidant and anti-inflammatory molecule responsible for its ability to dissolve metal salts and form stable nanoparticles [17, 19].
(c) *Zingiber officinale* (ginger): Gingerol and shogaol in ginger are strong reducing agents and possess anticancer activity [25].
(d) *Cymbopogon citratus* (lemongrass): Rich in citral and flavonoids, this plant has shown promising results in the synthesis of zinc oxide and silver nanoparticles [29].
(e) *Rosmarinus officinalis* (rosemary): Contains rosmarinic acid and carnosic acid as reducing and capping agents [32].
(f) *Mentha piperita* (peppermint): Its polyphenolic and its menthol contents contribute to nanoparticle stabilization [33].
(g) *Thymus vulgaris* (thyme): Thymol and carvacrol in thyme possess high antioxidant activity and are efficient in nanoparticle production [30].

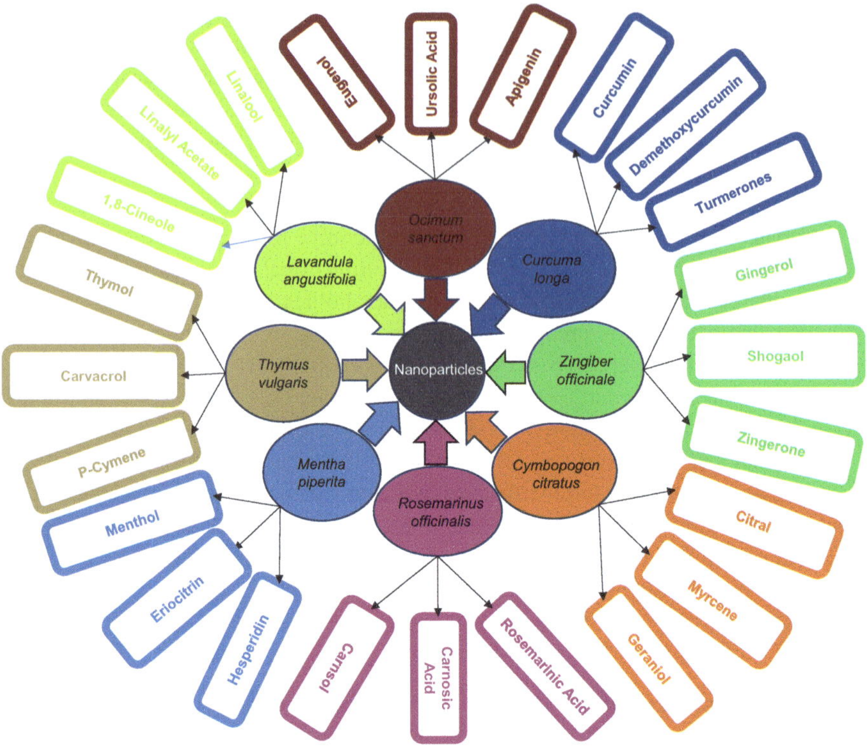

Fig. 1 Plants and their bioactive compounds in bio-nanoparticle synthesis

(h) *Lavandula angustifolia* (lavender): Lavender essential oils contain ingredients that are useful in stabilizing nanoparticles while providing therapeutic benefits [34].

These plants not only hold a place in ancient medicine but are studied in modern nanotechnology for green and sustainable production of nanoparticles for biomedical applications, primarily in cancer therapy.

3 General Methodology for Biosynthesis of Nanoparticles

The biosynthesis of nanoparticles with aromatic plant extracts is a systematic yet flexible process that combines classical phytochemistry with contemporary nanotechnology [14]. It typically commences with the making of plant extracts in fresh or dried tissues such as leaves, flowers, roots, or stems, which are washed, dried, pulverized, and subjected to solvent extraction with water, ethanol, methanol, or their combinations [15]. Harman extracts are commonly favored in biomedical applications for their safety and sustainability. The extract is filtered and stored at cold conditions until subsequent use. Simultaneously, a precursor solution for a metal is made using salts such as silver nitrate, gold chloride, zinc acetate, copper sulfate, or iron chloride with a selection based on a desired nanoparticle type [35].

When a plant extract is combined with a metal salt solution in a controlled setting such as stirring, pH, temperature, and extract-to-salt ratio, the phytochemicals become reducing and capping agents for initiating nanoparticle formation, which is frequently visualized with typical color transitions such as brown for Ag nanoparticles or ruby red for Au nanoparticles [36, 37]. Formed particles are separated by centrifuging away unreacted residues once or repeatedly washed with distilled water or ethanol and dried prior to preservation. Characterization then verifies the success of biosynthesis and analyzes the physicochemical attributes using UV-vis spectrophotometry, dynamic light scattering, scanning electron microscopy, transmission electron microscopy (SEM and TEM), Fourier-transform infrared spectroscopy (FTIR), and X-ray diffraction (XRD) [38, 39].

Phytochemical roles such as those played by flavonoids, alkaloids, phenolics, and terpenoids are paramount in reducing metal ionic forms and capping nanoparticles for their stability supplying their bioactivity [40]. Yet the synthesis result is extremely sensitive to parameters such as plant species, plant component, extraction solvent, developmental stage, and plant health, as well as process conditions such as pH in a reaction, temperature in a reaction, extract concentration in a reaction, and duration in a reaction [41]. Optimization of such factors facilitates reproducible preparations of nanoparticles of optimal size, morphology, stability, and functionality for applications in medicine.

4 Mechanisms of Anticancer Activity of Aromatic Plant-Based Nanoparticles

Aromatic medicinal plants are renowned for their several pharmacological actions, in particular their anticancer activity. When combined with nanotechnology, such plants not only serve as biological laboratories for nanoparticle construction but also as a reservoir for those bioactive molecules that enhance the therapeutic activity of thus-synthesized nanostructures. Nanostructures derived from aromatic plant extracts encompass a wide range of biological activity for cancer cells inasmuch as their synergetic effect is manifested in between their nanometric size and phytochemical capping [40]. Due to their ability to selectively kill cancer cells, modulate cell signaling pathways, and exhibit low toxicity toward normal tissues, aromatic medicinal plants bear a new promise in modern cancer medicine [42].

The mechanisms behind aromatic plant-sourced nanoparticles showing anticancer behavior are diverse and governed by several variables including nanoparticle type, phytochemicals used in their synthesis, and target cell line for cancer. Salient mechanistic attributes are cell death induction (programmed cell death or apoptosis), cell cycle disruption, reactive oxygen species (ROS) production, inhibition of

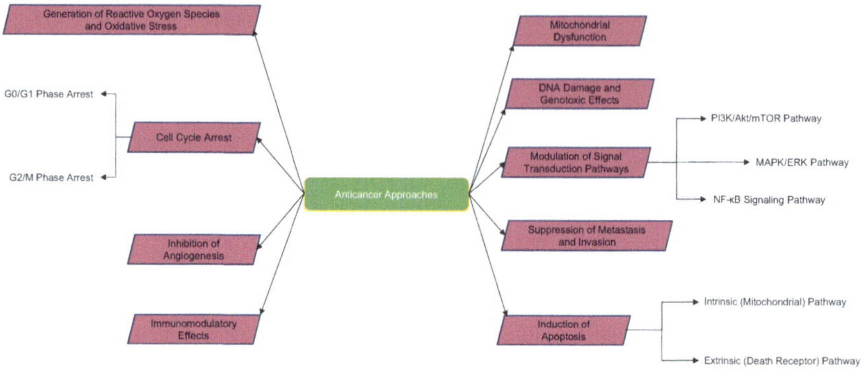

Fig. 2 Types of anticancer approaches

angiogenesis, disruption of mitochondrial function, and gene and protein expression modifications in tumor growth and metastasis [43] (Fig. 2).

4.1 Induction of Apoptosis

One of the best characterized anticancer mechanisms of aromatic plant nanoparticles is the induction of apoptosis, a synchronized and energy-demanding form of programmed cell death [44]. Unlike necrosis, which is a disease process, apoptosis is an orderly cell destruction program inherent in maintaining tissues in a state of homeostasis as well as in elimination of damaged or errant cells such as cancer cells. Nanoparticles prepared using aromatic plants tend to cause cell death by both intrinsic (mitochondrial) and extrinsic (death receptor-mediated) routes [21].

(a) Intrinsic (Mitochondrial) Pathway

The intrinsic pathway is provoked by internal cell stress such as oxidative stress, DNA damage, or mitochondrial disruption [45]. Aromatic plant nanoparticles can disturb mitochondrial membrane potential so as to secrete pro-apoptotic proteins such as cytochrome *c* into the cytosol. It ends up activating caspase-9 in an initiation cascade leading to effector caspase activation such as caspase-3 and caspase-7, with a net effect culminating in cell death. There are several plant-sourced molecules such as eugenol (from *Ocimum sanctum*), curcumin (from *Curcuma longa*), and citral (from *Cymbopogon citratus*) that are shown to modulate mitochondrial mechanisms by upregulation of pro-apoptotic proteins such as Bax while suppressing anti-apoptotic proteins such as Bcl-2 and Bcl-xL [46, 47].

(b) Extrinsic (Death Receptor) Pathway

Extrinsic pathway comprises cell-surface death receptor activation like Fas, TRAIL, or TNF-α receptors leading to recruitment in order to form a death-inducing signaling complex (DISC) as well as downstream activation of caspase-8 [48]. Bioactive phytochemical-capped nanoparticles are in a position to make cancer cells sensitive toward such signals so as to bring about apoptosis along this route. It is revealed in studies for nanoparticles made with *Zingiber officinale* or *Rosmarinus officinalis* for their capability to activate caspase-8 so as to assign a role for extrinsic path in their anticancer activity [49].

4.2 Generation of Reactive Oxygen Species and Oxidative Stress

Reactive oxygen species are reactive molecules generated by oxygen in a chemical reaction, including superoxide anions, hydrogen peroxide, and hydroxy radicals [50]. Even though physiological levels of ROS are a component of cell signaling and homeostasis, overproduction of ROS is destructive for cell macromolecules like DNA, proteins, and lipids leading to cell death. Nanoparticles made from aromatic plants are effective in producing ROS several times their tolerance level in cancer cells, generating an oxidative stress that leads to disruption in mitochondria, fragmentation in DNA, and cell death mechanisms [13]. Surface phytochemicals including flavonoids and phenolic acids can act as redox modulators in enhancing ROS production in the tumor microenvironment. In a typical experiment, silver or zinc oxide nanoparticles made with *Ocimum sanctum* or *Cymbopogon citratus* generate high levels of ROS in treated cancer cells observed using fluorescent ROS probes along with upregulated transcription in markers for oxidative stress [51]. Such pro-oxidant activity is a determining factor in their cytotoxicity in swiftly dividing cells that are predisposed severalfold toward oxidation damage. Surprisingly, phytochemicals' Janus-like activity as an antioxidant (in normal cells) as well as a prooxidant (under cancer cells) allows for selective toxicity favoring a predilection for cancerous tissues while sparing surrounding tissues [52].

4.3 Cell Cycle Arrest

Cancer cells demonstrate unrestrained proliferation due to cell cycle dysregulation. By breaking up key regulatory checkpoints, aromatic plant-based nanoparticles can halt cell cycle progression such that cell division is avoided as well as tumor expansion [53].

(a) G0/G1 Phase Arrest

Certain nanoparticles such as those made with *Curcuma longa* and *Mentha piperita* were found to inhibit cell cycle progression from G1 to S phase. It's brought about by inhibiting cyclins D and E and their respective cyclin-dependent kinases (CDKs) and activating CDK inhibitors such as p21 and p27

[54]. It inhibits DNA replication as well as cell proliferation at an early stage in cancer cells.

(b) G2/M Phase Arrest

These nanoparticles target the G2/M checkpoint for entering into mitosis. Zinc oxide nanoparticles formed using *Cymbopogon citratus* extract, for instance, induce G2/M arrest by suppressing cyclin B1 and CDK1 levels, leading to mitotic catastrophe and apoptosis [55]. Cell cycle disruption not only stops tumor progression but also enhances cancer cell vulnerability to chemotherapy agents and radiation therapy.

4.4 Inhibition of Angiogenesis

Angiogenesis, or creation of a new blood vessel network utilizing preexisting vasculature, is an indicator of cancer progression [56]. Tumors require a continuous oxygen and nutrient source in order to grow and metastasize, an activity encouraged by proteins such as vascular endothelial growth factor (VEGF) and its receptors. Plant aromatic nanoparticles are promising in suppressing cancer progression by inhibiting VEGF expression, blocking its receptor signaling, and inhibiting endothelial cell proliferation and migration. Plant molecules such as rosmarinic acid (from *Rosmarinus officinalis*) and curcumin are known anti-angiogenic agents [57]. Gold nanoparticles made with turmeric or basil extracts, for instance, were found to block VEGF expression in cancer cells, inhibiting neovascularization in model animals. Such anti-angiogenic activity can effectively starve tumors of essential nutrients, inhibiting their growth as well as their capacity for metastasis [58].

4.5 Mitochondrial Dysfunction

The mitochondria are key organelles in energy production, regulating redox, and cell death by apoptosis. Most anticancer approaches involve inhibition of mitochondrial pathways in order to affect cellular energy metabolism and initiate programmed cell death [59]. Plant-based nanoparticles can cause disruption in mitochondrial membrane potential ($\Delta\psi m$) such that pro-apoptotic proteins are released and the intrinsic apoptotic cascade gets initiated. Disruption is commonly monitored by staining with mitochondrial dyes such as JC-1 or Rhodamine-123 that indicate depolarization in treated cancer cells [60].

Silver nanoparticles stabilized with eugenol or gold nanoparticles functionalized with curcumin were found to perturb mitochondrial integrity in order to enhance apoptotic signals [61]. Further, mitochondrial DNA damage caused by oxidative stress adds its contribution toward cell loss in function and viability [62].

4.6 DNA Damage and Genotoxic Effects

Most anticancer agents exert their action by generating irreparable DNA lesions in cancer cells leading to cell cycle arrest and apoptosis [63]. Green-mediated nanoparticles are equally able to generate genotoxicity either by direct interaction with DNA or indirectly in generating ROS leading to genetic material damage [64]. Publications based on nanoparticles made with *Zingiber officinale* or *Ocimum sanctum* identified DNA strand breaks, chromatin condensation, and nuclear fragmentation in cancer cell lines [65]. These results are generally validated using means such as comet assay, TUNEL assay, and DAPI staining. It is further noted that certain nanoparticles slow down DNA repair mechanisms in such a way that genetic damage tends to accumulate more extensively while cancer cell survival is reduced [66].

4.7 Modulation of Signal Transduction Pathways

Cancer progression is marked by dysregulation of cell signaling pathways for proliferation, survival, invasion, and metastasis. Aromatic plant-derived nanoparticles are able to act on such pathways at multiple levels, thereby offering multi-targeted anticancer activity [3].

(a) PI3K/Akt/mTOR Pathway

It is a principal regulator of cell proliferation, metabolism, and survival. Hyperactivation of it is shared by most cancers [67]. Nanoparticles made with *Curcuma longa* and *Rosmarinus officinalis* were shown capable of inhibiting PI3K/Akt signaling with reduced cell proliferation and increased apoptosis [68].

(b) MAPK/ERK Pathway

The mitogen-activated protein kinase (MAPK) signaling regulates cell cycle and differentiation [69]. Plant nanoparticles can inhibit ERK phosphorylation and thereby stop downstream transcription of the growth-regulatory genes.

(c) NF-κB Signaling

NF-κB mediates inflammation, cell protection, and chemoresistance. Aromatic plant nanoparticles generally inhibit NF-κB activation by reducing anti-apoptotic and proliferative gene expression such as Bcl-2, COX-2, and IL-6 [70].

By modulating such signaling pathways, nanoparticles not only reduce tumor burden but also render cancer cells vulnerable to conventional therapies.

4.8 Suppression of Metastasis and Invasion

Metastasis is responsible for most cancer deaths and is composed of cancer cells migrating and invading away from their host organ into distant organs [71]. Metastasis is governed by matrix metalloproteinases (MMPs), epithelial-mesenchymal transition (EMT), and integrin signaling [72]. Aromatic plant nanoparticles can inhibit such metastatic procedures by downregulation of MMP-2 and MMP-9 involved in degrading extracellular matrix proteins; inhibition of EMT markers such as Snail, Slug, and Twist; and restoring E-cadherin expression, thereby enhancing cell-cell adherence. Silver nanoparticles synthesized with *Mentha piperita* and *Thymus vulgaris* showed anti-invasive behavior in vitro by reducing migratory potential of cancer cells in scratch wound migration assay as well as transwell migration test.

4.9 Immunomodulatory Effects

Besides direct killing of cancer cells, some nanoparticles based on plants are also capable of modulating the immune system in an attempt toward enhanced anticancer immunity. These actions include stimulation of NK cells, dendritic cells, and macrophages and modulation of cytokine production. Curcumin and eugenol are modifiers of immune responses, and their nanoparticle-attached forms are capable of maximizing bioavailability and stability in a manner whereby their immunomodulatory potential is enhanced [73]. As much as preclinical research in this aspect is in its nascent stages, it is nonetheless full of promise for integration of nanoparticle therapy with immunotherapy.

5 Selected Case Studies and In Vitro Findings

(a) *Taxus baccata* (Yew)—Silver Nanoparticles

Anisotropic silver nanoparticles (avg 75 nm) synthesized using aqueous *Taxus baccata* leaf extract had intense anticancer activity against MCF-7 breast cancer cells for an IC_{50} concentration of 0.25 μg/mL at 48 h, generating significant apoptosis and morphological transformations compared to ethanolic-extract NPs (5 μg/mL IC_{50}) [74].

(b) *Taxus baccata* (Yew)—Gold Nanoparticles

(c) Gold nanoparticles (<20 nm) made from Taxus bark extract reduced viability in cell lines MCF-7, HeLa, and Caov-4 by impacting caspase-independent mechanisms, revealing wide-spectrum anticancer activity [75].*Punica granatum* (Pomegranate)—Silver Nanoparticles

Silver NPs (6–45 nm) made with pomegranate peel extract inhibited A549 pulmonary cancer cell proliferation, induced cell cycle G1 arrest, and enhanced intracellular ROS without appreciable toxicity for normal cells [76].

(d) *Punica granatum* (Pomegranate)—Silver Nanoparticles

Plant extract-mediated AgNPs (35–60 nm) exhibited dose-dependent antiproliferative activity toward HepG2 liver cancer cells having $IC_{50} \approx 70$ µg/mL with DNA fragmentation confirming apoptosis [77].

(e) *Punica granatum* (Pomegranate)—Gold and Silver Nanoparticles

Pomegranate-derived Ag and Au NPs exhibited IC_{50} around 115 µg/mL for A549 and HCT116 cells, reducing viability by about 40–42%.

(f) *Elephantopus scaber*—Silver Nanoparticles

AgNPs (~59 nm) leaf mediated were assessed in MCF-7, A 549, and SCC 40 cells with $IC_{50} \approx 10$ µg/mL and exhibited notable anticancer activity [78].

(g) *Zingiber officinale* (Ginger)—Silver Nanoparticles

AgNPs (20–51 nm) made using ginger rhizome extract decreased HT 29 colon cancer proliferation by $IC_{50} \approx 150.8$ µg/mL [79].

(h) *Mentha arvensis* (Spearmint)—Silver Nanoparticles

Spherical AgNPs (<100 nm) from mint inhibited HCT116 colorectal carcinoma proliferation, showing $IC_{50} \approx 1.7$ µg/mL [80].

(i) *Moringa oleifera* (Drumstick Tree)—Silver Nanoparticles

AgNPs synthesized with leaf extract of *M. oleifera* induced apoptosis in HeLa cells, which holds great promise in treating cervical cancer [81].

(j) *Scutellaria barbata* (Barbat Skullcap)—Gold Nanoparticles

AuNPs generated using *S. barbata* leaf extract exhibited cell toxicity toward PANC-1 pancreatic cancer cells, although IC_{50} values were different, validating antineoplastic activity [82].

(k) *Juniperus polycarpos* (Persian Juniper)—Silver Nanoparticles

AgNPs (12–23 nm) from *J. polycarpos* needles induced apoptosis/necrosis in A549 cells through mitochondrial dysfunction and ROS elevation [83].

(l) *Catharanthus roseus* (Periwinkle)—Silver Nanoparticles

Leaf-extracted AgNPs (10–55 nm) reduced cell proliferation of A549 cells significantly with size- and exposure-dependent cytotoxicity [84].

(m) *Catharanthus roseus* (Periwinkle)—Zinc Oxide Nanoparticles

Quercetin-capped ZnO NPs from periwinkle leaves caused DNA binding and apoptosis in MCF-7 cells, for a potential nano-phytochemical synergism [46].

(n) *Morinda citrifolia* (Noni)—Zinc Oxide Nanoparticles

Green-synthesized zinc oxide nanoparticles using combined leaf extracts of *Catharanthus roseus* and *Morinda citrifolia* demonstrated strong stability, effective DNA-binding interactions, and significant anticancer activity against MCF-7 cells [85].

(o) *Pandanus odorifer* (Screw Pine)—Zinc Oxide Nanoparticles

P. odorifer-derived ZnO NPs (~50 nm) exhibited anticancer activity by reducing the viability of MCF-7, HepG2, and A549 cells to 60%, 62%, and 64%, respectively, at 100 µg/mL through apoptosis and necrosis [86].

(p) *Artemisia tournefortiana* (Mugwort)—Silver Nanoparticles
AgNPs (~40 to 50 nm) from mugwort extracts displayed moderate cytotoxicity against HT 29 colon cancer cells (IC_{50} ~40.7 µg/mL) [87].
(q) *Matricaria chamomilla* (Chamomile)—Silver Nanoparticles
Leaf-extract AgNPs (~42 nm) inhibited A549 cell growth with $IC_{50} \approx 42.4$ µg/mL, accompanied by apoptosis markers [88].
(r) *Cassia roxburghii* (Red Cassia)—Gold Nanoparticles
AuNPs (20–50 nm) prepared from *C. roxburghii* flower extracts provoked cell death in HepG2, HeLa, and MCF-7 lines, affecting DNA and enzyme activity [89].
(s) *Lonicera hypoglauca* (Honeysuckle)—Silver Nanoparticles
Variable size AgNPs made from flower extract induced apoptosis in MCF-7 cells but not in normal RAW 264.7 cells and thus exhibited selective cytotoxicity [90] (Table 1).

6 Advantages over Conventional Therapies

In the ongoing search for effective cancer treatments, conventional therapies such as chemotherapy, radiotherapy, and surgery continue to remain at the forefront of clinical practice. However, such therapies are often accompanied by severe side effects, nonspecific targetability, multidrug resistance, and toxicity to normal tissues [92]. The discovery of nanotechnology, namely, aromatic plant source-derived nanoparticles, introduced a new and possibly revolutionary means of anticancer therapy. Through the amalgamation of medicinal plants' therapeutic activity with nanoscale materials' precision and multifaceted functionalities, such systems promise a number of advancements over conventional therapies.

6.1 Enhanced Target Specificity and Reduced Side Effects

A major drawback for chemotherapy is its lack of specificity. Traditional chemotherapeutic agents are equally destructive to both normal and cancerous cells, which results in toxicity in critical organs like the liver, kidney, heart, and bone marrow. Plant-based nanoparticles can, however, be specifically engineered such that cancer cells are targeted more specifically either actively by functionalizing their surfaces with cancer cell receptor-specific ligands or passively by taking advantage of the enhanced permeability and retention (EPR) effect [93]. Examples are nanoparticles made out of *Taxus baccata* or *Punica granatum*, which were found to accumulate specifically in breast cancer cells and in lung cancer cells with minimal effect on normal cells [94]. Such a reduction in off-target toxicity translates into fewer side effects such as nausea, immunosuppression, and fatigue in such a manner that improves patients' quality of life while in therapy.

Table 1 Aromatic plants having anticancer activities

Aromatic plant species	Nanoparticle type	Cell lines used	Results	Key references
Taxus baccata	AgNPs	MCF-7	$IC_{50} \approx 0.25$ μg/mL; strong apoptosis	Abbasi Kajani et al. [74]
Taxus baccata	AuNPs	HeLa, MCF-7, Caov-4	Caspase-independent cytotoxicity	Abbasi Kajani et al. [75]
Punica granatum	AgNPs	A549	G1 arrest; ROS-mediated death	Annu et al. [76]
Punica granatum	AgNPs	HepG2	$IC_{50} \approx 70$ μg/mL; DNA fragmentation	
Elephantopus scaber	AgNPs	MCF-7, A549, SCC-40	$IC_{50} \approx 10$ μg/mL	Shinde and Mendhulkar [78]
Zingiber officinale	AgNPs	HT-29	$IC_{50} \approx 150.8$ μg/mL	Venkatadri et al. [79]
Mentha arvensis	AgNPs	HCT116	$IC_{50} \approx 1.7$ μg/mL	Javed and Mashwani [80]
Moringa oleifera	AgNPs	HeLa	Apoptosis induction	Vasanth et al. [81]
Scutellaria barbata	AuNPs	PANC-1	Moderate cytotoxicity	Wang et al. [82]
Juniperus polycarpos	AgNPs	A549	Mitochondrial dysfunction, ROS elevation	Noorbazargan et al. [83]
Catharanthus roseus	AgNPs	A549	Dose- and time-dependent cytotoxicity	Gumy et al. [84]
Catharanthus roseus	ZnO NPs	MCF-7	DNA binding, apoptosis	
Morinda citrifolia	ZnO NPs	MCF-7	ROS generation, apoptosis	Greeshma and Thamizselvi [85]
Pandanus odorifer	ZnO NPs	MCF-7, A549	Cell proliferation inhibition	Hussain et al. [86]
Artemisia tournefortiana	AgNPs	HT-29	$IC_{50} \approx 40.7$ μg/mL	Baghbani-Arani et al. [87]
Matricaria chamomilla	AgNPs	A549	$IC_{50} \approx 42.4$ μg/mL	Dadashpour et al. [88]
Cassia roxburghii	AuNPs	HepG2, HeLa, MCF-7	DNA damage, enzyme inhibition	Balashanmugam et al. [89]
Lonicera hypoglauca	AgNPs	MCF-7	Selective cytotoxicity	Xu et al. [90]
Ocimum sanctum	AgNPs	A549	Dose-dependent cytotoxicity	
Azadirachta indica	AuNPs	MCF-7	Apoptosis through mitochondrial pathway	
Cymbopogon citratus	ZnO NPs	A549	ROS-mediated apoptosis	
Curcuma longa	AuNPs	MDA-MB-231	Inhibited proliferation, induced apoptosis	Sheikh et al. [91]

(continued)

Table 1 (continued)

Aromatic plant species	Nanoparticle type	Cell lines used	Results	Key references
Rosmarinus officinalis	AgNPs	MCF-7, A549	Reduced viability, DNA damage	
Thymus vulgaris	AgNPs	A549	Mitochondrial damage, apoptosis	
Blumea eriantha	Ag/Fe NPs	MCF-7	DNA fragmentation, strong cytotoxicity	
Allium sativum	AgNPs	A549	Cytotoxicity (IC_{50} = 22 μg/mL)	
Camellia sinensis	AgNPs	MCF-7, HeLa	Dose-dependent cytotoxicity	Sulaiman et al. [49]
Terminalia arjuna	AgNPs	MCF-7	G2/M phase arrest, apoptosis	Mallikarjuna et al. (2014)
Astragalus fasciculifolius (Anzaroot)	AgNPs	MCF-7	Dose-dependent inhibition of cell proliferation, cytotoxicity	

6.2 Synergistic Effects of Phytochemicals and Nanocarriers

Aromatic plants are rich in bioactive phytochemicals, like flavonoids, alkaloids, terpenoids, tannins, and essential oils, that possess intrinsic anticancer activity. As used in nanoparticle synthesis, phytochemicals fulfill a dual mission: Nanoparticles are stabilized and reduced in green chemistry while offering additional therapeutic potential [13].

6.3 Overcoming Multidrug Resistance (MDR)

Another significant problem in cancer therapy is multidrug resistance (MDR), in which cancer cells become resistant to multiple chemotherapeutic compounds. Through mechanisms involving efflux pump overproduction (e.g., P-glycoprotein), drug target mutation, and enhanced DNA repair, tumor cells evade death and proliferation despite therapy [3]. Aromatic plant-derived nanoparticles are extremely promising in evading MDR. Due to their size (typically <100 nm), nanoparticles escape efflux pumps and accumulate within the cancer cells more efficiently [95]. Moreover, some phytochemical compounds like curcumin (*Curcuma longa*) and eugenol (*Ocimum sanctum*) are shown to modulate signaling processes involved in drug resistance in order to render cancer cells more sensitive to therapy.

Further, nanoparticles can also be co-loaded with standard drugs and phytomolecules such that combination treatments would act on multiple resistance mechanisms simultaneously [96]. Such multi-prong strategies enhance treatment outcomes and delay onset of resistance.

6.4 *Improved Drug Solubility and Bioavailability*

Most effective anticancer agents, such as plant-derived molecules like paclitaxel, are not very soluble in water, so their clinical uses are impaired. These translate into a need for toxic surfactants or solvents, which contribute to side effects [97]. Plant extract-synthesized nanoparticles form a stable biocompatible matrix for drug encapsulation for hydrophobic drugs, which improves their water-soluble characteristics along with their life of circulation [98]. Further, their nanoscale size leads to increased absorption and tumor entrapment based on the EPR effect. Gold nanoparticles based on *Scutellaria barbata* or *Cassia roxburghii* are examples where not only are there effective carriers for drug therapy but there is even increased penetration within solid tumors compared to bulk drug particles [99]. That improved pharmacokinetic profile guarantees there is an increased fraction of drug at its site of action.

6.5 *Eco-Friendly and Economical Synthesis*

Traditional nanoparticles are often synthesized with aggressive chemicals, excessive input energy, and sophisticated purification regimes, which are not desirable on environmental and economical fields. But green synthesis in aromatic plant extracts is an environment-friendly option, cost-effective, and scalable [14]. It includes the use of aqueous or ethanolic extracts of plants as reducing as well as stabilizing agents without use of any toxic reagents. This "green nanotechnology" is in line with global goals for sustainable development and is particularly valuable in low-resource settings where access is limited to advanced facilities [15].

6.6 *Versatility in Design and Functionalization*

Plant nanoparticles possess immense compositional flexibility in size, morphology, composition framework, and surface functionalization. Silver and gold nanoparticles along with zinc oxide, iron oxide, and selenium nanoparticles can be synthesized using variable extracts derived from aromatic plants (roots, bark, leaves, fruit pulp, flowers), while every constituent endows unique physicochemical and biological characteristics upon final end product [73]. Further, nanoparticles can also endow target ligands (antibodies, peptides), imaging probes (diagnostic purposes), or co-therapy payloads (double or multidrug entrapment) [22]. Such a multitude framework can permit simultaneous theranostic diagnosis along with therapy within an individual framework with promise for personalized patient-specific tumor profile-directed treatment solutions.

6.7 Induction of Multiple Cell Death Pathways

While standard chemotherapeutic agents primarily induce only apoptosis, most forms of cancer develop resistance to apoptosis in later stages. Plant nanoparticles are observed not just to induce apoptosis but also necrosis, autophagy, and immunogenic cell death, thereby evading mechanisms of resistance for complete eradication of tumors [100]. For example, AgNPs isolated from *Juniperus polycarpos* were shown to induce both necrosis and apoptosis by generating ROS and damaging mitochondria. Similarly, autophagy and DNA binding were promoted in ZnO NPs synthesized by using *Catharanthus roseus*, which indicated a potential in nanoparticles to activate several routes toward programmed cell death [65].

6.8 Biocompatibility and Safety

Biocompatibility is a key requirement for any drug. Even though efficient, synthetic nanoparticles are typically facing problems of gradual accumulation over a longer period of time, immunogenicity, and cell toxicity. Green-synthesized nanoparticles containing aromatic plants are endowed with a superior level of biocompatibility based on phytochemical capping on their surfaces that deters agglomeration as well as immune recognition [14]. Literature shows low hemolysis, low elevation in liver enzymes, and minimal histopathological change in animal models treated with nanoparticles based on *Rosmarinus officinalis* or *Matricaria chamomilla* [101]. Such a favorable safety profile enhances translatability into human uses and expedited approval by drug controlling agencies.

7 Challenges and Limitations of Aromatic Plant-Based Nanoparticles for Anticancer Activities

Although aromatic plant-mediated nanoparticles (NP)-based anticancer therapy is a field ripe with promise, it is imperative to note that it is fraught with a number of scientific, technical, regulatory, and translational issues. Even as green-synthesized NPs are beneficial in terms of eco-friendliness, biocompatibility, and drug synergy, there are several impediments that prevent their easy adoption in conventional cancer therapy. This section summarizes key issues and limitations with such an approach.

7.1 *Lack of Standardization in Plant Extracts and Synthesis Procedures*

Another key concern in the region is standardization in plant extract preparation for nanoparticle production. Aromatic and medicinal plants contain a large number of bioactive compounds such as terpenoids, flavonoids, alkaloids, and phenolic acids whose concentration varies significantly depending on factors such as geographic location, date of harvesting, extraction method, plant part used (leaf, root, flower, etc.), and seasonal variation [96]. The extract exhibits variability in its reducing and capping capability such that it results in inconsistent nanoparticle attributes (size, morphology, stability). Consequently, it is hard to reproduce results in different laboratories or even scale up for industrial applications [14].

7.2 *Unpredictable Nanoparticle Characteristics and Stability Issues*

Physical and chemical properties of nanoparticles such as size, morphology, charge at the surface, and crystallinity greatly affect their biological behavior. However, a fine-tuned control over such parameters is not feasible in plant-mediated synthesis since phytoconstituents of plants are highly intricate. Unlike chemical synthesis where a concentration of a reagent or a condition of a reaction can be set precisely, green synthesis is very often trial based. Moreover, long-term stability is another issue where most green-synthesized nanoparticles are susceptible to agglomeration, oxidation, or degradation in the longer term, particularly in physiological or storage environments [15]. Such instability impacts drug efficacy as well as safety, which is a severe limitation in clinical translation.

7.3 *Limited Understanding of Mechanisms of Action*

Although a number of in vitro reports indicate promising anticancer activity of aromatic plant-derived nanoparticles, their mechanisms are not understood. Whereas some nanoparticles induce reactive oxygen species (ROS)-dependent apoptosis, some modify mitochondrial function, cell cycle regulation, or autophagy [100].

7.4 Poor In Vivo Data and Lack of Clinical Trials

The majority of aromatic plant-based nanoparticles are explored only in vitro in cancer cell lines. Although encouraging outcomes are noted, in vitro results are not consistently converted into in vivo effectiveness due to complex physiological barriers in the form of immune clearance, enzyme-mediated destruction, and nonspecific disposition [102].

7.5 Potential for Unintended Toxicity

Although green synthesis is generally advocated as biocompatible and safe, toxicological profiling for plant nanoparticles is still in its infancy. Some compositions may end up in nontarget tissues, cause oxidation stress, or provoke immune responses. Long-term use, genotoxicity, and effects on reproductive or endocrine problems are still not extensively researched [103]. As such, full toxicological assessments involving acute, subacute, and chronic toxicity assessments are necessary prior to approval for use in humans.

7.6 Challenges in Drug Loading and Controlled Release

Another technical disadvantage is drug release control and inefficiency in drug loading. Even though plant nanoparticles are able to demonstrate inherent anticancer activity, co-loading with other drugs is in most situations needed for improved efficacy. Achieving maximum drug encapsulation, even distribution, and slow release is challenging due to variability in nanoparticle surface chemistry, inconsistent interactions between plant metabolites, drug therapies, and weak binding forces that lead to early leakage of drug. Controlled release preparations (enzyme sensitive, pH sensitive, or thermosensitive) are in development but yet require fine-tuning in their integration into green-synthesized NPs [96].

8 Future Perspectives

Aromatic plant-derived nanoparticles are a burgeoning frontier in cancer therapy in which ancient wisdom in medicinal plants is tied in with state-of-the-art nanotechnology innovation. Success is based upon converting lab outcomes into safe, standard, and clinically translatable applications. Standardization in extraction processes and reproducibility in nanoparticle preparations remain key in filling loops between experimental research and clinical translation. Integrations in artificial intelligence and machine learning remain full of promise in making nanoparticle design simpler, optimizing drug delivery routes, and predicting both therapy and toxicity outcomes. Screening for neglected medicinal plants based upon ethnobotanical leads and high-speed screening may identify new phytochemicals endowed with unique anticancer activity, while next-generation smart stimuli-responsive systems remain optimistically capable of optimizing site-specific delivery. Formulations in theranostic nanoparticles remain a dual platform for simultaneous diagnosis and therapy, which is particularly crucial in tracking aggressive and metastatic cancer.

Targeting cancer stem cells and metastatic corridors with plant-sourced nanomedicines may prevent relapse and resistance. However, systematic toxicological, pharmacokinetic, and immunogenicity profiling remains unavoidable in ensuring safety. Formulation in strong regulatory regimes, global quality control standards, and harmonized approval tracks remain paramount in clinical translation. Finally, collaboration in public-private partnership modes, industrial upscaling, and sustainable production modalities remain paramount in rapid integration in cancer treatment involving their full range as next-generation cancer therapeutics.

9 Conclusion

Cancer is still a global healthcare issue with millions affected annually and not many curative treatments for the majority of tumors. In recent years, there have been promising advances on the horizon with the fusion of nanotechnology and phytomedicine, and one of the most promising new advances has been that of employing aromatic plant-based nanoparticles as an anticancer treatment. This comprehensive review has charted the many-faced potential of these nanoformulations from their environment-friendly synthesis with pharmaceutical plant extracts to their mechanism of action, cytotoxicity profile, and clinical advantages. In light of selected results from broad swathes of in vitro and preclinical data, at this critical juncture now, with opportunity and challenge often sitting side by side, this last part integrates essential findings and supplies the rationale for converting this new platform into usable, clinical entities. At its very heart, there lies truly remarkable synergy: the centuries-long therapeutic wisdom of aromatic plants and the precision and flexibility of modern nanoscience. By utilizing bioactive molecules from extracts of plants as reducing and stabilizing agents, researchers are able to assemble nanoparticles not just sustainably sourced and biodegradable but also inherently therapeutic. Such nanoformulations generally are capable of inducing apoptosis, autophagy, or oxidative stress on cancer cells and, for the most part, possess low toxicity for normal cells. This type of double functionality with standard chemotherapeutics is not very frequently achievable and signifies the particular therapeutic promise of nanoparticulate derivatives of plants.

Competing Interests The authors declare that they have no competing interest.

Declaration of Generative AI and AI-Assisted Technologies in the Writing Process During the preparation of this work, the book chapter author(s) have not taken any assistance from AI, and the book chapter author(s) reviewed and edited the content as needed and take(s) full responsibility for the content of the publication.

References

1. Bray F, Laversanne M, Weiderpass E, Soerjomataram I (2020) The ever-increasing importance of cancer as a leading cause of premature death worldwide. Cancer 126(12):2919–2926
2. Hanahan D, Weinberg RA (2011) Hallmarks of cancer: the next generation. Cell 144(5):646–674
3. Holohan C, Van Schaeybroeck S, Longley DB, Johnston PG (2013) Cancer drug resistance: an evolving paradigm. Nat Rev Cancer 13(10):714–726
4. Paramanya A, Sharma S, Bagdat RB, Ali A (2020) Recent practices of medicinal and aromatic plants in nanotechnology. In: Nanomaterials for agriculture and forestry applications. Elsevier, Amsterdam, pp 435–467
5. Peer D, Karp JM, Hong S, Farokhzad OC, Margalit R, Langer R (2007) Nanocarriers as an emerging platform for cancer therapy. Nat Nanotechnol 2(12):751–760
6. Wagner V, Dullaart A, Bock AK, Zweck A (2006) The emerging nanomedicine landscape. Nat Biotechnol 24(10):1211–1217
7. Husen A (2019) Natural product-based fabrication of zinc oxide nanoparticles and their application. In: Husen A, Iqbal M (eds) Nanomaterials and plant potential. Springer International Publishing AG, Cham, pp 193–291. https://doi.org/10.1007/978-3-030-05569-1_7
8. Husen A (2020) Introduction and techniques in nanomaterials formulation. In: Husen A, Jawaid M (eds) Nanomaterials for agriculture and forestry applications. Elsevier, Cambridge, pp 1–14. https://doi.org/10.1016/B978-0-12-817852-2.00001-9
9. Husen A (2023) Secondary metabolites based green synthesis of nanomaterials and their applications. Springer Nature Singapore Pte Ltd, Singapore. https://doi.org/10.1007/978-981-99-0927-8
10. Siddiqi KS, Husen A (2016a) Fabrication of metal nanoparticles from fungi and metal salts: scope and application. Nanoscale Res Lett 11(98):1–15
11. Siddiqi KS, Husen A (2016b) Fabrication of metal and metal oxide nanoparticles by algae and their toxic effects. Nanoscale Res Lett 11(363):1–11
12. Husen A, Siddiqi KS (2014) Phytosynthesis of nanoparticles: concept, controversy and application. Nanoscale Res Lett 9(229):1–24
13. Husen A, Siddiqi KS (2023) Advances in smart nanomaterials and their applications. Elsevier, Cambridge. https://doi.org/10.1016/C2021-0-02202-1
14. Iravani S (2011) Green synthesis of metal nanoparticles using plants. Green Chem 13(10):2638–2650
15. Ahmed S, Ahmad M, Swami BL, Ikram S (2016) A review on plants extract mediated synthesis of silver nanoparticles for antimicrobial applications: a green expertise. J Adv Res 7(1):17–28
16. Balekundri A, Mannur V (2020) Quality control of the traditional herbs and herbal products: a review. Future J Pharm Sci 6:100037
17. Aggarwal BB, Sundaram C, Malani N, Ichikawa H (2007) Curcumin: the Indian solid gold. Adv Exp Med Biol 595:1–75

18. Prakash P, Gupta N (2005) Therapeutic uses of *Ocimum sanctum* Linn (Tulsi) with a note on eugenol and its pharmacological actions: a short review. Indian J Physiol Pharmacol 49(2):125–131
19. Prasad S, Tyagi AK, Aggarwal BB (2014) Recent developments in delivery, bioavailability, absorption and metabolism of curcumin: the golden pigment from golden spice. Cancer Res Treat 46(1):2–18
20. Quintero-Rincón P, Caballero-Gallardo K, Olivero-Verbel J (2025) Natural anticancer agents: prospection of medicinal and aromatic plants in modern chemoprevention and chemotherapy. Nat Prod Bioprospect 15:25
21. Singh P, Kim YJ, Zhang D, Yang DC (2018) Biological synthesis of nanoparticles from plants and their biomedical applications. Trends Biotechnol 34(7):588–599
22. Jeevanandam J, Barhoum A, Chan YS, Dufresne A, Danquah MK (2018) Review on nanoparticles and nanostructured materials: history, sources, toxicity and regulations. Beilstein J Nanotechnol 9:1050–1074
23. Zhang L, Gu FX, Chan JM, Wang AZ, Langer RS, Farokhzad OC (2012) Nanoparticles in medicine: therapeutic applications and developments. Clin Pharmacol Ther 83(5):761–769
24. Iravani S, Korbekandi H, Mirmohammadi SV, Zolfaghari B (2014) Synthesis of silver nanoparticles: chemical, physical and biological methods. Res Pharm Sci 9(6):385–406
25. Shanmugam MK, Kannaiyan R, Sethi G (2013) Targeting cell signaling and apoptotic pathways by dietary agents: role in the prevention and treatment of cancer. Nutr Cancer 63(2):161–173
26. Ghasemi F, Shafiee M, Banikazemi Z, Pourhanifeh MH, Khanbabaei H, Shamshirian A, Amiri Moghadam S, ArefNezhad R, Sahebkar A, Avan A, Mirzaei H (2019) Curcumin inhibits NF-κB and Wnt/β-catenin pathways in cervical cancer cells. Pathol Res Pract 215(10):152556
27. Shahbaz SK, Koushki K, Izadi O, Penson PE, Sukhorukov VN, Kesharwani P (2024) Advancements in curcumin-loaded PLGA nanoparticle delivery systems: progressive strategies in cancer therapy. J Drug Target 32(10):1207–1232
28. Siddiqi KS, Husen A (2016c) Fabrication of metal nanoparticles from fungi and metal salts: scope and applications. Nanoscale Res Lett 11(1):98
29. Chaudhuri SK, Malodia L (2017) Biosynthesis of zinc oxide nanoparticles using leaf extract of Aloe vera and their environmental applications. J Nanosci 2017:1–7
30. Sharifi-Rad J, Saheli B, Schnitzler P, Ayatollahi SA, Kobarfard F, Fathi M, Eisazadeh M, Sharifi-Rad M (2017) Susceptibility of herpes simplex virus type 1 to monoterpenes thymol, carvacrol, p-cymene and essential oils of *Sinapis* arvensis L, *Lallemantia royleana Benth* and *Pulicaria vulgaris* Gaertn. Cell Mol Biol (Noisy-le-grand) 63(8):42–47
31. Pradhan Y, Farah MA, Al-Anazi KM, Ali A (2023) Traditional knowledge, bioactive compounds, and antidiabetic and other potential response of holy basil (*Ocimum sanctum*). In: Antidiabetic medicinal plants and herbal treatments. CRC Press, Boca Raton, pp 295–307
32. Andrade JM, Faustino C, Gracia C, Laderias D, Reis CP, Rijo P (2018) Rosmarinus officinalis L.: an updated review of its phytochemistry and biological activity. Future Sci OA 4(4):FSO283
33. Gholamipourfard K, Salehi M, Banchio E (2021) *Mentha piperita* phytochemicals in agriculture, food industry and medicine: features and applications. S Afr J Bot 141:183–195
34. Belova MM, Shipunova VO, Kotelnikova PA, Babenyshev AV, Rogozhin EA, Cherednichenko MY, Deyev SM (2019) "Green" synthesis of cytotoxic silver nanoparticles based on secondary metabolites of *Lavandula angustifolia* Mill. Acta Nat 11(2):47–53
35. Patel A, Patel N, Ali A, Alim H (2023) Nanomaterials synthesis using saponins and their applications. In: Secondary metabolites based green synthesis of nanomaterials and their applications. Springer, Singapore, pp 141–157
36. Jha AK, Prasad K (2010) Green synthesis of silver nanoparticles using cycas leaf. Int J Green Nanotechnol Phys Chem 1(2):P110–P117

37. Thakkar KN, Mhatre SS, Parikh RY (2010) Biological synthesis of metallic nanoparticles. Nanomedicine 6(2):257–262
38. Chung IM, Park I, Seung-Hyun K, Thiruvengadam M, Rajakumar G (2016) Plant-mediated synthesis of silver nanoparticles: their characteristic properties and therapeutic applications. Nanoscale Res Lett 11(1):40
39. Shah M, Fawcett D, Sharma S, Tripathy SK, Poinern GEJ (2015) Green synthesis of metallic nanoparticles via biological entities. Materials (Basel) 8(11):7278–7308
40. Thatyana M, Dube NP, Kemboi D, Manicum ALE, Mokgalaka-Fleischmann NS, Tembu JV (2023) Advances in phytonanotechnology: a plant-mediated green synthesis of metal nanoparticles using *Phyllanthus* plant extracts and their antimicrobial and anticancer applications. Nanomaterials 13(19):2616
41. Eker F, Akdaşçi E, Duman H, Bechelany M, Karav S (2025) Green synthesis of silver nanoparticles using plant extracts: a comprehensive review of physicochemical properties and multifunctional applications. Int J Mol Sci 26(13):6222
42. Choudhari AS, Mandave PC, Deshpande M, Ranjekar P, Prakash O (2020) Phytochemicals in cancer treatment: from preclinical studies to clinical practice. Front Pharmacol 10:1614
43. Rao PV, Nallappan D, Madhavi K, Rahman S, Jun WL, Gan SH (2016) Phytochemicals and biogenic metallic nanoparticles as anticancer agents. Oxidative Med Cell Longev 2016:3685671
44. Elmore S (2007) Apoptosis: a review of programmed cell death. Toxicol Pathol 35(4):495–516
45. Kroemer G, Galluzzi L, Brenner C (2007) Mitochondrial membrane permeabilization in cell death. Physiol Rev 87(1):99–163
46. Gupta SC, Patchva S, Aggarwal BB (2021) Therapeutic roles of curcumin: lessons learned from clinical trials. AAPS J 23(2):22
47. Pavan AR, Silva GD, Jornada DH, Chiba DE, Fernandes GF, Man Chin C, Dos Santos JL (2016) Unraveling the anticancer effect of curcumin and resveratrol. Nutrients 8(11):628
48. Ashkenazi A, Dixit VM (1998) Death receptors: signaling and modulation. Science 281(5381):1305–1308
49. Sulaiman GM, Tawfeeq AT, Naji AS (2021) Biosynthesis of silver nanoparticles from ginger extract and evaluation of their anticancer potential. Mater Sci Eng C Mater Biol Appl 119:111625
50. Zahra KF, Lefter R, Ali A, Abdellah EC, Trus C, Ciobica A, Timofte D (2021) The involvement of the oxidative stress status in cancer pathology: a double view on the role of the antioxidants. Oxidative Med Cell Longev 2021:9965916
51. Shah STA, Awan SJ, Javed K (2019) ROS generation and anticancer activity of green-synthesized nanoparticles. Appl Nanosci 9(5):1247–1260
52. Scalbert A, Johnson IT, Saltmarsh M (2005) Polyphenols: antioxidants and beyond. Am J Clin Nutr 81(1):215S–217S
53. Vermeulen K, Van Bockstaele DR, Berneman ZN (2003) The cell cycle: a review of regulation, deregulation and therapeutic targets in cancer. Cell Prolif 36(3):131–149
54. Ravindran J, Prasad S, Aggarwal BB (2009) Curcumin and cancer cells: how many ways can curry kill tumor cells selectively? AAPS J 11(3):495–510
55. Patel DK, Singh A, Dutta S (2019) Green synthesis of ZnO nanoparticles using medicinal plants and evaluation of their anticancer potential. J Photochem Photobiol B 190:129–135
56. Folkman J (2007) Angiogenesis: an organizing principle for drug discovery? Nat Rev Drug Discov 6(4):273–286
57. Kumar S, Deepak, Prasad R (2020) Plant-derived nanoparticles as a novel platform for anti-angiogenic therapy. Front Pharmacol 11:562
58. Vankayala R, Lin CC, Kalluru P, Chiang CS, Hwang KC (2013) Gold nanoshells-mediated bimodal photodynamic and photothermal cancer treatment using ultra-low doses of near infra-red light. Biomaterials 34(28):6229–6238
59. Wallace DC (2012) Mitochondria and cancer. Nat Rev Cancer 12(10):685–698

60. Ly JD, Grubb DR, Lawen A (2003) The mitochondrial membrane potential ($\Delta\psi$m) in apoptosis. Apoptosis 8(2):115–128
61. Basu A, Jha S, Ghosh S (2018) Plant-mediated synthesis of metallic nanoparticles and their biological applications. Toxicol Res 34(4):281–291
62. Yu E (2012) Mitochondrial DNA damage and diseases. Biochim Biophys Acta Mol Basis Dis 1820(5):602–608
63. Jackson SP, Bartek J (2009) The DNA-damage response in human biology and disease. Nature 461(7267):1071–1078
64. Tanisha, Jadhav U, Alim H (2024) Genotoxicity of silver nanoparticles in plants and underlying mechanism. In: Plant response to silver nanoparticles: plant growth, development, production, and protection. Springer Nature Singapore, Singapore, pp 59–78
65. Rajeshkumar S, Bharath LV, Geetha M (2020) *Catharanthus roseus* extract-mediated zinc oxide nanoparticles induce cytotoxicity and autophagy in cancer cells. J Inorg Organomet Polym Mater 30(5):1711–1720
66. Helleday T, Lo J, van Gent DC, Engelward BP (2008) DNA double-strand break repair: from mechanistic understanding to cancer treatment. DNA Repair 7(4):605–613
67. Fruman DA, Rommel C (2014) PI3K and cancer: lessons, challenges and opportunities. Nat Rev Drug Discov 13(2):140–156
68. Agarwal R, Srivastava AK, Singh V, Raghuwanshi N (2019) Curcumin and rosmarinic acid-based nanoparticles targeting PI3K/Akt pathway in cancer. Phytomedicine 55:62–71
69. Dhillon AS, Hagan S, Rath O, Kolch W (2007) MAP kinase signalling pathways in cancer. Oncogene 26(22):3279–3290
70. Kawano M, Mabuchi S, Matsumoto Y (2015) The role of NF-κB in cancer: therapeutic implications. Int J Mol Sci 16(12):27897–27919
71. Chaffer CL, Weinberg RA (2011) A perspective on cancer cell metastasis. Science 331(6024):1559–1564
72. Lamouille S, Xu J, Derynck R (2014) Molecular mechanisms of epithelial–mesenchymal transition. Nat Rev Mol Cell Biol 15(3):178–196
73. Sharma G, Kumar A, Sharma S, Naushad M, Dwivedi RP, Alothman ZA, Mola GT (2019) Novel development of nanoparticles to bionanocomposites for cancer treatment: an overview. Mater Sci Eng C Mater Biol Appl 104:109921
74. Abbasi Kajani A, Bordbar AK, Zarkesh Esfahani SH, Khosropour AR, Razmjou A (2014) Green synthesis of anisotropic silver nanoparticles with potent anticancer activity using *Taxus baccata* extract. RSC Adv 4:61394–61403
75. Abbasi Kajani A, Bordbar AK, Zarkesh Esfahani SH, Razmjou A (2016) Gold nanoparticles as potent anticancer agent: green synthesis, characterization, and in vitro study. RSC Adv 6:63973–63983
76. Annu AS, Kaur G, Sharma P, Singh S, Ikram S (2018) Evaluation of the antioxidant, antibacterial and anticancer (lung cancer cell line A549) activity of *Punica granatum* mediated silver nanoparticles. Toxicol Res (Camb) 7(5):923–930
77. Saratale RG, Shin HS, Kumar G, Benelli G, Kim DS, Saratale GD (2017) Exploiting antidiabetic activity of silver nanoparticles synthesized using *Punica granatum* leaves and anticancer potential against human liver cancer cells (HepG2). Artif Cells Nanomed Biotechnol 46(Suppl 3):S211–S222
78. Shinde AS, Mendhulkar VD (2020) Antiproliferative activity of *Elephantopus scaber* mediated silver nanoparticles against MCF-7, A-549, SCC-40, and COLO-205 human cancer cell lines. Asian J Pharm Clin Res 13(2):163–167
79. Venkatadri B, Shanparvish E, Rameshkumar MR, Valan Arasu M, Al-Dhabi NA, Ponnusamy VK, Agastian P (2020) Green synthesis of silver nanoparticles using aqueous rhizome extract of *Zingiber officinale* and *Curcuma longa*: in vitro anticancer potential on human colon carcinoma HT 29 cells. Saudi J Biol Sci 27(11):2980–2986

80. Javed B, Mashwani ZR (2020) Synergistic effects of physicochemical parameters on bio-fabrication of mint silver nanoparticles: structural evaluation and action against HCT116 colon cancer cells. Int J Nanomedicine 15:3621–3637
81. Vasanth K, Priyadarshini S, Sakthivel N (2014) Biosynthesis of silver nanoparticles using *Moringa oleifera* and their cytotoxic effects on HeLa cells. J Nanomed Biother Discov 4(1):1–7
82. Wang L, Xu J, Yan Y, Liu H, Karunakaran T, Li F (2019) Green synthesis of gold nanoparticles from Scutellaria barbata and its anticancer activity in pancreatic cancer cell (PANC-1). Artif Cells Nanomed Biotechnol 47(1):1617–1627
83. Noorbazargan H, Amintehrani S, Dolatabadi A, Mashayekhi A, Khayam N, Moulavi P, Naghizadeh M, Mirzaie A, Mirzaei Rad F, Kavousi M (2021) Anti-cancer and anti-metastasis properties of bioorganic-capped silver nanoparticles fabricated from *Juniperus chinensis* extract against lung cancer cells. AMB Express 11(1):61
84. Gumy MN, Kanunfre CC, Padilha JP, Cruz L (2023) Silver nanoparticles green synthesis from Catharanthus roseus flowers and effect on A549 lung cancer cells. Braz Arch Biol Technol 66(spe):e2320989
85. Greeshma KP, Thamizselvi R (2022) Experimental and theoretical approach on green synthesized zinc oxide nanoparticles from combined leaf extracts of Catharanthus roseus and Morinda citrifolia for in vitro anti-cancer studies. J Mol Liq 351:118636
86. Hussain A, Oves M, Alajmi MF, Hussain I, Amir S, Ahmed J, Rehman MT, El-Seedif HR, Ali I (2019) Biogenesis of ZnO nanoparticles using Pandanus odorifer leaf extract: anticancer and antimicrobial activities. RSC Adv 9(27):15357–15369
87. Baghbani-Arani F, Movagharnia R, Sharifian A, Salehi S, Sadat Shandiz SA (2017) Photo-catalytic, anti-bacterial, and anti-cancer properties of phyto-mediated synthesis of silver nanoparticles from *Artemisia tournefortiana* Rchb extract. J Photochem Photobiol B Biol 173:640–649
88. Dadashpour M, Firouzi-Amandi A, Pourhassan-Moghaddam M, Maleki MJ, Soozangar N, Jeddi F, Nouri M, Zarghami N, Pilehvar-Soltanahmadi Y (2018) Biomimetic synthesis of silver nanoparticles using *Matricaria chamomilla* extract and their potential anticancer activity against human lung cancer cells. Mater Sci Eng C 92:902–912
89. Balashanmugam P, Gajendran V, Ravi M, Shanthi R (2016) Biosynthesis of gold nanoparticles using Cassia roxburghii flower extract and their cytotoxicity on HepG2, HeLa, and MCF 7 cells. J Nanosci Nanotechnol Mol Biol 3(2):42–48
90. Xu L, Qi X, Zhou Q, Wang L, Li S, Li M (2016) Lonicera hypoglauca-mediated silver nanoparticles exhibit selective cytotoxicity against MCF 7 breast cancer cells. J Nanopart Res 18(1):12
91. Sheikh E, Bhatt MLB, Tripathi M (2018) Bio-based synthesised and characterized monodispersed Curcuma longa silver nanoparticles induce targeted anticancer activity in breast cancer cells. Pharmacogn Mag 14(Suppl 3):S340–S345
92. Arruebo M, Fernández-Pacheco R, Ibarra MR, Santamaría J (2011) Magnetic nanoparticles for drug delivery. Nano Today 2(3):22–32
93. Danhier F (2016) To exploit the tumor microenvironment: since the EPR effect fails in the clinic, what is the future of nanomedicine. J Control Release 244(1):108–121
94. Sharma A, Goyal AK, Rath G, Jain S (2018) Green synthesis and characterization of silver nanoparticles using *Punica granatum* and their anticancer activity against MCF-7 cells. Pharm Nanotechnol 6(1):5–11
95. Wang H, Agarwal P, Zhao S (2021) Overcoming multidrug resistance in cancer with nanotechnology. Acta Pharm Sin B 11(9):2589–2619
96. Patra JK, Das G, Fraceto LF, Campos EVR, Rodriguez-Torres MP, Acosta-Torres LS, Diaz-Torres LA, Grillo R, Swamy MK, Sharma S, Habtemariam S, Shin H-S (2018) Nano based drug delivery systems: recent developments and future prospects. J Nanobiotechnol 16:71
97. Singh R, Lillard JW, Singh S (2017) Nanoparticle-based targeted drug delivery. Exp Mol Pathol 86(3):215–223

98. Bhattacharyya S, Kudgus RA, Bhattacharya R, Mukherjee P (2021) Inorganic nanoparticles in cancer therapy. Pharmacol Res 82:224–237
99. Dutta T, Paul S, Roy A, Roy M (2021) Phytochemical-assisted synthesis of gold nanoparticles and their applications in cancer therapy. Front Mol Biosci 8:635223
100. Siddiqi KS, Husen A (2017) Green synthesis, characterization and applications of nanoparticles. Nanoscale Res Lett 11(1):502
101. Miri A, Darroudi M, Kouhi M, Sarani M (2021) Green biosynthesis of silver nanoparticles using *Rosmarinus officinalis* and *Matricaria chamomilla* extracts: characterization and in vivo toxicity evaluation. Mater Sci Eng C Mater Biol Appl 118:111432
102. Mukherjee S, Chowdhury D, Kotcherlakota R, Patra S, Vinothkumar B, Bhadra MP, Sreedhar B, Patra CR (2014) Potential theranostics application of bio-synthesized silver nanoparticles (4-in-1 system). Theranostics 4(3):316–335
103. Vijayan R, Joseph S, Mathew B (2018) Green synthesis of silver nanoparticles using *Nervalia zeylanica* leaf extract and evaluation of their antioxidant, catalytic, and antimicrobial potentials. Mater Today Proc 5(1):809–819

Aromatic Plant-Based Nanoparticles for Wound Healing

Rafiq A. Rather and Mohd Saleem Dar

Abstract Wound healing is a precisely regulated biological process that progresses through four sequential phases—hemostasis, inflammation, proliferation, and remodeling—to restore tissue integrity and function. While acute wounds typically resolve efficiently, chronic wounds often fail to progress due to factors like infection, diabetes, or poor vascularization, necessitating advanced therapeutic approaches. Conventional wound care methods are limited in efficacy, driving interest in innovative strategies that combine nanotechnology and phytomedicine. Nanoparticles (NPs) derived from medicinal and aromatic plants (MAPs) offer a promising solution, merging the benefits of traditional herbal remedies with modern nanoscience. These plant-based NPs, abundant in bioactive constituents like flavonoids, terpenoids, and phenolics, exhibit strong antimicrobial, anti-inflammatory, antioxidant, and tissue regenerative activities. Green synthesis techniques enhance their safety, scalability, and ecological compatibility, while improving bioavailability and targeted delivery. Owing to their nanoscale size and high surface area, NPs enable enhanced tissue penetration and controlled release of therapeutic agents, promoting fibroblast activity, collagen deposition, and angiogenesis. This chapter examines the role of MAP-based NPs in wound healing, focusing on their synthesis, biological mechanisms, and incorporation into advanced wound dressings such as hydrogels and nanofibers. It also discusses current challenges, safety considerations, and future prospects, highlighting their potential to transform regenerative wound care.

Keywords Wound healing · Aromatic plants · Nanoparticles · Phytochemicals · Regenerative therapy

---

R. A. Rather (✉)
Department of Clinical Biochemistry, Government Degree College Shopian, Shopian, Jammu & Kashmir, India

M. S. Dar
National Cancer Institute, NIH, Bethesda, MD, USA

A. Husen (ed.), *Sustainable Uses of Aromatic Plants in Nanotechnology*, Smart Nanomaterials Technology, https://doi.org/10.1007/978-981-95-7409-4_10

1 Introduction

Wound healing is a highly coordinated and complex biological process that plays a crucial role in restoring tissue integrity and function following injury [1]. It progresses through four closely interconnected and overlapping phases: hemostasis, inflammation, proliferation, and remodeling [2]. Multiple factors are known to negatively influence this process—such as malnutrition, immunosuppression, hypoxia, chronic infection, stress, surgical interventions, certain medications, alcoholism, smoking, and chronic diseases—resulting in impaired or delayed wound healing [3]. While acute wounds typically heal without complication, chronic wounds often stall in the inflammatory phase due to intrinsic factors (e.g., diabetes) or extrinsic influences (e.g., infection), requiring advanced therapeutic interventions [4].

Traditional wound care practices, such as gauze dressings, foams, films, and topical antiseptics, are limited by poor moisture control, minimal antimicrobial properties, and insufficient delivery of bioactive agents [5]. In response, recent innovations have merged nanotechnology with phytomedicine particularly through the development of nanoparticles (NPs) derived from medicinal and aromatic plants (MAPs) to enhance wound healing outcomes [6].

Nanoparticles, generally ranging in size from 1 to 100 nm, exhibit distinctive physicochemical properties, including a high surface-area-to-volume ratio, adjustable surface functionality, and enhanced tissue penetration capabilities [7]. These properties of NPs facilitate the effective delivery of therapeutic molecules by enhancing target specificity and bioavailability, minimizing adverse effects, and promoting interaction with biological systems at the molecular level [7]. Nanoparticle-based scaffolds are often used to regenerate damaged tissues and organs [8]. When synthesized from plant extracts, NPs retain the bioactivity of their phytochemical precursors, offering a dual advantage of modern nanotechnology and traditional herbal efficacy [9].

The green synthesis of NPs using plant extracts primarily relies on the biogenic reduction of metal ions mediated by bioactive compounds present in the extract [10]. These phytochemicals act as reducing agents by donating electrons to metal ions, facilitating their transformation into NPs. Moreover, biomolecules such as proteins and polysaccharides function as capping and stabilizing agents, preventing particle aggregation and influencing the size and morphology of synthesized NPs [11]. Multiple factors such as pH, temperature, precursor concentration, and reaction time play critical roles in controlling nucleation and growth processes, thereby determining the final structural and physicochemical properties of the NPs [9].

Medicinal and aromatic plants such as *Lavandula angustifolia* (lavender), *Curcuma longa* (turmeric), *Rosmarinus officinalis* (rosemary), *Thymus vulgaris* (thyme), and *Azadirachta indica* (neem) are abundant sources of bioactive phytoconstituents such as alkaloids, flavonoids, phenolics, terpenoids, glucosinolates, cyanogenic glycosides, and essential oils [12]. These constituents exhibit strong antimicrobial, anti-inflammatory, antidiabetic, antioxidant, and regenerative activities, while also promoting digestion, pain relief, sedation, and immune modulation and enhancing memory and energy [12]. Leveraging green synthesis techniques, these phytochemicals serve as natural reducing and capping agents in NP fabrication, eliminating the need for hazardous chemicals and preserving biological activity. This method is eco-friendly, scalable, and cost-effective [13]. Moreover, owing to their biocompatibility, nanoscale dimensions, natural abundance, and favorable physicochemical characteristics, these plant-based NPs exhibit remarkable pharmacological properties, positioning them as highly potential candidates for medical applications such as wound healing [13].

Aromatic plant-based NPs, including metal-based (e.g., gold, silver), polymer-based (e.g., chitosan), and lipid-based (e.g., liposomes) systems, can be engineered for targeted therapeutic applications [14]. They promote wound healing by combating microbes, reducing oxidative stress, and modulating inflammation [15]. Phytochemicals like curcumin, thymol, carvacrol, eugenol, terpinen-4-ol, geraniol, cinnamaldehyde, and cineole, when encapsulated in NPs, exert antimicrobial effects by disrupting microbial membranes and interfering with vital enzymatic pathways [16, 17]. Their antioxidant and anti-inflammatory properties further mitigate oxidative damage and downregulate pro-inflammatory cytokines, including tumor necrosis factor (TNF)-α and interleukin (IL)-6. Beyond infection control and inflammation reduction, aromatic plant-based NPs also enhance tissue regeneration by stimulating fibroblast proliferation, collagen synthesis, angiogenesis, and keratinocyte migration—critical processes for wound closure and formation of new epithelium (epithelialization) [18]. Some NP formulations even activate fetal-like regenerative pathways, reducing scarring (fibrosis) and improving tissue architecture [19]. The efficacy of these NPs is further optimized when incorporated into advanced wound delivery systems such as hydrogels, nanofibers, films, and sprays [20]. These carriers maintain a moist, biocompatible wound environment, allow controlled release of therapeutic agents, and reduce dressing frequency and patient discomfort. For instance, nanofiber scaffolds mimic the extracellular matrix (ECM), supporting cellular attachment and growth, while hydrogels offer flexibility, permeability, and strong wound adherence [20].

In wound care management, NP-based delivery systems present advantages compared to conventional therapies. They improve drug solubility and stability, facilitate targeted delivery, and minimize systemic toxicity [21]. Despite promising preclinical results, the clinical translation of MAP-derived NPs in wound healing remains in its early stages [22]. Evidence from in vitro and animal studies highlights their potential to accelerate wound closure, prevent microbial infection, and improve tissue regeneration. However, robust clinical validation through human intervention trials is needed, alongside addressing regulatory issues concerning NP safety, reproducibility, and standardization [22]. Additional challenges include ensuring the scalability of green synthesis methods, achieving long-term stability, and optimizing dosage and delivery strategies.

Different classes of NPs contribute uniquely to wound therapy [23]. Metal NPs, especially silver and gold, provide strong antimicrobial effects and imaging capabilities [24]. Polymeric NPs and lipid-based carriers (such as phytosomes, SLNs, and NLCs) are ideal for encapsulating hydrophobic plant compounds, offering controlled release and biocompatibility [25]. Carbon-based NPs (e.g., nanotubes) and ceramic NPs offer structural strength and are being explored in biosensing applications, tissue scaffolding, and skin tissue care and management [26, 27]. Semiconductor NPs like quantum dots (QDs), although more relevant in diagnostics, highlight the versatility of NPs in biomedical innovation.

Green synthesis remains the cornerstone of aromatic plant-based NP production [28]. It typically involves the extraction of phytochemicals from plant parts (leaves, roots, flowers) using aqueous or organic solvents, followed by reaction with metal salt solutions (e.g., $AgNO_3$) to produce NPs [29]. Advanced methods such as microwave-, heat-, and ultrasonication-assisted synthesis improve reaction speed, uniformity, and stability. Beyond metallic NPs, phytochemical-rich extracts are also incorporated into biodegradable polymers like chitosan or PLGA, forming polymeric NPs with sustained-release profiles.

Aromatic plant-based NPs represent a promising interface between traditional herbal wisdom and modern nanotechnology [30]. Their multifunctional roles—including antimicrobial defense, anti-inflammatory action, antioxidant protection, and regenerative stimulation—make them ideal candidates for next-generation wound care [30]. While further research is needed to refine their synthesis, ensure clinical safety, and standardize therapeutic protocols, these novel agents hold significant potential to transform wound management and regenerative medicine.

2 Wounds and Mechanism of Wound Healing

A wound refers to any break, disruption, or injury of living tissue—including the skin, mucous membranes, or internal organs—that may occur suddenly due to direct trauma (mechanical, thermal, chemical, or radiation) or develop gradually as a result of chronic conditions such as diabetes mellitus, venous or arterial insufficiency, or immune disorders [31, 32]. Wounds are generally categorized as acute or chronic, depending on the duration since injury and their progression through the normal phases of healing process [33]. Both types can be further subcategorized based on several criteria: the cause of injury (surgical, traumatic, burns, diabetic ulcers, pressure ulcers), the extent and depth of tissue damage (superficial, partial-thickness, full-thickness), and the sterility of the wound bed (clean, clean-contaminated, contaminated, infected) [34]. The appearance and characteristics of wounds may vary significantly depending on factors such as their anatomical location, the mechanism and depth of injury, their chronicity (acute or chronic), and their sterility status (sterile or infected) [35]. Because treatment modalities depend on wound classification, a thorough evaluation by a qualified healthcare provider is crucial for appropriate management. To facilitate the evaluation, documentation, and management of wounds, various classification systems have been established. They include the Centers for Disease Control Surgical Wound Classification, the International Red Cross Wound Classification, the Tscherne classification, the Gustilo-Anderson classification for open fractures, and the AO Foundation's AO soft tissue grading system [36]. Each of these systems defines and categorizes wound healing differently according to its clinical context. In general terms, wound healing refers to the physiological process by which a living organism replaces damaged or destroyed tissue with newly formed tissue [1]. This involves a coordinated cascade of cellular and molecular mechanisms aimed at restoring the integrity and functional capacity of the skin or other affected tissues. Under normal physiological conditions, wound repair proceeds through four overlapping phases: hemostasis (clot formation), inflammation, cell proliferation (tissue regeneration), and tissue remodeling (maturation and cellular differentiation) [3]. In certain contexts, hemostasis (blood clotting) is regarded as part of the inflammatory phase rather than a distinct stage of healing. The rate and efficiency of wound repair can be affected by multiple factors including age, tissue oxygenation, underlying medical conditions, psychological or physiological stress, and the use of specific medications [3].

In healthy human skin, the outer epithelial layer (epidermis) and the deeper connective tissue layer (dermis) work together to form a protective barrier against external environmental insults [37]. When this barrier is disrupted, a well-orchestrated sequence of cellular and biochemical events is triggered to initiate tissue repair [1]. Cutaneous wound healing consists of four phases (Fig. 1), briefly described below:

Hemostasis (blood clotting): Immediately following an injury, blood platelets adhere to the damaged site and undergo a shape change into an irregular, amorphous form that facilitates clot formation. They also release chemical mediators that enhance the coagulation process [38]. This triggers the conversion of fibrinogen into fibrin, which creates a mesh-like network that binds the platelets together. The resulting clot seals the vessel injury and helps to minimize or stop further bleeding [39].

Inflammation: During the inflammatory phase, dead and damaged cells, along with pathogens (such as bacteria, fungi, and viruses) and other debris, are removed through phagocytosis—a process in which white blood cells engulf and eliminate the unwanted material. Platelet-derived growth factors are also released into the wound site, stimulating the migration and proliferation of fibroblasts and smooth muscle cells, thereby supporting the proliferative phase of healing [39].

Proliferation (growth of new tissue): During this phase, several important processes take place, including angiogenesis, collagen deposition, granulation tissue formation, epithelialization, and wound contraction. In angiogenesis, vascular endothelial cells generate blood vessels to restore oxygen and nutrient supply. During fibroplasia and granulation tissue formation, fibroblasts proliferate and secrete collagen and fibronectin to create ECM. Simultaneously, re-epithelialization of the epidermis occurs, in which epithelial cells multiply and migrate across the wound bed, covering the newly made tissue. Wound contraction is mediated by myofibroblasts, which pull the wound edges through a contraction mechanism similar to that of smooth muscle cells. Once their functions are complete, excess cells undergo apoptosis [39].

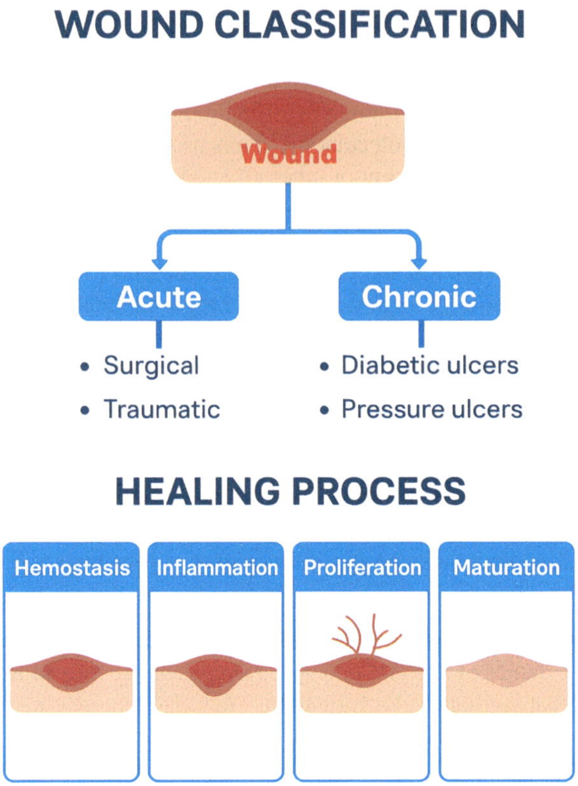

Fig. 1 Types of wounds and major steps in wound healing process

Maturation (remodeling): During this maturation and remodeling phase, collagen fibers are realigned along tension lines to enhance tissue strength, while surplus cells are removed through apoptosis, a regulated form of programmed cell death [39].

Wounds that do not progress through the normal stages of healing and remain unhealed for more than a month are categorized as chronic wounds [32]. These wounds have diverse causes, all of which pose a significant burden on the healthcare system. The coexistence of multiple chronic conditions is known as comorbidity, and such comorbidities often complicate the course of chronic wounds, making it difficult to define them as a single disease entity [32]. Individuals with diabetes or obesity are particularly susceptible to developing chronic wounds, and most people with persistent, non-healing wounds also experience other serious clinical conditions. Chronic wounds primarily affect the elderly, with 3% of Americans over 65 currently having open wounds. By 2060, the elderly population is projected to exceed 77 million, indicating a growing prevalence of chronic wounds. Overall, chronic wounds are estimated to affect about 2% of the US population, representing a growing global health concern [40]. When left untreated or poorly managed, non-healing wounds can lead to severe complications, such as infection, sepsis, limb amputation, and even death. Consequently, there is a pressing need to develop novel therapeutic agents that enhance wound healing and alleviate the increasing burden on the global healthcare system [30, 40]. In recent years, the integration of nanomaterials with unique physicochemical properties into therapeutic strategies has demonstrated promising outcomes in in vitro and animal model systems [41]. However, only a limited amount of research has been conducted to determine whether these nanomaterials significantly enhance the healing process.

3 Overview of Nanotechnology and Nanoparticles

Nanotechnology involves manipulating matter at the nanometer scale (1–100 nm), a range where materials exhibit unique quantum and surface phenomena not observed at larger dimensions [42]. Owing to their small size and high surface area, NPs possess superior physicochemical properties compared to bulk materials, making them highly useful in diagnostics, targeted drug delivery, medicine, and wound management [43]. The green synthesis of NPs using MAPs has gained increasing attention for its environmental sustainability, eco-friendliness, safety, and cost-effectiveness [28].

Nanoparticles are broadly classified into:

1. **Carbon-based NPs**: This group includes fullerenes, carbon nanotubes, graphene oxide, and QDs. Their structural versatility and high conductivity make them particularly suitable for drug delivery and wound therapy, especially due to their ability to carry cytoprotective agents to target sites with precision [26]. Carbon-based NPs have been integrated with tissue engineering strategies to create innovative approaches for wound healing, leading to accelerated recovery and effective tissue regeneration.
2. **Metallic NPs**: Made from metals like gold, silver, and palladium, these exhibit surface plasmon resonance, enabling applications in imaging, diagnostics, and radiosensitization in cancer therapy. Radiolabeled metal NPs also serve as theranostic agents. Metallic NPs (e.g., gold, silver, copper) have advanced wound treatment through their antimicrobial, drug-delivery, and healing-promoting properties [24]. Green synthesis further enhances their biocompatibility and safety by combining the effects of NPs with those of natural plant extracts [29].
3. **Ceramic NPs**: Composed of oxides and carbonates of metals like cerium and titanium. Known for their biocompatibility and heat resistance, they are used in implants, tissue engineering, photodynamic therapy, and cutaneous wound healing [44, 45]. Among them, cerium dioxide (CeO_2)-based nanomaterials have recently gained substantial attention as promising dermal substitutes, capable of enhancing fibroblast infiltration and tissue regeneration [44].
4. **Semiconductor NPs**: Include QDs made from materials like cadmium selenide. Due to their fluorescence and bioimaging capabilities, QDs find applications in

diagnostics, targeted drug delivery, studies on drug resistance, drug analysis, antimicrobial activity, tissue imaging, genetic assays, and photothermal therapy [46, 47].
5. **Polymeric NPs**: Formed from biocompatible polymers into nanospheres or nanocapsules. They enable controlled drug release and targeted delivery and are engineered to cross biological barriers [48].
6. **Lipid-based NPs**: Lipid NPs are effective carriers for drugs, particularly in oncotherapy. Their excellent biocompatibility and suitability for personalized medicine make them promising candidates for future oncology applications [49].

Medicinal and aromatic plant-based NPs have been shown to influence various phases of wound healing and are therefore used to treat wounds, promote healing, and prevent complications [15]. Their inherent nanoscale characteristics—such as ion release and the generation of reactive oxygen species—enable them to influence diverse molecular and cellular pathways, thereby promoting the transition of chronic, non-healing wounds into the active healing phase. Currently, NPs are used both as intrinsic antimicrobial agents and as delivery systems for antibiotics and other antimicrobial compounds [28].

Nanoparticles and other nanomaterials such as nanofibers and nanosheets exhibit distinctive physicochemical properties that make these particles extremely suitable for a range of applications in wound healing [20, 21]. These NPs can improve wound healing by serving as carriers for therapeutic compounds, including MAP-based antioxidant, anti-inflammatory, and antimicrobial agents [15]. In addition, NPs can be designed to interact with immune cells or fibroblasts, facilitating the targeted delivery of drugs to specific tissues [50]. Likewise, nanofibers used in wound dressing provide several advantages, such as high surface-area-to-volume ratio, strong mechanical properties, and tunable porosity. These features create a favorable healing microenvironment and mimic the ECM, supporting cell attachment and proliferation [20]. Similarly, polymer nanosheets have gained attention as wound dressing materials due to their ultrathin structure, large surface area, flexibility, and adhesive nature. These properties contribute to effective wound coverage, mechanical reinforcement, and barrier protection [31]. The demand for wound dressings with strong wet adhesion and antimicrobial capabilities remains particularly high for managing infected wounds.

Medicinal and aromatic plants and their bioactive phytoconstituents have long been utilized in the treatment and management of wounds [51]. Herbal medicines derived from MAPs constitute a significant fraction of the global pharmaceutical market. Despite extensive knowledge of their therapeutic potential, standardized protocols for the quality assessment of plant materials—encompassing phytochemical characterization, pharmacological evaluation, and therapeutic validation—remain insufficiently detailed in the current literature [52]. Establishing standardization practices for medicinal plants is essential to ensure product quality, phytochemical consistency, safety, and therapeutic efficacy [53]. Medicinal and aromatic plants have the potential to treat and cure various conditions such as ulcers, skin infections, inflammation, scabies, leprosy, venereal diseases, and wounds [30].

In wound management, these plants contribute through their disinfectant and debriding actions, as well as by maintaining a moist environment conducive to optimal natural healing [51].

Traditional healing therapies make extensive use of various plants, their extracts, decoctions, and pastes to treat burns, cuts, and wounds [51, 54]. However, modern wound healing therapies have notable limitations, especially in managing complex or chronic wounds. A major challenge is infection control, as traditional dressings (such as hydrocolloids, foams, films, gauze, and cotton) lack antimicrobial properties, leaving wounds susceptible to bacterial colonization—particularly in immunocompromised or diabetic patients [55]. These treatments also fail to maintain optimal moisture balance. Excess moisture can cause skin maceration, while dryness impairs tissue regeneration. Most basic dressings cannot effectively regulate this environment. Chronic wounds, including diabetic foot ulcers and pressure sores, frequently remain stalled in the inflammatory phase [56]. Traditional therapies offer little targeted support to reactivate healing and lack personalization to address issues like poor blood supply or growth factor imbalances [55]. Frequent dressing changes increase discomfort, healthcare costs, and the risk of wound disruption. Adherent dressings often cause pain and tissue damage during removal. Systemic antibiotics, commonly used in conventional care, contribute to antibiotic resistance—making infections harder to treat [57]. Moreover, traditional methods do not address wound oxygenation, with hypoxia hindering cell migration and collagen synthesis. Unregulated healing may also lead to excessive scarring, affecting function or appearance, especially in sensitive areas [57]. These limitations underscore the need for advanced, patient-specific wound care strategies.

In recent years, the combination of nanotechnology with plant-based therapeutics has emerged as a promising and biocompatible strategy for enhancing wound healing [15]. Medicinal and aromatic plant-based NPs are especially valuable due to their multifunctional roles. When synthesized through eco-friendly or green approaches using plant-derived bioactive compounds—such as flavonoids, terpenoids, alkaloids, and essential oils—these NPs serve both as therapeutic agents and as natural stabilizers, thereby eliminating the need for harmful chemicals [28]. Medicinal and aromatic plant-based NPs possess strong antimicrobial properties, making them effective against infections common in chronic wounds. Plants like *Azadirachta indica*, *Ocimum sanctum*, and *Curcuma longa* have been used to produce silver and zinc oxide NPs with potent antibacterial activity [58]. Additionally, these NPs exhibit antioxidant and anti-inflammatory effects by modulating cytokines and scavenging reactive oxygen species, which helps transition wounds from inflammation to regeneration [59]. These NPs further facilitate tissue regeneration by promoting fibroblast proliferation, collagen synthesis, and angiogenesis—especially when derived from plants like *Aloe vera* and *Centella asiatica* [60, 61]. When incorporated into microneedle array patches, these NPs can be delivered in a controlled and painless manner, enhancing treatment efficacy and patient compliance. While challenges remain—such as variability in plant extracts and standardization—MAP-based NPs, particularly in combination with advanced delivery systems, offer a sustainable, targeted, and effective strategy for managing chronic and complex wounds [62, 63].

4 Aromatic Plant-Based Nanoparticles and Their Mechanism of Action

These are nanoscale particles (~1 to 100 nm) synthesized using extracts from aromatic plants [64]. The synthesis is often green or eco-friendly, avoiding toxic chemicals and leveraging phytochemicals (e.g., alkaloids, flavonoids, phenols, tannins, terpenes) in the plants as reducing and stabilizing agents. Medicinal and aromatic plant-based NPs are typically categorized into metal-based (e.g., silver, gold), polymer-based (e.g., chitosan), or lipid-based (e.g., liposomes) types. These are commonly synthesized through green synthesis methods such as plant extract-mediated, microwave-assisted, or ultrasound-assisted synthesis [65, 66]. Medicinal and aromatic plant-based NPs can be incorporated into various delivery systems, including hydrogels, nanofiber mats, sprays, creams, and wound dressings [50]. Such delivery systems enable controlled drug release, improved bioavailability, and site-specific therapeutic action [50]. For example, hydrogel-based scaffolds offer a structural matrix that supports cellular adhesion, proliferation, and differentiation, ultimately facilitating tissue regeneration [50]. The concept of MAP-based NPs in wound healing centers on the synthesis and application of NPs derived from MAPs such as *Lavandula angustifolia* (lavender), *Rosmarinus officinalis* (rosemary), *Curcuma longa* (turmeric), *Thymus vulgaris* (thyme), and others that have antioxidant, antimicrobial, anti-inflammatory, and tissue-regenerative properties [62, 67]. Consequently, MAP-based NPs promote wound healing through multiple interrelated mechanisms (Table 1).

Table 1 Mechanisms of MAP-based NPs in wound healing

Mechanism	Role in healing	Key references
Antimicrobial	Prevents or treats infection by bacteria/fungi	Nandhini et al. [68], Tyavambiza et al. [69]
Antioxidant	Reduces oxidative stress at the wound site	Lakkim et al. [70]
Anti-inflammatory	Lowers pro-inflammatory cytokines (e.g., IL-6, TNF-α)	Hajialyani et al. [62]
Cell proliferation	Enhances fibroblast and keratinocyte activity	Nandhini et al. [68]
Collagen deposition	Supports tissue regeneration and ECM remodeling	Lakkim et al. [70]
Angiogenesis	Stimulates new blood vessel formation	Hajialyani et al. [62]

5 Synthesis of Aromatic Plant-Based NPs: An Overview

The synthesis of aromatic plant-based NPs is an emerging green nanotechnology approach that harnesses the rich phytochemical content of aromatic plants for the reduction and stabilization of NPs [65]. Aromatic plants such as thyme, basil, rosemary, oregano, lemongrass, and eucalyptus are rich in diverse bioactive compounds, including alkaloids, polyphenols, flavonoids, terpenoids, tannins, and essential oils [71]. These constituents not only impart the characteristic fragrance (aroma) of the plants but also exhibit strong redox activity, enabling them to function as natural reducing and capping agents during NP preparation [71].

The process typically begins with the preparation of a plant extract [72]. Selected plant parts—commonly leaves, buds, or flowers—are washed thoroughly to remove dust and other contaminants and then air-dried or oven-dried at a controlled temperature to preserve their bioactive components [73]. The dried material is finely ground and subjected to extraction using solvents such as water, ethanol, or methanol [72]. The extraction may involve boiling or cold maceration, depending on the nature of the target phytochemicals. After extraction, the solution is filtered to obtain a clear plant extract rich in phytoconstituents [73].

To synthesize NPs, the plant extract is mixed with a solution of a metal salt, such as silver nitrate for silver NPs (AgNPs) or chloroauric acid for gold NPs (AuNPs) [74]. Upon mixing, the phytochemicals in the extract reduce the metal ions to their corresponding metallic forms [75]. For instance, Ag^+ ions are reduced to elemental silver (Ag^0), resulting in the nucleation and subsequent growth of NPs [76]. The reaction is often conducted at room temperature or mildly elevated temperatures (typically 40–80 °C). A visual change in the color of the reaction mixture—such as the development of a brown or reddish hue—often serves as a preliminary indicator of NP formation.

Simultaneously, the same phytochemicals that reduce the metal ions also act as capping agents, adsorbing onto the surface of the forming NPs and stabilizing them against aggregation. This dual role of phytochemicals makes the green synthesis route particularly attractive, as it avoids the need for additional chemical stabilizers or surfactants [77]. After the synthesis is complete, the NPs are separated from the reaction mixture using techniques like centrifugation and are subsequently washed with distilled water or alcohol to remove any unbound plant residues [78].

The synthesized NPs are then subjected to a series of characterization techniques to confirm their formation; determine their size, shape, and crystalline nature; and identify the functional groups responsible for reduction and capping [79]. Commonly employed characterization techniques include ultraviolet-visible spectroscopy, scanning electron microscopy, transmission electron microscopy, X-ray diffraction, atomic force microscopy, and Fourier-transform infrared spectroscopy [79–81].

One notable example is the synthesis of AgNPs using clove extract [82, 83]. In this method, clove buds are boiled in distilled water to prepare the extract, which is subsequently mixed with silver nitrate solution. Gentle heating and continuous stirring lead to the appearance of a brown coloration, indicating the formation of

AgNPs. Characterization through HPLC and LC-MS revealed that the NPs are uniformly shaped with an average size of 20–25 nm. Moreover, these AgNPs exhibit enhanced antimicrobial activity, partly due to the intrinsic properties of clove phytochemicals such as eugenol [81, 83].

In summary, employing aromatic plant extracts for NP synthesis provides a sustainable, economical, and environmentally friendly alternative to conventional chemical methods [81]. The green method not only minimizes the use of hazardous reagents but also integrates the therapeutic and biofunctional properties of aromatic plants into the NPs, potentially enhancing their biomedical applications [81].

6 Role of Medicinal and Aromatic Plants in Nanoparticle Synthesis for Wound Healing

Aromatic plants, as their name implies, are those that emit a characteristic fragrance or aroma. These plants are frequently studied alongside medicinal plants, and together, they constitute a distinct group recognized by ethnobotanists as MAPs [84, 85]. Medicinal plants, also known as medicinal herbs, possess therapeutic properties or produce beneficial pharmacological effects in humans, animals, and experimental models. In contrast, aromatic plants are primarily valued for their essential oils, which are widely used as spices and flavorings and in perfumes and medicines. These plants also produce raw materials for various industrial uses. Although often grouped under the term "MAPs," a more accurate term might be "medicinal, aromatic, and cosmetic plants," as many of these species are also used in the cosmetic industry [85]. Aromatic plants are characterized by their essential oils—volatile, hydrophobic compounds with strong odors—extracted from various plant parts like flowers, leaves, seeds, bark, and roots [86]. These oils are complex mixtures of secondary metabolites, primarily low-boiling-point phenylpropanoids and terpenoids, which are responsible for their distinct fragrances and diverse biological activities [87]. They are widely used in aromatherapy, cosmetics, flavoring, and perfumery and even as pesticides and herbal beverages. In India, commonly produced essential oils include those from lemongrass, mentha, citronella, ajwain, eucalyptus, palmarosa, spearmint, patchouli, and others [86].

A wide range of botanicals with wound healing properties have been traditionally utilized across the Northern Hemisphere, including *Achillea millefolium* (yarrow), *Aloe vera* (aloe), *Althaea officinalis* (marshmallow), *Calendula officinalis* (calendula or pot marigold), *Matricaria chamomilla* (chamomile), *Curcuma longa* (turmeric), *Eucalyptus* spp. (eucalyptus), *Simmondsia chinensis* (jojoba), *Plantago major* (plantain), *Pinus* spp. (pine), *Camellia sinensis* (green tea), *Punica granatum* (pomegranate), and *Inula helenium* (elecampane). These medicinal plants have been employed since ancient times to promote the healing of cutaneous wounds, prevent infections, and reduce dependence on antibiotics that contribute to antibiotic resistance [88]. Research-based herbal approaches are increasingly important as alternatives to antibiotics, which are ineffective against viral infections. Plant families such as Lamiaceae, Apiaceae, and Zingiberaceae are extensively studied for their therapeutic potential owing to their strong antioxidant, anti-inflammatory, and antimicrobial properties [89]. Other notable medicinal plants include *Senegalia visco* (visco acacia), *Artemisia nilagirica* (Indian wormwood), *Vaccinium microcarpum* (small cranberry), and *Hypericum perforatum* (St. John's wort), all of which have shown efficacy in treating various skin disorders. These plants yield bioactive compounds such as lupeol, artemetin, hydroquinone, hypericin, epicatechin, and curcumin, which are effective against cardiovascular diseases, inflammatory disorders, skin cancer, and pigmentation disorders and may help reduce the risk of cancer [89]. Common active constituents in these plants include flavonoids (antioxidant, anti-inflammatory), tannins (astringent, promote tissue contraction), saponins (antimicrobial, promote cell proliferation), alkaloids (anti-inflammatory, analgesic), essential oils (antiseptic, anti-inflammatory), and polysaccharides (promote tissue hydration and repair) [90].

Medicinal and aromatic plants are gaining prominence in Indian agriculture for three main reasons. First, the rising popularity of traditional healthcare systems like Ayurveda, Siddha, and Unani has boosted demand for medicinal plant-based products due to their holistic, affordable, and low-side-effect treatments [91]. Second, overharvesting from wild sources has raised sustainability concerns, making cultivation essential to meet demand. Medicinal plants offer higher economic value than conventional crops due to their high worth and low volume [92]. Third, the Ministry of Environment, Forest and Climate Change has documented over 9500 MAP species important to the pharmaceutical industry, with 2000–2300 commonly used in traditional medicine and around 150 commercially cultivated at scale. Aromatic plants, containing essential oils or volatile bioactive compounds, are considered a subset of medicinal plants [93].

Aromatic plant-based NPs offer key benefits in wound healing by combating microbial infections, addressing antibiotic resistance with natural antimicrobial effects. Their antioxidant phytochemicals neutralize harmful free radicals, while anti-inflammatory actions reduce inflammatory cytokines such as TNF-α and IL-6, minimizing swelling and aiding healing [62]. These NPs promote fibroblast growth, collagen formation, epithelialization, and angiogenesis, essential for tissue regeneration. Though primarily supported by in vitro and animal studies, evidence shows faster healing, fewer infections, and better tissue repair than conventional treatments. Future work will focus on clinical safety, scalable production, and integrating these NPs into smart wound dressings for real-time wound monitoring and care (Table 2).

Recently, aromatic plant-based nanomaterials have been integrated into advanced wound dressings, representing a promising frontier in the development of multifunctional, bioactive, and sustainable wound care technologies. However, their practical application in clinical wound management has traditionally been limited by issues such as volatility, poor solubility, and instability under physiological conditions. Through nanoencapsulation, nanoemulsification, or the eco-friendly

Table 2 Aromatic plants used in NP synthesis for wound healing

Plant name	Botanical name	Type of NP synthesized	Wound healing benefits	Key references
Tulsi (holy basil)	*Ocimum sanctum*, *Ocimum canum*	Silver (AgNPs), gold (AuNPs)	Antibacterial, enhances wound closure, anti-inflammatory	Gautam et al. [94], Mohanta et al. [95], Rao et al. [96], Tailor et al. [97]
Neem	*Azadirachta indica*	Silver (AgNPs), zinc oxide (ZnO NPs)	Promotes fibroblast proliferation, reduces infection	Anjum et al. [98], Lakkim et al. [99]
Turmeric	*Curcuma longa*	Silver (AgNPs), gold (AuNPs)	Curcumin-stabilized NPs enhance tissue regeneration following injury or disease	Hamilton and Gilbert [100], Muniyappan et al. [101], Thirupathi et al. [102]
Lavender	*Lavandula angustifolia*	Silver (AgNPs)	Promotes angiogenesis and collagen deposition	Şuică-Bunghez et al. [103], Teymouri et al. [104]
Lemongrass	*Cymbopogon citratus*	Silver (AgNPs), zinc oxide (ZnO NPs)	Antimicrobial, speeds re-epithelialization	Abdelbaky et al. [105], Arsene et al. [106], Rakib-Uz-Zaman et al. [107]
Eucalyptus	*Eucalyptus globulus*	Silver (AgNPs)	Antibacterial, helps in infected wound healing	Balčiūnaitienė et al. [108], Elbhnsawi et al. [109]
Tea tree	*Melaleuca alternifolia*	Silver (AgNPs)	Antimicrobial and anti-inflammatory	Nova et al. [110], Ramadan et al. [111]
Clove	*Syzygium aromaticum*	Silver (AgNPs)	Eugenol-stabilized AgNPs promote wound contraction	Makky et al. [112], Sethuram et al. [113]
Cinnamon	*Cinnamomum verum*	Silver (AgNPs), gold (AuNPs)	Antioxidant, supports fibroblast migration, prevents obesity through gut microbiota reshaping	Elmitwalli et al. [114], Seyed Ahmadi et al. [115], Sharma et al. [116]
Rosemary	*Rosmarinus officinalis*	Silver (AgNPs)	Reduces oxidative stress and inflammation	Mohamed et al. [117]
Peppermint	*Mentha piperita*	Silver (AgNPs)	Refreshing effect; antimicrobial and promotes cell migration	Aftab et al. [118], Hudz et al. [119], Robles-Martínez et al. [120]

Table 2 (continued)

Plant name	Botanical name	Type of NP synthesized	Wound healing benefits	Key references
Basil	*Ocimum basilicum*	Silver (AgNPs)	Reduces inflammation and speeds closure	Pirtarighat et al. [121], Qamar et al. [122]
Sandalwood	*Santalum album*	Silver (AgNPs), gold (AuNPs)	Antiseptic, antioxidant, helps in tissue regeneration	Selvam et al. [123]
Chamomile	*Matricaria chamomilla*	Silver (AgNPs)	Anti-inflammatory, enhances granulation	Dogru et al. [124], El Mihyaoui et al. [125]
Geranium	*Pelargonium graveolens*	Silver (AgNPs)	Astringent, antimicrobial	Shankar et al. [126]
Ajwain	*Trachyspermum ammi*	Palladium (PdNPs)	Enhanced anticancer activity and protein interaction profiling	Goli et al. [127], MubarakAli et al. [128]
Ginger	*Zingiber officinale*	Nanofibrous mat	Antioxidant activity, prevent wound infection, accelerate wound healing	Maleki et al. [129]
Thyme	*Thymus vulgaris*	Nanofibers, silver (AgNPs)	Antioxidant activity, antimicrobial, accelerate wound healing	Maleki et al. [129], Tavukcuoglu et al. [130]
Long pepper	*Piper longum*	Silver (AgNPs)	Antimicrobial and anti-inflammatory	Imtiaz et al. [131]
Davana	*Artemisia annua* L.	Nanofibers	Anti-inflammatory, anti-bacterial and anti-microbial, promotes proliferation and attachment of the seeded fibroblast cells	Mirbehbahani et al. [132]
Black seed	*Nigella sativa*	Silver (AgNPs), nanoemulgels	Antioxidant activity; prevent wound infection; nanoemulgel of thymoquinone improves skin penetration and promotes faster wound healing	Algahtani et al. [133], Palanisamy et al. [134]
Indian camphorweed (Kutraj)	*Pluchea indica* L.	*P. indica* branch extract NPs	Enhanced proliferation and migration of keratinocytes and fibroblasts	Chiangnoon et al. [135]
Mint	*Mentha piperita*	Silver (AgNPs)	Significantly improved wound healing time in diabetic rats	Mojally et al. [136]

synthesis of metallic NPs using plant extracts, the therapeutic efficacy of aromatic compounds can be significantly enhanced [137].

Incorporating these nanomaterials into advanced wound dressing platforms—particularly hydrogels and electrospun nanofibers—further amplifies their clinical potential [137]. Hydrogels, owing to their high water content, biocompatibility, and tunable physicochemical properties, provide a moist environment conducive to healing and can be easily engineered to encapsulate plant-based NPs or nanoemulsions [138]. This integration enables the sustained release of growth factors and bioactive molecules, while maintaining optimal hydration and promoting tissue regeneration [137]. For example, novel lavender oil and AgNPs simultaneously loaded onto hydrophilic polyurethane nanofibers have demonstrated not only enhanced antimicrobial activity but also accelerated re-epithelialization and wound closure [139].

Similarly, electrospun nanofibers closely replicate the structural and functional characteristics of the ECM, making them highly suitable for skin regeneration applications [140]. These fibrous scaffolds possess a high surface-area-to-volume ratio, favorable mechanical strength, controllable topography, and adjustable porosity, along with the capacity for biochemical functionalization—all features that collectively promote oxygen exchange, nutrient diffusion, and cell migration [140]. When aromatic plant-derived nanomaterials are incorporated into these fibers—either by blending into the spinning solution or through surface functionalization—the resulting scaffolds exhibit synergistic wound healing effects [141]. Nanofibers loaded with zein/thyme essential oil, for instance, have demonstrated remarkable antimicrobial activity, while maintaining cytocompatibility with fibroblasts and keratinocytes [142].

Recent research has focused on developing smart wound dressings using these plant-based nanomaterials, including stimuli-responsive hydrogels and dual-drug releasing nanofiber systems. In vivo studies in animal models have further validated the efficacy of these systems, showing reduced infection, faster wound closure, and improved tissue regeneration compared to conventional dressings. The convergence of green nanotechnology and biomaterial engineering thus holds immense promise for creating next-generation wound care solutions that are not only effective but also aligned with principles of sustainability and biocompatibility [143]. By harnessing the intrinsic healing properties of nature and enhancing them through nanotechnology, it is possible to develop wound care systems that are safer, more effective, and responsive to the complex biological environment of chronic and acute wounds.

7 Safety Considerations and Future Prospects

The application of aromatic plant-based NPs in wound healing represents a novel and promising therapeutic approach; however, it also brings several safety considerations that must be carefully evaluated [21]. Aromatic plants are known for their bioactive agents—including essential oils, flavonoids, and terpenoids—that exhibit antimicrobial, antioxidant, anti-inflammatory, and tissue-regenerative properties beneficial for wound management. However, when these compounds are synthesized into NPs, their physicochemical properties change, potentially altering their biological behavior [144]. Key safety concerns include cytotoxicity to skin cells, allergic reactions, and unintended interactions with the immune system. The NP's size, surface charge, and composition can affect cellular uptake and distribution, raising concerns about systemic absorption, bioaccumulation, and long-term toxicity, especially if the NPs are not biodegradable [145].

Another critical aspect is the variability in plant-derived compounds depending on species, harvest conditions, and extraction methods, which may lead to inconsistent therapeutic effects or safety profiles [146]. Therefore, ensuring reproducibility, purity, and stability of the plant extracts used in NP formulations is essential. Furthermore, comprehensive in vitro and in vivo investigations are required to assess potential adverse effects and to define safe and effective dosage ranges [147]. Regulatory oversight remains limited in this area, particularly with products that are positioned as both therapeutic and cosmetic [147].

Despite these challenges, the future prospects of aromatic plant-based NPs in wound healing are encouraging. With ongoing advances in nanotechnology and green synthesis methods, it is increasingly possible to develop biocompatible and eco-friendly NPs with enhanced wound healing efficacy [68]. A preclinical murine study on *Catharanthus roseus* and *Azadirachta indica*–derived AgNPs showed potent antimicrobial effects against MDR bacteria and significantly enhanced wound closure (~94% and ~87% healing) [99]. Another study formulated *Cyperus rotundus*–mediated AgNP hydrogels, demonstrating strong antibacterial and antioxidant properties, with effective wound closure and minimal cytotoxicity [148]. Researchers are also exploring multifunctional NPs that combine antimicrobial action with regenerative properties and controlled drug release. Integration with smart wound dressings and biosensing materials could lead to next-generation therapies tailored to individual patient needs [149]. To fully realize this potential, interdisciplinary collaboration between botanists, material scientists, pharmacologists, and clinicians will be essential, alongside the development of standardized regulatory pathways for safe and effective clinical application.

8 Conclusion

Nanoparticles have revolutionized wound healing by offering targeted, controlled, and multifunctional therapeutic strategies that overcome the limitations of conventional wound care. Due to their nanoscale size, high surface-area-to-volume ratio, and ability to penetrate biological barriers, NPs can interact with cellular and molecular components of wounds more effectively than traditional treatments. When synthesized using aromatic plant extracts, NPs inherit bioactive compounds that possess antimicrobial, anti-inflammatory, antioxidant, and regenerative properties, critical for all stages of wound repair. These phytochemical-loaded NPs not only accelerate healing by promoting fibroblast proliferation, collagen synthesis, and angiogenesis but also provide an effective defense against infection and oxidative stress at the wound site. The green synthesis of such NPs is eco-friendly and biocompatible, making them ideal candidates for clinical applications. Their incorporation into delivery systems like hydrogels, nanofiber mats, and biodegradable films enhances localized action, minimizes systemic toxicity, and reduces the frequency of dressing changes. Although still in the early stages of clinical translation, plant-based NPs represent a promising advancement in regenerative medicine and wound management. Continued research, standardization, and clinical validation will be essential to fully harness their therapeutic potential. Overall, NPs especially those derived from MAPs offer a powerful and sustainable approach to improving wound healing outcomes and represent the future of smart, efficient, and patient-friendly wound care technologies.

Acknowledgments The corresponding author is grateful to the SERB-National Postdoctoral Fellowship program and Science and Engineering Research Board, Department of Science and Technology, New Delhi, India (grant number: PDF/2016/002730), for financial support.

Competing Interests The authors declare that they have no competing interest.

Declaration of Generative AI and AI-Assisted Technologies in the Writing Process During the preparation of this work, the book chapter author(s) have not taken any assistance from AI, and the book chapter author(s) reviewed and edited the content as needed and take(s) full responsibility for the content of the publication.

References

1. Bohling MW (2014) Chapter 17—Wound healing. In: Langley-Hobbs SJ, Demetriou JL, Ladlow JF (eds) Feline soft tissue and general surgery. W.B. Saunders, Philadelphia, pp 171–175. https://doi.org/10.1016/B978-0-7020-4336-9.00017-2
2. Singh S, Young A, McNaught C-E (2017) The physiology of wound healing. Surgery (Oxford) 35(9):473–477. https://doi.org/10.1016/j.mpsur.2017.06.004
3. Guo S, Dipietro LA (2010) Factors affecting wound healing. J Dent Res 89(3):219–229. https://doi.org/10.1177/0022034509359125
4. Schilrreff P, Alexiev U (2022) Chronic inflammation in non-healing skin wounds and promising natural bioactive compounds treatment. Int J Mol Sci 23(9):4928. https://doi.org/10.3390/ijms23094928
5. Lagoa T, Queiroga MC, Martins L (2024) An overview of wound dressing materials. Pharmaceuticals (Basel) 17(9):1110. https://doi.org/10.3390/ph17091110
6. Pereira RF, Bártolo PJ (2016) Traditional therapies for skin wound healing. Adv Wound Care (New Rochelle) 5(5):208–229. https://doi.org/10.1089/wound.2013.0506
7. Kurul F, Turkmen H, Cetin AE, Topkaya SN (2025) Nanomedicine: how nanomaterials are transforming drug delivery, bio-imaging, and diagnosis. Next Nanotechnol 7:100129. https://doi.org/10.1016/j.nxnano.2024.100129
8. Fathi-Achachelouei M, Knopf-Marques H, Ribeiro da Silva CE, Barthès J, Bat E, Tezcaner A, Vrana NE (2019) Use of nanoparticles in tissue engineering and regenerative medicine. Front Bioeng Biotechnol 7:113. https://doi.org/10.3389/fbioe.2019.00113

9. Kazemi S, Hosseingholian A, Gohari SD, Feirahi F, Moammeri F, Mesbahian G, Moghaddam ZS, Ren Q (2023) Recent advances in green synthesized nanoparticles: from production to application. Mater Today Sustain 24:100500. https://doi.org/10.1016/j.mtsust.2023.100500
10. Mittal AK, Chisti Y, Banerjee UC (2013) Synthesis of metallic nanoparticles using plant extracts. Biotechnol Adv 31(2):346–356. https://doi.org/10.1016/j.biotechadv.2013.01.003
11. El-Seedi HR, Omara MS, Omar AH, Elakshar MM, Shoukhba YM, Duman H, Karav S, Rashwan AK, El-Seedi AH, Altaleb HA, Gao H, Saeed A, Jefri OA, Guo Z, Khalifa SAM (2024) Updated review of metal nanoparticles fabricated by green chemistry using natural extracts: biosynthesis, mechanisms, and applications. Bioengineering 11(11):1095. https://www.mdpi.com/2306-5354/11/11/1095
12. Zamani S, Fathi M, Ebadi M-T, Máthé Á (2025) Global trade of medicinal and aromatic plants. A review. J Agric Food Res 21:101910. https://doi.org/10.1016/j.jafr.2025.101910
13. Singh H, Desimone MF, Pandya S, Jasani S, George N, Adnan M, Aldarhami A, Bazaid AS, Alderhami SA (2023b) Revisiting the green synthesis of nanoparticles: uncovering influences of plant extracts as reducing agents for enhanced synthesis efficiency and its biomedical applications. Int J Nanomedicine 18:4727–4750. https://doi.org/10.2147/ijn.S419369
14. Dikshit PK, Kumar J, Das AK, Sadhu S, Sharma S, Singh S, Gupta PK, Kim BS (2021) Green synthesis of metallic nanoparticles: applications and limitations. Catalysts 11(8):902. https://www.mdpi.com/2073-4344/11/8/902
15. Amutha Gokul T, Ramesh Kumar K, Venkatachalam K, Suresh Babu R, Veeramanikandan V, Sagadevan S, Balaji P (2024) Plant-based nanostructure for wound healing—an emerging paradigm for effective therapy. Inorg Chem Commun 162:112162. https://doi.org/10.1016/j.inoche.2024.112162
16. Kashi M, Noei M, Chegini Z, Shariati A (2024) Natural compounds in the fight against Staphylococcus aureus biofilms: a review of antibiofilm strategies [Review]. Front Pharmacol 15:1491363. https://doi.org/10.3389/fphar.2024.1491363
17. Shariati A, Didehdar M, Razavi S, Heidary M, Soroush F, Chegini Z (2022) Natural compounds: a hopeful promise as an antibiofilm agent against Candida species [Review]. Front Pharmacol 13:917787. https://doi.org/10.3389/fphar.2022.917787
18. Olteanu G, Neacșu SM, Joița FA, Musuc AM, Lupu EC, Ioniță-Mîndrican CB, Lupuliasa D, Mititelu M (2024) Advancements in regenerative hydrogels in skin wound treatment: a comprehensive review. Int J Mol Sci 25(7):3849. https://doi.org/10.3390/ijms25073849
19. Yates CC, Hebda P, Wells A (2012) Skin wound healing and scarring: fetal wounds and regenerative restitution. Birth Defects Res C Embryo Today 96(4):325–333. https://doi.org/10.1002/bdrc.21024
20. Awlqadr FH, Majeed KR, Altemimi AB, Hassan AM, Qadir SA, Saeed MN, Faraj AM, Salih TH, Abd Al-Manhel AJ, Najm MAA, Tsakali E, Van Impe JFM, Abd El-Maksoud AA, Abedelmaksoud TG (2025) Nanotechnology-based herbal medicine: preparation, synthesis, and applications in food and medicine. J Agric Food Res 19:101661. https://doi.org/10.1016/j.jafr.2025.101661
21. Kushwaha A, Goswami L, Kim BS (2022) Nanomaterial-based therapy for wound healing. Nanomaterials (Basel) 12(4):618. https://doi.org/10.3390/nano12040618
22. Nandhini J, Karthikeyan E, Rajeshkumar S (2024) Nanomaterials for wound healing: current status and futuristic frontier. Biomed Technol 6:26–45. https://doi.org/10.1016/j.bmt.2023.10.001
23. Wang F, Zhang W, Li H, Chen X, Feng S, Mei Z (2022) How effective are nano-based dressings in diabetic wound healing? A comprehensive review of literature. Int J Nanomedicine 17:2097–2119
24. Mendes C, Thirupathi A, Corrêa M, Gu Y, Silveira PCL (2022) The use of metallic nanoparticles in wound healing: new perspectives. Int J Mol Sci 23(23):15376. https://doi.org/10.3390/ijms232315376

25. Lu H, Zhang S, Wang J, Chen Q (2021) A review on polymer and lipid-based nanocarriers and its application to nano-pharmaceutical and food-based systems. Front Nutr 8:783831. https://doi.org/10.3389/fnut.2021.783831
26. Hajishoreh NK, Jamalpoor Z, Rasouli R, Asl AN, Sheervalilou R, Akbarzadeh A (2023) The recent development of carbon-based nanoparticles as a novel approach to skin tissue care and management—a review. Exp Cell Res 433(2):113821. https://doi.org/10.1016/j.yexcr.2023.113821
27. Tîlmaciu CM, Morris MC (2015) Carbon nanotube biosensors. Front Chem 3:59. https://doi.org/10.3389/fchem.2015.00059
28. Hano C, Abbasi BH (2022) Plant-based green synthesis of nanoparticles: production, characterization and applications. Biomolecules 12(1):31. https://www.mdpi.com/2218-273X/12/1/31
29. Vanlalveni C, Lallianrawna S, Biswas A, Selvaraj M, Changmai B, Rokhum SL (2021) Green synthesis of silver nanoparticles using plant extracts and their antimicrobial activities: a review of recent literature. RSC Adv 11(5):2804–2837. https://doi.org/10.1039/D0RA09941D
30. Puri A, Mohite P, Maitra S, Subramaniyan V, Kumarasamy V, Uti DE, Sayed AA, El-Demerdash FM, Algahtani M, El-kott AF, Shati AA, Albaik M, Abdel-Daim MM, Atangwho IJ (2024) From nature to nanotechnology: the interplay of traditional medicine, green chemistry, and biogenic metallic phytonanoparticles in modern healthcare innovation and sustainability. Biomed Pharmacother 170:116083. https://doi.org/10.1016/j.biopha.2023.116083
31. Afshar M, Rezaei A, Eghbali S, Nasirizadeh S, Alemzadeh E, Alemzadeh E, Shadi M, Sedighi M (2024) Nanomaterial strategies in wound healing: a comprehensive review of nanoparticles, nanofibres and nanosheets. Int Wound J 21(7):e14953. https://doi.org/10.1111/iwj.14953
32. Kujath P, Michelsen A (2008) Wounds—from physiology to wound dressing. Dtsch Arztebl Int 105(13):239–248. https://doi.org/10.3238/arztebl.2008.0239
33. Raziyeva K, Kim Y, Zharkinbekov Z, Kassymbek K, Jimi S, Saparov A (2021) Immunology of acute and chronic wound healing. Biomolecules 11(5):700. https://doi.org/10.3390/biom11050700
34. Whitney JD (2005) Overview: acute and chronic wounds. Nurs Clin 40(2):191–205. https://doi.org/10.1016/j.cnur.2004.09.002
35. Pollock PJ, Schumacher J (2012) Chapter 23—Principles of wound management. In: Mair TS, Love S, Schumacher J, Smith RKW, Frazer G (eds) Equine medicine, surgery and reproduction, 2nd edn. W.B. Saunders, Philadelphia, pp 469–487. https://doi.org/10.1016/B978-0-7020-2801-4.00023-7
36. Herman T, Popowicz P, Bordoni B (2024) Wound classification [Updated 2023 Aug 17]
37. Salazar J, Carmona T, Zacconi FC, Venegas-Yazigi D, Cabello-Verrugio C, Il Choi W, Vilos C (2022) The human dermis as a target of nanoparticles for treating skin conditions. Pharmaceutics 15(1):10. https://doi.org/10.3390/pharmaceutics15010010
38. Golebiewska EM, Poole AW (2015) Platelet secretion: from haemostasis to wound healing and beyond. Blood Rev 29(3):153–162. https://doi.org/10.1016/j.blre.2014.10.003
39. Reinke JM, Sorg H (2012) Wound repair and regeneration. Eur Surg Res 49(1):35–43. https://doi.org/10.1159/000339613
40. Sen CK (2023) Human wound and its burden: updated 2022 compendium of estimates. Adv Wound Care (New Rochelle) 12(12):657–670. https://doi.org/10.1089/wound.2023.0150
41. Sangnim T, Puri V, Dheer D, Venkatesh DN, Huanbutta K, Sharma A (2024) Nanomaterials in the wound healing process: new insights and advancements. Pharmaceutics 16(3):300. https://doi.org/10.3390/pharmaceutics16030300
42. Logothetidis S (2011) Nanotechnology: principles and applications. In: Nanostructured materials and their applications. Springer, Cham, pp 1–22
43. Blanco-Fernandez B, Castano O, Mateos-Timoneda MÁ, Engel E, Pérez-Amodio S (2021) Nanotechnology approaches in chronic wound healing. Adv Wound Care (New Rochelle) 10(5):234–256

44. Sanmugam A, Sellappan LK, Manoharan S, Rameshkumar A, Kumar RS, Almansour AI, Arumugam N, Kim HS, Vikraman D (2024) Development of chitosan-based cerium and titanium oxide loaded polycaprolactone for cutaneous wound healing and antibacterial applications. Int J Biol Macromol 256(Pt 1):128458. https://doi.org/10.1016/j.ijbiomac.2023.128458
45. Skriabin AS, Shakurov AV, Vesnin VR, Lukina YS, Tsygankov PA, Bionyshev-Abramov LL, Serejnikova NB, Vorob'ev EV (2022) Titanium membranes with hydroxyapatite/titania bioactive ceramic coatings: characterization and in vivo biocompatibility testing. ACS Omega 7(51):47880–47891. https://doi.org/10.1021/acsomega.2c05718
46. Biju V, Itoh T, Anas A, Sujith A, Ishikawa M (2008) Semiconductor quantum dots and metal nanoparticles: syntheses, optical properties, and biological applications. Anal Bioanal Chem 391:2469–2495
47. Singh D, Thapa S, Singh KRB, Verma R, Singh RP, Singh J (2023a) Cadmium selenide quantum dots and its biomedical applications. Mater Lett X 18:100200. https://doi.org/10.1016/j.mlblux.2023.100200
48. Begines B, Ortiz T, Pérez-Aranda M, Martínez G, Merinero M, Argüelles-Arias F, Alcudia A (2020) Polymeric nanoparticles for drug delivery: recent developments and future prospects. Nanomaterials (Basel) 10(7):1403. https://doi.org/10.3390/nano10071403
49. Waheed I, Ali A, Tabassum H, Khatoon N, Lai WF, Zhou X (2024) Lipid-based nanoparticles as drug delivery carriers for cancer therapy. Front Oncol 14:1296091. https://doi.org/10.3389/fonc.2024.1296091
50. Ahmad N, Bukhari SNA, Hussain MA, Ejaz H, Munir MU, Amjad MW (2024) Nanoparticles incorporated hydrogels for delivery of antimicrobial agents: developments and trends. RSC Adv 14(19):13535–13564
51. Sharma A, Khanna S, Kaur G, Singh I (2021) Medicinal plants and their components for wound healing applications. Future J Pharm Sci 7(1):53. https://doi.org/10.1186/s43094-021-00202-w
52. Klein-Junior LC, de Souza MR, Viaene J, Bresolin TMB, de Gasper AL, Henriques AT, Heyden YV (2021) Quality control of herbal medicines: from traditional techniques to state-of-the-art approaches. Planta Med 87(12–13):964–988. https://doi.org/10.1055/a-1529-8339
53. Li S, Han Q, Qiao C, Song J, Lung Cheng C, Xu H (2008) Chemical markers for the quality control of herbal medicines: an overview. Chin Med 3(1):7. https://doi.org/10.1186/1749-8546-3-7
54. Kumar B, Vijayakumar M, Govindarajan R, Pushpangadan P (2007) Ethnopharmacological approaches to wound healing—exploring medicinal plants of India. J Ethnopharmacol 114(2):103–113. https://doi.org/10.1016/j.jep.2007.08.010
55. Nur MG, Rahman M, Dip TM, Hossain MH, Hossain NB, Baratchi S, Padhye R, Houshyar S (2025) Recent advances in bioactive wound dressings. Wound Repair Regen 33(1):e13233. https://doi.org/10.1111/wrr.13233
56. Andleeb A, Khan H, Andleeb A, Khan M, Tariq M (2024) Advances in chronic wound management: from conventional treatment to novel therapies and biological dressings. Crit Rev Biomed Eng 52(5):29–62. https://doi.org/10.1615/CritRevBiomedEng.2024053066
57. Lindholm C, Searle R (2016) Wound management for the 21st century: combining effectiveness and efficiency. Int Wound J 13 Suppl 2(Suppl 2):5–15. https://doi.org/10.1111/iwj.12623
58. Priyadarshini S, Sulava S, Bhol R, Jena S (2019) Green synthesis of silver nanoparticles using Azadirachta indica and Ocimum sanctum leaf extract. Curr Sci 117(8):1300–1307
59. Chae SY, Park R, Hong SW (2022) Surface-mediated high antioxidant and anti-inflammatory effects of astaxanthin-loaded ultrathin graphene oxide film that inhibits the overproduction of intracellular reactive oxygen species. Biomater Res 26(1):30. https://doi.org/10.1186/s40824-022-00276-4
60. Arribas-López E, Zand N, Ojo O, Snowden MJ, Kochhar T (2022) A systematic review of the effect of Centella asiatica on wound healing. Int J Environ Res Public Health 19(6):3266. https://www.mdpi.com/1660-4601/19/6/3266

61. Iosageanu A, Mihai E, Seciu-Grama AM, Utoiu E, Gaspar-Pintiliescu A, Gatea F, Cimpean A, Craciunescu O (2024) In vitro wound-healing potential of phenolic and polysaccharide extracts of aloe vera gel. J Funct Biomater 15(9):266. https://doi.org/10.3390/jfb15090266
62. Hajialyani M, Tewari D, Sobarzo-Sánchez E, Nabavi SM, Farzaei MH, Abdollahi M (2018) Natural product-based nanomedicines for wound healing purposes: therapeutic targets and drug delivery systems. Int J Nanomedicine 13:5023–5043. https://doi.org/10.2147/ijn.S174072
63. Zhao L, Sabri AHB, McGuckin MB, Li L, Wang Z, Larrañeta E, Donnelly RF (2025) Implantable microarray patch: engineering at the nano and macro scale for sustained therapeutic release via synthetic biodegradable polymers. Adv Mater Technol 10(6):2400346
64. Soundhararajan R, Mohamed Sheik Meeran SAF, Prakash SP, Mohammad W, Kandasamy R, Srinivasan H (2020) Multi potent aromatic nano colloid: synthesis, characterization and applications. AMB Express 10(1):168. https://doi.org/10.1186/s13568-020-01104-5
65. Kshtriya V, Koshti B, Gour N (2021) Chapter Four—Green synthesized nanoparticles: classification, synthesis, characterization, and applications. In: Verma SK, Das AK (eds) Comprehensive analytical chemistry, vol 94. Elsevier, Amsterdam, pp 173–222. https://doi.org/10.1016/bs.coac.2020.12.009
66. Mobarak MB, Sikder MF, Muntaha KS, Islam S, Rabbi SF, Chowdhury F (2025) Plant extract-mediated green synthesized CuO nanoparticles for environmental and microbial remediation: a review covering basic understandings to mechanistic study. Nanoscale Adv 7(9):2418–2445
67. Tang X, Li L, You G, Li X, Kang J (2023) Metallic elements combine with herbal compounds upload in microneedles to promote wound healing: a review [Review]. Front Bioeng Biotechnol 11:1283771. https://doi.org/10.3389/fbioe.2023.1283771
68. Nandhini SN, Sisubalan N, Vijayan A, Karthikeyan C, Gnanaraj M, Gideon DAM, Jebastin T, Varaprasad K, Sadiku R (2023) Recent advances in green synthesized nanoparticles for bactericidal and wound healing applications. Heliyon 9(2):e13128. https://doi.org/10.1016/j.heliyon.2023.e13128
69. Tyavambiza C, Dube P, Goboza M, Meyer S, Madiehe AM, Meyer M (2021) Wound healing activities and potential of selected African medicinal plants and their synthesized biogenic nanoparticles. Plants (Basel) 10(12):2635. https://www.mdpi.com/2223-7747/10/12/2635
70. Lakkim V, Reddy MC, Lekkala VVV, Lebaka VR, Korivi M, Lomada D (2023) Antioxidant efficacy of green-synthesized silver nanoparticles promotes wound healing in mice. Pharmaceutics 15(5):1517. https://doi.org/10.3390/pharmaceutics15051517
71. Grigore-Gurgu L, Dumitrașcu L, Aprodu I (2025) Aromatic herbs as a source of bioactive compounds: an overview of their antioxidant capacity, antimicrobial activity, and major applications. Molecules 30(6):1304. https://doi.org/10.3390/molecules30061304
72. Sharma P, Navneet, Kaushal A (2022) Green nanoparticle formation toward wound healing, and its application in drug delivery approaches. Eur J Med Chem Rep 6:100088. https://doi.org/10.1016/j.ejmcr.2022.100088
73. Azmir J, Zaidul ISM, Rahman MM, Sharif K, Mohamed A, Sahena F, Jahurul M, Ghafoor K, Norulaini N, Omar, A (2013) Techniques for extraction of bioactive compounds from plant materials: a review. J Food Eng 117(4):426–436
74. Kumar S, Basumatary IB, Sudhani HPK, Bajpai VK, Chen L, Shukla S, Mukherjee A (2021) Plant extract mediated silver nanoparticles and their applications as antimicrobials and in sustainable food packaging: a state-of-the-art review. Trends Food Sci Technol 112:651–666. https://doi.org/10.1016/j.tifs.2021.04.031
75. Fahim M, Shahzaib A, Nishat N, Jahan A, Bhat TA, Inam A (2024) Green synthesis of silver nanoparticles: a comprehensive review of methods, influencing factors, and applications. JCIS Open 16:100125. https://doi.org/10.1016/j.jciso.2024.100125
76. Thomas S, Gonsalves RA, Jose J, Zyoud SH, Prasad AR, Garvasis J (2024) Plant-based synthesis, characterization approaches, applications and toxicity of silver nanoparticles: a comprehensive review. J Biotechnol 394:135–149. https://doi.org/10.1016/j.jbiotec.2024.08.009

77. Lithi IJ, Nakib KIA, Chowdhury AS, Hossain MS (2025) A review on the green synthesis of metal (Ag, Cu, and Au) and metal oxide (ZnO, MgO, Co3O4, and TiO2) nanoparticles using plant extracts for developing antimicrobial properties. Nanoscale Adv 7(9):2446–2473
78. Priya N, Kaur K, Sidhu AK (2021) Green synthesis: an eco-friendly route for the synthesis of iron oxide nanoparticles. Front Nanotechnol 3. https://doi.org/10.3389/fnano.2021.655062
79. Moosavy MH, de la Guardia M, Mokhtarzadeh A, Khatibi SA, Hosseinzadeh N, Hajipour N (2023) Green synthesis, characterization, and biological evaluation of gold and silver nanoparticles using Mentha spicata essential oil. Sci Rep 13(1):7230. https://doi.org/10.1038/s41598-023-33632-y
80. Chaudhari RK, Shah PA, Shrivastav PS (2023) Green synthesis of silver nanoparticles using Adhatoda vasica leaf extract and its application in photocatalytic degradation of dyes. Discov Nano 18(1):135. https://doi.org/10.1186/s11671-023-03914-5
81. Lakhan MN, Chen R, Shar AH, Chand K, Shah AH, Ahmed M, Ali I, Ahmed R, Liu J, Takahashi K, Wang J (2020) Eco-friendly green synthesis of clove buds extract functionalized silver nanoparticles and evaluation of antibacterial and antidiatom activity. J Microbiol Methods 173:105934. https://doi.org/10.1016/j.mimet.2020.105934
82. Ajitha B, Reddy YAK, Lee Y, Kim MJ, Ahn CW (2019) Biomimetic synthesis of silver nanoparticles using Syzygium aromaticum (clove) extract: catalytic and antimicrobial effects. Appl Organomet Chem 33(5):e4867
83. Tekin V, Kozgus Guldu O, Dervis E, Yurt Kilcar A, Uygur E, Biber Muftuler FZ (2019) Green synthesis of silver nanoparticles by using eugenol and evaluation of antimicrobial potential. Appl Organomet Chem 33(7):e4969
84. Devika M (2021) An overview of medicinal and aromatic plants. Med Aromat Plants (Los Angeles) 10(5):368
85. Pandey AK, Kumar P, Saxena MJ, Maurya P (2020) Chapter 6—Distribution of aromatic plants in the world and their properties. In: Florou-Paneri P, Christaki E, Giannenas I (eds) Feed additives. Academic Press, New York, pp 89–114. https://doi.org/10.1016/B978-0-12-814700-9.00006-6
86. Thomas J, Joy P, Mathew S, Skaria BP (2000) Plant sources of aroma chemicals and medicines in India, pp 104–108
87. Kliszcz A, Danel A, Puła J, Barabasz-Krasny B, Możdżeń K (2021) Fleeting beauty—the world of plant fragrances and their application. Molecules 26(9):2473
88. Albahri G, Badran A, Hijazi A, Daou A, Baydoun E, Nasser M, Merah O (2023) The therapeutic wound healing bioactivities of various medicinal plants. Life (Basel) 13(2):317. https://doi.org/10.3390/life13020317
89. Škrovánková S, Mišurcová L, Machů L (2012) Antioxidant activity and protecting health effects of common medicinal plants. Adv Food Nutr Res 67:75–139. https://doi.org/10.1016/b978-0-12-394598-3.00003-4
90. Fierascu RC, Fierascu I, Baroi AM, Ortan A (2021) Selected aspects related to medicinal and aromatic plants as alternative sources of bioactive compounds. Int J Mol Sci 22(4):1521. https://doi.org/10.3390/ijms22041521
91. Riaz U, Iqbal S, Sohail MI, Samreen T, Ashraf M, Akmal F, Siddiqui A, Ahmad I, Naveed M, Khan NI (2021) A comprehensive review on emerging importance and economical potential of medicinal and aromatic plants (MAPs) in current scenario. Pak J Agric Res 34(2)
92. Kala CP, Dhyani PP, Sajwan BS (2006) Developing the medicinal plants sector in northern India: challenges and opportunities. J Ethnobiol Ethnomed 2:32. http://europepmc.org/abstract/PMC/PMC1562365
93. Kumar A, Jnanesha AC (2016) Medicinal and aromatic plants biodiversity in India and their future prospects: a review. Ind J Unani Med 9(1):10–17
94. Gautam D, Dolma KG, Khandelwal B, Gupta M, Singh M, Mahboob T, Teotia A, Thota P, Bhattacharya J, Goyal R, Oliveira SMR, Pereira ML, Wiart C, Wilairatana P, Eawsakul K, Rahmatullah M, Saravanabhavan SS, Nissapatorn V (2023) Green synthesis of silver

nanoparticles using Ocimum sanctum Linn. and its antibacterial activity against multidrug resistant Acinetobacter baumannii. PeerJ 11:e15590. https://doi.org/10.7717/peerj.15590

95. Mohanta YK, Biswas K, Mishra AK, Patra B, Mishra B, Panda J, Avula SK, Varma RS, Panda BP, Nayak D (2024) Amelioration of gold nanoparticles mediated through Ocimum oil extracts induces reactive oxygen species and mitochondrial instability against MCF-7 breast carcinoma. RSC Adv 14(38):27816–27830
96. Rao YS, Kotakadi VS, Prasad T, Reddy AV, Gopal D (2013) Green synthesis and spectral characterization of silver nanoparticles from Lakshmi tulasi (Ocimum sanctum) leaf extract. Spectrochim Acta A Mol Biomol Spectrosc 103:156–159
97. Tailor G, Yadav BL, Chaudhary J, Joshi M, Suvalka C (2020) Green synthesis of silver nanoparticles using Ocimum canum and their anti-bacterial activity. Biochem Biophys Rep 24:100848. https://doi.org/10.1016/j.bbrep.2020.100848
98. Anjum MR, Islam R, Hasan MM, Rashid B, Parvez MMM, Sarkar S, Ali MH, Asha MEZ (2023) Effect of neem extract with zinc oxide nanoparticles for burn wound treatment. J Appl Life Sci Int 26(1):17–23
99. Lakkim V, Reddy MC, Pallavali RR, Reddy KR, Reddy CV, Inamuddin, Bilgrami AL, Lomada D (2020) Green synthesis of silver nanoparticles and evaluation of their antibacterial activity against multidrug-resistant bacteria and wound healing efficacy using a murine model. Antibiotics (Basel) 9(12):902. https://doi.org/10.3390/antibiotics9120902
100. Hamilton AE, Gilbert RJ (2023) Curcumin release from biomaterials for enhanced tissue regeneration following injury or disease. Bioengineering (Basel) 10(2):262. https://www.mdpi.com/2306-5354/10/2/262
101. Muniyappan N, Pandeeswaran M, Amalraj A (2021) Green synthesis of gold nanoparticles using Curcuma pseudomontana isolated curcumin: its characterization, antimicrobial, antioxidant and anti-inflammatory activities. Environ Chem Ecotoxicol 3:117–124. https://doi.org/10.1016/j.enceco.2021.01.002
102. Thirupathi A, Guzzatti MFM, Corrêa MEAB, Venturini LM, de Roch Casagrande L, Lima IR, Da Costa C, De Pieri E, Tietbohl LTW, Feuser PE, Machado-de-Ávila RA, Gu Y, Silveira PCL (2023) Green synthesis of gold nanoparticles with curcumin or açai in the tissue repair of palatal wounds. Antioxidants 12(8):1574. https://www.mdpi.com/2076-3921/12/8/1574
103. Şuică-Bunghez IR, Senin RM, Sorescu AA, Ganciarov M, Răut I, Firincă C, Constantin M, Gifu IC, Stoica R, Fierăscu I, Fierăscu RC (2024) Application of Lavandula angustifolia Mill. extracts for the phytosynthesis of silver nanoparticles: characterization and biomedical potential. Plants (Basel) 13(3):333. https://doi.org/10.3390/plants13030333
104. Teymouri H, Mohammadimehr M, Ahanjan M, Sheidaei S, Saeedi M, Mellati A (2024) Effect of collagen hydrogel containing Lavandula officinalis essential oil nanoemulsion in wound healing of infectious burn. Iran J Microbiol 16(3):376–388. https://doi.org/10.18502/ijm.v16i3.15795
105. Abdelbaky AS, Mohamed AMHA, Sharaky M, Mohamed NA, Diab YM (2023) Green approach for the synthesis of ZnO nanoparticles using Cymbopogon citratus aqueous leaf extract: characterization and evaluation of their biological activities. Chem Biol Technol Agric 10(1):63. https://doi.org/10.1186/s40538-023-00432-5
106. Arsene MMJ, Davares AKL, Goriainov S, Viktorovna PI, Parfait K, Andreevna SL, Vyacheslavovna YN, Aleksandrovna VE, Zdislavovna EI, Sulikoevich KZ, Alekseevna KA, Nikolaïevna BM (2024) Synthesis of green-engineered silver nanoparticles using Cymbopogon citratus (lemongrass) and its antibacterial activity against clinical Pseudomonas aeruginosa. Int J One Health 10(1):100–113
107. Rakib-Uz-Zaman SM, Hoque Apu E, Muntasir MN, Mowna SA, Khanom MG, Jahan SS, Akter N, Khan MAR, Shuborna NS, Shams SM, Khan K (2022) Biosynthesis of silver nanoparticles from Cymbopogon citratus leaf extract and evaluation of their antimicrobial properties. Challenges 13(1):18. https://www.mdpi.com/2078-1547/13/1/18
108. Balčiūnaitienė A, Liaudanskas M, Puzerytė V, Viškelis J, Janulis V, Viškelis P, Griškonis E, Jankauskaitė VJ (2022) Eucalyptus globulus and Salvia officinalis extracts mediated green

synthesis of silver nanoparticles and their application as an antioxidant and antimicrobial agent. Plants 11(8):1085

109. Elbhnsawi NA, Elwakil BH, Hassanin AH, Shehata N, Elshewemi SS, Hagar M, Olama ZA (2023) Nano-chitosan/eucalyptus oil/cellulose acetate nanofibers: manufacturing, antibacterial and wound healing activities. Membranes (Basel) 13(6):604. https://www.mdpi.com/2077-0375/13/6/604
110. Nova BGV, dos Santos Silva L, da Silva Andrade M, de Santana AVS, da Silva LCT, Sá GC, Zafred IF, de Aguiar Moreira PH, Monteiro CA, da Silva LCN (2024) The essential oil of Melaleuca alternifolia incorporated into hydrogel induces antimicrobial and anti-inflammatory effects on infected wounds by Staphylococcus aureus. Biomed Pharmacother 173:116389
111. Ramadan MA, Shawkey AE, Rabeh MA, Abdellatif AO (2020) Promising antimicrobial activities of oil and silver nanoparticles obtained from Melaleuca alternifolia leaves against selected skin-infecting pathogens. J Herb Med 20:100289
112. Makky S, Rezk N, Abdelsattar AS, Hussein AH, Eid A, Essam K, Kamel AG, Fayez MS, Azzam M, Agwa MM, El-Shibiny A (2023) Characterization of the biosynthesized Syzygium aromaticum-mediated silver nanoparticles and its antibacterial and antibiofilm activity in combination with bacteriophage. Results Chem 5:100686. https://doi.org/10.1016/j.rechem.2022.100686
113. Sethuram L, Thomas J, Mukherjee A, Chandrasekaran N (2021) Eugenol micro-emulsion reinforced with silver nanocomposite electrospun mats for wound dressing strategies. Mater Adv 2(9):2971–2988
114. Elmitwalli OSMMS, Kassim DAK, Algahiny AT, Henari FZ (2025) Green synthesis of metal nanoparticles using Cinnamomum-based extracts and their applications. Nanotechnol Sci Appl 18:93–114
115. Seyed Ahmadi SG, Farahpour MR, Hamishehkar H (2019) Topical application of Cinnamon verum essential oil accelerates infected wound healing process by increasing tissue antioxidant capacity and keratin biosynthesis. Kaohsiung J Med Sci 35(11):686–694. https://doi.org/10.1002/kjm2.12120
116. Sharma VK, Prateeksha, Gupta SC, Singh BN, Rao CV, Barik SK (2022) Cinnamomum verum-derived bioactives-functionalized gold nanoparticles for prevention of obesity through gut microbiota reshaping. Mater Today Bio 13:100204. https://doi.org/10.1016/j.mtbio.2022.100204
117. Mohamed A, Dayo M, Alahmadi S, Ali S (2024) Anti-inflammatory and antimicrobial activity of silver nanoparticles green-synthesized using extracts of different plants. Nanomaterials (Basel) 14(17):1383. https://doi.org/10.3390/nano14171383
118. Aftab R, Akbar F, Afroz A, Asif A, Khan MR, Rehman N, Zeeshan N (2024) Mentha piperita silver nanoparticle-loaded hydrocolloid film for enhanced diabetic wound healing in rats. J Wound Care 33(Suppl 3a):xlviii–lx. https://doi.org/10.12968/jowc.2024.33.Sup3a.xlviii
119. Hudz N, Kobylinska L, Pokajewicz K, Horčinová Sedláčková V, Fedin R, Voloshyn M, Myskiv I, Brindza J, Wieczorek PP, Lipok J (2023) Mentha piperita: essential oil and extracts, their biological activities, and perspectives on the development of new medicinal and cosmetic products. Molecules 28(21):7444. https://doi.org/10.3390/molecules28217444
120. Robles-Martínez M, Patiño-Herrera R, Pérez-Vázquez FJ, Montejano-Carrizales JM, González JFC, Pérez E (2020) Mentha piperita as a natural support for silver nanoparticles: a new anti-Candida albicans treatment. Colloid Interface Sci Commun 35:100253. https://doi.org/10.1016/j.colcom.2020.100253
121. Pirtarighat S, Ghannadnia M, Baghshahi S (2019) Biosynthesis of silver nanoparticles using Ocimum basilicum cultured under controlled conditions for bactericidal application. Mater Sci Eng C Mater Biol Appl 98:250–255. https://doi.org/10.1016/j.msec.2018.12.090

122. Qamar SUR, Virijević K, Arsenijević D, Avdović E, Živanović M, Filipović N, Ćirić A, Petrović I (2024) Silver nanoparticles from Ocimum basilicum L. tea: a green route with potent anticancer efficacy. Colloid Interface Sci Commun 59:100771. https://doi.org/10.1016/j.colcom.2024.100771
123. Selvam K, Sudhakar C, Ragu Prasath A (2024) Green synthesis and characterization of silver nanoparticles from sandalwood (Santalum album L.) extract for efficient catalytic reduction, antioxidant and antibacterial activity. Biocatal Agric Biotechnol 57:103094. https://doi.org/10.1016/j.bcab.2024.103094
124. Dogru E, Demirbas A, Altinsoy B, Duman F, Ocsoy I (2017) Formation of Matricaria chamomilla extract-incorporated Ag nanoparticles and size-dependent enhanced antimicrobial property. J Photochem Photobiol B 174:78–83. https://doi.org/10.1016/j.jphotobiol.2017.07.024
125. El Mihyaoui A, Esteves da Silva JCG, Charfi S, Candela Castillo ME, Lamarti A, Arnao MB (2022) Chamomile (Matricaria chamomilla L.): a review of ethnomedicinal use, phytochemistry and pharmacological uses. Life (Basel) 12(4):479. https://doi.org/10.3390/life12040479
126. Shankar SS, Ahmad A, Sastry M (2003) Geranium leaf assisted biosynthesis of silver nanoparticles. Biotechnol Prog 19(6):1627–1631. https://doi.org/10.1021/bp034070w
127. Goli F, Divsalar A, Rasouli M, Gholami H (2024) Ajwain-assisted synthesis of oxalipalladium nanoparticles for colorectal cancer treatment: enhanced anticancer activity and protein interaction profiling. New J Chem 48(35):15620–15632
128. MubarakAli D, Kim H, Venkatesh PS, Kim JW, Lee SY (2023) A systemic review on the synthesis, characterization, and applications of palladium nanoparticles in biomedicine. Appl Biochem Biotechnol 195(6):3699–3718. https://doi.org/10.1007/s12010-022-03840-9
129. Maleki H, Doostan M, Khoshnevisan K, Baharifar H, Maleki SA, Fatahi MA (2024) Zingiber officinale and thymus vulgaris extracts co-loaded polyvinyl alcohol and chitosan electrospun nanofibers for tackling infection and wound healing promotion. Heliyon 10(1):e23719. https://doi.org/10.1016/j.heliyon.2023.e23719
130. Tavukcuoglu O, Evcimen Duygulu N, Altinbay A, Ciftci F (2024) Green synthesis of silver nanoparticles from Thymus vulgaris and Sambucus nigra extracts in poly(vinyl alcohol) nanofiber matrix: in vitro evaluation. Ind Crop Prod 222:119825. https://doi.org/10.1016/j.indcrop.2024.119825
131. Imtiaz T, Priyadharshini R, Rajeshkumar S, Sinduja P (2021) Green synthesis and characterization of silver nanoparticles synthesized using piper longum and its antioxidant activity. J Pharm Res Int 33(51A):342–352
132. Mirbehbahani FS, Hejazi F, Najmoddin N, Asefnejad A (2020) Artemisia annua L. as a promising medicinal plant for powerful wound healing applications. Prog Biomater 9(3):139–151. https://doi.org/10.1007/s40204-020-00138-z
133. Algahtani MS, Ahmad MZ, Shaikh IA, Abdel-Wahab BA, Nourein IH, Ahmad J (2021) Thymoquinone loaded topical nanoemulgel for wound healing: formulation design and in-vivo evaluation. Molecules 26(13):3863. https://doi.org/10.3390/molecules26133863
134. Palanisamy CP, Poompradub S, Sansanaphongpricha K, Jayaraman S, Subramani K, Sonsudin F (2025) Green synthesis of Nigella sativa-mediated silver nanoparticles for enhanced antibacterial activity and wound healing: mechanistic insights and biomedical applications. Environ Nanotechnol Monitor Manag 24:101085. https://doi.org/10.1016/j.enmm.2025.101085
135. Chiangnoon R, Samee W, Uttayarat P, Jittachai W, Ruksiriwanich W, Sommano SR, Athikomkulchai S, Chittasupho C (2022) Phytochemical analysis, antioxidant, and wound healing activity of Pluchea indica L. (less) branch extract nanoparticles. Molecules 27(3):635. https://doi.org/10.3390/molecules27030635
136. Mojally M, Sharmin E, Alhindi Y, Obaid NA, Almaimani R, Althubiti M, Idris S, Abdelghany AH, Refaat B, Al-Amodi HS, Abdalla AN, Kamel HFM (2022) Hydrogel

films of methanolic Mentha piperita extract and silver nanoparticles enhance wound healing in rats with diabetes Type I. J Taibah Univ Sci 16(1):308–316. https://doi.org/10.1080/16583655.2022.2054607
137. Ataide JA, Zanchetta B, Santos ÉM, Fava ALM, Alves TFR, Cefali LC, Chaud MV, Oliveira-Nascimento L, Souto EB, Mazzola PG (2022) Nanotechnology-based dressings for wound management. Pharmaceuticals (Basel) 15(10):1286. https://doi.org/10.3390/ph15101286
138. Kaur H, Gogoi B, Sharma I, Das DK, Azad MA, Pramanik DD, Pramanik A (2024) Hydrogels as a potential biomaterial for multimodal therapeutic applications. Mol Pharm 21(10):4827–4848. https://doi.org/10.1021/acs.molpharmaceut.4c00595
139. Sofi HS, Akram T, Tamboli AH, Majeed A, Shabir N, Sheikh FA (2019) Novel lavender oil and silver nanoparticles simultaneously loaded onto polyurethane nanofibers for wound-healing applications. Int J Pharm 569:118590. https://doi.org/10.1016/j.ijpharm.2019.118590
140. Su Y, Toftdal MS, Le Friec A, Dong M, Han X, Chen M (2021) 3D electrospun synthetic extracellular matrix for tissue regeneration. Small Sci 1(7):2100003. https://doi.org/10.1002/smsc.202100003
141. Liu Q, Luo S, Peng J, Chang R (2024) Electrospun nanofibers from plant natural products: a new approach toward efficient wound healing. Int J Nanomedicine 19:13973–13990. https://doi.org/10.2147/ijn.S501970
142. Liu JX, Dong WH, Mou XJ, Liu GS, Huang XW, Yan X, Zhou CF, Jiang S, Long YZ (2020) In situ electrospun zein/thyme essential oil-based membranes as an effective antibacterial wound dressing. ACS Appl Bio Mater 3(1):302–307. https://doi.org/10.1021/acsabm.9b00823
143. Afonso IS, Cardoso B, Nobrega G, Minas G, Ribeiro JE, Lima RA (2024) Green synthesis of nanoparticles from olive oil waste for environmental and health applications: a review. J Environ Chem Eng 12(5):114022. https://doi.org/10.1016/j.jece.2024.114022
144. Velgosova O, Dolinská S, Podolská H, Mačák L, Čižmárová E (2024) Impact of plant extract phytochemicals on the synthesis of silver nanoparticles. Materials (Basel) 17(10):2252. https://www.mdpi.com/1996-1944/17/10/2252
145. Sharma A, Madhunapantula SV, Robertson GP (2012) Toxicological considerations when creating nanoparticle-based drugs and drug delivery systems. Expert Opin Drug Metab Toxicol 8(1):47–69. https://doi.org/10.1517/17425255.2012.637916
146. Shipkowski KA, Betz JM, Birnbaum LS, Bucher JR, Coates PM, Hopp DC, MacKay D, Oketch-Rabah H, Walker NJ, Welch C, Rider CV (2018) Naturally complex: perspectives and challenges associated with Botanical Dietary Supplement Safety assessment. Food Chem Toxicol 118:963–971. https://doi.org/10.1016/j.fct.2018.04.007
147. Gupta PD, Daswani PG, Birdi TJ (2014) Approaches in fostering quality parameters for medicinal botanicals in the Indian context. Indian J Pharm 46(4):363–371. https://doi.org/10.4103/0253-7613.135946
148. Sharma S, Bose A, Biswas S, Sen S, Roy I (2025) Cyperus rotundus mediated green synthesis of silver nanoparticles for antibacterial wound dressing applications. Sci Rep 15(1):18394. https://doi.org/10.1038/s41598-025-03555-x
149. Khalid A, Bai D, Abraham AN, Jadhav A, Linklater D, Matusica A, Nguyen D, Murdoch BJ, Zakhartchouk N, Dekiwadia C et al (2020) Electrospun nanodiamond–silk fibroin membranes: a multifunctional platform for biosensing and wound-healing applications. ACS Appl Mater Interfaces 12(43):48408–48419

Aromatic Plant-Based Nanoparticles for Improving Skin and Their Restoration Process

Loushambam Samananda Singh, Ajay Kumare Saw, Chabungbam Satyananda Singh, Yungkham Rajeevkumar Singh, Waikhom Somraj Singh, Thounaojam Premlata, and Lautambam Sanathoiba Singha

Abstract Aromatic flora represents abundant reservoirs of bioactive constituents, particularly polyphenolic compounds and essential oils, that manifest antimicrobial, anti-inflammatory, antioxidant, and regenerative characteristics advantageous for dermal health. However, traditional formulations frequently encounter obstacles such as inadequate solubility, instability, and restricted penetration, which constrain their therapeutic efficacy. Nanotechnology offers a proficient approach to surmount these impediments by increasing solubility, stability, bioavailability, and controlled release of phytochemicals, while facilitating targeted delivery to specific cutaneous strata. Aromatic plant-derived nanoparticles including lipid-based systems, polymeric carriers, liposomes, and eco-friendly synthesized metallic nanoparticles exhibit improved encapsulation efficiency, dermal permeation, and sustained release. Mechanistically, these nanoformulations modulate oxidative stress, attenuate pro-inflammatory cytokines, inhibit matrix metalloproteinases, and stimulate angiogenesis and re-epithelialization, culminating in expedited wound healing, enhanced photoprotection, diminished hyperpigmentation, and mitigation of acne and inflammatory dermatoses. While preclinical and nascent translational investigations bolster their therapeutic potential, significant challenges persist in standardizing botanical sources, ensuring formulation stability, executing thorough safety

L. S. Singh (✉) · W. S. Singh · L. S. Singha
Institute of Pharmacy, Assam Don Bosco University, Sonapur, Assam, India

A. K. Saw (✉)
Department of Genomic Medicine, University of Texas, MD Anderson Cancer Center, Houston, TX, USA
e-mail: aksaw@mdanderson.org

C. S. Singh
Department of Physics, Assam Don Bosco University, Sonapur, Assam, India

Y. R. Singh
Department of Zoology, Assam Don Bosco University, Sonapur, Assam, India

T. Premlata
Department of Botany, Assam Don Bosco University, Sonapur, Assam, India

A. Husen (ed.), *Sustainable Uses of Aromatic Plants in Nanotechnology*, Smart Nanomaterials Technology, https://doi.org/10.1007/978-981-95-7409-4_11

assessments, and producing robust clinical evidence. Future trajectories entail the integration of green synthesis, stimuli-responsive delivery, and rational carrier design within cohesive regulatory frameworks. Collectively, aromatic plant-derived nanoparticles epitomize a sustainable and adaptable platform with substantial potential to revolutionize dermatological care and skin restoration therapies.

Keywords Aromatic plants · Nanoparticles · Phytochemicals · Essential oils · Dermal drug delivery

1 Introduction

Aromatic plant-based nanoparticles represent a rapidly advancing frontier in dermatological science, offering innovative solutions for skin improvement and restoration by leveraging the unique properties of plant-derived bioactives and the enhanced delivery capabilities of nanotechnology. Nanoparticle preparation using extracts from aromatic plants such as dittany, sage, sea buckthorn, and calendula has demonstrated significant potential in topical applications, primarily due to the high content of phenolic and flavonoid compounds that impart strong antioxidant activity to the nanoparticles' surfaces, which is crucial for combating oxidative stress and supporting skin health. These plant-based nanoparticles, particularly silver nanoparticles (AgNPs) synthesized via green methods, exhibit favorable physicochemical characteristics, including controlled morphology, stability, and safety, as evidenced by their retention in the upper layers of the stratum corneum without deep skin penetration, thus minimizing systemic exposure and potential toxicity during short-term use [1].

The integration of plant extracts into nanostructures addresses several limitations inherent to traditional herbal therapies, such as poor solubility, instability, and limited skin permeability, by enhancing the bioavailability, photostability, and controlled release of active phytochemicals directly to the target skin layers [2, 3]. In the context of wound healing and skin restoration, plant-based nanomaterials have been shown to modulate key biological processes, including the suppression of inflammatory cytokines, reduction of oxidative factors, and promotion of neovascularization through upregulation of growth factors like VEGF, FGF, and PDGF, thereby accelerating tissue repair and regeneration [4]. Furthermore, encapsulation of plant polyphenols and flavonoids in nanoparticles has proven particularly effective in treating inflammatory skin conditions such as acne, as these nanoformulations enhance the stability, solubility, and targeted delivery of bioactives, resulting in improved anti-inflammatory, antioxidant, and antibacterial effects with fewer adverse effects than conventional therapies [5]. Lipid-based nanoparticles, such as nanostructured lipid carriers (NLCs) and solid lipid nanoparticles (SLNs), have emerged as especially promising carriers for essential oils and herbal extracts, offering high encapsulation efficiency, sustained release, and improved skin permeation, which collectively contribute to superior therapeutic outcomes and user compliance [3].

Despite these advances, several challenges remain, including the need for standardization of plant sources, comprehensive characterization of phytochemical profiles, formulation stability, and thorough toxicity assessments, as well as the translation of promising in vitro and in vivo findings into robust clinical evidence [2]. Addressing these gaps will require interdisciplinary collaboration to optimize extraction methods, nanoparticle design, and delivery systems, ensuring that the full therapeutic potential of aromatic plant-based nanoparticles is realized in skin care and restoration.

2 Aromatic Plants and Their Bioactive Components

Plants have served as a primary source of medicinal agents since ancient times, with aromatic plants occupying a prominent position due to their therapeutic potential. These species are rich reservoirs of bioactive compounds, particularly polyphenols and essential oils, which demonstrate a broad spectrum of health-promoting activities, including antimicrobial, antioxidant, and anti-inflammatory effects [6–10]. Notable examples include oregano (*Origanum vulgare*), lavender (*Lavandula* spp.), rosemary (*Rosmarinus officinalis*), chamomile (*Matricaria chamomilla*), tea tree (*Melaleuca alternifolia*), and basil (*Ocimum basilicum*), all of which have been extensively utilized in traditional medicine and food preservation [11]. In aromatic plants, essential oils are predominantly biosynthesized through two primary biochemical pathways: the 2-C-methylerythritol-4-phosphate and the mevalonic acid pathway. These plants also produce an array of secondary metabolites, including terpenoids, glycosides, and alkaloids, each possessing significant pharmacological properties. Their natural origin, environmental compatibility, and wide range of industrial applications contribute to their increasing global demand [12]. Hydrodistillation remains an environmentally sustainable and economically viable method for extracting volatile constituents from essential oils. These oils and plant extracts have gained considerable attention as natural growth promoters, often demonstrating superior efficacy compared to conventional antibiotics [13]. Their applications span the food, feed, pharmaceutical, and cosmetic sectors, where they enhance health and nutrition in humans and animals and serve as effective adjuvants to synthetic drugs. Bioactive constituents from aromatic plants also protect biological systems against oxidative stress by quenching singlet oxygen and modulating enzymatic activity, including cytochrome-related pathways [14]. Furthermore, certain aromatic species exhibit high adaptability for cultivation in degraded or contaminated soils. They can produce essential oils free from toxic metal contaminants while simultaneously stabilizing or removing soil pollutants, thereby offering promising applications in phytoremediation [15].

2.1 Key Aromatic Plants Used for Skin Health

Aromatic plants, with their rich phytochemical profiles, offer a wide range of therapeutic benefits and have long been recognized for their significant role in promoting skin health. Their antimicrobial, antioxidant, and anti-inflammatory properties make them particularly valuable in cosmeceuticals, supporting the management of various dermatological conditions [16]. Many indigenous communities have traditionally utilized local aromatic plants valued for their wound healing and anti-inflammatory effects. While the use of these herbs in cosmetics is well established, there is an increasing emphasis on the need for standardized formulations and clinical validation to ensure their safety and efficacy in dermatological applications. Overall, aromatic plants are extensively employed in skin care for their anti-inflammatory, antioxidant, antimicrobial, and rejuvenating activities. The key aromatic plants, along with their major bioactive compounds and associated skin health benefits, are presented in Table 1.

Table 1 Major aromatic plants, compounds, and skin benefits

Plant name (family)	Main bioactive compounds	Skin health benefits	Key references
Origanum vulgare (Lamiaceae)	Carvacrol, thymol, linalool	Antimicrobial, anti-inflammatory, anti-aging, wound healing, anti-hyperpigmentation	Bora et al. [17], Lombrea et al. [18], Olivero-Verbel et al. [19]
Calendula officinalis (Asteraceae)	Flavonoids, triterpenoids	Anti-inflammatory, wound healing, soothing, tissue regeneration	Olivero-Verbel et al. [19]
Matricaria chamomilla (Asteraceae)	Bisabolol, chamazulene	Anti-inflammatory, soothing, anti-allergic, skin barrier repair	Olivero-Verbel et al. [19]
Thymus vulgaris (Lamiaceae)	Thymol, carvacrol	Antimicrobial, antioxidant, anti-inflammatory	Olivero-Verbel et al. [19]
Melaleuca alternifolia (Myrtaceae)	Terpinen-4-ol, α-terpineol	Antimicrobial (notably anti-acne), anti-inflammatory	Michalak [20], Olivero-Verbel et al. [19]
Cymbopogon citratus (Poaceae)	Citral, geraniol	Antimicrobial, antioxidant, soothing	Olivero-Verbel et al. [19]
Foeniculum vulgare (Apiaceae)	Anethole, fenchone	Antioxidant, anti-inflammatory, skin rejuvenation	Olivero-Verbel et al. [19]
Piper nigrum (Piperaceae)	Piperine, β-caryophyllene	Antioxidant, antimicrobial, wound healing	Michalak [20], Salehi et al. [21], Sounouvou et al. [22]
Camellia sinensis (Theaceae)	Catechins, polyphenols	Antioxidant, anti-aging, photoprotective, skin lightening	Olivero-Verbel et al. [19]
Citrus sinensis (Rutaceae)	Limonene, flavonoids	Antioxidant, skin brightening, anti-inflammatory	Olivero-Verbel et al. [19]
Althaea officinalis (Malvaceae)	Mucilage, flavonoids	Soothing, anti-inflammatory, moisturizing	Olivero-Verbel et al. [19]
Sambucus nigra (Adoxaceae)	Anthocyanins, flavonoids	Anti-inflammatory, photoprotective, tissue regeneration	Olivero-Verbel et al. [19]

3 Nanotechnology in Skin Therapy

Nanotechnology has revolutionized skin therapy by introducing nanoparticles—ultrafine particles generally measuring between 1 and 100 nm—that offer unique physicochemical properties and significant therapeutic advantages. Nanoparticles are broadly grouped into inorganic (such as gold, silver, and metal oxides), polymer-based, and lipid-based types, each with distinct characteristics that influence their interaction with skin tissues and their suitability for various therapeutic applications [23]. Inorganic nanoparticles, like gold and silver, are valued for their stability, ease of functionalization, and potential intrinsic therapeutic effects, while polymeric nanoparticles provide biocompatibility, controlled degradation, and versatility in drug encapsulation. Lipid-based nanoparticles, such as SLNs and liposome nanoparticles, mimic biological membranes, enhancing compatibility and facilitating the delivery of both hydrophilic and lipophilic drugs. "The small size and large surface area-to-volume ratio of nanoparticles enable them to traverse the skin's formidable barrier the stratum corneum more effectively than conventional formulations, allowing for enhanced penetration into deeper skin layers or even targeted delivery to specific skin structures such as hair follicles" [24]. This property is particularly advantageous for treating localized skin diseases, including skin cancers, chronic wounds, and inflammatory conditions, where precise delivery is crucial for efficacy and minimizing systemic side effects. Nanoparticles also enable controlled and sustained drug release, maintaining therapeutic concentrations at the target site over extended periods and reducing the frequency of application, which facilitates greater patient compliance and diminishes adverse reactions arising from peak concentrations [25].

Furthermore, nanoparticles can protect encapsulated drugs from degradation by enzymes or environmental factors, thereby enhancing drug stability and bioavailability. Functionalization of nanoparticle surfaces with ligands or polymers can further improve targeting, prolong circulation time, and enable stimuli-responsive release, tailoring therapy to the pathological microenvironment of skin diseases [26]. In the context of skin cancer, nanoparticles enable passive tumor targeting through enhanced permeability and retention effect, allowing for higher drug accumulation in tumor tissues, while sparing healthy skin, and can be engineered to co-deliver multiple therapeutic agents for synergistic effects or to combine therapy with diagnostic imaging (theranostics) [23]. The versatility of nanoparticles extends to gene therapy, where they serve as carriers for nucleic acids such as siRNA, overcoming the skin's barrier and intracellular obstacles to achieve efficient gene silencing in dermatological disorders [27]. Despite these advances, several challenges persist, such as the requirement for deeper understanding of skin penetration mechanisms, minimizing toxicity, and ensuring scalability and regulatory compliance for clinical translation [24].

4 Fabrication of Aromatic Plant-Based Nanoparticles

Aromatic plant-based nanoparticles are increasingly fabricated using green synthesis approaches that leverage the rich phytochemical content of aromatic plants, offering an eco-friendly, sustainable, and cost-effective alternative to conventional physical and chemical nanoparticle synthesis methods, which are usually associated with the use of toxic reagents and the production of hazardous residues. In green synthesis, plant extracts rich in flavonoids, terpenoids, polyphenols, sugars, and other bioactive molecules act as natural reducing, stabilizing, and capping agents, facilitating the reduction of metal ions and the formation of nanoparticles characterized by uniform size and defined morphology [28–34]. The process typically involves mixing plant extracts with metal salt solutions under controlled conditions (such as temperature, pH, and reaction time), leading to nucleation, growth, and stabilization of nanoparticles, with the specific phytochemical profile of the aromatic plant influencing the physicochemical properties of the resulting nanomaterials [34]. Characterization of these nanoparticles is essential to ensure their suitability for various applications, and a suite of analytical techniques is employed: scanning electron microscopy and transmission electron microscopy provide detailed information on particle size and shape; dynamic light scattering measures hydrodynamic diameter and size distribution; X-ray diffraction reveals crystalline structure; Fourier-transform infrared spectroscopy detects surface functional groups and confirms the presence of plant-derived capping agents; and zeta potential analysis assesses surface charge and colloidal stability [32, 35]. These characterization methods collectively enable precise control and optimization of nanoparticle properties, which is crucial for their performance in biomedical, environmental, and agricultural applications [35].

Encapsulation of aromatic plant-derived bioactives or essential oils within nanocarriers further enhances their stability, bioavailability, and targeted delivery, and several encapsulation strategies are widely used: nanoemulsions, which represent thermodynamically stable mixtures of oil and water maintained by surfactants, are particularly effective for encapsulating hydrophobic essential oils and improving their solubility and controlled release; liposomes, consisting of concentric phospholipid bilayers, can encapsulate both hydrophobic and hydrophilic compounds, offering biocompatibility and protection of sensitive plant actives; SLNs provide a solid lipid matrix for encapsulation, combining the advantages of controlled release and high physical stability; and polymeric nanoparticles, formed from natural or synthetic polymers, allow for tunable release profiles and surface functionalization for targeted delivery [36].

The integration of green synthesis with advanced encapsulation techniques not only maximizes the therapeutic and functional potential of aromatic plant-based nanoparticles but also aligns with the principles of sustainability and safety, minimizing environmental impact and toxicity concerns [32, 35, 36]. These plant-based nanomaterials have demonstrated promising applications in drug delivery, antimicrobial therapy, biosensing, food preservation, and agriculture, owing to their unique physicochemical properties, biocompatibility, and the inherent bioactivity of aromatic plant constituents. However, challenges remain in scaling up production, achieving uniformity in nanoparticle characteristics, and ensuring regulatory compliance, necessitating further research into process optimization, mechanistic understanding of plant-mediated synthesis, and comprehensive safety assessments.

5 Mechanisms of Action: Enhancing Skin Delivery with Aromatic Plant-Based Nanoparticles

Aromatic phytochemical nanoparticles represent advanced dermal delivery systems, wherein their nanoscale dimensions facilitate enhanced penetration across the stratum corneum, and controlled release kinetics ensure prolonged therapeutic action. These nanoformulations exhibit multifunctional bioactivities including antioxidant protection, anti-inflammatory regulation, antimicrobial properties, and reinforcement of the cutaneous barrier, which collectively contribute to improved skin hydration, elasticity, and resilience against environmental insults. The underlying mechanisms are depicted in Fig. 1.

5.1 Antioxidant Mechanisms in Skin Protection

Nanoparticles synthesized from plant extracts enhance the stability, solubility, and skin penetration of bioactive compounds, enabling controlled delivery and maximizing therapeutic efficacy. The primary mechanism by which these nanoparticles

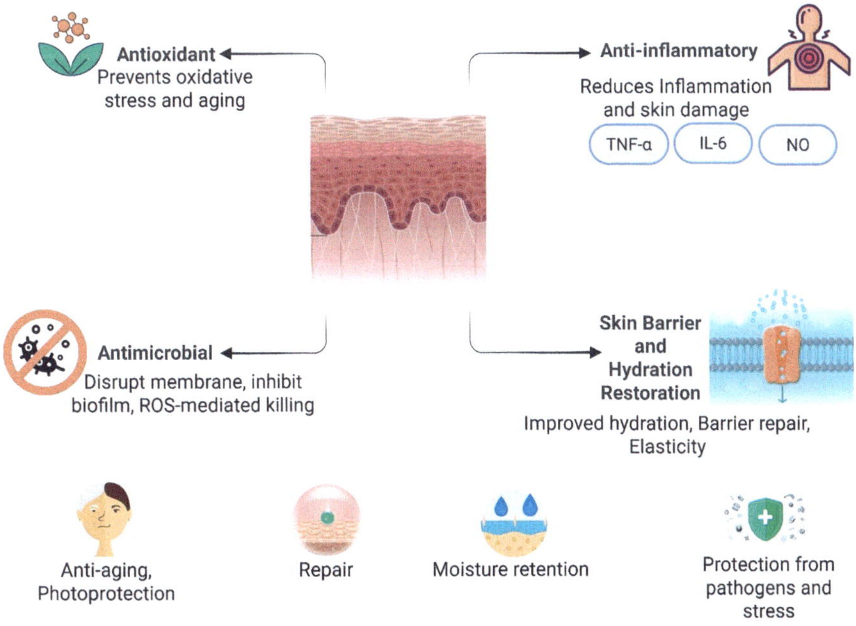

Fig. 1 Mechanisms of aromatic plant-based nanoparticles in skin protection and restoration

improve skin health is through the neutralization of reactive oxygen species (ROS), which are generated by environmental stressors like ultraviolet (UV) radiation and pollution. Excessive ROS may adversely affect vital cellular structures, accelerate skin aging, and trigger inflammatory responses. For instance, ginseng root-derived exosome-like nanoparticles have been shown to protect skin cells from UV-induced oxidative stress by reducing ROS production, downregulating proapoptotic and aging-related genes, and suppressing pro-inflammatory signaling pathways such as activator protein-1, ultimately preventing cell death and promoting skin resilience [37]. Similarly, nanoparticles loaded with plant polyphenols and flavonoids exhibit superior antioxidant and anti-inflammatory effects compared to their non-encapsulated counterparts, as the nanoformulation enhances their stability, bioavailability, and skin permeability, leading to more effective free radical scavenging and alleviation of oxidative damage [2, 5]. The improved solubility and controlled release properties of these nanoparticles ensure sustained antioxidant activity, which is crucial for ongoing protection against daily oxidative insults [3, 38]. Additionally, the use of plant extracts in the green synthesis of metallic nanoparticles, such as silver or gold, further augments their antioxidant capacity due to the presence of surface-bound phenolic and flavonoid compounds, which not only confer strong free radical scavenging activity but also contribute to the nanoparticles' stability and safety for topical application [39]. These antioxidant mechanisms are complemented by the nanoparticles' ability to modulate key molecular pathways involved in skin aging and inflammation, such as the inhibition of matrix metalloproteinases that degrade collagen and the suppression of nuclear factor-kappa B (NF-κB) and mitogen-activated protein kinase (MAPK) signaling, which are key mediators of inflammatory and stress responses [37]. Furthermore, the enhanced skin penetration achieved by reducing particle size and increasing surface area allows these nanoparticles to deliver active compounds into deeper skin layers, providing comprehensive protection and repair at the cellular level [3]. The cumulative effect of these mechanisms is reflected in improved skin hydration, elasticity, and photoprotection, as well as accelerated wound healing and reduced signs of aging, such as wrinkles and hyperpigmentation [2]. Despite these promising outcomes, ongoing research emphasizes the need for standardized plant sourcing, thorough characterization of nanoparticle formulations, and comprehensive safety assessments to ensure consistent efficacy and minimize potential toxicity.

5.2 *Anti-inflammatory Pathways and Immune Modulation*

Aromatic plant-based nanoparticles offer a promising approach to anti-inflammatory therapy by improving the therapeutic efficacy, absorption, and site-specific delivery of phytochemicals. Their mechanisms involve direct modulation of inflammatory mediators and regulation of cellular signaling pathways. At the molecular level, nanoparticles synthesized from aromatic plant extracts rich in terpenoids, flavonoids, polyphenols, tannins, saponins, alkaloids, and essential oil constituents can inhibit the production of pro-inflammatory cytokines such as IL-6, TNF-α, and nitric oxide, as demonstrated by astragaloside IV-chitosan nanoparticles, which significantly reduced these mediators in LPS-stimulated macrophages by acting on metabolic pathways including arginine and proline, glutathione, sphingolipid, and glycerophospholipid metabolism, as well as amino acid metabolism [40, 41]. These nanoparticles also interfere with the dynamic balance between pro- and anti-inflammatory responses, helping to restore immune homeostasis by modulating the cellular immune microenvironment and suppressing excessive inflammatory cascades. Plant-derived exosome-like nanoparticles further contribute by transferring proteins, lipids, and nucleic acids from plant cells to mammalian cells, mediating interspecies communication and regulating immune responses at the cellular level [42]. The high surface area-to-volume ratio of nanoparticles enhances their cellular uptake and ligand-binding properties, allowing for more efficient delivery and interaction with intracellular targets, which is crucial for the inhibition of key inflammatory signaling pathways such as NF-κB, MAPK, and JAK/STAT [43, 44]. Additionally, the encapsulation of phytochemicals within nanoparticles overcomes challenges of poor solubility, instability, and limited bioavailability, ensuring sustained release and improved pharmacokinetics, which is particularly important for chronic inflammatory conditions where long-term modulation of inflammation is required [44, 45]. Plant-mediated green synthesis of nanoparticles delivers an eco-friendly and sustainable approach but also imparts multifunctional biomedical properties, including antioxidant, anti-apoptotic, and anti-angiogenic effects, which synergistically contribute to the suppression of inflammation and the prevention of tissue damage [46]. Furthermore, these nanoparticles can target specific tissues or cells, reduce off-target effects, and minimize toxicity compared to conventional anti-inflammatory drugs [45].

5.3 Antimicrobial Actions Against Pathogens

Aromatic plant-based nanoparticles, particularly those synthesized using plant extracts rich in aromatic phytochemicals, have emerged as potent antimicrobial agents due to their multifaceted mechanisms of action that address the growing challenge of multidrug-resistant pathogens. These nanoparticles, often metallic or metal oxide in nature (such as silver, gold, zinc oxide, and copper oxide), leverage the unique physicochemical properties conferred by their nanoscale size and the bioactive compounds from aromatic plants to exert broad-spectrum antimicrobial effects. The primary mechanism involves direct interaction with microbial cell walls and membranes, where nanoparticles attach to and disrupt the structural integrity of these barriers, leading to increased permeability, leakage of cellular contents, and ultimately cell death; this process is largely independent of traditional antibiotic resistance mechanisms, as it does not require penetration into the cell or specific molecular targets, thereby circumventing common resistance pathways such as efflux pumps and enzymatic degradation [47, 48]. Additionally, these nanoparticles induce oxidative stress within microbial cells by generating ROS, which damage vital biomolecules including DNA, proteins, and lipids, further compromising cell viability and function. The release of metal ions from the nanoparticles can also interfere with microbial metabolic processes and enzyme activity, disrupt gene expression, and inhibit ribosomal function, collectively impairing the microorganism's ability to survive and replicate [47]. Another significant mechanism is the inhibition of biofilm formation, a key factor in microbial resistance and persistence; nanoparticles prevent the initial adhesion of microbes to surfaces and disrupt established biofilms, making pathogens more susceptible to host defenses and other antimicrobial agents [47, 48]. Furthermore, aromatic plant-based nanoparticles can modulate quorum sensing, the bacterial communication system that regulates virulence and resistance traits, thereby attenuating pathogenicity and reducing the likelihood of resistance development. The synergistic effects observed when nanoparticles are combined with plant-derived phytochemicals or conventional antibiotics further enhance their antimicrobial efficacy, often resulting in lower required doses and reduced toxicity [49]. Importantly, the biogenic synthesis of these nanoparticles using aromatic plant extracts not only imparts additional antimicrobial properties from the plant's secondary metabolites but also improves biocompatibility and reduces environmental toxicity compared to chemically synthesized counterparts [50]. Despite these promising attributes, the application of aromatic plant-based nanoparticles in clinical and agricultural settings necessitates careful consideration of potential risks, including toxicity to non-target organisms and impacts on beneficial microbial communities, underscoring the need for further research to optimize their safety and efficacy profiles [47, 51].

5.4 Skin Barrier Restoration and Hydration Enhancement

Aromatic plant-based nanoparticles have emerged as innovative agents for skin barrier restoration and hydration enhancement, leveraging both the bioactivity of plant-derived compounds and the advanced delivery capabilities of nanotechnology. These nanoparticles, often formulated as lipid-based carriers such as SLNs, NLCs, or nanoemulsions, encapsulate essential oils, polyphenols, flavonoids, or ceramides to overcome the inherent limitations of plant actives, such as poor solubility, instability, and low skin permeability, thereby significantly improving their bioavailability and efficacy for topical applications [3]. The mechanisms underlying their action are multifaceted: first, the nanoscale size of these carriers facilitates deeper penetration through the stratum corneum, the skin's outermost barrier, allowing for more efficient delivery of active compounds to target sites within the epidermis and dermis [52]. For example, rosemary essential oil-loaded NLCs have demonstrated superior skin hydration and elasticity compared to free essential oil, attributed to the nanoparticles' ability to enhance the retention and controlled release of bioactives, as well as their interaction with skin lipids to reinforce barrier function [53]. Similarly, nanoemulsions containing polyphenolic-rich extracts, such as those from

Elaeis guineensis leaves, provide controlled and sustained release of antioxidants like catechin, which, according to in silico modeling, can bind to aquaporin-3 proteins in the stratum corneum, thereby promoting water retention and improving skin hydration [54]. The inclusion of penetration enhancers (e.g., glycerin) and optimization of particle size further augment the delivery and efficacy of encapsulated actives, as seen in ceramide-loaded oil-in-water nanoemulsions that restore barrier integrity and moisture content in dermatitis models by efficiently replenishing depleted skin lipids [52]. Moreover, the encapsulation of plant polyphenols and flavonoids in nanoparticles not only stabilizes these compounds but also enables their controlled release and targeted delivery, enhancing their anti-inflammatory, antioxidant, and antibacterial effects, which are crucial for repairing damaged skin and maintaining hydration [5]. Among nanocarriers, SLNs and NLCs stand out for their high encapsulation efficiency, stability, and favorable sensory characteristics, positioning them as promising platforms for dermatological and cosmetic use; however, continued research is essential to identify novel bioactives and optimize formulations for targeted skin conditions [3].

6 Applications in Skin Restoration

Aromatic phytogenic nanoparticles have emerged as advanced modalities in dermal restoration, attributable to their ability to deliver bioactive constituents with superior stability, enhanced permeability, and sustained therapeutic efficacy. Their clinical and biomedical applications span diverse domains, including wound healing and tissue regeneration, wherein they accelerate re-epithelialization and promote angiogenesis; anti-aging interventions, through stimulation of collagen biosynthesis and attenuation of oxidative stress; and the management of dermatological disorders such as acne, eczema, and psoriasis via barrier reinforcement and immunomodulatory activity. In addition, their demonstrated photoprotective capacity against UV-induced oxidative stress and DNA damage further underscores their potential as versatile nanoplatforms for both therapeutic and cosmetic dermatological applications (Fig. 2).

6.1 *Wound Healing and Tissue Regeneration*

Wound healing and tissue regeneration in skin restoration is a complex, multi-phase biological process involving coordinated cellular and molecular events, hemostasis, inflammation, proliferation, and remodeling regulated by diverse cells and signaling molecules to restore skin integrity and function [55]. While small wounds in healthy adult skin can heal rapidly, larger or chronic wounds, such as those resulting from burns, diabetes, or vascular insufficiency, often face impaired healing, leading to excessive scarring or non-healing ulcers, which underscores the need for advanced

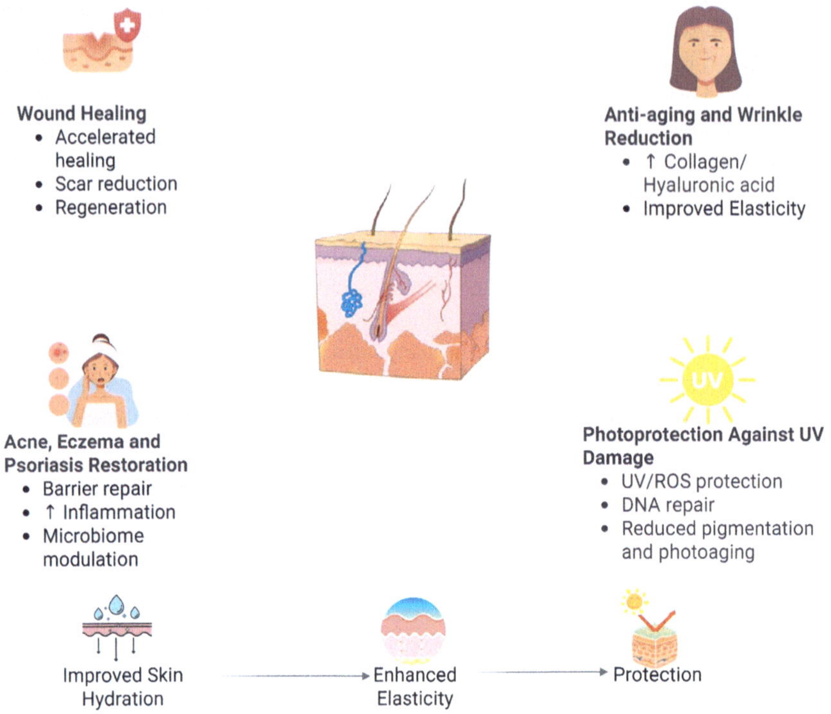

Fig. 2 Applications of aromatic plant-based nanoparticles in skin restoration

therapeutic strategies [56]. Traditional approaches, including wound dressings, skin grafts, and topical agents, have evolved with the advent of innovative technologies such as bioengineered skin substitutes, 3D bioprinting, in situ forming hydrogels, and nanotherapeutics, which aim to recapitulate the structure and function of native skin, accelerate healing, and minimize scarring [57]. Engineered skin substitutes, incorporating scaffolds seeded with autologous or allogeneic cells, growth factors, or bioactive molecules, have shown promise in promoting re-epithelialization, vascularization, and even regeneration of skin appendages like hair follicles and sebaceous glands, moving closer to true tissue regeneration rather than mere repair [58]. Cellular therapies, particularly those utilizing mesenchymal stem/stromal cells (MSCs) and their derivatives (such as exosomes), have demonstrated paracrine effects that modulate inflammation, enhance angiogenesis, and stimulate resident cell proliferation, offering new avenues for treating chronic wounds and reducing pathological fibrosis [59]. The integration of smart biomaterials—such as hydrogels, protein fiber-coated matrices, and marine-derived macromolecules—provides instructive microenvironments that deliver biochemical and biophysical cues, regulate cell behavior, and support endogenous tissue regeneration, while advanced wound dressings with antimicrobial and immunomodulatory properties further optimize the healing milieu [58]. Recent research also highlights the critical role of immune regulation, particularly by regulatory T cells, in balancing inflammation and promoting regenerative outcomes, as well as the influence of specific cytokines like IL-22, which can drive both tissue repair and, if dysregulated, pathological processes [60]. Despite these advances, challenges remain, including the need for standardized, cost-effective, and scalable therapies, improved understanding of the mechanisms underlying scarless healing (as observed in fetal skin and certain animal models), and the translation of experimental techniques into clinically effective, personalized treatments [57].

6.2 Anti-aging and Wrinkle Reduction

Anti-aging and wrinkle reduction are central applications in skin restoration, driven by the desire to counteract the visible and functional decline associated with intrinsic aging and environmental factors, particularly UV exposure. Skin aging manifests as loss of elasticity, increased dryness, thinning, and the formation of wrinkles, primarily due to the degradation of structural proteins like collagen and elastin, reduced hyaluronic acid (HA) content, and impaired skin barrier function [61, 62]. Modern anti-aging strategies leverage a multifaceted approach, including topical agents, nutraceuticals, injectables, and advanced biomaterials, each targeting specific molecular pathways to restore skin integrity and appearance. Hyaluronic acid, a key molecule for skin hydration and volume, is widely used in creams, serums, and injectable fillers; its topical and injectable application replenishes moisture, enhances elasticity, and visibly reduces fine lines and wrinkles, with clinical studies confirming significant improvements in skin hydration and wrinkle depth after regular use [62]. Innovations such as RNAi-based therapies delivered via adipose-derived stem cells have demonstrated the ability to modulate endogenous HA metabolism, protecting against UV-induced HA degradation, restoring water content, and smoothing deep wrinkles, thus offering a promising biological strategy for facial rejuvenation [63]. Collagen peptides, administered orally, have shown substantial reductions in wrinkle volume, area, and depth, alongside increased skin moisturization, supporting their role as effective nutraceuticals for anti-aging [64]. Peptides like Tetrapeptide-68 and Argireline, which mimic neuromodulatory effects like Botox, have emerged as non-invasive alternatives, achieving up to 52% wrinkle reduction within weeks by inhibiting neurotransmitter release and promoting muscle relaxation, although their efficacy is sometimes limited by skin absorption challenges [65]. Botanical extracts and antioxidants, such as astaxanthin, oleocanthal, and oleacein, contribute to anti-aging by mitigating oxidative stress, improving skin moisture, and reducing wrinkle formation, with clinical trials highlighting their potential, especially in older adults [66]. Additionally, innovative delivery systems like microneedles and molecular modifications are being explored to enhance the penetration and efficacy of anti-wrinkle agents, addressing limitations of conventional topical applications [67]. Regenerative cosmetics, including tissue engineering and the use of stem cells, are at the forefront of skin restoration, aiming to stimulate collagen synthesis, repair damaged tissues, and rejuvenate aging skin at a cellular level, thus offering long-term improvements in skin structure and function [61]. Despite the proliferation of anti-aging products, the effectiveness of many botanical and vitamin-based formulations remains variable, often complicated using multiple active ingredients and the need for more rigorous, long-term clinical studies to validate their claims [68].

6.3 Skin Restoration in Acne, Eczema, and Psoriasis

The treatment of acne, eczema, and psoriasis for skin restoration involves a multifaceted approach that targets both the underlying inflammation and the repair of the skin barrier, with recent advances highlighting the importance of barrier-supportive therapies, microbiome modulation, and integration of both conventional and alternative treatments. In acne management, topical agents such as retinoids and benzoyl peroxide are standard, but these can exacerbate barrier damage, leading to dryness, erythema, and increased healing time; adjunctive use of ceramide-containing cleansers and moisturizers has been shown to significantly reduce these adverse effects, accelerate barrier repair, and improve overall skin function, as evidenced by lower transepidermal water loss and reduced inflammatory lesion counts compared to basic skincare routines [69]. For eczema (atopic dermatitis), the restoration and maintenance of the lipid barrier is a cornerstone of therapy, with ceramide-based products and topical agents like linoleic acid and amitriptyline offering promising results in reducing dryness, itch, and inflammation without the side effects associated with long-term corticosteroid use, thus supporting a more personalized and sustainable approach to skin restoration [70]. In psoriasis, innovative therapies such as immunomodulatory hydrogel microspheres containing *Cutibacterium acnes*-derived extracellular vesicles have demonstrated the ability to restore skin barrier function, reduce pro-inflammatory gene expression, and rebalance the skin microbiota, which is often disrupted in psoriatic lesions; these approaches not only alleviate symptoms but also promote long-term skin homeostasis by modulating immune cell populations and enhancing resistance to pathogenic bacteria. Across all three conditions, the role of the skin microbiome has gained prominence, with both probiotics and postbiotics emerging as adjunctive therapies that can modulate immune responses, improve barrier integrity, and reduce recurrence rates, as seen in reductions in inflammatory markers and improvements in skin quality in patients with acne, eczema, and psoriasis [71]. Complementary and alternative medicine (CAM), including herbal remedies such as tea tree oil, lavender, turmeric, aloe vera, and sandalwood oil, has shown anti-inflammatory, antimicrobial, and healing properties, with some clinical evidence supporting their efficacy in reducing symptoms and promoting skin restoration, although the quality of evidence varies and further well-designed studies are needed to confirm their long-term benefits and safety [72]. Patient preference for natural therapies is often driven by the perception of fewer side effects and greater control over their condition, particularly in acne, where CAM use is associated with improved psychological outcomes, although confidence in efficacy is lower for eczema and psoriasis. Despite the growing popularity of CAM, systematic reviews indicate that the evidence base remains insufficient for strong recommendations, highlighting the need for individualized treatment plans that integrate both conventional and alternative modalities under medical supervision [73]. The integration of Ayurvedic and naturopathic therapies, focusing on holistic balance and lifestyle modification, further expands the therapeutic landscape, offering external and internal interventions aimed at restoring skin health and systemic equilibrium.

6.4 Photoprotection Against UV Damage

Photoprotection against UV damage is a cornerstone in the prevention and restoration of skin health, given the well-established role of UV radiation in causing acute and chronic skin disorders such as erythema, photoaging, hyperpigmentation, and skin cancers. The skin's natural defense, primarily melanin, provides a degree of intrinsic photoprotection by absorbing and scattering UV radiation and neutralizing ROS, but this endogenous shield is often insufficient, especially in lighter phototypes or under intense sun exposure, necessitating the development and application of exogenous photoprotective strategies [74]. Traditional sunscreens, containing organic and inorganic UV filters, remain the frontline defense, with organic agents absorbing UV and inorganic agents reflecting and scattering both UV and visible light; however, concerns about allergenicity and regulatory limitations on synthetic molecules have driven a surge in interest toward natural photoprotective compounds derived from plants, algae, and marine organisms, which not only absorb UV but also exhibit antioxidant, anti-inflammatory, and immunomodulatory properties, thereby offering multifaceted protection and supporting skin restoration [75]. Recent advances have expanded the photoprotection paradigm beyond UVB and UVA to include visible light and infrared radiation, both of which contribute to oxidative stress and photoaging, particularly in individuals with skin of color, highlighting the need for broad-spectrum protection that incorporates agents like iron oxide, titanium dioxide, and biologically relevant antioxidants into sunscreen formulations [74]. The integration of DNA repair enzymes, such as photolyase, into topical products represents a significant innovation, as these enzymes can actively repair UV-induced DNA lesions like cyclobutane pyrimidine dimers, thus not only preventing but also reversing some of the molecular damage that underlies photoaging and carcinogenesis, and clinical studies suggest that such combined photoprotection and photorepair strategies are effective in reducing new skin lesions in photodamaged skin [76]. Additionally, systemic photoprotection through dietary antioxidants such as carotenoids, vitamins C and E, polyphenols, and specific plant extracts has demonstrated efficacy in enhancing the skin's resilience to UV-induced oxidative stress, with intervention studies showing that long-term consumption of these micronutrients can reduce erythema and DNA damage, and their combination may offer synergistic benefits [77]. Probiotic-derived HA has emerged as another promising agent, with studies showing that HA produced by certain bacterial strains can significantly improve the viability of UVB-irradiated keratinocytes, suggesting a role in both photoprotection and post-exposure skin restoration [78]. Furthermore, novel synthetic and natural compounds, such as benzylidene derivatives of glycyrrhetinic acid and rosemary diterpenes, have shown potent anti-oxidant, anti-apoptotic, and collagen-preserving effects in cellular models, paving the way for their development as next-generation photoprotective agents [79]. The importance of tailored photoprotection is underscored by the variability in skin phototypes, DNA repair capacity, and susceptibility to pigmentary disorders, especially in populations with skin of color, where visible light-induced damage is more pronounced and sunscreen use is often suboptimal, necessitating culturally sensitive education and product development [80]. In addition to topical and systemic approaches, the use of skin-derived precursor cells and stimulation of endogenous protective pathways, such as the Nrf2 antioxidant response, have shown promise in experimental models for both preventing and repairing UV-induced skin damage, further broadening the arsenal for skin restoration [81]. Despite these advances, key challenges persist, such as the need for rigorous clinical trials to validate the efficacy of novel agents, the creation of formulations offering uniform protection across the solar spectrum, and the evaluation of interactions between different light wavelengths that may impact overall photoprotective effectiveness.

7 Factors Influencing Efficacy

The efficacy of bioactive compound delivery systems is governed by a complex interplay of factors, notably particle size, surface charge, concentration of bioactive compounds, stability, and compatibility with the delivery system. Particle size is a critical determinant, as it influences cellular uptake, biodistribution, and bioactivity. Smaller particles generally exhibit higher cellular uptake and bioactivity due to their enhanced ability to cross biological barriers, but may be cleared more rapidly from the body, reducing retention time and potentially limiting therapeutic efficacy; conversely, larger particles may offer improved stability and retention but can face challenges in traversing biological membranes and may accumulate in organs such as the liver and spleen, as seen with gold nanoparticles and nanoemulsions [82]. The surface charge of particles further modulates their interaction with biological systems: positively charged nanoparticles often demonstrate increased cellular uptake and improved delivery to target tissues, such as the brain, or within biofilms, but may also induce higher cytotoxicity and hemolysis, whereas negatively charged or slightly negative particles tend to have better biocompatibility and mucus penetration, with reduced cellular uptake and toxicity [83].

The concentration of bioactive compounds within the carrier affects encapsulation efficiency, loading capacity, and the release profile; higher concentrations can enhance therapeutic outcomes but may also risk instability or phase separation, as observed in quercetin-loaded egg yolk granules [84]. Stability encompassing thermal, photochemical, and storage stability ensures that bioactive compounds remain effective until they reach their target; this is influenced by both particle size and surface charge, with intermediate sizes often providing a balance between protection and bioaccessibility and surface modifications (e.g., multilayer emulsions, surface coatings) enhancing resistance to environmental stressors [84]. Delivery system compatibility is essential for maintaining the integrity and function of the carrier and its payload; compatibility is shaped by the physicochemical properties of both the bioactive compound and the carrier, including hydrophobicity, flexibility, and surface chemistry, which together determine the system's ability to encapsulate, protect, and release the compound effectively in the desired biological context [83, 84]. The interplay of these factors is further complicated by the route of administration (oral, intravenous, inhaled, etc.), the target tissue or organ, and the specific biological barriers encountered, necessitating a tailored approach to the design of delivery systems. For instance, in vaccine delivery, modulation of particle size and surface charge can enhance immune responses, while in drug delivery to the brain or tumors, optimizing these parameters can improve targeting and bioavailability [85]. Additionally, the use of advanced engineering techniques, such as ultrasonic treatment or nanoprecipitation, allows for precise control over particle characteristics, further enhancing efficacy [84].

8 Safety, Toxicity, and Regulatory Aspects

Plant-based nanoparticles have emerged as promising alternatives to conventional nanomaterials due to their eco-friendly synthesis, biocompatibility, and diverse applications in medicine, agriculture, and environmental remediation. However, their safety, toxicity, and regulatory aspects remain critical concerns for clinical and environmental deployment. Cytotoxicity studies reveal that while plant-mediated nanoparticles often exhibit reduced toxicity compared to chemically synthesized counterparts, they are not inherently risk-free; their effects depend on factors such as size, concentration, surface charge, and the specific plant metabolites involved in synthesis, which can influence cellular uptake and biological interactions [86]. In vitro and in vivo assays have demonstrated that plant-based silver nanoparticles (PB-AgNPs), for example, can induce oxidative stress, DNA damage, and organ-specific toxicities (e.g., hepatotoxicity, neurotoxicity) in non-target organisms, with lethal and sublethal effects observed particularly in aquatic models, raising concerns about their environmental impact and potential for bioaccumulation in food chains [87]. Biocompatibility assessments generally support the safer profile of plant-based nanoparticles, as plant-derived capping agents and reducing biomolecules can enhance stability and reduce cytotoxicity, making them suitable for biomedical applications such as drug delivery, imaging, and antimicrobial therapies [86]. Nonetheless, the biocompatibility is not universal; some plant-based nanoparticles may still disrupt cellular homeostasis, trigger immune responses, or interfere with physiological processes in plants and animals, necessitating comprehensive toxicological evaluations tailored to each nanoparticle type and application [88]. Regulatory challenges for clinical translation are substantial, as current frameworks are not fully equipped to address the unique properties of plant-based nanoparticles, including their complex composition, batch-to-batch variability, and interactions with biological systems [89]. Standardized protocols for synthesis, characterization, and toxicity testing are lacking, complicating risk assessment and approval processes. Furthermore, the absence of long-term safety data, the limited understanding of nanoparticle fate in vivo and in the environment, and the need for harmonized international guidelines hinder regulatory acceptance and commercialization [87, 89].

9 Challenges and Future Perspectives

Personalized skincare using nanoformulations offers targeted delivery, enhanced skin penetration, and improved stability of active ingredients but faces major research gaps. Most studies remain in vitro or preclinical, with limited in vivo and clinical evidence to confirm safety, efficacy, and long-term effects in diverse populations. Key challenges include precise drug release control, biocompatibility, large-scale production, and unresolved safety concerns over nanomaterial toxicity and systemic effects [90]. Clinical translation is further limited by skin disease complexity, individual physiological variability, and unclear regulations. Future progress depends on robust in vivo studies, personalized approaches integrating genetic and lifestyle factors, and advances in smart, stimuli-responsive nanocarriers. A growing interest in natural and herbal nanoformulations also demands research into their unique properties and skin interactions.

10 Conclusion

Aromatic plant-derived nanoparticles signify a noteworthy evolution in dermatological treatment by amalgamating the bioactivity of botanically sourced compounds with the delivery benefits of nanotechnology. These systems augment solubility, stability, and dermal penetration, while facilitating controlled and targeted release of phytochemicals, culminating in enhanced outcomes in wound repair, anti-aging, pigmentation abnormalities, acne, and barrier restoration. Their mechanisms encompassing antioxidant, anti-inflammatory, antimicrobial, and skin-regenerative properties underscore their potential as secure and efficacious cosmeceutical and therapeutic agents. Notwithstanding these advancements, hurdles persist in standardizing botanical sources, guaranteeing formulation stability, ascertaining long-term safety, and producing robust clinical evidence. Future advancements will hinge on synchronized protocols for extraction and characterization, sustainable green synthesis, scalable manufacturing, and meticulously designed clinical trials. With these foundations in place, aromatic plant-derived nanoformulations can progress from experimental innovations to clinically substantiated, sustainable solutions for dermal health and rejuvenation.

Competing Interests The authors declare that they have no competing interest.

Declaration of Generative AI and AI-Assisted Technologies in the Writing Process During the preparation of this work, the book chapter author(s) have not taken any assistance from AI, and the book chapter author(s) reviewed and edited the content as needed and take(s) full responsibility for the content of the publication.

References

1. Siakavella IK, Lamari F, Papoulis D, Orkoula M, Gkolfi P, Lykouras M, Avgoustakis K, Hatziantoniou S (2020) Effect of plant extracts on the characteristics of silver nanoparticles for topical application. Pharmaceutics 12:1244
2. Reolon JB, Sari MHM, Marchiori C, Dallabrida KG, dos Santos JAR, de Fátima Ramos de Almeida I, Alves FMS, Bonini JS, Ferreira LM (2023) Herbal drugs-loaded soft nanoparticles for treating skin disorders: where do we stand? Ind Crops Prod 206:117602
3. Safta DA, Bogdan C, Moldovan M-L (2024) SLNs and NLCs for skin applications: enhancing the bioavailability of natural bioactives. Pharmaceutics 16:1270
4. Hajialyani M, Tewari D, Sobarzo-Sánchez E, Nabavi SM, Farzaei MH, Abdollahi M (2018) Natural product-based nanomedicines for wound healing purposes: therapeutic targets and drug delivery systems. Int J Nanomedicine 13:5023–5043
5. Puspadewi R, Milanda T, Muhaimin M, Chaerunisaa AY (2025) Nanoparticle-encapsulated plant polyphenols and flavonoids as an enhanced delivery system for anti-acne therapy. Pharmaceuticals 18:209
6. Husen A (2022) Herbs, shrubs and trees of potential medicinal benefits. CRC Press, Boca Raton. https://doi.org/10.1201/9781003205067
7. Husen A (2023) Antidiabetic medicinal plants and herbal treatments. CRC Press, Boca Raton, pp 33487–32742. https://doi.org/10.1201/b23347
8. Husen A (2023) Exploring poisonous plants: medicinal values, toxicity response and other potential uses. CRC Press, Boca Raton, pp 33487–32742. https://doi.org/10.1201/b23017
9. Husen A (2024) Antimalarial medicinal plants. Taylor & Francis Group, LLC, Boca Raton. https://doi.org/10.1201/9781003378396
10. Husen A (2024) Medicinal spices and condiment crops. Taylor & Francis Group, LLC, Boca Raton. https://doi.org/10.1201/9781003387046
11. Caballero-Gallardo K, Quintero-Rincón P, Olivero-Verbel J (2025) Aromatherapy and essential oils: holistic strategies in complementary and alternative medicine for integral wellbeing. Plants 14:400
12. Naji E, Abdulfatah HF, AlZawi K (2024) Plant secondary metabolites, their classification and biological roles: a review. J Univ Anbar Pure Sci 18:106–115
13. Rafeeq M, Bilal RM, Alagawany M, Batool F, Yameen K, Farag MR, Ali S, Elnesr SS, El-Shall NA (2022) The use of some herbal plants as effective alternatives to antibiotic growth enhancers in poultry nutrition. Worlds Poult Sci J 78:1067–1085
14. Gupta I, Singh R, Muthusamy S, Sharma M, Grewal K, Singh HP, Batish DR (2023) Plant essential oils as biopesticides: applications, mechanisms, innovations, and constraints. Plants 12:2916

15. Li X, Tan Q, Zhou Y, Chen Q, Sun P, Shen G, Ma L (2023) Synergic remediation of polycyclic aromatic hydrocarbon-contaminated soil by a combined system of persulfate oxidation activated by biochar and phytoremediation with basil: a compatible, robust, and sustainable approach. Chem Eng J 452:139502
16. Budzianowska A, Banaś K, Budzianowski J, Kikowska M (2025) Antioxidants to defend healthy and youthful skin—current trends and future directions in cosmetology. Appl Sci 15:2571
17. Bora L, Avram S, Pavel IZ, Muntean D, Liga S, Buda V, Gurgus D, Danciu C (2022) An up-to-date review regarding cutaneous benefits of Origanum vulgare L. essential oil. Antibiotics 11:549
18. Lombrea A, Antal D, Ardelean F, Avram S, Pavel IZ, Vlaia L, Mut A-M, Diaconeasa Z, Dehelean CA, Soica C, Danciu C (2020) A recent insight regarding the phytochemistry and bioactivity of Origanum vulgare L. essential oil. Int J Mol Sci 21:9653
19. Olivero-Verbel J, Quintero-Rincón P, Caballero-Gallardo K (2024) Aromatic plants as cosmeceuticals: benefits and applications for skin health. Planta 260:132
20. Michalak M (2022) Plant-derived antioxidants: significance in skin health and the ageing process. Int J Mol Sci 23:585
21. Salehi B, Zakaria ZA, Gyawali R, Ibrahim SA, Rajkovic J, Shinwari ZK, Khan T, Sharifi-Rad J, Ozleyen A, Turkdonmez E, Valussi M, Tumer TB, Monzote Fidalgo L, Martorell M, Setzer WN (2019) Piper species: a comprehensive review on their phytochemistry, biological activities and applications. Molecules 24:1364
22. Sounouvou HT, Toukourou H, Catteau L, Toukourou F, Evrard B, Van Bambeke F, Gbaguidi F, Quetin-Leclercq J (2021) Antimicrobial potentials of essential oils extracted from West African aromatic plants on common skin infections. Sci Afr 11:e00706
23. Zeng L, Gowda BHJ, Ahmed MG, Abourehab MAS, Chen Z-S, Zhang C, Li J, Kesharwani P (2023) Advancements in nanoparticle-based treatment approaches for skin cancer therapy. Mol Cancer 22:10
24. Liu L, Zhao W, Ma Q, Gao Y, Wang W, Zhang X, Dong Y, Zhang T, Liang Y, Han S, Cao J, Wang X, Sun W, Ma H, Sun Y (2023) Functional nano-systems for transdermal drug delivery and skin therapy. Nanoscale Adv 5:1527–1558
25. Sghier K, Mur M, Veiga F, Paiva-Santos AC, Pires PC (2024) Novel therapeutic hybrid systems using hydrogels and nanotechnology: a focus on nanoemulgels for the treatment of skin diseases. Gels 10:45
26. Moni SS, Moshi JM, Matou-Nasri S, Alotaibi S, Hawsawi YM, Elmobark ME, Hakami AMS, Jeraiby MA, Sulayli AA, Moafa HN (2025) Advances in materials science for precision melanoma therapy: nanotechnology-enhanced drug delivery systems. Pharmaceutics 17:296
27. Silvestrini AVP, Morais MF, Debiasi BW, Praça FG, Bentley MVLB (2024) Nanotechnology strategies to address challenges in topical and cellular delivery of SiRNAs in skin disease therapy. Adv Drug Deliv Rev 207:115198
28. Husen A (2017) Gold nanoparticles from plant system: synthesis, characterization and their application. In: Ghorbanpourn M, Manika K, Varma A (eds) Nanoscience and plant–soil systems, vol 48. Springer International Publishing AG, Cham, pp 455–479. https://doi.org/10.1007/978-3-319-46835-8_17
29. Husen A (2019) Natural product-based fabrication of zinc oxide nanoparticles and their application. In: Husen A, Iqbal M (eds) Nanomaterials and plant potential. Springer International Publishing AG, Cham, pp 193–291. https://doi.org/10.1007/978-3-030-05569-1_7
30. Husen A (2023) Secondary metabolites based green synthesis of nanomaterials and their applications. Springer Nature Singapore Pte Ltd, Singapore. https://doi.org/10.1007/978-981-99-0927-8
31. Husen A, Rahman QI, Iqbal M, Yassin MO (2019) Plant-mediated fabrication of gold nanoparticles and their applications. In: Husen A, Iqbal M (eds) Nanomaterials and plant potential. Springer International Publishing AG, Cham, pp 71–110. https://doi.org/10.1007/978-3-030-05569-1_3

32. Khan F, Shariq M, Asif M, Siddiqui MA, Malan P, Ahmad F (2022) Green nanotechnology: plant-mediated nanoparticle synthesis and application. Nanomaterials 12:673
33. Mishra VK, Husen A, Rahman QI, Iqbal M, Sohrab SS, Yassin MO, Bachheti RK (2019) Plant-based fabrication of silver nanoparticles and their application. In: Husen A, Iqbal M (eds) Nanomaterials and plant potential. Springer International Publishing AG, Cham, pp 135–175. https://doi.org/10.1007/978-3-030-05569-1_5
34. Rani N, Singh P, Kumar S, Kumar P, Bhankar V, Kumar K (2023) Plant-mediated synthesis of nanoparticles and their applications: a review. Mater Res Bull 163:112233
35. Nguyen NTT, Nguyen LM, Nguyen TTT, Nguyen TT, Nguyen DTC, Van Tran T (2022) Formation, antimicrobial activity, and biomedical performance of plant-based nanoparticles: a review. Environ Chem Lett 20:2531–2571
36. Barhoum A, Jeevanandam J, Rastogi A, Samyn P, Boluk Y, Dufresne A, Danquah MK, Bechelany M (2020) Plant celluloses, hemicelluloses, lignins, and volatile oils for the synthesis of nanoparticles and nanostructured materials. Nanoscale 12:22845–22890
37. Choi W, Cho JH, Park SH, Kim DS, Lee HP, Kim D, Kim HS, Kim JH, Cho JY (2024) Ginseng root-derived exosome-like nanoparticles protect skin from UV irradiation and oxidative stress by suppressing activator protein-1 signaling and limiting the generation of reactive oxygen species. J Ginseng Res 48:211–219
38. Pozos-Nonato S, Domínguez-Delgado CL, Campos-Santander KA, Benavides AA, Pacheco-Ortin SM, Higuera-Piedrahita RI, Resendiz-González G, Molina-Trinidad EM (2023) Novel nanotechnological strategies for skin anti-aging. Curr Pharm Biotechnol 24:1397–1419
39. Xu XY, Moon S-K, Kim J-K, Kim WJ, Kim Y-J, Kim H (2022) Structural properties and anti-dermatitis effects of flavonoids-loaded gold nanoparticles prepared by Eupatorium japonicum. Front Pharmacol 13:1055378
40. Gonfa YH, Tessema FB, Bachheti A, Rai N, Tadesse MG, Nasser Singab A, Chaubey KK, Bachheti RK (2023) Anti-inflammatory activity of phytochemicals from medicinal plants and their nanoparticles: a review. Curr Res Biotechnol 6:100152
41. Qu T, Suo S, Yan L, Wang L, Liang T (2025) Anti-inflammatory effects of astragaloside IV–chitosan nanoparticles and associated anti-inflammatory mechanisms based on liquid chromatography-mass spectrometry metabolomic analysis of RAW264.7 cells. Chem Biodivers 22:e202402234
42. Yi Q, Xu Z, Thakur A, Zhang K, Liang Q, Liu Y, Yan Y (2023) Current understanding of plant-derived exosome-like nanoparticles in regulating the inflammatory response and immune system microenvironment. Pharmacol Res 190:106733
43. Agarwal H, Nakara A, Shanmugam VK (2019) Anti-inflammatory mechanism of various metal and metal oxide nanoparticles synthesized using plant extracts: a review. Biomed Pharmacother 109:2561–2572
44. Conte R, Marturano V, Peluso G, Calarco A, Cerruti P (2017) Recent advances in nanoparticle-mediated delivery of anti-inflammatory phytocompounds. Int J Mol Sci 18:709
45. Jogpal V, Sanduja M, Dutt R, Garg V, Tinku (2022) Advancement of nanomedicines in chronic inflammatory disorders. Inflammopharmacology 30:355–368
46. Naseer F, Ahmed M, Majid A, Kamal W, Phull AR (2022) Green nanoparticles as multifunctional nanomedicines: insights into anti-inflammatory effects, growth signaling and apoptosis mechanism in cancer. Semin Cancer Biol 86:310–324
47. Kaushal P, Maity D, Awasthi R (2024) Nano-green: harnessing the potential of plant extracts for sustainable antimicrobial metallic nanoparticles. J Drug Deliv Sci Technol 94:105488
48. Ozdal M, Gurkok S (2022) A recent advances in nanoparticles as antibacterial agent. ADMET DMPK 10(2):115–129
49. Baptista PV, McCusker MP, Carvalho A, Ferreira DA, Mohan NM, Martins M, Fernandes AR (2018) Nano-strategies to fight multidrug resistant bacteria—"a battle of the titans". Front Microbiol 9:1441
50. Susanti D, Haris MS, Taher M, Khotib J (2022) Natural products-based metallic nanoparticles as antimicrobial agents. Front Pharmacol 13:895616

51. Ali MA, Ahmed T, Wu W, Hossain A, Hafeez R, Islam Masum MM, Wang Y, An Q, Sun G, Li B (2020) Advancements in plant and microbe-based synthesis of metallic nanoparticles and their antimicrobial activity against plant pathogens. Nanomaterials 10:1146
52. Zhou Y, Wu L, Zhang Y, Hu J, Fardous J, Ikegami Y, Ijima H (2025) Topical delivery of ceramide by oil-in-water nanoemulsion to retain epidermal moisture content in dermatitis. Biomolecules 15:608
53. Montenegro L, Pasquinucci L, Zappalà A, Chiechio S, Turnaturi R, Parenti C (2017) Rosemary essential oil-loaded lipid nanoparticles: in vivo topical activity from gel vehicles. Pharmaceutics 9:48
54. Romes NB, Abdul Wahab R, Abdul Hamid M, Oyewusi HA, Huda N, Kobun R (2021) Thermodynamic stability, in-vitro permeability, and in-silico molecular modeling of the optimal Elaeis guineensis leaves extract water-in-oil nanoemulsion. Sci Rep 11:20851
55. Peña OA, Martin P (2024) Cellular and molecular mechanisms of skin wound healing. Nat Rev Mol Cell Biol 25:599–616
56. Kolimi P, Narala S, Nyavanandi D, Youssef AAA, Dudhipala N (2022) Innovative treatment strategies to accelerate wound healing: trajectory and recent advancements. Cells 11:2439
57. Tottoli EM, Dorati R, Genta I, Chiesa E, Pisani S, Conti B (2020) Skin wound healing process and new emerging technologies for skin wound care and regeneration. Pharmaceutics 12:735
58. He C, Yin M, Zhou H, Qin J, Wu S, Liu H, Yu X, Chen J, Zhang H, Zhang L, Wang Y (2025) Magnetic nanoactuator-protein fiber coated hydrogel dressing for well-balanced skin wound healing and tissue regeneration. ACS Nano 19:1713–1731
59. Guillamat-Prats R (2021) The role of MSC in wound healing, scarring and regeneration. Cells 10:1729
60. Knoedler S, Knoedler L, Kauke-Navarro M, Rinkevich Y, Hundeshagen G, Harhaus L, Kneser U, Pomahac B, Orgill DP, Panayi AC (2023) Regulatory T cells in skin regeneration and wound healing. Mil Med Res 10:49
61. Dan X, Li S, Chen H, Xue P, Liu B, Ju Y, Lei L, Li Y, Fan X (2024) Tailoring biomaterials for skin anti-aging. Mater Today Bio 28:101210
62. Shang L, Li M, Xu A, Zhuo F (2024) Recent applications and molecular mechanisms of hyaluronic acid in skin aging and wound healing. Med Novel Technol Devices 23:100320
63. Sun B, He Y, Zhang L, Liu S, Chen M, Pan J, Fang J, Wang Y, Jiang H, Liu X, Zhang C-Y, Li J (2025) Protecting and rejuvenating ageing skin by regulating endogenous hyaluronan metabolism using adipose-derived stem cell-secreted SiRNAs. Front Med 12:1529936
64. Carrillo-Norte JA, García-Mir B, Quintana L, Buracchio B, Guerrero-Bonmatty R (2024) Anti-aging effects of low-molecular-weight collagen peptide supplementation on facial wrinkles and skin hydration: outcomes from a six-week randomized, double-blind, placebo-controlled trial. Cosmetics 11:137
65. Lee S-G, Kang S-M, Kang H (2024) Wrinkle reduction using tetrapeptide-68 contained in an O/W formulation: a randomized double-blind placebo-controlled study. Pharmaceutics 16:987
66. Zhou X, Cao Q, Orfila C, Zhao J, Zhang L (2021) Systematic review and meta-analysis on the effects of astaxanthin on human skin ageing. Nutrients 13:2917
67. Islam H, Poly TS, Tisha ZT, Rahman S, Naveed AIJ, Ahmed A, Ahmed SN, Hassan J, Uddin MJ, Das DB (2023) 3D printed hollow microneedles for treating skin wrinkles using different anti-wrinkle agents: a possible futuristic approach. Cosmetics 10:41
68. Hunt KJ, Hung SK, Ernst E (2010) Botanical extracts as anti-aging preparations for the skin. Drugs Aging 27:973–985
69. Draelos ZD, Baalbaki N, Colon G, Dreno B (2023) Ceramide-containing adjunctive skin care for skin barrier restoration during acne vulgaris treatment. J Drugs Dermatol 22:554–558
70. Blaess M, Deigner H-P (2019) Derailed ceramide metabolism in atopic dermatitis (AD): a causal starting point for a personalized (basic) therapy. Int J Mol Sci 20:3967
71. Prajapati SK, Lekkala L, Yadav D, Jain S, Yadav H (2025) Microbiome and postbiotics in skin health. Biomedicine 13:791

72. Dontje AEWK, Schuiling-Veninga CCM, van Hunsel FPAM, Ekhart C, Demirci F, Woerdenbag HJ (2024) The therapeutic potential of essential oils in managing inflammatory skin conditions: a scoping review. Pharmaceuticals 17:571
73. Jones VA, Patel PM, Wilson C, Wang H, Ashack KA (2021) Complementary and alternative medicine treatments for common skin diseases: a systematic review and meta-analysis. JAAD Int 2:76–93
74. Rigel DS, Lim HW, Draelos ZD, Weber TM, Taylor SC (2022) Photoprotection for all: current gaps and opportunities. J Am Acad Dermatol 86:S18–S26
75. Solano F (2020) Photoprotection and skin pigmentation: melanin-related molecules and some other new agents obtained from natural sources. Molecules 25:1537
76. Garcia-Mouronte E, Pérez-González LA, Naharro-Rodriguez J, Fernández Guarino M (2024) Understanding active photoprotection: DNA-repair enzymes and antioxidants. Life 14:822
77. Hyeraci M, Papanikolau ES, Grimaldi M, Ricci F, Pallotta S, Monetta R, Minafò YA, Di Lella G, Galdo G, Abeni D, Fania L, Dellambra E (2023) Systemic photoprotection in melanoma and non-melanoma skin cancer. Biomolecules 13:1067
78. Shaheen AE, Gebreel HM, Moussa LA, Zakaria AE, Nemr WA (2023) Photoprotection against UV-induced skin damage using hyaluronic acid produced by Lactiplantibacillus plantarum and Enterococcus durans. Curr Microbiol 80:262
79. Kumar A, Archo S, Singh CP, Naikoo SH, Singh B, Kaur S, Tasduq SA (2022) Photoprotective effect of 18β-glycyrrhetinic acid derivatives against ultra violet (UV)-B-induced skin aging. Bioorg Med Chem Lett 76:128984
80. Tsai J, Chien AL (2022) Photoprotection for skin of color. Am J Clin Dermatol 23:195–205
81. Xian D, Gao X, Xiong X, Xu J, Yang L, Pan L, Zhong J (2017) Photoprotection against UV-induced damage by skin-derived precursors in hairless mice. J Photochem Photobiol B Biol 175:73–82
82. Li T, Yao J, Everett DW, Hou Y, Pan Z, Su H, Fu Y (2024) Recombined egg yolk granules with controllable particle sizes for quercetin delivery—effects of particle size on physicochemical stability, bioaccessibility, and bioactivity of quercetin. Food Biophys 19:46–57
83. Yasuno G, Koide H, Hirata S, Okamoto T, Watanabe M, Saito K, Sato K, Matayoshi K, Yonezawa S, Asai T (2025) Influence of size, flexibility, hydrophobicity, surface charge, and surface chemistry on the biodistribution of orally administered polymer nanoparticles. Biol Pharm Bull 48:399–409
84. Shi Y, Liu Y, Sun Y, Zhong M, Rashid A, Qayum A, Liang Q, Rehman A, Ma H, Ren X (2024) Interfacial multilayer self-assembly of protein and polysaccharides: ultrasonic regulation, stability and application in delivery lutein. Int J Biol Macromol 272:132880
85. Ali MM, Shoukri RA, Yousry C (2024) Effect of particle size versus surface charge on the brain targeting behavior of elastic nanovesicles: in-vitro characterization, comparison between I-optimal and D-optimal statistical optimization and in-vivo pharmacokinetic evaluation. J Drug Deliv Sci Technol 96:105693
86. Karunakaran G, Sudha KG, Ali S, Cho E-B (2023) Biosynthesis of nanoparticles from various biological sources and its biomedical applications. Molecules 28:4527
87. Silva Brito R, João Bebianno M, Rocha TL (2024) Plant-based silver nanoparticles ecotoxicity: perspectives about green technologies in the One Health Context. Crit Rev Environ Sci Technol 54:1218–1235
88. Gao M, Chang J, Wang Z, Zhang H, Wang T (2023) Advances in transport and toxicity of nanoparticles in plants. J Nanobiotechnol 21:75
89. Zulfiqar Z, Khan RRM, Summer M, Saeed Z, Pervaiz M, Rasheed S, Shehzad B, Kabir F, Ishaq S (2024) Plant-mediated green synthesis of silver nanoparticles: synthesis, characterization, biological applications, and toxicological considerations: a review. Biocatal Agric Biotechnol 57:103121
90. Kim J (2025) Smart nanocarriers in cosmeceuticals through advanced delivery systems. Biomimetics 10:217

Aromatic Plant-Based Nanoemulsions or Gels and Their Biomedical Applications: A Microbiological and Pharmaceutical Botany Perspective

Gülnihan Selim, Elif Özlem Arslan Aydoğdu, Mine Koçyiğit, and Nihal Doğruöz Güngör

Abstract Nanotechnology has become a game-changer in the biomedical sciences, providing new ways to improve drug delivery, control infections, and heal wounds. In this context, nanoemulsions and hydrogels have attracted significant interest, owing to their distinctive physicochemical properties and biocompatibility. Nanoemulsions, which are made up of tiny droplets of liquids that don't mix with each other and are held together by surfactants, have better solubility, stability, and permeability. This makes them great for use in the mouth, on the skin, in the eyes, in the nose, and in the veins. Hydrogels, on the other hand, are polymeric networks that are hydrophilic, hold a lot of water, are structurally similar to natural tissues, and are very biocompatible. They create a wet environment for tissue regeneration and let drugs be released in a controlled way, which makes the treatment more effective. Chitosan-based hydrogels are very useful because they stick to the skin and last a long time. Combining these two systems into hybrid formulations called nanoemulgels makes them even more effective by making them thicker, able to hold more drugs, and able to stay on the skin longer. Experimental studies underscore their biomedical potential: lavender nanoemulsions combined with chitosan hydrogels exhibited enhanced antiviral efficacy against herpes simplex virus type 1; psoralen formulations incorporating fennel and clove oil nanoemulsions improved transdermal delivery in psoriasis and vitiligo treatment; *Pituranthos tortuosus* nanoemulsion gels expedited wound healing; and clove-cinnamon oil nanoemulgels demonstrated significant antimicrobial and antibiofilm activity against resistant pathogens. These results underscore the function of nanoemulsions and gels not

G. Selim
Institute of Graduate Studies in Sciences, Istanbul University, Istanbul, Turkey

E. Ö. Arslan Aydoğdu (✉) · N. Doğruöz Güngör
Faculty of Science, Department of Biology, Istanbul University, Istanbul, Turkey
e-mail: eoarslan@istanbul.edu.tr

M. Koçyiğit
Faculty of Pharmacy, Department of Pharmaceutical Botany, Istanbul University, Istanbul, Turkey

A. Husen (ed.), *Sustainable Uses of Aromatic Plants in Nanotechnology*, Smart Nanomaterials Technology, https://doi.org/10.1007/978-981-95-7409-4_12

merely as delivery mechanisms but also as dynamic therapeutic platforms that utilize the inherent bioactivity of essential oils derived from aromatic plants. These systems combine traditional herbal medicine with modern nanotechnology to provide solutions for treating infectious diseases, speeding up wound healing, and speeding up tissue regeneration. In general, aromatic plant-based nanoemulsions and hydrogels are promising new approaches in microbiology and pharmaceutical botany that connect natural resources with biomedical innovation.

Keywords Antimicrobial activity · Aromatic plants · Biomedical applications · Gel formulations · Nanoemulsion

1 Introduction

Aromatic plants have been widely used in traditional medicine for thousands of years for human health. Today, thanks to modern science and technology, these plants are valuable natural resources with new applications in the biomedical field. The traditional use of aromatic plants in the treatment of many diseases and the effective results obtained are generally due to organic substances called essential oils found in these plants. These secondary metabolites found in aromatic plants have antioxidant, antibacterial, antifungal, and other properties [1–9]. The field of drug development has its origins in plants. Plants remain a source of information for effective drugs used in new clinical treatments and against new diseases. This is because plant-based sources generally have fewer side effects, and the synergistic effect of the compounds enables better results to be achieved [10–15]. However, traditional herbal treatments have limitations such as low water solubility, lack of stability, low bioavailability, volatility, susceptibility to oxidation, and non-targeted effects [8, 9, 16, 17].

Nanotechnology has emerged as a new approach to maximizing the biomedical and several other potentials of aromatic plant components by overcoming these challenges [16, 18–27]. Nanotechnology enables the use of essential oils and extracts derived from aromatic plants in nanoemulsion and nanogel formulations. These systems can provide targeted and controlled release to diseased areas, thanks to the high encapsulation capacity, homogeneity, low toxicity, improved stability, and bioavailability of the active ingredients [16, 18, 28–31].

Nanoemulsions are ultrafine, thermodynamically stable or semi-stable products containing a surfactant alongside an oil phase and a water phase within a size range of 20–200 nm. These products overcome fundamental challenges such as low water solubility, oxidation tendency, and volatility that inhibit the effectiveness of essential oils. Nanoemulsions are obtained by converting nanoemulsions into a gel form and increase treatment efficacy and offer ease of application to the user by enabling the appropriate consistency and longer-lasting release in topical applications [19, 25, 32–34]. Nanoformulations based on volatile oils have very large surface areas due to their nano-size. Therefore, they have the potential to overcome biological barriers that drugs normally cannot pass through. These nano-sizes also increase the absorption of active ingredients and their delivery to target tissues. Stability and, consequently, storage life are also increased with nanoemulsion formulations. Thanks to nanoemulsions, even very low doses of active ingredients can provide high efficacy, thereby reducing overall toxicity and minimizing the side effects of the drugs used [35–38].

This section focuses on the efficacy of essential oils and bioactive compounds derived from aromatic plants in nanoemulsion and gel forms for biomedical applications. The morphological and chemical properties and pharmacological potential of plants will be evaluated in pharmaceutical botany, while the antimicrobial activity and resistance mechanisms of nanoformulations on microorganisms will be evaluated in microbiology. Pharmacokinetic properties, toxicological safety, clinical application potential, and future research areas will also be discussed. This section lies at the intersection of botany, microbiology, pharmaceutical technology, and biomedical sciences.

2 Aromatic Medicinal Plants: Pharmaceutical Botanical Overview

2.1 Olfactory System: How Do We Smell?

Olfaction, or the sense of smell, is a unique chemical process that lets living things find and understand volatile chemicals in their surroundings. The process starts when airborne odorant molecules dissolve in the mucus layer of the nasal cavity and reach the olfactory epithelium, which is at the top of the nasal cavity. There are millions of olfactory receptor neurons (ORNs) in this epithelium. Each one only expresses one type of olfactory receptor, and these receptors are part of the large family of G protein-coupled receptors. When an odorant binds to its specific receptor, it starts a chain of events inside the cell that involves olfactory G protein (GFP), adenylyl cyclase, and cyclic AMP (cAMP). The rise in cAMP opens cyclic nucleotide-gated ion channels that let sodium and calcium ions in and chloride ions out. This depolarizes the receptor neuron and starts an action potential [39]. These action potentials travel along the axons of ORNs, which come together to make the olfactory nerve and send signals to the olfactory bulb. In the olfactory bulb, the axons connect to mitral and tufted cells in structures called glomeruli. This is where signals from ORNs that express the same receptor type are combined (Fig. 1). This arrangement of space helps to encode the identity of smells at first. The processed signals are then sent to higher brain areas, such as the piriform cortex, the orbitofrontal cortex, and limbic structures like the amygdala and hippocampus. These connections elucidate not only the uniqueness of odors but also their capacity to elicit emotional and mnemonic responses [40].

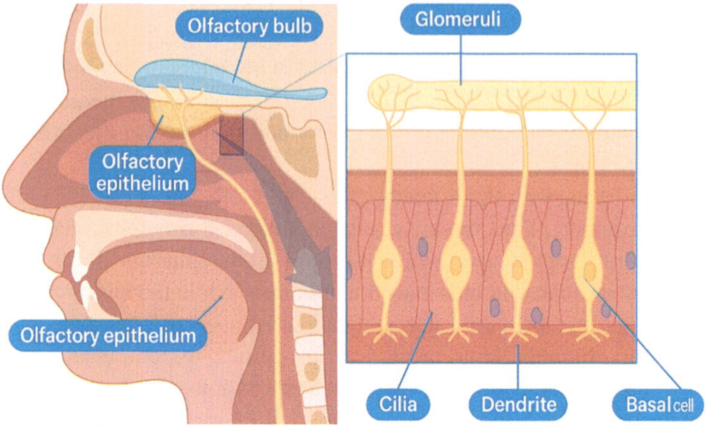

Fig. 1 Mechanism of the olfactory system

In short, the sense of smell is a complicated system that turns chemical information from the outside world into electrical signals and useful perceptions. Unique neural pathways that go around the thalamus show how important it was for evolution in terms of survival, behavior, and social interactions.

2.2 *Secretory Tissues in Plants: Why and Where Do Plants Produce Aromas or Scents?*

Plants produce a wide variety of metabolites as a result of their physiological and biochemical activities. Some of these metabolites are stored in specialized structures without undergoing any changes, while others are secreted directly onto the plant surface or into the external environment. These substances are called secretions, and the structures that produce them are called secretory tissues. Secretions include water, enzymes, nectar, mucilage, latex, resin, essential oils, and tannins. Secretory tissues are generally classified into two main groups: intracellular secretions and extracellular secretions. Intracellular secretions accumulate in the cytoplasm or vacuoles of specialized secretory cells. They store the secretion products internally. These can be found in solid form (e.g., calcium oxalate crystals) or liquid form (e.g., essential oils) [41]. The volatile oil cells in *Laurus nobilis* (bay laurel) leaves are a typical example. Bay laurel oil is widely used in cosmetic products as a fragrance and in skin care due to its antimicrobial properties [42]. Milk ducts (laticifer system) consist of cells with cellulose walls and contain a secretion called latex. Latex may contain components such as sugar, tannin, glycoside, alkaloid, and rubber. There are jointless and jointed types. Extracellular secretions are not stored inside the cell but are secreted directly outside. They are grouped into three categories:

1. *Secretions discharged into intercellular spaces*: secretion pockets are oval or round cavities. They are found in the Rutaceae, Apiaceae, and Hypericaceae families. Many species in these families are widely used in cosmetics (e.g., citrus oils, fennel oil, St. John's wort extracts) due to their volatile oils. Secretory ducts, on the other hand, are long structures formed by overlapping intercellular spaces [43]. Resin ducts in *Pinus* species are an example. Resins are used in perfumery and dermocosmetic products [44].
2. *Secretions produced by epidermal cells between the cell wall and the cuticle*: these include glandular trichomes (Fig. 2). Glandular trichomes secrete pure water, saline solutions, mucilage, enzymes, and especially essential oils. In these structures, which consist of a stalk and a head, the secretion accumulates under the cuticle. Secretory hairs found in the Lamiaceae family (e.g., lavender, rosemary, mint) are the fundamental structures that enable the widespread use of these plants in cosmetics [45].
3. *Secretions released directly into the external environment*: hydathodes, especially in plants in humid environments, secrete water through guttation. Nectaries secrete nectar and contribute to pollination (Fig. 2). Honey and honeydew are valued as natural moisturizers in cosmetic products. Digestive glands secrete acids and enzymes that aid digestion in insectivorous plants. Salt glands secrete salt in halophytes, which crystallizes on the leaf surface [46].

Essential oils are one of the most valuable secondary metabolites synthesized and stored in the secretory cells, pockets, channels, and hairs of plants. They are used in the cosmetics industry for their fragrance and antimicrobial, antioxidant,

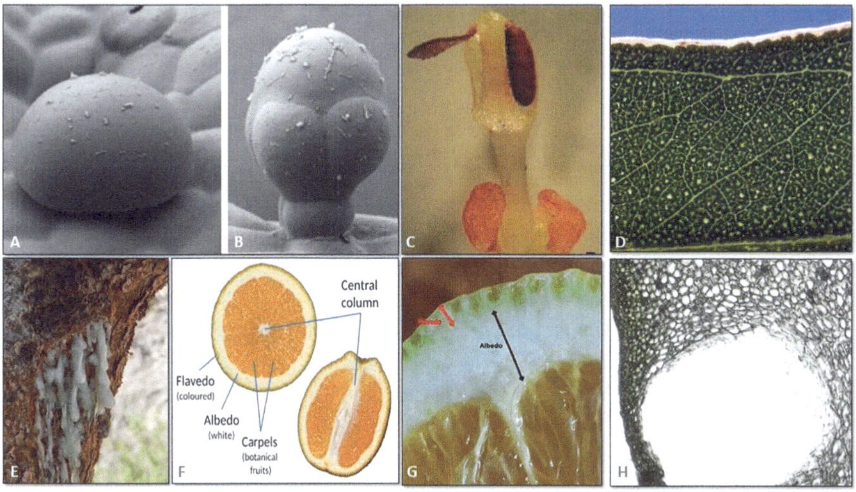

Fig. 2 (**a**, **b**) Glandular hairs observed in the Lamiaceae family. (**c**) Nectariums attached to the filament of *Laurus nobilis*. (**d**) Secretion pockets (transparent dots) on the leaves of *Eucalyptus* species. (**e**) Resin flowing from the stem of *Boswellia* species. (**f–h**) Lysigenous secretion canal found in the flavedo section of *Citrus* fruits

anti-inflammatory, and protective properties. The Lamiaceae, Rutaceae, Apiaceae, and Myrtaceae families, in particular, stand out in perfumery, skin care, and aromatherapy products due to their richness in essential oils.

2.3 The Most Commonly Used Plants Due to Their Essential Oils

Essential oils represent highly valued natural products due to their unique chemical complexity and broad spectrum of biological activities. Their phytochemical constituents, including terpenes, phenols, aldehydes, and esters, exhibit antimicrobial, antioxidant, anti-inflammatory, and wound healing properties, making them attractive candidates for integration into modern pharmaceutical and microbiological applications. However, challenges such as volatility, hydrophobicity, and instability under environmental conditions have limited their direct use [47]. Recent advances in formulation science, particularly nanoemulsions and gel-based delivery systems, have provided effective solutions to these limitations by improving solubility, stability, and bioavailability, while enabling controlled release. From a pharmaceutical botany perspective, plant-derived essential oils serve as both therapeutic agents and natural preservatives, bridging traditional medicinal knowledge with contemporary biomedical innovation [48]. Their applications extend across dermatology, infectious disease management, wound care, and even neurological support, where their bioactive profiles play a decisive role. The therapeutic potential of each essential oil is strongly linked to its phytochemical composition, which also determines its commercial value [49]. Table 1 presents the most widely used aromatic plants, their commercial names, and their main biomedical applications.

3 Essential Oils and Bioactive Compounds

Essential oils, phenolic compounds, and sterols derived from aromatic plants are the main components responsible for therapeutic activity. Their antioxidant, anti-inflammatory, and antimicrobial properties, in particular, offer an alternative to synthetic additives, especially in the food, cosmetic, and pharmaceutical industries. Essential oils contain monoterpenes, sesquiterpenes, and terpenoids, as well as some phenolic monoterpenes such as thymol and carvacrol, which provide therapeutic efficacy. Phenolic compounds, on the other hand, are mostly found in non-volatile polar extracts such as phenolic acids, flavonoids, and tannins. These compounds provide antioxidant and anti-inflammatory activity. Terpenes such as β-sitosterol, ursolic acid, and oleanolic acid are non-volatile and exhibit anti-inflammatory and antioxidant effects like phenolic compounds [52, 53] (Table 2).

Table 1 The most commonly used plants due to their essential oils, their families, trade names, and intended uses (selected from [47–51])

Scientific name (with authority)	Family	Commercial name	Main use
Anthemis nobilis L.	Asteraceae	Roman chamomile oil	Soothing for sensitive skin; relieves stress, mild sedative
Boswellia carterii Birdw.	Burseraceae	Frankincense oil	Anti-aging, wound healing; anti-inflammatory, respiratory health
Calendula officinalis L.	Asteraceae	Calendula oil	Sensitive skin care; wound healing, anti-inflammatory
Cananga odorata (Lam.) Hook.f. & Thomson	Annonaceae	Ylang-ylang oil	Perfume, hair care; reduces stress, regulates blood pressure
Chamomilla recutita (L.) Rauschert	Asteraceae	German chamomile oil	Skin soothing; anti-inflammatory, digestive aid, mild sedative
Citrus aurantium L. var. *amara*	Rutaceae	Neroli oil	Perfume, skin rejuvenation; anxiolytic, digestive stimulant
Citrus bergamia Risso & Poit.	Rutaceae	Bergamot oil	Perfume, acne control; mood enhancer, antimicrobial
Citrus limon (L.) Osbeck	Rutaceae	Lemon oil	Skin brightening; antioxidant, immune booster, antiseptic
Citrus sinensis (L.) Osbeck	Rutaceae	Sweet orange oil	Aromatherapy, perfume; stress relief, digestive support
Commiphora myrrha (Nees) Engl.	Burseraceae	Myrrh oil	Oral and skin care; antimicrobial, wound healing, anti-inflammatory
Cymbopogon citratus (DC.) Stapf	Poaceae	Lemongrass oil	Antibacterial, deodorant; digestive tonic, antimicrobial
Cymbopogon nardus (L.) Rendle	Poaceae	Citronella oil	Insect repellent; antifungal, mild anti-inflammatory
Eucalyptus globulus Labill.	Myrtaceae	Eucalyptus oil	Deodorant; respiratory relief, expectorant, antimicrobial
Helichrysum italicum (Roth) G.Don	Asteraceae	Immortelle oil	Anti-aging; wound and scar healing, anti-inflammatory
Jasminum grandiflorum L.	Oleaceae	Jasmine absolute	Perfume, skin softener; antidepressant, aphrodisiac
Jasminum sambac (L.) Aiton	Oleaceae	Arabian jasmine oil	Perfume, aphrodisiac; relaxation, skin care
Lavandula angustifolia Mill.	Lamiaceae	Lavender oil	Skin soothing, perfume; anxiolytic, sedative, wound healing
Melaleuca alternifolia (Maiden & Betche) Cheel	Myrtaceae	Tea tree oil	Acne treatment; antifungal, antibacterial, wound healing

(continued)

Table 1 (continued)

Scientific name (with authority)	Family	Commercial name	Main use
Mentha × piperita L.	Lamiaceae	Peppermint oil	Cooling, shampoo; digestive aid, analgesic, decongestant
Origanum vulgare L.	Lamiaceae	Oregano oil	Antiseptic; antimicrobial, antifungal, immune stimulant
Pelargonium graveolens L'Hér. ex Aiton	Geraniaceae	Geranium oil	Skin balancing; antimicrobial, stress relief, hormone regulation
Pogostemon cablin (Blanco) Benth.	Lamiaceae	Patchouli oil	Perfume fixative; anti-inflammatory, wound healing, stress relief
Rosa centifolia L.	Rosaceae	Rose absolute	Perfume; antioxidant, mood enhancer
Rosa damascena Mill.	Rosaceae	Rose oil/rose water	Moisturizer, perfume; antidepressant, anti-inflammatory
Rosmarinus officinalis L. (syn. *Salvia rosmarinus Spenn.*)	Lamiaceae	Rosemary oil	Hair tonic; improves circulation, memory booster, antimicrobial
Salvia officinalis L.	Lamiaceae	Sage oil	Deodorant; antimicrobial, memory enhancer, hormonal balance
Santalum album L.	Santalaceae	Sandalwood oil	Perfume, moisturizer; anti-inflammatory, anxiolytic
Styrax benzoin Dryand.	Styracaceae	Benzoin resin/oil	Perfume fixative; wound healing, respiratory relief
Thymus vulgaris L.	Lamiaceae	Thyme oil	Skin cleanser; antimicrobial, expectorant, immune support
Vetiveria zizanioides (L.) Nash (syn. *Chrysopogon zizanioides* (L.) *Roberty*)	Poaceae	Vetiver oil	Perfume base; calming, grounding, anti-inflammatory

The essential oils and phenolic compounds of aromatic plants integrate into the cell membrane and can increase the cell's permeability, leading to disruption of ion balance and leakage of cytoplasmic contents outside the cell. Additionally, these compounds exert a broad-spectrum antimicrobial activity by negatively affecting the protein and DNA structures of microorganisms, regulating oxidative stress genes, weakening the defense mechanisms of microorganisms, and inhibiting quorum sensing [52, 71, 72]. The polyphenols and terpenoids in aromatic plants contribute to combating oxidative stress and chronic diseases caused by oxidative stress, such as diabetes, cardiovascular diseases, neurodegenerative disorders, and cancer, by exhibiting antioxidant activity through free radical scavenging, metal chelation, and induction of endogenous antioxidant enzymes [71, 73, 74]. The bioactive compounds in aromatic plants inhibit inflammation by suppressing COX-2

Table 2 Bioactive properties of some plant essential oils and extracts

Category/plant (example)	Main compounds/ examples	Bioactive properties (antioxidant, anti-inflammatory, antimicrobial)	Key references
Pelargonium graveolens L'Hér. (rose geranium EO)	Geraniol, β-citronellol, geranyl tiglate	Strong antioxidant (DPPH/FRAP, etc.); anti-inflammatory enzyme inhibition; antibacterial/antifungal activity including *S. aureus*/MRSA and *Candida* spp.; membrane/biofilm disruption reported	M'hamdi et al. [54], Draiaia et al. [55]
Thymus L. spp. (thyme EO)	Carvacrol, thymol, *p*-cymene, linalool	Potent antioxidant and ROS scavenging activity; anti-inflammatory effects (thymol/carvacrol modulates cytokines, NF-κB); broad antibacterial and antifungal activity (including MRSA, *P. aeruginosa*); antibiofilm effects	Chroho et al. [56], Vassiliou et al. [57]
Ammoides pusilla (Brot.) Breistr. (APEO)	Carvacrol, d-limonene (major components reported)	Antioxidant and dermatoprotective activities; anti-elastase/anti-tyrosinase (skin-protective); antimicrobial activity vs. dermatopathogens (e.g., *S. aureus*, *Candida* spp.)	Belaiba et al. [58], Taibi et al. [59]
Polyphenols (EGCG, quercetin, kaempferol, flavonoids)	EGCG (green tea), quercetin, kaempferol, stilbenes, tannins	Strong ROS scavengers (high antioxidant capacity); anti-inflammatory via NF-κB inhibition; demonstrated antimicrobial/antibiofilm effects in dermatological contexts and wound models	Mokra et al. [60], Capasso et al. [61]
Terpenoids/ monoterpenes	Borneol, thymol, α-terpineol, d-limonene; carotenoids (β-carotene)	Widely reported antioxidant + anti-inflammatory activities; many terpenoids show antibacterial/ antifungal activity and membrane-targeting mechanisms	Masyita et al. [62], Siddiqui et al. [63]
Allium sativum L. (garlic)	Allicin and sulfur compounds	Moderate to strong antioxidant activity; immunomodulatory/ anti-inflammatory effects; broad antimicrobial spectrum (bacteria and fungi) attributed largely to allicin	Verma et al. [64], Sahidur et al. [65]
Trigonella foenum-graecum L. (fenugreek)	Thymoquinone/other phenolics, timol in some extracts	Antioxidant (DPPH, etc.), anti-inflammatory actions, and broad-spectrum antibacterial activity reported in vitro	Akhtar et al. [66]
Aloe vera (L.) Burm.f. (gel/ extracts)	Polysaccharides, anthraquinones, aloin, vitamins	Antioxidant enzyme activation; anti-inflammatory and wound healing properties; antimicrobial effects (e.g., *S. aureus*, *C. albicans*)—used frequently in wound/dressing formulations	Chelu et al. [67], Catalano et al. [68]
Humulus lupulus L. (hop)	Humulones (α-acids), lupulones (β-acids), xanthohumol (polyphenol)	Moderate antioxidant activity; anti-inflammatory actions in keratinocytes; notable antibacterial and antibiofilm activity vs. *S. aureus*, *Cutibacterium acnes*, and others (hops extracts/CO_2 extracts)	Weber et al. [69], Kolenc et al. [70]

and 5-LOX enzymes, reducing prostaglandin and leukotriene synthesis, suppressing iNOS and lowering NO production, and regulating pro-inflammatory cytokines such as TNF-α, IL-1β, and IL-6 [52, 71–74].

4 Nanoemulsions and Gels as Drug Delivery Systems

Nanoemulsions hold great potential in next-generation drug systems. They are readily soluble, non-toxic, biocompatible, and stable emulsion droplets, making them suitable candidates for use in drug delivery systems (Fig. 3). Nanoemulsions with these properties can be used as topical, nasal, oral, ocular, or intravenous drug delivery systems [75–77].

Hydrogels are hydrophilic macromolecules composed of natural or synthetic polymers with high water holding capacity. Alginate, chitosan polypeptides, Pluronic/polyethylene glycol conjugates, and hyaluronic acid are the most frequently studied hydrogels [78]. Their high water content, biocompatibility, and porous and soft structures provide properties similar to natural tissues. They are becoming a component of drug delivery systems in tissue engineering, wound healing, and biomedical applications. Various therapeutic agents can be dissolved in hydrogels, enabling their use as part of a drug delivery system. The design of hydrogels allows them to release drugs in a controlled manner, consistent with therapeutic needs, and sustained for extended periods [79].

Chitosan is one of the most suitable polymers for topical drug application. Chitosan is a semisynthetic polymer that exhibits bioadhesive properties because it can trap large amounts of water molecules. Furthermore, chitosan-containing hydrogels have properties that allow prolong retention on the skin and promote drug absorption [80].

A 2021 study demonstrated promising results in the treatment of herpes simplex virus by developing a nanoemulsified drug delivery system using *Lavandula* plant optimized with chitosan hydrogel. Herpes simplex virus type 1 is a painful skin lesion typically found in the oral region. Permeability studies in sheep buccal mucosa revealed that the resulting matrix was more effective than the antiviral penciclovir used in the treatment of herpes simplex virus type 1 [81, 82].

Psoralens are a sensitive chemical agent used in diseases such as psoriasis and vitiligo that require UV light during treatment. While generally used topically or orally, they are known to have various side effects related to their use. Inadequate skin absorption of some topical medications can occur. It has also been reported that psoralens can reach the eyes and cause adverse effects in some cases. In one study, 8-methoxypsoralen (8-MOP), a type of psoralen, was used in combination with nanoemulsions containing essential oils from sweet fennel and clove plants and chitosan-containing hydrogels. The results showed that this combination increased transdermal drug delivery and skin persistence [80].

Nanometric spheres in nanoemulsions facilitate drug delivery through their enhanced stability and solubility [83]. Furthermore, the novel properties of nanostructures play a significant role in clinical studies due to their effects on targeted

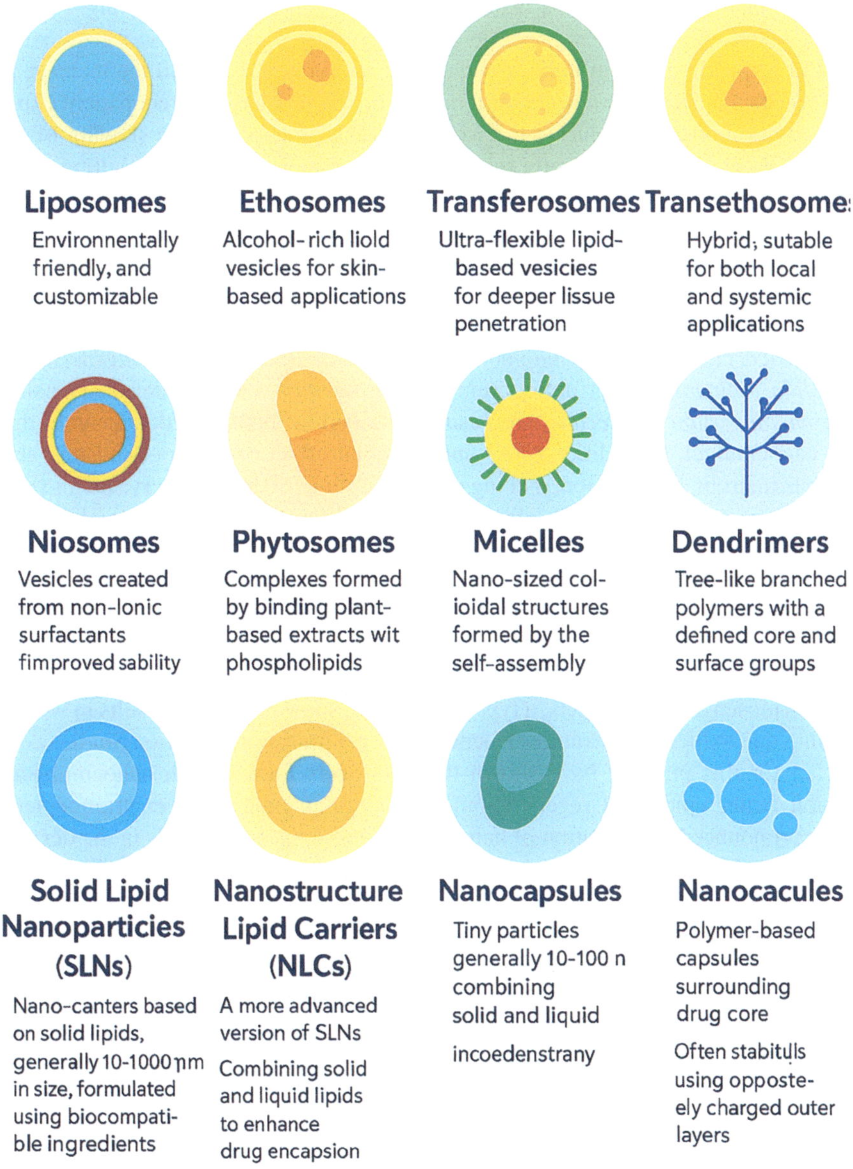

Fig. 3 Nanotechnological systems for plant-derived drugs

drug delivery, tissue penetration, and enhanced cellular response. This allows for both the treatment of bacterial infections and the inhibition of the growth of pathogenic bacteria. Therefore, nanoemulsions are also effective in drug delivery systems [84]. Nanoemulsions derived from the aromatic plant *Pituranthos tortuosus* were gelled with a nonionic polymer to investigate their wound healing effects. The study found that the nanoemulgel was much more effective in accelerating the healing process than healing creams within 10 days, with a wound healing rate of 85.87% [84].

The acquisition of antibiotic resistance by microorganisms, and the increase in this resistance over time, also leads to global health problems. Direct patient exposure to microorganisms during the wound healing process and the slowing of the healing process are also directly related to this situation. Carbopol, a gelling agent composed of acrylic acid polymers cross-linked with polyalkenyl ethers or divinyl glycol, has an important place in topical applications due to its non-toxicity and low irritancy. Gelation of essential oils from clove (*Syzygium aromaticum* (L.) Merr. & L.M.Perry) and cinnamon (*Cinnamomum verum* J.Presl) (syn., *C. zeylanicum*) with Carbopol demonstrated antimicrobial activity against *Staphylococcus epidermidis*, *Staphylococcus aureus*, *Pseudomonas aeruginosa*, *Bacillus chungangensis*, and *Candida albicans*. Potent antiquorum sensing activity was also detected against *Chromobacterium violaceum*. The results obtained in this study demonstrate that the nanoemulgel, in addition to its antimicrobial effect on microorganisms, also inhibits microbial interaction, making it a good option for drug delivery studies [83].

Dressings used in wound healing (gauze, bandages, etc.) can both stop bleeding and protect the wound from microbial infections, but they lack antimicrobial activity. Hydrogels, on the other hand, hold a place in wound healing studies due to their properties such as high water capacity, biocompatibility, wound moisturizing, oxygen permeability, absorption of excess exudate, and resemblance to the extracellular matrix [85, 86].

Thermosensitive hydrogels, a type of hydrogel, show promising results in wound healing due to their temperature-dependent sol-gel transition behavior. Studies conducted for this purpose have shown that hydrogels formed with nanoemulsions exhibit better epidermal regeneration, more regular collagen fiber arrangement, more pronounced regeneration of hair follicles and sebaceous glands in the dermis, and reduced inflammatory cell infiltration compared to control organisms. The significant effects observed here are thought to be in addition to the significant effects of aromatic plants (regulation of inflammatory response, increased cell proliferation, etc.), as well as the prolonged duration of action of the hydrogels, which provides a stable release and thus enhances their therapeutic efficacy [85]. Nanoemulsion gels offer more effective treatment options compared to nanoemulsion approaches alone due to their prolonged contact, improved infusion capability, higher drug loading capacity, and permeability [83].

5 Microbiological Aspects and Antimicrobial Potential

The majority of aromatic plants are used in traditional medicine to treat various diseases or infections. Past and present studies have demonstrated that the potential of aromatic plants can be further developed. *Cinnamomum cassia* (L.) J.Presl is an aromatic plant known for its antioxidant, antibacterial, anti-inflammatory, and anticancer properties. A study determined that nanoemulsions derived from *C. cassia* essential oils had a minimum inhibitory concentration of 0.025% and a minimum bactericidal concentration of 0.05% against the bacterium *Klebsiella pneumoniae* (MTCC 8911). These nanoemulsions also demonstrated antioxidant properties in addition to their anticancer activity against the alveolar lung adenocarcinoma cell line (A549). While the essential oils of aromatic plants possess various bioactive properties, their use as nanoemulsions has been reported to provide both more stable and more effective results [87].

Another study conducted with various aromatic plants found that essential oils obtained from these plants possessed antibacterial activity. In a study using the disk diffusion method, plants such as *Punica granatum* L. and *Thymus hirtus* Willd. were found to be effective against various pathogens, while the pathogens mentioned included *Escherichia coli*, *Staphylococcus aureus*, and *Pseudomonas aeruginosa*, demonstrating the potential of these essential oils. Furthermore, the study determined that *P. granatum* extract had higher antibacterial activity compared to the antibiotic streptomycin (10 μg/μL) [88]. Similarly, nanoemulsions of essential oils from *Deverra denudata* (Viv.) Pfisterer & Podlech (syn., *Pituranthos denudatus* Viv.) and *Deverra tortuosa* (Desf.) DC. (syn., *Pituranthos tortuosus* Benth. & Hook.f. ex Asch. & Schweinf.) were also found to have strong antimicrobial activity against *Streptococcus pyogenes*, *Enterobacter aerogenes*, and *S. aureus* using the disk diffusion method [89].

Foodborne pathogenic bacteria such as *Salmonella* cause significant risks to food safety. Contamination of food with these foodborne pathogens and subsequent consumption of these foods carry the risk of illnesses that can lead to death. *Salmonella typhimurium* is one of the bacteria known to cause infection due to its high concentration in foods such as fruit juice. One study demonstrated that adding *Litsea cubeba* (Lour.) Pers. nanoemulsions to kiwifruit juice containing *S. typhimurium* and incubating them significantly reduced the number of pathogens present in the environment. This study demonstrates the potential of aromatic plant-derived nanoemulsions for improving food safety [90].

Nanoemulsions of *Lavandula spica* exhibited antibacterial activity at effective MIC values against *S. typhimurium* ATCC 1402 and *S. aureus* ATCC 6538 bacteria, while also exhibiting high inhibitory activity against *C. albicans*. It was also found that the growth of mycelia of *Aspergillus flavus* and *Aspergillus niger* fungi was inhibited. Nanoemulsions exhibited higher antimicrobial properties compared to crude lavender oil. This study demonstrates promising results in infections affected by multiple pathogens [81]. *Melaleuca citrina* (Curtis) Dum.Cours. (syn., *Callistemon citrinus* (Curtis) Skeels) nanoemulsions exhibited antimicrobial activity against the pathogens *Fusarium oxysporum* (MIC 62.5 mg/mL) and *E. coli* (MIC 0.256 mg/mL), while also exhibiting antilarvicidal activity against *Chilo suppressalis* larvae [19]. Similarly, *M. citrina* nanoemulsions were found to exhibit antifungal activity against the fungal *Fusarium proliferatum* [91].

Timur plant (*Zanthoxylum armatum* DC.) nanoemulsion and Carbopol hydrogel were determined to have antifungal activity against *C. albicans* in disk diffusion tests, while no signs of erythema and toxicity were observed on the skin of rats in animal experiments, thus confirming their safety. The obtained formulation may be an alternative treatment option for topical use, especially in the treatment of resistant fungal infections [92]. *Coriandrum sativum* L. nanoemulsion Carbopol hydrogel was found to have activity against *P. aeruginosa*, *K. pneumoniae*, and methicillin-resistant *S. aureus* (MRSA) bacteria in agar diffusion tests, with MIC values determined as 2.3, 3.75, and 6.5 μg/mL [93]. In one study, it was observed that nanoemulsions of the *Ocimum basilicum* L. plant had both antimicrobial and antibiofilm activities when matrixed with a gellan gum-based hydrogel [94]. When alginate hydrogels of nanoemulsions of thyme and clove plants were tested using the disk diffusion method, it was determined that they exhibited strong antibacterial activity against *E. coli* and *Bacillus cereus* bacteria [95]. The study demonstrated that the thermosensitive hydrogel prepared with essential oils obtained from the aromatic plant *Nepeta annua* Pall. (syn., *Schizonepeta annua* (Pall.) Schischk.) exhibited antimicrobial activity against MRSA ATCC 43300, *S. aureus* ATCC 25923, *E. coli* ATCC 43895, and *C. albicans* ATCC 10231 microorganisms, with an MIC value of 3906 μg/mL. Obtaining such an effect against common wound pathogens suggests that the nanoemulsion-hydrogel combination may provide advantages in terms of infection control and tissue repair [85].

Studies have shown that aromatic plants possess antimicrobial activity, but systems formed in the form of nanoemulsions or gels have much higher antimicrobial activity. Nanostructures offer smaller sizes and greater penetration. Gels, on the other hand, allow for a longer release of the antimicrobial agent due to their structure. Combining these two properties results in much more effective antimicrobial applications. Nanoemulsions and gels hold significant potential in the treatment of many infections, especially those where resistance is present [81, 93].

6 Biomedical Applications of Aromatic Nanoemulsions and Gels

The integration of phenolic and terpenoids with anti-inflammatory, antimicrobial, antioxidant, and wound healing properties, into nanoemulsion and gel-based carriers enhances their advantages in biomedical applications. These carrier systems facilitate penetration or delivery to the application site, while also improving half-life and stability. Furthermore, these materials are enabling the slow release of aromatics and limit high-dose exposure and the associated indications [96, 97].

Aromatic gels and nanoemulsions are frequently used in wound and burn treatments. For example, Miastkowska et al. [98] tested lavender oil nanoemulsions to accelerate wound healing, and they found that collagen type 1/2 gene expression increased in damaged tissue and that it also had a pro-regenerative effect, alleviating the inflammatory response. Nanoemulsions containing *Schizonepeta annua* basil oil, essential oil, and clove essential oil are other examples compelling in wound and burn healing [99]. Another example is tea tree oil, which has anti-edematous effects and the potential to improve skin regeneration. A study demonstrated that clindamycin hydrochloride supplemented with a nanoemulsion containing propolis and tea tree oil supported wound healing through epithelialization and collagen production, in addition to its antimicrobial effect [100].

A study conducted on gels containing sea buckthorn oil has shown that it can be an effective wound dressing for burn and wound healing. It has been observed that gels containing this oil, which has antioxidant and anti-inflammatory effects, promote skin regeneration after burns, form a protective layer during this process, and help optimize post-wound tissue deformation by increasing hydration [101]. Another nanocarrier system that promotes the healing rate of burn tissue is sandalwood oil-based nanoemulsions. This material supports tissue regeneration, while exhibiting anti-inflammatory and antioxidant activity, suppressing both the sensation of pain and the free radicals associated with deformation [102].

Aromatic nanoemulsion applications can also be seen for acne treatment. Tea tree (*Melaleuca alternifolia* (Maiden & Betche) Cheel) nanoemulsions have been shown to be quite successful in reducing comedones and follicular hyperkeratinization. *Origanum vulgare* L. essential oil nanoemulsions are also used in topical acne treatments. Isotretinoin, adapalene, nadifloxacin, and retinol, included in nanoemulsion formulations, enhance acne treatment [103–105].

Aromatic nanoemulsions are among the innovative treatments for atopic dermatitis, as the innovative treatments is aromatic nanoemulsions. For example, a nanoemulsion formulation containing *Linum usitatissimum* L. seed oil (flaxseed oil) has been found to have high potential [106]. Essential oils, such as calendula, chamomile, tea tree oil, and lavender, are known to be effective in the treatment of atopic dermatitis and are used in natural topical creams. However, one of the most significant problems is that these oils lose their effectiveness quickly. As a solution, successful results have been achieved by loading these essential oils into nanocarrier systems [107]. Excessive application of fatty acids such as limonene, linalool, and geraniol, which are used to support the skin barrier, can cause skin irritation, burning sensation, and even allergic reactions. One way to prevent this is to deliver these active compounds using nanoemulsions, enabling slow release and application at low concentrations. For example, when Muta et al. [108] developed a lipid-based nanoformulation of tea tree oil, they were able to create a gel formulation that allowed for high stability and slow release of essential oils. Another study demonstrated that a microemulsion-based hydrogel containing tea tree oil and azelaic acid could form a stable structure that reduces the challenges associated with using these compounds [109]. Limonene-based syringic acid nanoemulsions can also be used in the treatment of psoriasis. Experiments showed that this high anti-inflammatory nanomaterial is successful in treating psoriasis in rats [110]. Topical application of methotrexate-loaded nanoemulsions of clobetasol propionate, calcipotriol, curcumin, chaulmoogra, and tacrolimus oil has also been proven successful in treating psoriasis [111].

Another condition with potential for medical use of aromatic nanocarriers is hypermelanosis. In the treatment of this disease, applications such as chemical peeling or laser therapy, as well as tyrosinase inhibitors, which can cause irritation and side effects, are preferred. Quercetin and olive oil, which are alternatives to these, can be used in the form of nanoemulsions due to their poor water solubility. In a study on quercetin-bound nanoemulsions, Silva et al. [112] obtained a formula with high stability and homogeneous emulsion, as well as high antioxidant and tyrosinase inhibition effects. The researchers also found that this nanomaterial, which can be used to neutralize free radicals and balance pigment distribution in the skin, does not irritate. In another study, kojic monooleate, another tyrosinase inhibitor, was integrated into a nanocarrier system to improve its penetration into the skin, resulting in a highly stable, elastic anti-hyperpigmentation formulation [113].

The potential for using nanoemulsions containing volatile oils in antiseptic formulations is also relatively high. For example, nanoemulsions of clove and thyme essential oils, with their antibacterial, antifungal, and antibiofilm properties, can be used in applications such as surgical field cleaning and hand sanitizers. Basil oil nanoemulsion hydrogels are effective against both Gram-negative and Gram-positive bacteria, such as *E. coli*, *Staphylococcus aureus*, and *Bacillus subtilis*, and also exhibit both antimicrobial and antibiofilm effects with cytotoxicity against *C. albicans*. Eugenol nanoemulsions also have potential for use in antiseptic gels due to their antibacterial effects [94, 114, 115]. Antiseptic and also oral gels such as nanoemulsion gels based on clove and cinnamon essential oils and mucosal gels such as nanoemulsions with herbal ingredients for the treatment of vaginal candidiasis can be produced. These gels reduce the chemical content of drugs, while enhancing the mucosal penetration of essential oils [83, 116].

Due to their anti-inflammatory effects, aromatic nanoemulsions and gels have high potential in topical applications to alleviate musculoskeletal pain and local inflammation and may help resolve arthritic problems by regulating the cyclooxygenase/lipoxygenase pathways in the skin and subcutaneous tissue [117]. Nanoemulsions of *Aniba canelilla* (Kunth) Mez, curcuma, thyme, and ginger extracts or oils have been found to possess anti-inflammatory effects [118, 119]. Additionally, limonene-, coumaric acid-, and syringic acid-enhanced acetic acid nanoemulsions also possess anti-inflammatory potential [120, 121]. The integration of aromatherapeutic agents with nanoemulsions and gels has been a focus of recent research. Inhalation of these agents has a calming effect on the nervous system. When integrated with nanocarriers, topical and local applications are also possible. These applications aim to achieve the impact at low doses. For example, lavender oil, known for its anxiety-reducing and sleep-promoting effects, is an effective phytotherapeutic agent. However, the fact that this oil evaporates rapidly even at low temperatures causes difficulties in its application. As with most fatty acids, the use of nanocarriers is being considered to solve this problem in lavender oil. However, studies are still insufficient to reach a definitive conclusion [122].

Drugs that are large enough to cross the blood-brain barrier and are lipophilic in nature are produced to deliver a drug to the central nervous system. Therefore, nanocarriers are also required for the increased medical applications of natural neuromodulators. These nanocarriers are highly branched three-dimensional dendrimers, nanogels, and carbon nanoparticles. When delivered in this manner, eucalyptus oil has provided an analgesic effect in rats. It has been determined that a cream containing bergamot oil nanoemulsion has an analgesic effect, and another formulation provides agitation control in dementia patients [123].

A study on the use of rosemary oil-loaded nanoemulsions for Alzheimer's disease has revealed that they cause acetylcholinesterase inhibition and thus have the potential to support cognitive function [124]. In another study, quercetin- and vitamin E-loaded nanoemulsions were used intranasally to manage oxidative stress in a Parkinson's disease model. The effects of a developed vitamin E-loaded naringenin nanoemulsion administered via the nasal route were examined in rats using different behavioral models, including a muscle coordination test, narrow beam test, forced swim test, grip strength test, and akinesia test. The study concluded that the nanoemulsion reversed 6-OHDA-induced Parkinson's symptoms in rats when administered together with levodopa [125].

7 Microbiological Safety, Sterility, and Preservation

Plant-based raw materials are exposed to environmentally derived microbial contamination during cultivation, harvesting, transportation, and storage. Their organic nature makes these products highly attractive to microorganisms. Plant-based nanoemulsions and gels are also at serious risk in this regard. Microbial contamination can shorten the shelf life of the product and also cause biosafety vulnerabilities. In plant products, spoilage caused by microorganism groups such as Firmicutes, Actinobacteria, and Proteobacteria can cause deterioration in texture, color, and odor. In this case, the product becomes unusable. The most frequently reported microbial contaminants include *Bacillus* spp., which are relatively difficult to combat because they can form spores; *Pseudomonas aeruginosa*, which can multiply rapidly in humid environments and form biofilms; and various mold species such as *Aspergillus*, *Penicillium*, and *Fusarium* spp. [126, 127]. Molds and yeasts are particularly important because they are commonly found in storage environments. Although many aromatic compounds exhibit antifungal properties, molds pose a greater threat due to the toxins they release into the environment.

Spore-forming bacteria are more resistant to disinfectants than vegetative forms. The presence of spore-forming Gram-positive bacteria, particularly *Bacillus subtilis*, in plant-based nanoemulsions and gels can be suppressed by the antimicrobial properties of the bioactive compounds in the emulsion. For example, the inhibitory effect of chamomile essential oil nanoemulsion on *B. subtilis* has been reported [128]. A similar situation applies to *P. aeruginosa*. This contaminant, which can originate from organic fertilizers, can be inhibited by most antimicrobial essential oils such as mint, myrtle, clove, and tea tree oil [129, 130].

There are several monitoring parameters used to control microbial contamination in pharmaceutical and cosmetic products. Total aerobic microbial count and total yeast and mold count, along with the identification of *S. aureus*, *E. coli*, *P. aeruginosa*, and *Candida albicans*, are the preferred basic tests. The total microbial load helps predict both how well the production process can suppress the microbial load during raw material processing and the formulation's success in protecting the product. In addition to these tests, the efficacy tests of the preservatives used must also be studied during the formulation development stage. These tests will also enable the determination of the amount of chemical preservative to be added to the product if it contains an aromatic compound, which is one of the components of the nanoemulsion, and if its antimicrobial efficacy has been detected. To ensure the sustained efficacy and increased stability of the prepared nanocarrier-aromatic complex, it is important to test the final product for its antimicrobial limits, as well as to perform direct testing. The efficacy of the nanomaterial to be used as a raw material is generally examined using minimum inhibitory concentration and minimum bactericidal concentration determination tests. In these tests, the target pathogen selection requires the clear identification of the product's intended use and possible contaminating pathogens. Various microorganisms such as *E. coli*, *S. aureus*, *Pseudomonas* sp., *Listeria monocytogenes*, *Klebsiella pneumoniae*, *Bacillus subtilis*, *Proteus vulgaris*, *Streptococcus pyogenes*, and *C. albicans* are used in these tests. Another group of antimicrobial efficacy tests are agar diffusion tests. Although many nanoemulsions do not pose a diffusion problem due to their small size, these tests are less preferred because they do not provide results as accurate as dilution-based tests [128, 131–134].

In addition to their antimicrobial activities, the cytotoxicity and mutagenic effects of aromatic nanoemulsions and gels are also being investigated. Different cell lines can be used in the studies, and various parameters such as oxidative stress, cellular deformation, and rapid cell death are examined after a specific period of exposure. Cytotoxicity tests can be conducted in two different ways: in vivo and in vitro. In vitro, MTT tests, which frequently test mitochondrial function and cellular oxidoreductase enzymes, are preferred. Additionally, lactate dehydrogenase, hemolysis, and plasma membrane damage are other commonly preferred tests [135–137]. In vivo tests generally use animal models. These experiments examine parameters such as the efficacy duration, penetration, and bioavailability of the material obtained, while also evaluating side effects such as cellular damage, irritation, and allergy [138, 139].

Microbial validation of raw materials for pharmaceutical and cosmetic products, as well as stability tests that determine product safety, quality, and shelf life, is another factor to consider. Findings related to the long-term stability and effectiveness of nanoemulsions focus on parameters such as droplet size, polydispersity index, zeta potential, and the ability to maintain effectiveness at different pH values. Standards such as Good Manufacturing Practices (GMP) and ISO 22716 are also important for products containing these materials. These GMP include a series of measures and application protocols designed to avoid microbial, physical, and chemical contamination. Manufacturing in compliance with specific standards and protocols offers advantages in various aspects, such as product standardization, protection of user health, and protection of the business against unforeseen harm [129, 140].

8 Regulatory and Pharmacopeial Considerations

Throughout human history, the search for healing began with the plants offered by nature, and this knowledge has been passed down from generation to generation, reaching the present day. Today, herbal treatments are not only a part of cultural heritage but also a crucial component of modern healthcare systems. Approximately 85% of the world's population currently uses herbal medicine, and this rate exceeds 90% in Africa. Almost half (46%) of new drugs approved between 1981 and 2019 were derived from natural compounds. The effectiveness of traditional Chinese formulas in symptom control during the COVID-19 pandemic has once again demonstrated the current value of this heritage [141]. The roots of herbal pharmacopoeias date back to the second century AD with the Chinese work *Shennong Bencaojing*, where drugs were divided into three classes based on their toxicity and therapeutic power. The Romanian pharmacopoeia began in the mid-nineteenth century with 180 plant monographs, but the herbal content gradually decreased over the following century, making way for synthetic drugs. Russia, a plant-associated society, survived on wild plants even during wartime [142]. These historical examples demonstrate the importance of herbal pharmacopoeias not only for treatment but also for cultural identity and social resilience. Currently, major references such as the European Pharmacopoeia (Ph. Eur.), the United States Pharmacopoeia (USP), the Chinese Pharmacopoeia (ChP), and the African Herbal Pharmacopoeia (AfrHP) provide varying levels of standards to ensure the quality of herbal medicines. The European Pharmacopoeia represents the most stringent system with binding rules, while the AfrHP provides a more flexible framework with voluntary standards. The Chinese Pharmacopoeia, on the other hand, contains approximately 1000 monographs on traditional Chinese medicine, making it one of the most comprehensive resources [143, 144].

The most significant limitation of herbal medicines is their low bioavailability. Nanotechnology stands out as a powerful tool to overcome this limitation. Systems such as liposomes, phytosomes, nanoemulsions, nanocapsules, and dendrimers increase the solubility and efficacy of herbal compounds. Products such as Meriva® (curcumin-phosphatidylcholine complex) and Legalon® SIL (silibinin succinate phytosome complex) are examples approved for clinical use. However, the toxicity, bioaccumulation, and environmental impacts of nanoparticles remain controversial. A notable shortcoming is the lack of monographs specific to nanoformulations in pharmacopoeias.

Another issue as important as nanotechnology is the ethical framework. The Nagoya Protocol mandates equitable benefit sharing in the use of biological resources and traditional knowledge. Hoodia and rooibos in South Africa are successful examples of sharing the economic benefits derived from the knowledge and resources of local communities. *Hoodia gordonii* (Masson) Sweet ex Decne. is a succulent plant native to South Africa and Namibia. It has been used for centuries by the San (Bushmen) people as an appetite suppressant and thirst reducer during long hunts in desert conditions. In the 1990s, the South African Council for Scientific and Industrial Research (CSIR) isolated a molecule called P57 from *Hoodia* as an appetite suppressant and licensed it to major pharmaceutical companies (Pfizer, Unilever, etc.). Initially, the use of traditional knowledge by San communities was not acknowledged, but after international pressure, an agreement was reached in 2003, granting the San people a share of the commercial revenues generated from *Hoodia*. This is cited as an example of the principle of "fair benefit sharing" envisaged by the Nagoya Protocol [145]. Another example is rooibos (*Aspalathus linearis* (Burm.f.) R.Dahlgren), a plant native to South Africa, whose leaves have been consumed as tea by the local Khoi and San people for centuries. Its antioxidant properties and health benefits have made it popular worldwide as "rooibos tea." The global trade in rooibos has created a multimillion-dollar industry. However, in the process, the communities that first used traditional knowledge were denied any rights. In 2019, the South African government mandated a benefit-sharing agreement between the rooibos industry and the local Khoi and San communities, with the community receiving a percentage of the revenue from rooibos sales [146]. This approach aims to ensure not only medical benefits but also social justice.

Adverse reaction reporting rates for herbal medicines range from 0.03% to 30%. This discrepancy demonstrates the inadequacy of safety monitoring systems. Artificial intelligence and big data-based approaches (e.g., WHO VigiBase, India AYUSH practices) have the potential to more accurately and earlier detect safety signals specific to herbal medicines [147]. DNA barcoding, metabolomics, and chemoprofiling methods also open new horizons in preventing counterfeiting and ensuring standardization [148].

Herbal pharmacopoeias are no longer legacies of the past but rather living structures, merging with nanotechnology, artificial intelligence, and global regulatory systems. This transformation is not only a scientific advancement but also a necessity for health equity and sustainability. The combination of traditional knowledge and modern innovation stands out as the strongest candidate for a just, accessible, and effective healthcare model that can address the health inequities of the twenty-first century [149].

The US Food and Drug Administration (FDA) and the European Medicines Agency (EMA) are responsible for ensuring the effectiveness, safety, and standard of medicines. Both internationally recognized regulatory agencies have strict Good Manufacturing Practices (GMP) guidelines [150].

8.1 FDA Requirements

- The FDA's Current Good Manufacturing Practices (cGMPs) cover manufacturing, packaging, labeling, and quality control. FDA inspections are conducted to ensure that pharmaceutical products entering the US market comply with the highest quality standards.

- The US pharmaceutical manufacturing standards (21 CFR Part 211) are the rules and regulations that encompass cGMPs for the production and preparation of finished drug products for human and animal use [151, 152].

Laboratory practices in pharmaceutical manufacturing include rigorous controls over analytical methodologies, specifications, sampling, and testing procedures for raw materials, intermediates, finished products, and packaging components under section 211, "Laboratory Controls." Pharmaceutical companies must validate analytical procedures, conduct stability testing to determine product shelf life, develop product standards, and maintain detailed records of laboratory activities [152].

Any drug that does not comply with cGMP regulations is considered "adult." The FDA also provides guidelines for the selection and testing of systems used to protect products from contamination, spoilage, and external factors that could compromise product quality. It also mandates cleaning validation to prevent cross-contamination and ensure that the equipment used is thoroughly cleaned [153].

8.2 EMA Requirements

- Pharmaceutical production in the European Union is governed by the EMA's GMP rules and includes key aspects of manufacturing, quality control, and documentation. Compliance with the EMA's GMP rules is required for obtaining EudraGMP certification and commercializing pharmaceutical products within the European Union [154].
- The European Pharmacopoeia (Ph. Eur.) provides quality standards for the production and control of medicines in Europe. These quality standards cover active pharmaceutical ingredients and excipients in their original form and in the form of pharmaceutical preparations and are legally binding for all 40 signatories [155].

8.3 Preclinical Requirements for Efficacy and Safety

The FDA ensures safety and efficacy criteria before products reach consumers and even implements post-market monitoring measures [156, 157].

- With cGMP, quality is integrated into every step of the design and manufacturing process, ensuring the efficacy and safety of drugs [153].
- Good laboratory practices regulate the planning, execution, monitoring, recording, reporting, and archiving of laboratory studies. This ensures the reliability and integrity of results [158].
- The International Council on Safety of Pharmaceutical Excipients (SCIPEC) creates safety tests for excipients based on their pharmacological effects, exposure circumstances, and physical and chemical characteristics [159].

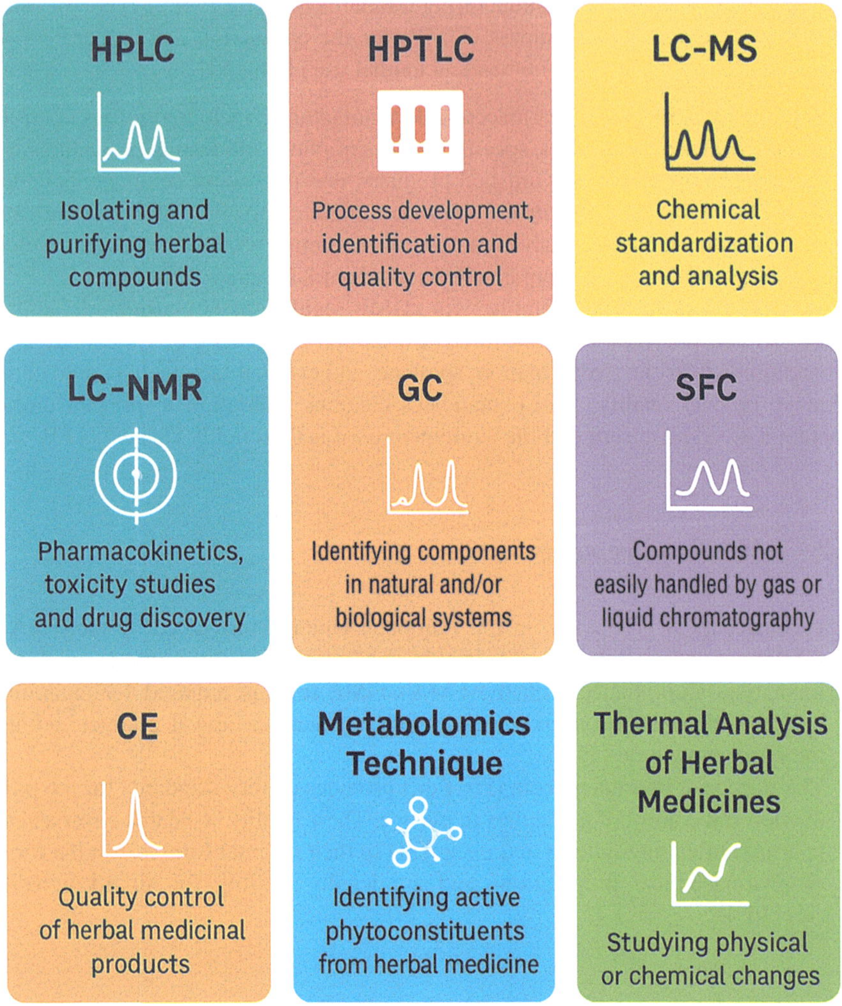

Fig. 4 Techniques in herbal drug identification and characterization

– Analytical techniques must be validated to show that they satisfy requirements for robustness, accuracy, precision, and specificity (Fig. 4) [158].

8.4 *Labeling, Shelf Life, and Preservative Use Limits*

Labeling, shelf life, and preservative use limits for pharmaceutical products are strictly regulated to ensure quality, safety, and patient compliance. Pharmacopoeia standards are officially recognized by a country or region and define the quality standards of pharmaceuticals. These standards include the European Pharmacopoeia (Ph. Eur.), the American Pharmacopoeia (USP), the Japanese Pharmacopoeia (JP), and the Turkish Pharmacopoeia (TF). These standards include packaging and labeling requirements. The goal is to ensure that products and their components are of a standard quality. Listing the excipients used by all pharmaceutical manufacturers will help determine the overall frequency and extent of problems (bioequivalence, toxicity, etc.) that these substances may cause in the population. In the United States, the FDA recognizes the American Pharmacopoeia (USP-NF) as the official and binding standard for quality standards and in Europe the EMA Ph. Eur. In Turkey, its standards are accepted by the Turkish Pharmaceuticals and Medical Devices Agency (TİTCK) and the Turkish Pharmacopoeia (TF). Pharmacopoeia standards define quality, while the FDA (or institutions such as EMA and TİTCK) monitors compliance with these standards and issues licenses [150, 155].

Stability studies are conducted to determine the shelf life and storage conditions of drug products. These studies evaluate the long-term stability of pharmaceutical products under various conditions, such as temperature, humidity, and light exposure. Additionally, accelerated stability tests are also available. These tests utilize high temperature and humidity conditions to assess the long-term stability of pharmaceutical formulations over shorter periods of time. Consequently, the shelf life of pharmaceutical products is determined. Shelf life analyses help determine drug expiration dates [158].

References do not directly specify specific numerical values for preservative use limits. However, it has been noted that non-sterile products can be preserved by adding chemical compounds to prevent microbial growth without compromising product quality. When performing microbial load tests, samples containing preservatives or antimicrobial agents may produce false-negative results. Therefore, the agents used to neutralize these substances should be validated. It has been emphasized that preservative-free drugs and safer diluents should be developed for sensitive populations such as low-birth-weight babies, asthma patients, etc., as they react to excipients added to formulations to maintain product stability, increase bioavailability, and optimize manufacturing processes [160]

9 Challenges and Future Perspectives

Aromatic plants are natural resources used in traditional medicine for thousands of years, but their use in modern biomedical applications requires overcoming certain challenges. While the aromatic plant-based nanoemulsions and gels discussed in this section lie at the intersection of nanotechnology and phytotherapy, there is still some way to go to improve the efficacy and safety of these systems.

One of the most important challenges to be addressed in the biomedical use of aromatic plant-based nanoemulsions and nanogels is homogeneity and stability. Low water solubility, volatility, susceptibility to oxidation, and low bioavailability limit the effectiveness of bioactive compounds derived from plants in biomedical applications. Another challenge is formulation challenges. The variation in aromatic plants makes it difficult to achieve standardization in standardized products. Furthermore, the volatile nature and oxidation tendency of the essential oils used, as mentioned above, pose a significant challenge to long-term storage stability and the availability of stable formulations. However, by selecting appropriate surfactants such as Tween 80, using co-surfactants, and adding polymeric stabilizers such as chitosan and Carbopol 934 to the formulation, stable formulations can be achieved for 6 months or more. Furthermore, a low polydispersity index ensures homogeneous size distribution and long-term physical stability, while a high zeta potential ensures stability through the surface charge and electrostatic repulsion of the particles.

Transitioning nanoemulsions and nanogels from small-scale laboratory production to large-scale industrial production presents both cost and technological challenges. Production innovations such as continuous flow reactors and micromixer technologies can offer solutions to overcome these challenges.

Nanotechnology is also seen as a solution to the increasing resistance to antibiotics, one of the most important problems of our time. Nanoemulsion and nanogel products will increase the solubility, stability, bioavailability, and targeted release of essential oils and bioactive compounds, making them an effective alternative to synthetic antibiotics for future antimicrobial strategies. While the combination of aromatic plants and nanotechnology promises to yield novel, potent, and effective therapeutic products, standardization and regulation are necessary before these products can be introduced to the pharmaceutical market. Despite standards established by regulatory bodies such as the FDA and EMA, product development is slowed by the lack of standardized protocols, safety assessment criteria, and clinical studies for aromatic plant-based nanoemulsions and gels. Furthermore, comprehensive studies and determination of the biocompatibility and toxicological properties of the nanomaterials used are essential. The toxicity of the concentrations of surfactants used in the production of nanoemulsions must be determined. Furthermore, good laboratory practice toxicology studies are crucial before their use in humans.

10 Conclusion

The essential oils and bioactive compounds of aromatic plants offer great potential for the pharmaceutical and cosmetic industries, thanks to their natural origin and broad spectrum of biological activity. Their antioxidant, anti-inflammatory, and antimicrobial properties make them key components in next-generation therapeutic and product development.

Aromatic plant-based nanoemulsions and gels have the potential to overcome fundamental challenges of active ingredients, such as low water solubility, volatility, susceptibility to oxidation, and low bioavailability. Nanoemulsions in the

20–200 nm size range significantly increase skin penetration, stability, and bioavailability, while also providing controlled and targeted release.

The intersection of aromatic plant-based systems and nanotechnology could enable the overcoming of challenges such as drug resistance and formulation limitations. Standardization will focus on addressing safety assessment issues by standardizing protocols and completing long-term toxicity studies, as well as intelligent controlled release systems responsive to environmental cues such as pH or temperature; hybrid carriers combining nanoemulsions with other nanocarrier systems such as liposomes, nanocapsules, dendrimers, and solid lipid nanoparticles; and sustainable and environmentally friendly manufacturing approaches. This will provide patients with more effective, safe, and sustainable treatment solutions. Realizing all of this potential requires close collaboration between experts in pharmaceutical botany, microbiology, nanotechnology, pharmacology, toxicology, and regulatory science.

Competing Interests The authors declare that they have no competing interest.

Declaration of Generative AI and AI-Assisted Technologies in the Writing Process During the preparation of this work, the book chapter author(s) have not taken any assistance from AI, and the book chapter author(s) reviewed and edited the content as needed and take(s) full responsibility for the content of the publication.

References

1. Dabravolski SA, Sukhorukov VN, Melnichenko AA, Khotina VA, Orekhov AN (2023) Potential application of the plant-derived essential oils for atherosclerosis treatment: molecular mechanisms and therapeutic potential. Molecules 28:5673
2. Husen A (2022) Herbs, shrubs and trees of potential medicinal benefits. CRC Press, Boca Raton. https://doi.org/10.1201/9781003205067
3. Husen A (2023a) Antidiabetic medicinal plants and herbal treatments. CRC Press, Boca Raton. https://doi.org/10.1201/b23347
4. Husen A (2023b) Exploring poisonous plants: medicinal values, toxicity response and other potential uses. CRC Press, Boca Raton. https://doi.org/10.1201/b23017
5. Husen A (2023c) Secondary metabolites based green synthesis of nanomaterials and their applications. Springer Nature Singapore Pte Ltd, Singapore. https://doi.org/10.1007/978-981-99-0927-8
6. Husen A (2024a) Antimalarial medicinal plants. Taylor & Francis Group, LLC, Boca Raton. https://doi.org/10.1201/9781003378396
7. Husen A (2024b) Medicinal spices and condiment crops. Taylor & Francis Group, LLC, Boca Raton. https://doi.org/10.1201/9781003387046
8. Perveen S, Kanwal S, Muneer S, Aslam S, ul ane Gillani Q, Qadeer I et al (2025) A application of herbal medicine and essential oils in new era. In: Medicinal plants and aromatics: a holistic health perspective, vol 109. Unique Scientific Publishers, Faisalabad
9. Tanwar SN, Parauha YR, There Y, Swart HC, Dhoble SJ (2024) Plant-based biosynthesis of metal and metal oxide nanoparticles: an update on antimicrobial and anticancer activity. ChemBioEng Rev 11:e202400012

10. Che CT, George V, Ijinu TP, Pushpangadan P, Andrae-Marobela K (2024) Traditional medicine. In: Pharmacognosy. Academic Press, New York, pp 11–28
11. Das M, Roy S, Guha C, Saha AK, Singh M (2021) In vitro evaluation of antioxidant and antibacterial properties of supercritical CO2 extracted essential oil from clove bud (Syzygium aromaticum). J Plant Biochem Biotechnol 30:387–391
12. Giri VP, Pandey S, Shukla P, Gupta SC, Srivastava M, Rao CV et al (2024) Facile fabrication of sandalwood oil-based nanoemulsion to intensify the fatty acid composition in burned and rough skin. ACS Omega 9(6):6305–6315
13. Gupta AK, Yadav R, Sharma S, Sawale J, Kahwa I (2023) Anti-ageing activity and antioxidant activity of methanol extracts of leaves and flowers of Salvia officinalis and Rosmarinus officinalis. J Young Pharm 15:448–455
14. Nokhodchi A, Chavan S, Ghafourian T (2023) In vitro dissolution and permeability testing of inhalation products: challenges and advances. Pharmaceutics 15:983
15. Singh A, Ansari VA, Mahmood T, Ahsan F, Wasim R, Maheshwari S (2022) Development and evaluation of topical liposomal gel of lemon oil and orange oil for synergistic effect to enhance skin beauty. In: Recent research transition in life sciences, p 34. ISBN: 978-93-93810-10-6
16. Gupta S, Gupta J, Anand A, Ojha S (2022) Review on nanogel/emulgel formulations of traditional medicines. Biol Sci 2:269–278
17. Rasheed S, Zahid S, Jamil I, Rafi U, Bano E, Aslam M et al (2023) A comprehensive review on biomedical application of plant-derived essential oils. Mater Chem Mech 1:1–23
18. Al-Balushi RA, Haque A, Saeed M, Al-Harthy T, Al-Hinaai M, Al-Hashmi S (2024) Unlocking the anticancer potential of frankincense essential oils (FEOs) through nanotechnology: a review. Mol Biotechnol 66:3013–3024
19. Haghbayan H, Moghimi R, Sarrafi Y, Taleghani A, Hosseinzadeh R (2024) Enhancing bioactivity of Callistemon citrinus (Curtis) essential oil through novel nanoemulsion formulation. J Biomater Sci Polym Ed 35:2660–2681
20. Husen A (2023d) Nanomaterials from agricultural and horticultural products. Springer Nature Singapore Pte Ltd, Singapore. https://link.springer.com/book/9789819934348
21. Husen A, Siddiqi KS (2023) Advances in smart nanomaterials and their applications. Elsevier, Cambridge. https://doi.org/10.1016/C2021-0-02202-1
22. Jin-Chul K, Madhusudhan A, Husen A (2021) Smart nanomaterials in biomedical applications. Springer Nature Switzerland AG, Cham https://doi.org/10.1007/978-3-030-84262-8
23. Madhusudhan A, Purohit SD, Prasad R, Husen A (2024) Functional smart nanomaterials and their theranostics approaches. Springer Nature Singapore Pte Ltd, Singapore
24. Mandal AK, Ghorai S, Husen A (2023) Functionalized smart nanomaterials for point-of-care testing. Springer Nature Singapore Pte Ltd., Singapore
25. Pavlačková J, Egner P, Mokrejš P, Janalíková M (2024) Formulating sustainable emulsions: mandelic acid and essential oils as natural preservatives. Molecules 29:4510
26. Rajagopal M, Paul AK, Lee MT, Joykin AR, Por CS, Mahboob T et al (2022) Phytochemicals and nano-phytopharmaceuticals use in skin, urogenital and locomotor disorders: are we there? Plants 11:1265
27. Rath SK, Dwivedi V, Husen A, Akhtar N (2024) Smart nanomaterials for drug delivery and neurological diseases management. Springer Nature Singapore Pte Ltd, Singapore. https://doi.org/10.1007/978-981-97-0308-1
28. Kaplan A (2022) The nanocomposites designs of phytomolecules from medicinal and aromatic plants: promising anticancer-antiviral applications. Beni-Suef Univ J Basic Appl Sci 11:17
29. Kaur N, Bains A, Kaushik R, Dhull SB, Melinda F, Chawla P (2021) A review on antifungal efficiency of plant extracts entrenched polysaccharide-based nanohydrogels. Nutrients 13:2055
30. Ozay C, Karpuz M (2024) Phytocompounds and lipid-based drug delivery system for neurodegenerative diseases. Asian Pac J Trop Biomed 14:417–426

31. Rajamma SS, Krishnaswami V, Prabu SL, Kandasamy R (2022) Geophila repens phytosome-loaded intranasal gel with improved nasal permeation for the effective treatment of Alzheimer's disease. J Drug Deliv Sci Technol 69:103087
32. Iskandar B, Liu TW, Mei HC, Kuo IC, Surboyo MDC, Lin HM, Lee CK (2024) Herbal nanoemulsions in cosmetic science: a comprehensive review of design, preparation, formulation, and characterization. J Food Drug Anal 32:428
33. Javed S, Mangla B, Salawi A, Sultan MH, Almoshari Y, Ahsan W (2024) Essential oils as dermocosmetic agents, their mechanism of action and nanolipidic formulations for maximized skincare. Cosmetics 11:210
34. Omidian H, Cubeddu LX, Gill EJ (2025) Harnessing nanotechnology to enhance essential oil applications. Molecules 30:520
35. Govindaswamy B, Ravikrishnan B, Panda D, Perumal S (2024) In vitro anti cancer, anti inflammatory and anti oxidant activity of Celery oil (Apium graveolens)-Myristic acid based microemulsion system: characterization and biomedical application. bioRxiv 2024:11
36. Hu F, Sun DS, Wang KL, Shang DY (2022) Nanomedicine of plant origin for the treatment of metabolic disorders. Front Bioeng Biotechnol 9:811917
37. Kaspute G, Ramanavicius A, Prentice U (2025) Natural drug delivery systems for the treatment of neurodegenerative diseases. Mol Biol Rep 52:217
38. Serri C, Cruz-Maya I, Bonadies I, Rassu G, Giunchedi P, Gavini E, Guarino V (2023) Green routes for bio-fabrication in biomedical and pharmaceutical applications. Pharmaceutics 15:1744
39. Okada K, Shimatani K (2024) Effect of olfactory stimulation from aromatherapy on the autonomic nervous activity during aerobic exercises. Sci Rep 14(1):11198
40. Choi NY, Wu YT, Park SA (2022) Effects of olfactory stimulation with aroma oils on psychophysiological responses of female adults. Int J Environ Res Public Health 19(9):5196
41. Lo MM, Benfodda Z, Molinié R, Meffre P (2024) Volatile organic compounds emitted by flowers: ecological roles, production by plants, extraction, and identification. Plants 13(3):417
42. Sardarova AS, Ibadullayeva SJ (2025) Ecological anatomical study of structural-plastic response reactions of the medicinally important species Laurus nobilis (Lauraceae) in various ecological conditions. Plant Fungal Res 8(1):23–35
43. Shunmugam S, Roslia NS, Manickamb S, Yusoffa NF, Alitheena NB, Namasivayama P (2021) Morphology and anatomy of leaf, stem and petiole of Luvunga crassifolia Tanaka (Rutaceae). Mal J Fund Appl Sci 17(6):818–828
44. Tulik M, Jura-Morawiec J (2023) An arrangement of secretory cells involved in the formation and storage of resin in tracheid-based secondary xylem of arborescent plants. Front Plant Sci 14:1268643
45. Gostin IN (2023) Glandular and non-glandular trichomes from Phlomis herba-venti subsp. pungens leaves: light, confocal, and scanning electron microscopy and histochemistry of the secretory products. Plants 12(13):2423
46. Chatt EC, Mahalim SN, Mohd-Fadzil NA, Roy R, Klinkenberg PM, Horner HT, Hampton M, Carter CJ, Nikolau BJ (2021) Nectar biosynthesis is conserved among floral and extrafloral nectaries. Plant Physiol 185(4):1595–1616
47. Caballero-Gallardo K, Quintero-Rincón P, Olivero-Verbel J (2025) Aromatherapy and essential oils: holistic strategies in complementary and alternative medicine for integral wellbeing. Plants 14(3):400
48. Nayak S, Hegde RB, Rao AS, Poojary R (2025) Unlocking the potential of essential oils in aromatic plants: a guide to recovery, modern innovations, regulation and AI integration. Planta 262(1):6
49. Ali SS, Al-Tohamy R, Al-Zahrani M, Badr A, Sun J (2025) Essential oils and plant-derived bioactive compounds: a comprehensive review of their therapeutic potential, mechanisms of action, and advances in extraction technologies. Phytochem Rev 14:1–49
50. Bolouri P, Salami R, Kouhi S, Kordi M, Asgari Lajayer B, Hadian J, Astatkie T (2022) Applications of essential oils and plant extracts in different industries. Molecules 27(24):8999

51. Worwood VA (2016) The complete book of essential oils and aromatherapy, revised and expanded: over 800 natural, nontoxic, and fragrant recipes to create health, beauty, and safe home and work environments. New World Library, Toronto
52. Dong X (2024) Pharmacological effects of aromatic medicinal plants: comprehensive analysis of active ingredients and mechanisms of action. J Med Plant Res 14:11–30
53. Soleimani M, Arzani A, Arzani V, Roberts TH (2022) Phenolic compounds and antimicrobial properties of mint and thyme. J Herb Med 36:100604
54. M'hamdi Z, Davì F, Elhourri M, Amechrouq A, Mondello F, Cacciola F et al (2024) Phytochemical investigations, antioxidant and insecticidal properties of essential oil and extracts from the aerial parts of Pelargonium graveolens from Morocco. Molecules 29:4036
55. Draiaia R, Amri A, Boubsil S, Necib A, Ketfi L, Mohamadi N (2025) GC/MS analysis, antioxidant and anti-inflammatory activity of Pelargonium graveolens. Bol latinoam Caribe plantas med aromát 24:199–211
56. Chroho M, Rouphael Y, Petropoulos SA, Bouissane L (2024) Carvacrol and thymol content affects the antioxidant and antibacterial activity of Origanum compactum and Thymus zygis essential oils. Antibiotics 13:139
57. Vassiliou E, Awoleye O, Davis A, Mishra S (2023) Anti-inflammatory and antimicrobial properties of thyme oil and its main constituents. Int J Mol Sci 24:6936
58. Belaiba M, Saoudi MM, Abedrabba M, Bouajila J (2024) Ammoides pusilla aerial part: GC-MS profiling and evaluation of in vitro antioxidant and biological activities. Processes 12:1274
59. Taibi M, Elbouzidi A, Haddou M, Yahyaoui MI, Baraich A, Allali A et al (2024) Exploring dermatological applications of Ammoides pusilla essential oil: evaluation of anti-aging, dermatoprotective and antimicrobial properties against dermatopathogenic microorganisms. Discov Life 54:19
60. Mokra D, Joskova M, Mokry J (2022) Therapeutic effects of green tea polyphenol (−)-Epigallocatechin-3-Gallate (EGCG) in relation to molecular pathways controlling inflammation, oxidative stress, and apoptosis. Int J Mol Sci 24:340
61. Capasso L, De Masi L, Sirignano C, Maresca V, Basile A, Nebbioso A et al (2025) Epigallocatechin gallate (EGCG): pharmacological properties, biological activities and therapeutic potential. Molecules 30:654
62. Masyita A, Sari RM, Astuti AD, Yasir B, Rumata NR, Emran TB et al (2022) Terpenes and terpenoids as main bioactive compounds of essential oils, their roles in human health and potential application as natural food preservatives. Food Chem: X 13:100217
63. Siddiqui T, Khan MU, Sharma V, Gupta K (2024) Terpenoids in essential oils: chemistry, classification, and potential impact on human health and industry. Phytomedicine Plus 4:100549
64. Verma T, Aggarwal A, Dey P, Chauhan AK, Rashid S, Chen KT, Sharma R (2023) Medicinal and therapeutic properties of garlic, garlic essential oil, and garlic-based snack food: an updated review. Front Nutr 10:1120377
65. Sahidur MR, Islam S, Jahurul MHA (2023) Garlic (Allium sativum) as a natural antidote or a protective agent against diseases and toxicities: a critical review. Food Chem Adv 3:100353
66. Akhtar H, Ali YA, Wei CR, Albassam RS, Ahmed F, Yasmin A et al (2025) Bioactive potential and health benefits of Trigonella foenum-graecum L.: a comprehensive review. Food Sci Nutr 13:e70887
67. Chelu M, Musuc AM, Popa M, Calderon Moreno J (2023) Aloe vera-based hydrogels for wound healing: properties and therapeutic effects. Gels 9:539
68. Catalano A, Ceramella J, Iacopetta D, Marra M, Conforti F, Lupi FR et al (2024) Aloe vera—an extensive review focused on recent studies. Foods 13:2155
69. Weber N, Biehler K, Schwabe K, Haarhaus B, Quirin KW, Frank U et al (2019) Hop extract acts as an antioxidant with antimicrobial effects against Propionibacterium acnes and Staphylococcus aureus. Molecules 24:223
70. Kolenc Z, Langerholc T, Hostnik G, Ocvirk M, Štumpf S, Pintarič M et al (2022) Antimicrobial properties of different hop (Humulus lupulus) genotypes. Plants 12:120

71. Liang J, Huang X, Ma G (2022) Antimicrobial activities and mechanisms of extract and components of herbs in East Asia. RSC Adv 12:29197–29213
72. Wang W, Chen YF, Wei ZF, Jiang JJ, Peng JQ, He QT, Liu HM (2022) Microemulsion of cinnamon essential oil formulated with tea polyphenols, gallic acid, and tween 80: antimicrobial properties, stability and mechanism of action. Microorganisms 11:2
73. Dobros N, Zawada KD, Paradowska K (2022) Phytochemical profiling, antioxidant and anti-inflammatory activity of plants belonging to the Lavandula genus. Molecules 28:256
74. Ţicolea M, Pop RM, Pârvu M, Usatiuc LO, Uifălean A, Ranga F, Pârvu AE (2024) Phytochemical composition antioxidant and anti-inflammatory activity of Artemisia dracunculus and Artemisia abrotanum. Antioxidants 13(8):1016
75. Souto EB, Cano A, Martins-Gomes C, Coutinho TE, Zielińska A, Silva AM (2022) Microemulsions and nanoemulsions in skin drug delivery. Bioengineering 9(4):158
76. Sutradhar KB, Amin ML (2013) Nanoemulsions: increasing possibilities in drug delivery. Eur J Nanomedicine 5(2):97–110
77. Wilson RJ, Li Y, Yang G, Zhao CX (2022) Nanoemulsions for drug delivery. Particuology 64:85–97
78. Mikhail AS, Morhard R, Mauda-Havakuk M, Kassin M, Arrichiello A, Wood BJ (2023) Hydrogel drug delivery systems for minimally invasive local immunotherapy of cancer. Adv Drug Deliv Rev 202:115083
79. Thang NH, Chien TB, Cuong DX (2023) Polymer-based hydrogels applied in drug delivery: an overview. Gels 9(7):523
80. Barradas TN, Senna JP, Cardoso SA, Nicoli S, Padula C, Santi P et al (2017) Hydrogel-thickened nanoemulsions based on essential oils for topical delivery of psoralen: permeation and stability studies. Eur J Pharm Biopharm 116:38–50
81. Badr MM, Badawy ME, Taktak NE (2021) Characterization, antimicrobial activity, and antioxidant activity of the nanoemulsions of Lavandula spica essential oil and its main monoterpenes. J Drug Deliv Sci Technol 65:102732
82. Hosny KM, Sindi AM, Alkhalidi HM, Kurakula M, Alruwaili NK, Alhakamy NA et al (2021) Oral gel loaded with penciclovir–lavender oil nanoemulsion to enhance bioavailability and alleviate pain associated with herpes labialis. Drug Deliv 28(1):1043–1054
83. Ullah N, Amin A, Farid A, Selim S, Rashid SA, Aziz MI, Mohtasheemul Hasan M (2023) Development and evaluation of essential oil-based nanoemulgel formulation for the treatment of oral bacterial infections. Gels 9:252
84. Bahloul B, Ben Bnina E, Hamdi A, Castillo Henríquez L, Baccar D, Kalboussi N et al (2024) Investigating the wound-healing potential of a nanoemulsion–gel formulation of Pituranthos tortuosus essential oil. Gels 10(3):155
85. Feng C, Xin W, Reyanggu A, Xueping M, Sodik N, Tao W et al (2025) Development and evaluation of a thermosensitive nanoemulsion hydrogel loaded with Schizonepeta annua essential oil for enhanced wound healing. Ind Crops Prod 233:121371
86. Gu R, Zhou H, Zhang Z, Lv Y, Pan Y, Li Q et al (2023) Research progress related to thermosensitive hydrogel dressings in wound healing: a review. Nanoscale Adv 5(22):6017–6037
87. Alam A, Ansari MJ, Alqarni MH, Salkini MA, Raish M (2023) Antioxidant, antibacterial, and anticancer activity of ultrasonic nanoemulsion of Cinnamomum cassia L. essential oil. Plants 12(4):834
88. Fatma G, Sami BHA, Ahmed L (2017) Investigation of extracts from Tunisian ethnomedicinal plants as antioxidants, cytotoxins, and antimicrobials. Biomed Environ Sci 30(11):811–824
89. Mighri H, Sabri K, Eljeni H, Neffati M, Akrout A (2015) Chemical composition and antimicrobial activity of Pituranthos chloranthus (Benth.) hook and Pituranthos tortu-osus (Coss.) maire essential oils from southern Tunisia. Adv Biol Chem 5(7):273–278
90. Sun R, Li Y, Su R, Cai X, Kong Y, Jiang T et al (2025) Antibacterial effect of ultrasound combined with Litsea cubeba essential oil nanoemulsion on Salmonella typhimurium in kiwifruit juice. Int J Food Microbiol 426:110898

91. Ling CQ, Liao HX, Wen JR, Nie HY, Zhang LY, Xu FR et al (2024) Investigation of the inhibitory effects of Illicium verum essential oil nanoemulsion on Fusarium proliferatum via combined transcriptomics and metabolomics analysis. Curr Microbiol 81(7):182
92. Noor A, Jamil S, Sadeq TW, Mohammed Ameen MS, Kohli K (2023) Development and evaluation of nanoformulations containing timur oil and rosemary oil for treatment of topical fungal infections. Gels 9(7):516
93. Eid AM, Issa L, Al-Kharouf O, Jaber R, Hreash F (2021) Development of Coriandrum sativum oil nanoemulgel and evaluation of its antimicrobial and anticancer activity. BioMed Res Int 2021(1):5247816
94. Chinnaiyan SK, Pandiyan R, Natesan S, Chindam S, Gouti AK, Sugumaran A (2022) Fabrication of basil oil nanoemulsion loaded gellan gum hydrogel—evaluation of its antibacterial and anti-biofilm potential. J Drug Deliv Sci Technol 68:103129
95. Mostaghimi M, Majdinasab M, Hosseini SMH (2022) Characterization of alginate hydrogel beads loaded with thyme and clove essential oils nanoemulsions. J Polym Environ 30(4):1647–1661
96. Abdollahi A, Fereydouni N, Moradi H, Karimivaselabadi A, Zarenezhad E, Osanloo M (2024) Nanoformulated herbal compounds: enhanced antibacterial efficacy of camphor and thymol-loaded nanogels. BMC Complement Med Ther 24:138
97. Azadi S, Osanloo M, Zarenezhad E, Farjam M, Jalali A, Ghanbariasad A (2023) Nano-scaled emulsion and nanogel containing Mentha pulegium essential oil: cytotoxicity on human melanoma cells and effects on apoptosis regulator genes. BMC Complement Med Ther 23:6
98. Miastkowska M, Sikora E, Kulawik-Pióro A, Kantyka T, Bielecka E, Kałucka U, Staniszewska-Ślęzak E (2023) Bioactive Lavandula angustifolia essential oil-loaded nanoemulsion dressing for burn wound healing: in vitro and in vivo studies. Biomater Adv 148:213362
99. Chelu M (2024) Hydrogels with essential oils: recent advances in designs and applications. Gels 10:636
100. Flores FC, De Lima JA, Da Silva CR, Benvegnú D, Ferreira J, Burger ME, Da Silva CDB (2015) Hydrogels containing nanocapsules and nanoemulsions of tea tree oil provide antiedematogenic effect and improved skin wound healing. J Nanosci Nanotechnol 15:800–809
101. Miastkowska M, Kulawik-Pióro A, Szczurek M (2020) Nanoemulsion gel formulation optimization for burn wounds: analysis of rheological and sensory properties. Processes 8:1416
102. Meghani N, Mansuri A, Chaudhari R, Kumar A (2024) Fabrication, assessment, and potential anti-bacterial activity of sandalwood oil nanoemulsion and its hand rub sanitizer. Environ Sci Pollut Res 31:46546–46557
103. Najafi-Taher R, Ghaemi B, Amani A (2018) Delivery of adapalene using a novel topical gel based on tea tree oil nano-emulsion: permeation, antibacterial and safety assessments. Eur J Pharm Sci 120:142–151
104. Taleb MH, Abdeltawab NF, Shamma RN, Abdelgayed SS, Mohamed SS, Farag MA, Ramadan MA (2018) Origanum vulgare essential oil as a potential anti-acne topical nanoemulsion: in vitro and in vivo study. Molecules 23:2164
105. Yousefpoor Y, Baharifar H, Esnaashari SS, Gheybi F, Mehrabi M, Osanloo M, Amani A (2024) A review of topical micro- and nanoemulsions for common skin diseases. *Nanomed J 11*(3): 205–221
106. Kildaci I, Budama-Kilinc Y, Kecel-Gunduz S, Altuntas E (2021) Linseed oil nanoemulsions for treatment of atopic dermatitis: formulation, characterization, in vitro and in silico evaluations. J Drug Deliv Sci Technol 64:102652
107. Damiani G, Eggenhöffner R, Pigatto PDM, Bragazzi NL (2019) Nanotechnology meets atopic dermatitis: current solutions, challenges and future prospects. Bioact Mater 4:380–386
108. Muta T, Parikh A, Kathawala K, Haidari H, Song Y, Thomas J, Garg S (2020) Quality-by-design approach for the development of nano-sized tea tree oil formulation-impregnated biocompatible gel with antimicrobial properties. Pharmaceutics 12:1091

109. Bisht A, Hemrajani C, Rathore C, Dhiman T, Rolta R, Upadhyay N, Tambuwala MM (2022) Hydrogel composite containing azelaic acid and tea tree essential oil as a therapeutic strategy for Propionibacterium and testosterone-induced acne. Drug Deliv Transl Res 12:2501–2517
110. Assalem N, Abd-Allah H, Ragaie MH, Ahmed SS, Elmowafy E (2024) Therapeutic potential of limonene-based syringic acid nanoemulsion: enhanced ex-vivo cutaneous deposition and clinical anti-psoriatic efficacy. Int J Pharm 660:124376
111. Latif MS, Nawaz A, Asmari M, Uddin J, Ullah H, Ahmad S (2022) Formulation development and in vitro/in vivo characterization of methotrexate-loaded nanoemulsion gel formulations for enhanced topical delivery. Gels 9:3
112. Silva CC, Benati RB, Massaro TN, Pereira KC, Gaspar LR, Marcato PD (2023) Antioxidant and anti-tyrosinase activities of quercetin-loaded olive oil nanoemulsion as potential formulation for skin hyperpigmentation. J Dispers Sci Technol 44:2628–2638
113. Syed Azhar SNA, Ashari SE, Salim N (2018) Development of a kojic monooleate-enriched oil-in-water nanoemulsion as a potential carrier for hyperpigmentation treatment. Int J Nanomedicine 13:6465–6479
114. Fu X, Gao Y, Yan W, Zhang Z, Sarker S, Yin Y, Chen J (2022) Preparation of eugenol nanoemulsions for antibacterial activities. RSC Adv 12:3180–3190
115. Hashem AH, Doghish AS, Ismail A, Hassanin MM, Okla MK, Saleh IA, Shehabeldine AM (2024) A novel nanoemulsion based on clove and thyme essential oils: characterization, antibacterial, antibiofilm and anticancer activities. Electron J Biotechnol 68:20–30
116. Negi P, Singh A, Pundir S, Parashar A, Upadhyay N, Agarwal S, Chauhan R, Tambuwala MM (2023) Essential oil and nanocarrier-based formulation approaches for vaginal candidiasis. Ther Deliv 14:207–225
117. Buriti BMAB, Figueiredo PLB, Passos MF, da Silva JKR (2024) Polymer-based wound dressings loaded with essential oil for the treatment of wounds: a review. Pharmaceuticals 17:897
118. Kreutz T, Carneiro SB, Soares KD, Limberger RP, Apel MA, Veiga-Junior VF, Koester LS (2021) Aniba canelilla essential oil-loaded nanoemulsion: improved stability of the main constituents and in vitro antichemotactic activity. Ind Crop Prod 171:113949
119. Wuttikul K, Sainakham M (2022) In vitro bioactivities and preparation of nanoemulsion from coconut oil loaded Curcuma aromatica extracts for cosmeceutical delivery systems. Saudi J Biol Sci 29:103435
120. Ekhtiar M, Ghasemi-Dehnoo M, Mirzaei Y, Azadegan-Dehkordi F, Amini-Khoei H, Lorigooini Z, Bagheri N (2023) The coumaric acid and syringic acid ameliorate acetic acid-induced ulcerative colitis in rats via modulation of Nrf2/HO-1 and pro-inflammatory cytokines. Int Immunopharmacol 120:110309
121. Kummer R, Fachini-Queiroz FC, Estevão-Silva CF, Grespan R, Silva EL, Bersani-Amado CA, Cuman RKN (2013) Evaluation of anti-inflammatory activity of Citrus latifolia essential oil and limonene in experimental mouse models. Evid Based Complement Alternat Med 2013:859083
122. Vora LK, Gholap AD, Hatvate NT, Naren P, Khan S, Chavda VP, Khatri DK (2024) Essential oils for clinical aromatherapy: a comprehensive review. J Ethnopharmacol 330:118180
123. Soares GABE, Bhattacharya T, Chakrabarti T, Tagde P, Cavalu S (2021) Exploring pharmacological mechanisms of essential oils on the central nervous system. Plants 11:21
124. Vasdev N, Handa M, Kesharwani P, Shukla R (2022) Rosemary oil low energy nanoemulsion: optimization, µrheology, in silico, in vitro, and ex vivo characterization. J Biomater Sci Polym Ed 33:1901–1923
125. Gaba B, Khan T, Haider MF, Alam T, Baboota S, Parvez S, Ali J (2019) Vitamin E loaded naringenin nanoemulsion via intranasal delivery for the management of oxidative stress in a 6-OHDA Parkinson's disease model. Biomed Res Int 2019:2382563
126. Kosalec I, Cvek J, Tomic S (2009) Contaminants of medicinal herbs and herbal products. Arh Hig Rada Toksikol 60:485
127. Steinhoff B (2021) Challenges in the quality of herbal medicinal products with a specific focus on contaminants. Phytochem Anal 32:117–123

128. Jamir Y, Bhushan M, Sanjukta R, Singh LR (2024) Plant-based essential oil encapsulated in nanoemulsions and their enhanced therapeutic applications: an overview. Biotechnol Bioeng 121:415–433
129. Drais HK (2024) Development and evaluation essential oils nanoemulgel as human skin sanitizer using novel method. Turk J Pharm Sci 21:456
130. Sevinç-Özakar R, Seyret E, Özakar E, Adıgüzel MC (2022) Nanoemulsion-based hydrogels and organogels containing propolis and dexpanthenol: preparation, characterization, and comparative evaluation of stability, antimicrobial, and cytotoxic properties. Gels 8:578
131. Ashaolu TJ (2021) Nanoemulsions for health, food, and cosmetics: a review. Environ Chem Lett 19:3381–3395
132. Hadian M, Rajaei A, Mohsenifar A, Tabatabaei M (2017) Encapsulation of Rosmarinus officinalis essential oils in chitosan-benzoic acid nanogel with enhanced antibacterial activity in beef cutlet against Salmonella typhimurium during refrigerated storage. LWT 84:394–401
133. Luciano WA, Pimentel TC, Bezerril FF, Barão CE, Marcolino VA, Carvalho RDSF, Magnani M (2023) Effect of citral nanoemulsion on the inactivation of Listeria monocytogenes and sensory properties of fresh-cut melon and papaya during storage. Int J Food Microbiol 384:109959
134. Wang W, Zhao D, Xiang Q, Li K, Wang B, Bai Y (2021) Effect of cinnamon essential oil nanoemulsions on microbiological safety and quality properties of chicken breast fillets during refrigerated storage. LWT 152:112376
135. de Godoi SN, Quatrin PM, Sagrillo MR, Nascimento K, Wagner R, Klein B, Ourique AF (2017) Evaluation of stability and in vitro security of nanoemulsions containing Eucalyptus globulus oil. Biomed Res Int 2017:2723418
136. Mota Ferreira L, Gehrcke M, Ferrari Cervi V, Bitencourt PER, da Silveira EF, Azambuja JH, Cruz L (2016) Pomegranate seed oil nanoemulsions with selective antiglioma activity: optimization and evaluation of cytotoxicity, genotoxicity and oxidative effects on mononuclear cells. Pharm Biol 54:2968–2977
137. Yoon HJ, Zhang X, Kang MG, Kim GJ, Shin SY, Baek SH, Bae H (2018) Cytotoxicity evaluation of turmeric extract incorporated oil-in-water nanoemulsion. Int J Mol Sci 19:280
138. Hort MA, Alves BS, Ramires Júnior OV, Falkembach MC, Araújo GMS, Fernandes CLF, Tavella RA, Bidone J, Dora CL, Silva Júnior FMR (2021) In vivo toxicity evaluation of nanoemulsions for drug delivery. Drug Chem Toxicol 44:585–594
139. Yousefpoor Y, Esnaashari SS, Baharifar H, Mehrabi M, Amani A (2023) Current challenges ahead in preparation, characterization, and pharmaceutical applications of nanoemulsions. WIREs Nanomed Nanobiotechnol 15:e1920
140. Azmi NAN, Elgharbawy AA, Motlagh SR, Samsudin N, Salleh HM (2019) Nanoemulsions: factory for food, pharmaceutical and cosmetics. Processes 7:617
141. Simmonds MS (2023) Plants and medicine. In: A cultural history of plants in the nineteenth century, vol 5, p 105
142. Choudhury PA, Singh N, Bajwa S, Dash P, Bisht (2023) Pharmacovigilance of herbal medicines: concerns and future prospects. J Ethnopharmacol 309:116383
143. Fursenco C, Drăgălin EA, Calalb T, Uncu L (2023) Comparative assessment of pharmacopoeia requirements regarding the standardization of herbal drugs. In: Direcții de reformare a sistemului farmaceutic din perspectiva cursului european al Republicii Moldova, pp 104–116
144. Zellner L, Wierer M, Himmelsbach M, Schatzmann B, Thalhamer B, Klampfl CW (2025) Determination of withanolides and withanosides in ashwagandha based products using HPLC-drift-tube-ion-mobility quadrupole time-of-flight mass spectrometry. Electrophoresis 46(5–6):340–346
145. Shomope I, Percival KM, Abdel Jabbar NM, Husseini GA (2024) Predicting calcein release from ultrasound-targeted liposomes: a comparative analysis of random forest and support vector machine. Technol Cancer Res Treat 23:15330338241296725
146. Shree D, Patra CN, Sahoo BM (2022) Novel herbal nanocarriers for treatment of dermatological disorders. Pharm Nanotechnol 10(4):246–256

147. Farooq S, Mehmood Z, Qais FA, Khan MS, Ahmad I (2019) Nanoparticles in ayurvedic medicine: potential and prospects. In: New look to phytomedicine. Academic Press, New York, pp 581–596
148. Sharma K (2025) Herbal pharmacopeias: bridging ancient traditions, nanotechnological innovation, and global regulatory cohesion for equitable healthcare. Pharmacol Res Nat Prod 8:100301
149. Kumari R, Kotecha M (2016) A review on the standardization of herbal medicines. Int J Pharma Sci Res 7(2):97–106
150. Good Manufacturing Practice—European Medicines Agency (2023). https://www.ema.europa.eu/en/human-regulatory/research-development/compliance/good-manufacturing-practice. Accessed 12 Oct 2023
151. Akram A, Suryawanshi SJ, Sharma N, Shukla VK, Arora S (2025) US Pharma Manufacturing Standards: CFR 210 & 211. Understanding pharmaceutical standards and regulations, pp 1–19
152. ECFR: 21 CFR Part 211—Current Good Manufacturing Practice for Finished Pharmaceuticals (2023). https://www.ecfr.gov/current/title-21/chapter-I/subchapter-C/part-211. Accessed 13 Oct 2023
153. Current Good Manufacturing Practice (CGMP) Regulations—FDA (2023). https://www.fda.gov/drugs/pharmaceutical-quality-resources/currentgood-manufacturing-practice-cgmp-regulations. Accessed 12 Oct 2023
154. Eudra GMP—Public Layout (2023). https://eudragmdp.ema.europa.eu/inspections/display-welcome.do. Accessed 13 Oct 2023
155. Sarma N, Upton R, Rose U, Guo DA, Marles R, Khan I, Giancaspro G (2023) Pharmacopeial standards for the quality control of botanical dietary supplements in the United States. J Diet Suppl 20:485–504
156. The FDA's Drug Review Process: Ensuring Drugs Are Safe and Effective—FDA (2023). https://www.fda.gov/drugs/information-consumersand-patients-drugs/fdas-drug-review-process-ensuring-drugs-are-safe-and-effective. Accessed 13 Oct 2023
157. What We Do—FDA (2023). https://www.fda.gov/about-fda/what-we-do. Accessed 12 Oct 2023
158. Akash MSH, Rehman K (2025) Comprehensive insights into pharmaceutical analysis. In: Essentials of pharmaceutical analysis. Springer Nature Singapore, Singapore, pp 1–62
159. About Warning and Close-Out Letters—FDA (2024). https://www.fda.gov/inspections-compliance-enforcement-and-criminalinvestigations/warning-letters/about-warning-and-close-out-letters. Accessed 27 Mar 2024
160. Barbosa R, Sampaio C, Sousa L, Fraga S, Soares P, Baltazar F. et al (2026) Exposure to potentially harmful excipients in neonates admitted to intensive care units using compounded medicines. *EJHP, 33*(2):137–141.

Aromatic Plant-Based Nanoemulsions in the Food Industry

Weria Weisany and Shima Yousefi

Abstract The incorporation of aromatic plant-based nanoemulsions in the food industry marks a significant advancement in natural preservation, flavor enhancement, and clean-label formulation. Rich in essential oils (EOs) and bioactive compounds, aromatic plants such as thyme, oregano, clove, and rosemary possess potent antimicrobial and antioxidant properties. However, their direct application in food systems is constrained by poor water solubility, volatility, and instability under processing conditions. Nanoemulsion technology overcomes these limitations by encapsulating hydrophobic plant bioactives into nanoscale oil droplets, enhancing their stability, solubility, bioavailability, and controlled release. This chapter provides a comprehensive overview of the formulation strategies, functional benefits, and applications of aromatic nanoemulsions across various food sectors including dairy, beverages, baked goods, ready-to-eat meals, and active packaging. Key aspects such as green emulsification methods, interaction with food matrices, regulatory challenges, and emerging trends in smart delivery and sustainability are also discussed. By leveraging the synergistic power of nature and nanotechnology, aromatic plant-based nanoemulsions are poised to transform modern food preservation and sensory enhancement in alignment with consumer demand for safe, functional, and environmentally responsible products.

Keywords Aromatic plants · Nanoemulsion · Essential oils · Food preservation · Functional food systems

W. Weisany (✉) · S. Yousefi
Department of Agriculture Science and Engineering, SR.C., Islamic Azad University, Tehran, Iran
e-mail: w.weisany@iau.ac.ir

A. Husen (ed.), *Sustainable Uses of Aromatic Plants in Nanotechnology*, Smart Nanomaterials Technology, https://doi.org/10.1007/978-981-95-7409-4_13

1 Introduction

The global food industry is undergoing a paradigm shift driven by the growing demand for safer, more natural, and functionally enhanced food products. Consumers today are increasingly aware of the health implications of synthetic preservatives, flavoring agents, and chemical stabilizers. As a result, there is mounting pressure on food manufacturers to replace artificial additives with natural alternatives that are not only safe but also effective in preserving food quality and extending shelf life. In this context, aromatic plant-based nanoemulsions have emerged as a promising solution at the interface of nature and nanotechnology. Aromatic plants, such as thyme, oregano, clove, cinnamon, rosemary, and basil, are rich sources of EOs and bioactive compounds known for their antimicrobial, antioxidant, and flavor-enhancing properties [1]. These secondary metabolites, especially phenolic compounds, terpenoids, and aldehydes, have long been used in traditional medicine and food preservation. However, their application in modern food systems has been limited by poor water solubility, volatility, strong aroma, and instability under processing conditions. Nanotechnology offers innovative strategies to overcome these limitations through the formulation of nanoemulsions, which encapsulate these bioactives in nanometer-sized droplets, thereby improving their dispersibility, stability, and functional performance [2, 3].

Nanoemulsions are thermodynamically unstable, kinetically stable systems typically composed of oil, water, surfactants, and sometimes cosurfactants. Their droplet size generally ranges between 20 and 200 nm, giving them a transparent or translucent appearance and high physical stability against creaming, sedimentation, and phase separation [4]. Unlike conventional emulsions, nanoemulsions can deliver hydrophobic bioactives in aqueous food matrices with improved solubility and bioavailability. In the food industry, this has opened new opportunities for incorporating EOs into beverages, dairy products, baked goods, sauces, and active packaging films [5].

One of the most attractive features of plant-based nanoemulsions is their dual functionality; they act as natural preservatives while also enhancing organoleptic qualities [6]. The antimicrobial efficacy of nanoemulsions has been shown to outperform their bulk counterparts due to increased surface area and greater interaction with microbial membranes. Additionally, nanoemulsions offer controlled release mechanisms, which ensure sustained delivery of bioactives during storage or digestion, thus prolonging their efficacy. These attributes are particularly valuable in tackling microbial spoilage and oxidative degradation, two major contributors to food waste and quality loss [7].

The green synthesis of nanoemulsions using plant-derived emulsifiers, such as saponins, lecithin, proteins, and polysaccharides, also aligns with sustainability goals. Unlike synthetic surfactants, natural emulsifiers are biodegradable and nontoxic and often provide synergistic health benefits [8, 9]. Furthermore, recent advances in high-energy (ultrasonication, high-pressure homogenization) and low-energy (phase inversion temperature, spontaneous emulsification) methods have enabled the scalable production of food-grade nanoemulsions with well-defined physicochemical characteristics [9, 10]. In addition to product enhancement, aromatic nanoemulsions play a vital role in food packaging technologies. Incorporating nanoemulsions into edible coatings, biopolymer-based films, and intelligent packaging systems enhances microbial protection, reduces oxidation, and even improves moisture barrier properties [11].

Despite the promising attributes, challenges remain in translating laboratory-scale nanoemulsion technologies to commercial food systems. Issues such as scalability, ingredient interactions, regulatory compliance, consumer perception, and cost-effectiveness must be addressed to ensure widespread adoption [12]. Regulatory agencies such as the FDA and EFSA are still developing clear frameworks for the approval of nanomaterials in food, particularly regarding labeling and toxicological safety.

This chapter aims to provide a comprehensive overview of the scientific principles, functional benefits, and practical applications of aromatic plant-based nanoemulsions in the food industry. It begins by exploring the phytochemical richness of aromatic plants and their potential as functional ingredients. Next, it delves into nanoemulsion formulation techniques, focusing on methods suitable for food applications. The chapter then highlights key functional roles—antimicrobial action, oxidative stability, and sensory enhancement—followed by real-world applications across various food sectors, including dairy, beverages, bakery, meat, and packaging. Lastly, it discusses safety considerations, regulatory challenges, and emerging trends that will shape the future of nano-enabled food systems. This represents a transformative approach to food formulation and preservation. By leveraging the strengths of both natural compounds and nanoscience, the food industry can develop clean-label, functional products that meet modern consumer expectations while addressing critical issues like food spoilage, waste reduction, and sustainability. As research and innovation continue to evolve, aromatic plant-based nanoemulsions are poised to become a cornerstone of next-generation food systems.

2 Nanoemulsion Technology in Food Formulation

Nanoemulsion technology has emerged as a cutting-edge tool for the encapsulation, protection, and delivery of bioactive compounds in modern food systems. Unlike conventional emulsions, nanoemulsions consist of extremely small droplets—typically ranging from 20 to 200 nm—dispersed within another immiscible liquid, such as oil-in-water (O/W) or water-in-oil (W/O) systems [13]. This nanoscale architecture offers significant advantages in terms of optical clarity, long-term stability, enhanced bioavailability, and controlled release, making nanoemulsions particularly attractive for formulating functional foods, beverages, and active packaging materials [14].

In the context of aromatic plant-based bioactives, such as EOs and hydrophobic phenolics, nanoemulsions serve as efficient carriers that address key limitations like low solubility, poor stability, volatility, and strong odor [15]. These compounds, although potent in their antimicrobial and antioxidant properties, are often chemically unstable and prone to rapid degradation when exposed to oxygen, heat, or light [16]. By encapsulating them within nano-sized droplets, nanoemulsion systems protect sensitive bioactives from environmental stressors while enhancing their dispersion in aqueous food matrices [17, 18].

A typical nanoemulsion designed for food applications contains three primary components: the oil phase (dispersed phase), the water phase (continuous phase), and emulsifying agents. The oil phase includes the active bioactive ingredient, most often EOs such as thymol, carvacrol, or eugenol, derived from aromatic plants. The physicochemical properties of the chosen oil, including polarity and interfacial tension, directly influence droplet formation, emulsion stability, and delivery behavior [19]. The continuous aqueous phase is generally composed of water or hydrophilic solutions, and its characteristics such as pH, ionic strength, and viscosity can affect droplet dynamics and emulsion stability [20]. Emulsifiers play a critical role in maintaining droplet dispersion and preventing coalescence. Food-grade emulsifiers such as lecithin, Tween 80, saponins, whey protein isolate, and gum Arabic are commonly employed [21]. The growing consumer preference for clean-label formulations has promoted the use of natural emulsifiers over synthetic alternatives. Additionally, cosurfactants like ethanol or glycerol, as well as antioxidants, may be incorporated to further improve system performance, enhance interfacial stability, and prolong shelf life [22].

Various fabrication methods can be employed to prepare nanoemulsions, broadly categorized as high-energy and low-energy techniques. High-energy methods, such as high-pressure homogenization, ultrasonication, and microfluidization, rely on mechanical forces to reduce droplet size and achieve uniformity. High-pressure homogenization involves forcing the mixture through narrow gaps at pressures ranging from 100 to 1000 bar, producing fine and stable emulsions [23]. Ultrasonication utilizes high-frequency sound waves to create cavitation and shear forces that break down droplets, making it particularly suitable for temperature-sensitive plant-derived compounds. Microfluidization, which pushes fluids through micro-channels at high velocity, is another efficient method for generating nano-sized emulsions with narrow size distributions [24]. Alternatively, low-energy methods such as spontaneous emulsification and phase inversion temperature (PIT) techniques depend on the intrinsic physicochemical properties of the system [25]. Spontaneous emulsification exploits composition or temperature changes to drive the formation of nanoemulsions without mechanical input, while PIT involves temperature shifts that induce emulsion inversion, enabling fine droplet formation [25]. Regardless of the method, food-grade nanoemulsions must be prepared using nontoxic solvents, food-approved emulsifiers, and gentle processing conditions to preserve the activity and integrity of plant-based compounds.

Nanoemulsions provide multiple functional and sensory benefits when incorporated into food systems. One of the most significant advantages is the improved dispersion of hydrophobic bioactives such as EOs and terpenes in water-based foods, including juices, soups, and dairy products. The small droplet size dramatically increases the surface area, leading to enhanced solubility and uniform distribution in the product matrix [26]. This also contributes to improved bioavailability, as the nanoscale droplets facilitate better absorption of active compounds in the gastrointestinal tract. Moreover, nanoemulsions can be engineered for targeted delivery and controlled release, responding to triggers such as pH, temperature, or enzymatic activity during digestion or storage [27]. Their optical transparency ensures minimal visual alteration of the final product, while encapsulation masks strong odors or flavors of potent oils, resulting in improved sensory acceptability for consumers [28, 29].

However, ensuring the physical and chemical stability of nanoemulsions remains a critical consideration. Several destabilization phenomena can occur during processing or storage. Coalescence, where droplets merge due to insufficient emulsifier coverage, leads to phase separation. Ostwald ripening, driven by solubility differences, causes small droplets to shrink while larger ones grow, ultimately destabilizing the emulsion [30]. Creaming or sedimentation may also occur due to density differences between the oil and aqueous phases. These issues can be mitigated by optimizing emulsifier concentration, adjusting pH and ionic strength, and incorporating ripening inhibitors such as medium-chain triglycerides [31]. Selecting appropriate formulation strategies based on the intended application is essential for achieving long-term stability and functional performance.

Despite the wide potential of nanoemulsions in food applications, there are notable challenges that need to be addressed. Scalability and economic feasibility are primary concerns, especially for small- and medium-scale manufacturers, as high-energy equipment can be cost-prohibitive. Regulatory uncertainty also poses a barrier; nano-enabled food technologies remain under scrutiny by regulatory agencies due to concerns over safety, bioaccumulation, and labeling transparency. Rigorous safety assessments and clear guidelines are necessary to support industry adoption and consumer trust. Furthermore, interactions between nanoemulsions and complex food matrices may affect their functionality. Proteins, polysaccharides, salts, and lipids present in the matrix can influence droplet behavior, bioactive release, and overall efficacy [32] (Fig. 1).

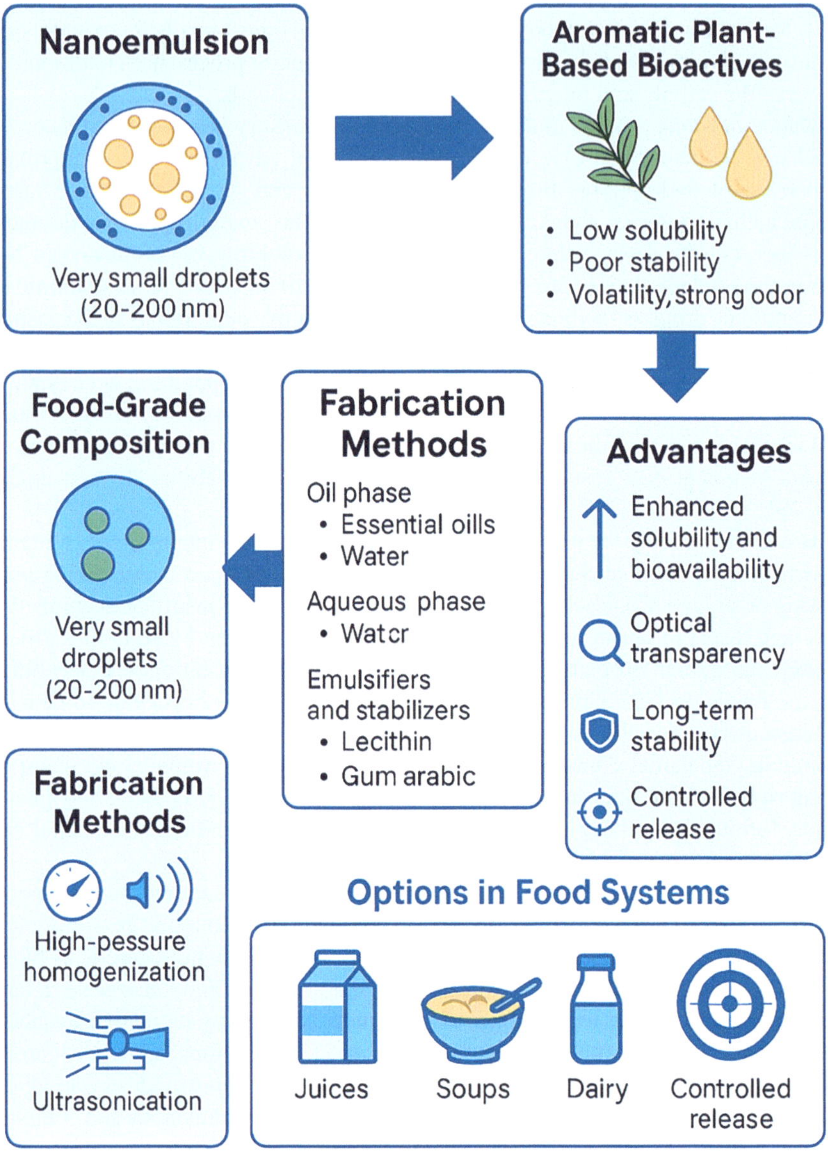

Fig. 1 Nanoemulsions, composed of nano-sized droplets (20–200 nm), encapsulate aromatic plant-based bioactives such as EOs to overcome issues of low solubility, instability, and volatility. The system includes a dispersed oil phase, an aqueous continuous phase, and food-grade emulsifiers (lecithin, gum Arabic). Fabrication methods include high-pressure homogenization and ultrasonication. Benefits include improved solubility, enhanced bioavailability, optical transparency, long-term stability, and controlled release. Applications span across various food systems, including juices, soups, dairy, and functional delivery platforms

3 Aromatic Plants as Natural Sources of Functional Nanoemulsions

Aromatic plants have played an integral role in food preservation, traditional medicine, and flavor enhancement for centuries. Their richness in EOs, phenolic compounds, terpenes, and aldehydes makes them a compelling natural source of bioactive agents with potent antimicrobial, antioxidant, and sensory-modulating effects. In recent years, these phytochemicals have gained renewed scientific interest for their potential application in nanoemulsion systems, which offer a means to stabilize, protect, and efficiently deliver these functional compounds in food formulations [33]. The transition from traditional use to nanoscale application represents a major advancement in maximizing the efficacy and versatility of aromatic plant extracts in the food industry.

Among the most commonly studied aromatic plants for nanoemulsion applications are oregano (*Origanum vulgare*) [34], thyme (*Thymus vulgaris*) [35], clove (*Syzygium aromaticum*) [36], cinnamon (*Cinnamomum zeylanicum*) [37], rosemary (*Rosmarinus officinalis*) [38], basil (*Ocimum basilicum*) [39], lemongrass (*Cymbopogon citratus*) [40], and peppermint (*Mentha piperita*) [17]. These species are renowned for their EO content, which includes key volatile compounds such as thymol, carvacrol, eugenol, cinnamaldehyde, citral, and menthol. These molecules are largely responsible for the antimicrobial and antioxidant properties of the oils and are the primary active components targeted for encapsulation in nanoemulsion systems.

The mode of action of these compounds is often linked to their ability to disrupt microbial cell membranes, inhibit enzymatic activity, and scavenge reactive oxygen species (ROS). For instance, thymol and carvacrol, found abundantly in oregano and thyme oils, are known to integrate into the lipid bilayer of microbial membranes, leading to increased permeability and cell lysis [41]. Eugenol, a major component of clove oil, exhibits strong radical-scavenging activity and can chelate metal ions that catalyze oxidative reactions [42]. Cinnamaldehyde, from cinnamon bark, shows broad-spectrum antimicrobial activity and has been effective against foodborne pathogens such as *Escherichia coli*, *Salmonella enterica*, and *Listeria monocytogenes*. These natural properties, when preserved within nanoemulsion systems, make aromatic plant extracts ideal candidates for active food components or additives [43].

The use of aromatic plant-derived compounds in their native, unencapsulated form, however, is often restricted by challenges such as poor aqueous solubility, volatility, instability under processing conditions, and intense flavor that may not be acceptable at high concentrations. Nanoemulsification offers a practical and efficient approach to overcoming these limitations [44]. Encapsulation in oil-in-water (O/W) nanoemulsions allows these hydrophobic bioactives to be dispersed in water-based food matrices, stabilizing them against environmental degradation and enabling their sustained release over time [3, 18]. In addition, the use of natural emulsifiers derived from the same or compatible botanical sources—such as saponins from *Quillaja saponaria*, proteins from legumes, or polysaccharides from guar and gum Arabic—further supports a clean-label and biocompatible approach in food-grade nanoemulsion design.

Another important advantage of aromatic plant-based nanoemulsions lies in their multifunctionality. Unlike synthetic preservatives, which often target a single preservation mechanism, plant-derived bioactives offer synergistic effects. For instance, a nanoemulsion containing both citral and limonene, from lemongrass and citrus peels, respectively, can simultaneously inhibit microbial growth and delay lipid oxidation in food emulsions such as mayonnaise or salad dressings [45]. Moreover, many aromatic plants also exhibit insecticidal, antifungal, and even immunomodulatory properties, which may open avenues for their use in food packaging or nutraceutical delivery systems.

The variability in chemical composition among EOs, however, remains a challenge in formulation. The yield and profile of EOs can be significantly influenced by factors such as plant cultivar, growth environment, extraction method, and harvest timing. Standardization is essential to ensure batch-to-batch consistency, particularly for industrial applications. Furthermore, encapsulating entire EOs as opposed to isolated compounds is often more effective, as the synergistic action of the whole oil tends to outperform that of individual constituents [46]. This has been demonstrated in nanoemulsion systems containing full-spectrum rosemary or thyme oils, which showed enhanced efficacy in inhibiting lipid oxidation and microbial proliferation in meat and dairy products [47].

In addition to EOs, aromatic plants are a rich source of nonvolatile bioactives such as flavonoids (quercetin, apigenin), phenolic acids (caffeic acid, rosmarinic acid), and alkaloids, which also contribute significantly to the functional properties of nanoemulsions. These compounds can be co-encapsulated or used as stabilizing agents that enhance antioxidant properties or modulate the release kinetics of volatile actives. For example, rosmarinic acid from rosemary not only exhibits potent antioxidant activity but can also serve as a cosurfactant in nanoemulsion systems, further stabilizing the emulsion and enhancing bioactivity [48].

From a technological standpoint, the integration of aromatic plant-based nanoemulsions into food systems requires careful consideration of their interaction with the food matrix. Factors such as pH, ionic strength, temperature, and the presence of other macromolecules (proteins, starches, fats) can significantly influence the stability and functionality of the nanoemulsion. Nevertheless, recent studies have demonstrated successful applications in various food systems, including salad dressings, flavored waters, dairy drinks, bakery products, and active edible coatings for fruits and vegetables (Fig. 2).

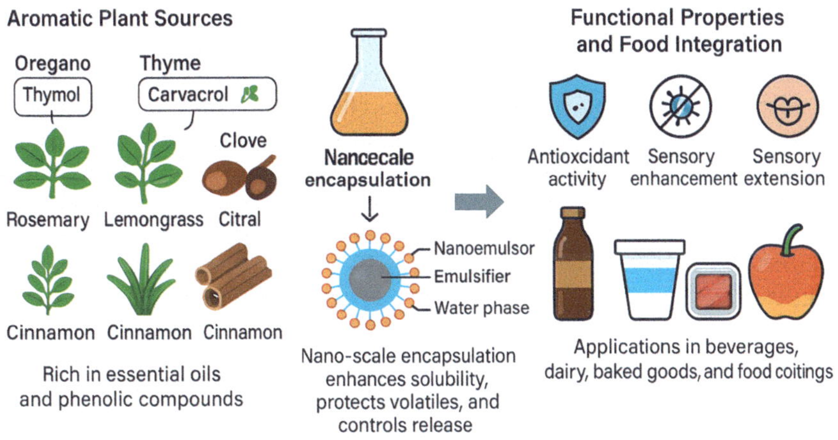

Fig. 2 Key essential oil-rich plants such as oregano, thyme, clove, rosemary, and lemongrass provide bioactive compounds like thymol, carvacrol, eugenol, and citral. These are encapsulated in nanoscale oil droplets using food-grade emulsifiers, enhancing solubility, protecting volatile compounds, and enabling controlled release. Nanoemulsions contribute antioxidant activity, microbial inhibition, and sensory improvements in various food products including beverages, dairy, meat, and fresh produce coatings

4 Functional Properties of Aromatic Nanoemulsions in Food Products

Aromatic plant-based nanoemulsions have gained considerable attention in food science due to their multifunctional properties, particularly in improving food safety, extending shelf life, and enhancing sensory attributes. These nanoscale delivery systems serve as efficient carriers for EOs and other bioactive compounds, which are known for their potent antimicrobial, antioxidant, and organoleptic characteristics. The encapsulation of these volatile and hydrophobic compounds within nanoemulsions not only enhances their dispersibility in aqueous food matrices but also protects them from environmental degradation, facilitates controlled release, and amplifies their functional performance across a variety of food systems [49].

One of the most important functional properties of aromatic nanoemulsions is their antimicrobial activity, which is primarily attributed to the EO constituents such as thymol, carvacrol, eugenol, cinnamaldehyde, and citral. These bioactives exhibit strong inhibitory effects against a wide range of foodborne pathogens and spoilage microorganisms, including *Listeria monocytogenes*, *Escherichia coli*, *Salmonella enterica*, and *Staphylococcus aureus* [50]. The nanoemulsion format significantly enhances the interaction of these compounds with microbial cell membranes due to the increased surface area and improved dispersion. Mechanistically, the EOs disrupt microbial membrane integrity, alter intracellular pH, inhibit enzyme function, and cause leakage of cellular contents. This mechanism is further amplified in nanoemulsions where the nanoscale droplets can penetrate microbial biofilms and enhance the antimicrobial efficacy at lower concentrations compared to bulk EOs [51]. Thus, aromatic nanoemulsions represent a promising natural alternative to synthetic preservatives, especially in minimally processed and ready-to-eat food products.

Another critical functionality of aromatic nanoemulsions lies in their antioxidant capacity, which helps prevent oxidative degradation of food components such as lipids, proteins, and vitamins. Oxidation leads to undesirable changes in flavor, color, and nutritional value, especially in high-fat and protein-rich foods like meat, dairy, and bakery products. Many EOs derived from aromatic plants contain phenolic compounds with excellent radical-scavenging and metal-chelating abilities [52]. When encapsulated in nanoemulsion systems, these antioxidants exhibit improved stability and sustained release, offering prolonged protection against oxidative stress during food processing and storage [3].

Beyond their preservative functions, aromatic nanoemulsions also contribute to sensory enhancement in food products. EOs are rich in volatile compounds that impart distinctive aromas and flavors, which can be leveraged to improve the overall sensory appeal of food. However, in their free form, EOs may be too pungent or volatile, often overpowering the product or degrading under heat and light [53]. Nanoemulsion and encapsulation allow for better control of flavor release, masking undesirable notes, and providing a more consistent and prolonged sensory experience. For example, lemon or mint EOs encapsulated in nanoemulsions can be incorporated into beverages and confectionery to impart a pleasant aroma and refreshing aftertaste while preserving product clarity and homogeneity [3].

Furthermore, nanoemulsions enable targeted and controlled release of encapsulated compounds in response to specific environmental stimuli such as pH, temperature, and enzymatic activity. This is particularly useful in food systems with complex matrices or during gastrointestinal transit, where the release of active agents needs to be timed for maximum efficacy. For instance, nanoemulsions designed with pH-sensitive emulsifiers or coatings can protect sensitive compounds in the acidic environment of the stomach and release them in the more neutral pH of the intestines [2, 3].

Another emerging application of aromatic nanoemulsions is in active and intelligent food packaging. Incorporating nanoemulsions into edible coatings or biodegradable films can create packaging materials with intrinsic antimicrobial and antioxidant properties. These functional films can reduce microbial load on the surface of fresh produce, meat, and dairy products, delay oxidation, and even serve as carriers for flavor compounds or indicators of spoilage [11, 54].

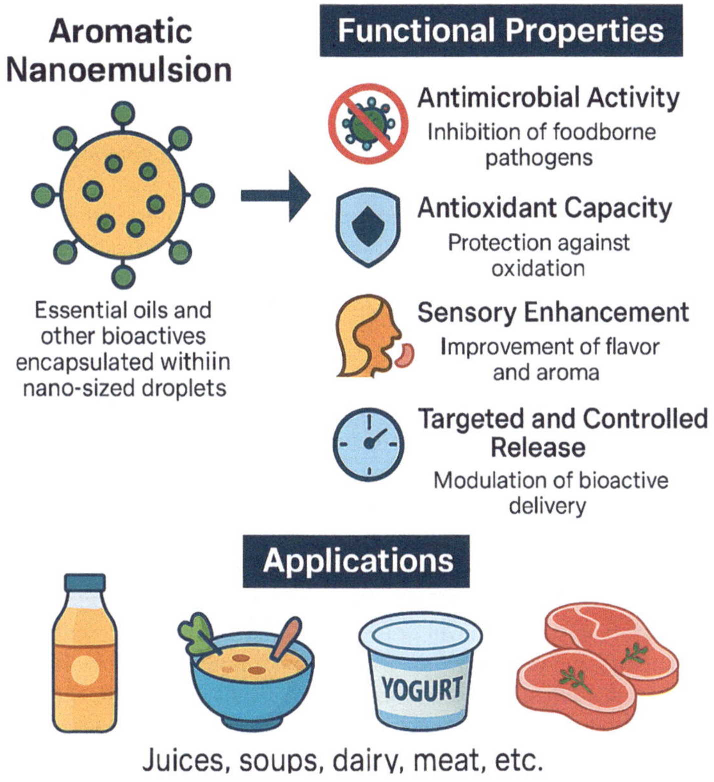

Fig. 3 Essential oils and plant-derived bioactives are encapsulated within nano-sized droplets to enhance stability and delivery. Key functional properties include antimicrobial activity against foodborne pathogens, antioxidant capacity to prevent oxidation, sensory enhancement of flavor and aroma, and targeted or controlled release of actives. These nanoemulsions are applied across a variety of food systems such as juices, soups, dairy products, and meat, contributing to improved food quality, safety, and shelf life

Despite these advantages, the functionality of nanoemulsions in real food systems depends on several matrix interactions and formulation parameters. Factors such as droplet size, surface charge, emulsifier type, and oil polarity can influence how nanoemulsions interact with proteins, carbohydrates, and other food constituents. Additionally, processing conditions such as heat treatment, freezing, and pH shifts can affect nanoemulsion stability and release behavior. Therefore, a thorough understanding of formulation-food matrix compatibility is essential to optimize functional performance [55] (Fig. 3).

5 Integration in Food Processing and Preservation

The integration of aromatic plant-based nanoemulsions into food processing and preservation strategies represents a promising advancement in sustainable and functional food innovation. These nano-sized delivery systems, composed of EOs or plant bioactives encapsulated in emulsified droplets, are increasingly being applied at various stages of food production to enhance product safety, quality, and shelf life. Their physicochemical properties—particularly high surface area, enhanced solubility, and controlled release—make them ideally suited for preserving perishable foods, inhibiting spoilage, and replacing synthetic additives in modern food processing lines [7].

Aromatic nanoemulsions have been successfully incorporated into a range of thermal and nonthermal food processing methods. In thermal processes such as pasteurization, baking, or cooking, nanoemulsions act as reservoirs that shield volatile and heat-sensitive bioactives from degradation. Encapsulation protects EO compounds like thymol, eugenol, or citral—otherwise prone to thermal loss—allowing their bioactivity to be retained even under high temperatures [18]. This is particularly beneficial in baked goods, where rosemary or cinnamon oil-based nanoemulsions can be added to dough or batter to inhibit mold growth and oxidative rancidity without negatively impacting flavor or texture [56]. Likewise, dairy and plant-based beverages that undergo pasteurization have shown improved microbial stability and oxidative resistance when fortified with nanoemulsions of clove or oregano oil. In nonthermal processing systems, such as cold storage, vacuum packaging, high-pressure processing (HPP), and UV treatment, nanoemulsions provide an additive layer of protection by continuously releasing antimicrobial or antioxidant agents throughout storage [57]. The slow and sustained release of active compounds ensures longer shelf-life and better-quality retention.

The application of nanoemulsions in post-processing stages—especially in surface treatments and edible coatings—is particularly impactful. In fresh-cut fruits, vegetables, and minimally processed foods, nanoemulsions are applied as dipping solutions, sprays, or components of edible films to provide direct antimicrobial contact and barrier functionality. This is often achieved using biopolymer-based matrices such as chitosan, alginate, or starch, into which aromatic nanoemulsions are dispersed [11, 54]. These coatings not only reduce microbial load but also regulate moisture transfer and gas exchange, delaying physiological spoilage and discoloration. For example, strawberries treated with a chitosan coating containing oregano oil nanoemulsion have shown delayed fungal growth and extended postharvest life under refrigeration [58].

Another significant area of integration is in the preservation of high-protein foods, such as meats, seafood, dairy, and plant-based analogs, which are highly susceptible to microbial spoilage and lipid oxidation. In meat and poultry products, nanoemulsions serve as both marinade additives and packaging components. Their antimicrobial action can suppress the growth of spoilage organisms and foodborne pathogens, while antioxidant activity slows the formation of malondialdehyde (MDA) and other oxidative markers [59]. In seafood, EO nanoemulsions help control off-odor development and lipid deterioration, which are major quality-limiting factors in chilled fish storage. For plant-based meat alternatives, integrating nanoemulsions during formulation helps improve microbial safety and oxidative stability without resorting to synthetic preservatives, aligning with clean-label consumer expectations.

Importantly, nanoemulsions can be tailored to the specific needs of food systems by manipulating droplet size, emulsifier type, oil concentration, and release kinetics. This level of customization allows food processors to control not only the level of microbial inhibition or antioxidant protection but also the sensory impact on the final product. A major advantage of nanoemulsions over traditional emulsions or free EOs is their minimal impact on taste, appearance, and texture when used at functional levels. This is particularly valuable in products where strong odors or flavors from EOs could be off-putting to consumers, such as neutral-flavored beverages, dairy, or plant-based snacks [60].

Moreover, aromatic nanoemulsions are being explored in synergistic preservation strategies, where they are combined with other hurdles such as mild heat, UV-C light, or natural antimicrobial peptides. These multi-hurdle approaches have shown promising results in reducing microbial counts more effectively than individual treatments, enabling milder processing with higher nutritional and sensory retention. In some studies, combining nanoemulsions of thyme oil with mild thermal treatment has resulted in enhanced inactivation of *L. monocytogenes* on fresh produce compared to either treatment alone, pointing toward broader industrial potential [61].

The integration of aromatic nanoemulsions into food processing and preservation systems offers a powerful and adaptable tool for the industry. Whether used during thermal treatment, as nonthermal preservatives, or in active coatings, these nanoscale systems deliver multifunctional benefits—including microbial control, oxidative stability, and quality retention—without compromising sensory attributes. As the food industry moves toward cleaner, more sustainable, and minimally processed products, nanoemulsion technology is poised to play a pivotal role in next-generation preservation strategies.

6 Active Food Packaging with Aromatic Nanoemulsions

The development of active food packaging systems represents a major advancement in food preservation, aligning with industry goals of extending shelf life, ensuring safety, and reducing dependency on synthetic additives. Active packaging goes beyond passive containment, incorporating functional agents that interact with the packaged food or its environment to inhibit spoilage and maintain quality. Among the most promising innovations in this field is the incorporation of aromatic plant-based nanoemulsions, which deliver natural antimicrobial and antioxidant effects directly into the packaging material or surface coating. By leveraging the high surface area, controlled release properties, and compatibility with biopolymers, nanoemulsions have enabled the next generation of smart, sustainable, and efficient food packaging systems [62].

Aromatic nanoemulsions are typically embedded into edible films, biodegradable coatings, or multilayer polymer matrices that serve as physical barriers while simultaneously releasing active compounds [28]. These nanoemulsions, composed of EOs such as thyme, clove, oregano, or cinnamon, encapsulate bioactives within oil droplets stabilized by natural emulsifiers like lecithin, whey protein, or gum Arabic. The nanoscale dispersion of droplets allows for uniform distribution within the packaging matrix and facilitates sustained release of volatile components over time, enhancing their preservative action while avoiding sensory overload [63].

The flexibility and biocompatibility of nanoemulsion-based packaging systems support their application in various formats. Edible coatings applied by dipping or spraying are commonly used for fresh produce, nuts, and bakery items, forming a thin, transparent barrier that enhances moisture retention and reduces microbial contamination. Composite films for refrigerated meat and dairy packaging can be engineered with multilayer structures, where the active layer is enriched with nanoemulsions [64]. These materials are typically based on biodegradable polymers such as chitosan, polylactic acid (PLA), starch, or pectin—contributing to the overall sustainability of the packaging system. In addition to functionality, these films exhibit favorable mechanical, optical, and barrier properties when properly plasticized and formulated.

Incorporating aromatic nanoemulsions also supports the development of multifunctional and clean-label packaging. Unlike synthetic preservatives and coatings, EO nanoemulsions are derived from natural sources and can often be labeled as GRAS (generally recognized as safe) ingredients [65]. This addresses growing consumer demand for clean, eco-friendly packaging materials that are not only functional but also safe and biodegradable. Furthermore, by reducing the need for direct chemical preservatives in the food product itself, these packaging systems support reformulation efforts aimed at simplifying ingredient lists and improving transparency [66].

Despite these promising attributes, the commercial implementation of nanoemulsion-based active packaging systems faces several technical and regulatory challenges. Ensuring homogenous dispersion of nanoemulsions within film-forming solutions, maintaining stability during film drying or extrusion, and scaling up production processes without loss of function are critical concerns. Additionally, regulatory approval of nanomaterials in packaging, particularly regarding migration limits and labeling, varies between jurisdictions and requires thorough toxicological evaluation. Addressing these barriers will be essential to move from experimental success to real-world adoption (Fig. 4).

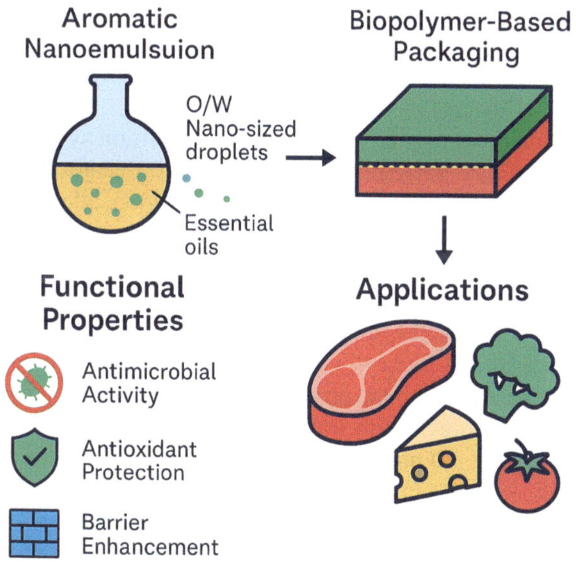

Fig. 4 Essential oils are encapsulated in oil-in-water (O/W) nano-sized droplets, forming the core of the nanoemulsion. These are incorporated into biopolymer-based packaging materials, enabling controlled release and enhancing functional performance. Key properties include antimicrobial activity, antioxidant protection, and barrier enhancement. Applications span across various food products, including meat, dairy, fruits, and vegetables, supporting improved food safety, shelf-life extension, and quality preservation within sustainable packaging frameworks

7 Enhancing Flavor and Sensory Appeal Using Aromatic Nanoemulsions

Beyond their established roles in food preservation and safety, aromatic plant-based nanoemulsions are increasingly being recognized for their potential to enhance the flavor and sensory appeal of food products. Sensory quality—comprising taste, aroma, mouthfeel, and visual appearance—is a critical determinant of consumer acceptance. Given the volatility, intensity, and instability of aromatic compounds in their free form, nanoemulsification offers an innovative strategy to optimize their delivery, control their release, and refine their sensory impact within complex food matrices.

Aromatic plants such as mint, basil, rosemary, lemongrass, thyme, and citrus varieties are rich in volatile terpenes, aldehydes, esters, and phenolics. These compounds impart distinctive flavors and aromas that are highly desirable in various food categories including beverages, confectionery, dairy, sauces, and bakery products [67]. However, many EO components are hydrophobic, chemically unstable, or highly potent, which can pose formulation challenges and lead to off-flavors, rapid volatilization, or degradation under thermal processing [3]. Incorporating these compounds into nanoemulsion systems addresses these limitations by encapsulating the active compounds within nano-sized droplets, providing protection, improved dispersibility, and modulated release.

One of the primary sensory advantages of nanoemulsion systems is their ability to preserve and modulate flavor release over time [68]. Encapsulation shields volatile compounds from premature oxidation and evaporation, allowing for their gradual and sustained liberation during storage, processing, or consumption. This slow release mimics the natural flavor perception curve, enhancing product enjoyment and reducing the need for excessive flavoring dosages [3, 18].

Another major benefit is the improved homogeneity and stability of flavor compounds in aqueous or semisolid food matrices. Free EOs often separate, create cloudiness, or concentrate in fat phases, leading to uneven flavor distribution. Nanoemulsions, by contrast, form translucent or transparent dispersions that are stable over time and exhibit uniform flavor intensity across the entire product [69]. This is especially valuable in products such as salad dressings, dairy drinks, or emulsified sauces where consistent sensory quality is essential.

Taste masking is another application where nanoemulsions contribute significantly. Certain bioactive compounds, including plant phenolics or functional oils (garlic, turmeric), can possess pungent, bitter, or astringent characteristics that limit their use in functional foods. Nanoemulsification can help mask these undesirable notes by entrapping them within the oil droplets and reducing their direct interaction with taste receptors. This not only broadens the potential for incorporating health-promoting ingredients but also improves the palatability of functional or fortified foods [70].

Furthermore, nanoemulsions can be tailored to enhance mouthfeel and textural perception. The small droplet size of nanoemulsions allows them to integrate seamlessly into food matrices, contributing to creaminess, smoothness, or lubrication without the addition of fats or synthetic emulsifiers [69]. In low-fat or plant-based dairy alternatives, EO nanoemulsions can serve a dual function: contributing to flavor richness and improving texture. Another emerging application involves multisensory delivery systems, where nanoemulsions are used alongside visual and textural cues to create enhanced flavor perception. The combination of color-stable nanoemulsions with aroma-active oils, delivered through edible films or capsules, is being explored in confectionery, chewing gum, and snack coatings. These systems can be engineered for burst release upon chewing or temperature change, offering consumers a dynamic flavor profile and greater enjoyment [71].

Despite these advantages, there are challenges to the widespread use of aromatic nanoemulsions for sensory enhancement. Stability under processing conditions such as heating, freezing, or pH changes must be ensured to preserve the flavor integrity. Additionally, overuse of EOs, even when nanoencapsulated, can still lead to overpowering or off notes, especially in delicate matrices. Regulatory and labeling considerations are also important, as some EOs are subject to concentration limits or specific approval for food use (Fig. 5).

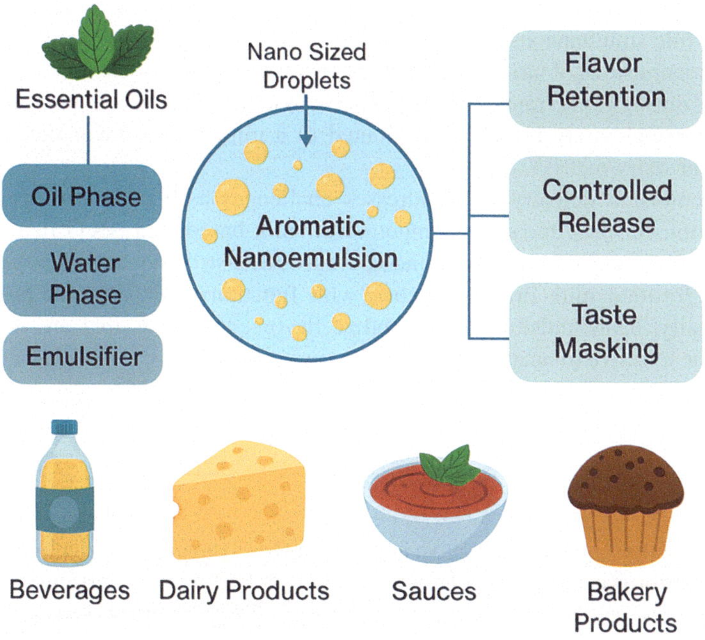

Fig. 5 Essential oils are encapsulated within nano-sized oil-in-water droplets formed by combining an oil phase, water phase, and food-grade emulsifier. This nanoemulsion system enables improved flavor retention, controlled release of aroma compounds, and effective taste masking. Applications include a wide range of food categories such as beverages, dairy products, sauces, and bakery items, where sensory quality is critical to consumer acceptance

8 Applications in Dairy, Bakery, Beverages, and Ready-to-Eat Foods

The growing demand for clean-label, functional, and minimally processed food products has fueled interest in incorporating aromatic plant-based nanoemulsions into a wide range of food matrices. These nanoemulsions, composed of EO bioactives stabilized in oil-in-water systems, offer multifunctional benefits including microbial inhibition, oxidative stability, flavor enhancement, and shelf-life extension. Their adaptability and compatibility with both aqueous and semisolid systems make them especially relevant in four major categories of the food industry: dairy, bakery, beverages, and ready-to-eat (RTE) foods.

8.1 Dairy Products

Dairy systems—including yogurt, cheese, milk-based beverages, and dairy desserts—are ideal platforms for nanoemulsion integration due to their colloidal nature and susceptibility to microbial spoilage and oxidation. Essential oils such as thyme, oregano, clove, and mint have been nanoencapsulated to function as natural preservatives while simultaneously enhancing flavor. In yogurt and fermented milk products, nanoemulsions of tarragon oil have been used to inhibit *L. monocytogenes* and *S. aureus* during refrigerated storage [72]. These nanoemulsions not only improved microbiological safety but also contributed to a mild herbal flavor that complemented the dairy matrix.

Nanoemulsion delivery also addresses challenges associated with dispersing hydrophobic flavor or bioactive compounds in milk-based matrices [73]. Their fine droplet size ensures optical clarity and homogenous distribution, making them suitable for fortified milk drinks enriched with functional oils or plant phenolics. Additionally, nanoemulsions help stabilize flavor compounds that are otherwise volatile or sensitive to heat treatment during pasteurization.

8.2 Beverages

Beverages represent one of the most active areas for nanoemulsion deployment due to the need for transparent, stable incorporation of hydrophobic bioactives such as EOs, vitamins, and flavonoids. Functional drinks, flavored waters, teas, juices, and plant-based milks can all benefit from the enhanced solubility and bioavailability provided by nanoemulsions. Aromatic nanoemulsions of lemon, orange, or mint oils are commonly used in citrus-based drinks to improve flavor stability and freshness perception. These systems provide optical clarity and eliminate issues such as phase separation, sedimentation, or flavor degradation. Moreover, nanoemulsions enhance the sensory delivery of volatile aroma compounds that would otherwise be lost during thermal pasteurization or prolonged storage.

In addition to flavor enhancement, nanoemulsions serve as delivery vehicles for natural preservatives and antioxidants in beverage formulations. For example, incorporation of rosemary oil nanoemulsions into fruit juices has been shown to inhibit microbial growth while preserving color and vitamin content. Their presence can also support reduced-sugar formulations by compensating for the flavor loss often associated with sweetener reduction [74].

8.3 Ready-to-Eat (RTE) Foods

Ready-to-eat foods, including salads, deli meats, precooked meals, and convenience snacks, are highly vulnerable to microbial contamination and oxidative spoilage due to their minimal processing and extended shelf life. Aromatic nanoemulsions offer an effective means of protection without compromising the freshness, taste, or visual appeal of these products. In RTE meat and poultry products, thyme or oregano oil nanoemulsions incorporated into marinades or surface coatings have been shown to significantly reduce bacterial counts during refrigerated storage. When used in combination with modified atmosphere packaging (MAP), these systems enhance safety while extending shelf life. In plant-based or vegetarian RTE meals, nanoemulsions provide both preservative and sensory functions. For instance, incorporating basil or garlic oil nanoemulsions into pasta sauces or hummus improves flavor profile while inhibiting microbial growth. Nanoemulsions have also been explored in salad dressings, where they improve emulsion stability, flavor dispersion, and oxidative protection of lipids and herbs.

9 Future Directions and Industry Perspectives

The integration of aromatic plant-based nanoemulsions into food systems represents a convergence of natural product chemistry, nanotechnology, and consumer-driven innovation. As the food industry continues to seek clean-label, multifunctional, and sustainable solutions, nanoemulsion technology is positioned as a transformative tool in both ingredient formulation and packaging innovation. However, for nanoemulsions to achieve widespread adoption, future developments must address technical optimization, cost-efficiency, regulatory clarity, and consumer trust.

9.1 Technological Innovations and Research Priorities

Future research will likely focus on next-generation nanoemulsion systems that go beyond static delivery to achieve dynamic, responsive functionality. Smart nanoemulsions capable of responding to pH, temperature, or enzymatic triggers could allow for site-specific release of bioactives during digestion or under storage conditions. This opens the door to personalized nutrition, active release packaging, and functional foods with tailored health benefits.

Emerging trends also include the use of green emulsification technologies. Replacing synthetic surfactants with fully natural or biodegradable alternatives such as protein-polysaccharide complexes or saponins aligns with consumer demand and reduces environmental impact. Low-energy emulsification methods, such as spontaneous emulsification or membrane emulsification, are gaining traction for their scalability, reduced energy input, and compatibility with heat-sensitive compounds [10].

Multifunctional nanoemulsions will also be of interest those that can simultaneously deliver flavor, antimicrobial activity, and nutritional benefits. Co-encapsulation of EOs with other bioactives such as polyphenols, vitamins, or probiotics may yield synergistic effects that enhance food safety and functional value. Finally, advanced analytical techniques such as cryo-TEM, atomic force microscopy, and nano-tracking analysis will be increasingly applied to understand nanoemulsion behavior, interfacial dynamics, and in vivo fate at the molecular level. Such insight will improve formulation accuracy and safety predictions.

9.2 Commercialization and Scale-Up Challenges

Despite strong scientific validation, the translation of nanoemulsion technology from lab to market remains limited by production scalability and cost constraints. High-energy homogenization, ultrasonication, and microfluidization require capital investment and may not be viable for small- and medium-sized enterprises. Future directions will focus on cost-effective, modular manufacturing systems that can be adapted to various production scales and food categories. Stability during processing and storage is another bottleneck, particularly in complex food matrices or high-moisture products. Collaborations between formulation scientists and process engineers will be essential to develop matrix-compatible systems that retain efficacy throughout the product life cycle.

Moreover, the integration of nanoemulsions into existing food supply chains requires alignment with food-grade standards, packaging protocols, and supply chain logistics. For example, incorporating nanoemulsions into packaging films or spray-coating systems for ready-to-eat meals demands compatibility with packaging lines and food-contact regulations.

9.3 Regulatory Readiness and Labeling

The future success of nanoemulsion-based products will depend heavily on regulatory harmonization across jurisdictions. Clear guidelines for nanoscale materials in food, including definition, risk assessment, and labeling requirements, are needed to reduce ambiguity and facilitate market entry. Regulatory bodies such as the FDA, EFSA, and *Codex Alimentarius* must continue developing frameworks that are science based yet innovation-friendly.

In parallel, transparent communication and labeling strategies will be vital. Using terminology that reflects safety and natural origin ("encapsulated plant extracts" or "botanical oil delivery system") may enhance consumer confidence compared to technical terms like "nanotechnology." Regulatory alignment must also consider international trade, export barriers, and harmonized ingredient listings.

9.4 Market Trends and Industry Perspectives

From an industry standpoint, nanoemulsion technology supports multiple emerging trends:

- Clean-label innovation: Natural EOs encapsulated in safe, food-grade carriers allow for reduced use of synthetic preservatives and flavorings.

- Functional food development: Nanoemulsions improve bioavailability and targeted delivery of active ingredients, supporting health claims in fortified foods and beverages.
- Plant-based and low-fat formulations: Nanoemulsions enhance flavor and mouthfeel in nondairy, vegan, or reduced-fat products, filling sensory gaps left by traditional fat or dairy components.
- Sustainable packaging: Active films incorporating nanoemulsions contribute to shelf-life extension and waste reduction while aligning with biodegradable material trends.

Major food companies are beginning to invest in nanotech-enabled ingredients and packaging solutions, often through partnerships with research institutions or startups. For example, active packaging films containing EO nanoemulsions are under development for fresh produce, meat, and dairy sectors. Similarly, beverage manufacturers are testing nanoemulsion-based flavor delivery systems to reduce sugar content without compromising taste.

To maintain momentum, public-private collaboration is essential. Industry consortia, academic partnerships, and government funding initiatives can support pilot-scale validation, safety data collection, and consumer education campaigns. These alliances will be instrumental in bridging the gap between bench-scale innovation and commercial viability.

Competing Interests The authors declare that they have no competing interest.

Declaration of Generative AI and AI-Assisted Technologies in the Writing Process During the preparation of this work, the book chapter author(s) have not taken any assistance from AI, and the book chapter author(s) reviewed and edited the content as needed and take(s) full responsibility for the content of the publication.

References

1. Weisany W, Amini J, Samadi S, Hossaini S, Yousefi S, Struik PC (2019a) Nano silver-encapsulation of thymus daenensis and Anethum graveolens essential oils enhances antifungal potential against strawberry anthracnose. Ind Crops Prod 141:111808
2. Weisany W, Yousefi S, Soufiani SP, Pashang D, McClements DJ, Ghasemlou M (2024b) Mesoporous silica nanoparticles: a versatile platform for encapsulation and delivery of essential oils for food applications. Adv Colloid Interf Sci 325:103116
3. Weisany W, Yousefi S, Tahir NA-R, Golestanehzadeh N, McClements DJ, Adhikari B, Ghasemlou M (2022b) Targeted delivery and controlled released of essential oils using nanoencapsulation: a review. Adv Colloid Interface Sci 303:102655
4. Dasgupta N, Ranjan S, Dasgupta N, Ranjan S (2018) Food nanoemulsions: stability, benefits and applications. In: An introduction to food grade nanoemulsions. Springer, Cham, pp 19–48
5. Chudasama M, Goyary J (2024) Nanostructured materials in food science: current progress and future prospects. Next Mater 5:100206

6. McClements DJ, Das AK, Dhar P, Nanda PK, Chatterjee N (2021) Nanoemulsion-based technologies for delivering natural plant-based antimicrobials in foods. Front Sustain Food Syst 5:643208
7. Sahoo M, Vishwakarma S, Panigrahi C, Kumar J (2021) Nanotechnology: current applications and future scope in food. Food Front 2:3–22
8. Hunde Y, Bachheti A, Chaubey KK, Husen A, Bachheti RK (2023) Nanoemulsions from essential oils: preparation, characterization, and their applications. In: Husen A, Bachheti RK, Bachheti A (eds) current trends in green Nano-emulsions. Springer Nature Singapore Pte Ltd., Singapore, pp 21–38. https://doi.org/10.1007/978-981-99-5398-1_2
9. Tamang N, Shrestha P, Khadka B, Mondal MH, Saha B, Bhattarai A (2021) A review of biopolymers' utility as emulsion stabilizers. Polymers 14:127
10. Mohamadyan W, Yousefi S, Weisany W (2024) Development of edible nanoemulsions containing vitamin E using a low-energy method: evaluation of particle size and physicochemical properties for food and beverage applications. Heliyon 10:e32415
11. Daeialiakbar M, Yousefi S, Weisany W (2025) Enhanced properties of chitosan-PVA nanocomposite films with lemongrass oil microcapsules. Carbohydr Polym Technol Appl 9:100668
12. Gddoa A-SST, Kareem NA, Verma DK, Pawan P, Patel AR, Mamta T, Singh S (2025) Applications of green synthesis of nanoparticles using microorganisms in food and dairy products. Processes 13:1560
13. McClements DJ (2012) Nanoemulsions versus microemulsions: terminology, differences, and similarities. Soft Matter 8:1719–1729
14. Aguilar-Pérez KM, Ruiz-Pulido G, Medina DI, Parra-Saldivar R, Iqbal HM (2023) Insight of nanotechnological processing for nano-fortified functional foods and nutraceutical—opportunities, challenges, and future scope in food for better health. Crit Rev Food Sci Nutr 63:4618–4635
15. Butani N, Bhatt MD, Parmar P, Jobanputra J, Dobriyal AK, Bhatt D (2020) Nanoemulsions of plant-based bioactive compounds: synthesis, properties, and applications. In: Nanotechnological approaches in food microbiology. CRC Press, Boca Raton, pp 187–226
16. Weisany W, Samadi S, Amini J, Hossaini S, Yousefi S, Maggi F (2019b) Enhancement of the antifungal activity of thyme and dill essential oils against Colletotrichum nymphaeae by nano-encapsulation with copper NPs. Ind Crop Prod 132:213–225
17. Weisany W, Samadi S, Tahir NA-r, Amini J, Hossaini S (2022a) Nano-encapsulated with mesoporous silica enhanced the antifungal activity of essential oil against Botrytis cinerea (Helotiales; Sclerotiniaceae) and Colletotrichum nymphaeae (Glomerellales; Glomerellaceae). Physiol Mol Plant Pathol 122:101902
18. Yousefi S, Weisany W, Hosseini SE, Ghasemlou M (2024) Mechanisms of nanoencapsulation to boost the antimicrobial efficacy of essential oils: a review. Food Hydrocoll 150:109655
19. Zolfaghari R, Fakhru'l-Razi A, Abdullah LC, Elnashaie SS, Pendashteh A (2016) Demulsification techniques of water-in-oil and oil-in-water emulsions in petroleum industry. Sep Purif Technol 170:377–407
20. Tadros TF (2013) Emulsion formation, stability, and rheology. In: Emulsion formation and stability. Wiley, New York, pp 1–75
21. Norn V (2015) Emulsifiers in food technology. Wiley, New York
22. Islam RU, Khan T, Junaid PM, Naseem S (2024) Recent developments in sustainable bio-based wax Nanoemulsions as edible coatings for food commodities: a comprehensive review. Packag Technol Sci 37:1175–1194
23. Rajasekaran B, Singh A, Nilsuwan K, Ma L, Nazeer RA, Benjakul S (2024) Shrimp oil nanoemulsions prepared by microfluidization and ultrasonication: characteristics and stability. RSC Adv 14:6135–6145
24. Modarres-Gheisari SMM, Gavagsaz-Ghoachani R, Malaki M, Safarpour P, Zandi M (2019) Ultrasonic nano-emulsification–a review. Ultrason Sonochem 52:88–105
25. Jintapattanakit A (2018) Preparation of nanoemulsions by phase inversion temperature (PIT) method. Pharmaceut Sci Asia 45:1–12

26. Kumari L, Choudhari Y, Patel P, Gupta GD, Singh D, Rosenholm JM, Bansal KK, Kurmi BD (2023) Advancement in solubilization approaches: a step towards bioavailability enhancement of poorly soluble drugs. Life 13:1099
27. Li L, Zeng Y, Chen M, Liu G (2022) Application of nanomicelles in enhancing bioavailability and biological efficacy of bioactive nutrients. Polymers 14:3278
28. Soufiani SP, Yousefi S, Honarvar M, Weisany W, Asadi G (2025) Polyvinyl alcohol films incorporating cellulose nanocrystals, chitosan, and mesoporous silica nanoparticles for enhanced stability and controlled release of beet extracts. Carbohydr Polym Technol Appl 10:100740
29. Szente L, Sohajda T, Fenyvesi É (2021) Encapsulation for masking off-flavor and off-tasting in food production. In: Functionality of cyclodextrins in encapsulation for food applications. Springer, Cham, pp 223–253
30. Ravera F, Dziza K, Santini E, Cristofolini L, Liggieri L (2021) Emulsification and emulsion stability: the role of the interfacial properties. Adv Colloid Interf Sci 288:102344
31. Alam MS, Akhtar A, Ahmad J, Nollet LM (2021) Stability perspectives of Nanoemulsions. In: Nanoemulsions in food technology. CRC Press, Boca Raton, pp 89–110
32. Isibor, P.O., Kayode-Edwards, I.I., Oyewole, O.A., Agbontaen, D.O., Onwaeze, O.O., Paul, E., Oyegbade, S.A., Ilyasu, U.S., Chimbekujwo, K.I., 2024. Risk assessment and management in nanotoxicology, environmental nanotoxicology: combatting the minute contaminants. Springer, Cham pp. 267–293
33. Zuccari G, Alfei S (2023) Development of phytochemical delivery systems by nano-suspension and nano-emulsion techniques. Int J Mol Sci 24:9824
34. Saffarian H, Rahimi E, Khamesipour F, Hashemi Dehkordi SM (2024) Antioxidant and antimicrobial effect of sodium alginate nanoemulsion coating enriched with oregano essential oil (Origanum vulgare L.) and Trachyspermum ammi oil (Carum cupticum) on food pathogenic bacteria. Food Sci Nutr 12:2985–2997
35. Doghish AS, Shehabeldine AM, El-Mahdy HA, Hassanin MM, Al-Askar AA, Marey SA, AbdElgawad H, Hashem AH (2023) Thymus vulgaris oil nanoemulsion: synthesis, characterization, antimicrobial and anticancer activities. Molecules 28:6910
36. Pandey VK, Srivastava S, Dash KK, Singh R, Dar AH, Singh T, Farooqui A, Shaikh AM, Kovacs B (2024) Bioactive properties of clove (Syzygium aromaticum) essential oil nanoemulsion: a comprehensive review. Heliyon 10:e22437
37. Badr MM, Badawy ME, Taktak NE (2022) Preparation, characterization, and antimicrobial activity of cinnamon essential oil and cinnamaldehyde nanoemulsions. J Essent Oil Res 34:544–558
38. Sharma AD, Chhabra R, Jain P, Kaur I, Chauhan A, Rani R (2023) Preparation, characterization, and biological potential of Nanoemulsion from Rosmarinus officinalis L. Essent Oil BioNanoSci 13:1955–1975
39. Ahmad I, Al-Dolaimy F, Kzar MH, Kareem AT, Mizal TL, Omran AA, Alazbjee AAA, Obaidur Rab S, Eskandar M, Alawadi AH (2024) Microfluidic-based nanoemulsion of Ocimum basilicum extract: constituents, stability, characterization, and potential biomedical applications for improved antimicrobial and anticancer properties. Microsc Res Tech 87:411–423
40. Balasubramanian D, Girigoswami A, Girigoswami K (2022) Antimicrobial, pesticidal and food preservative applications of lemongrass oil nanoemulsion: a mini-review. Recent Adv Food Nutr Agric 13:51–58
41. Gago C, Serralheiro A, da Graça Miguel M (2025) Anti-inflammatory activity of thymol and thymol-rich essential oils: mechanisms, applications, and recent findings. Molecules 30:2450
42. Kowalewska A, Majewska-Smolarek K (2023) Eugenol-based polymeric materials—antibacterial activity and applications. Antibiotics 12:1570
43. Suchi TA, Mishu ID, Akhter MZ, Hoque MM (2023) Comparative antimicrobial efficacy, kinetic destruction pattern and microbial inactivation dynamics of extracted cinnamon essential oil and commercial cinnamaldehyde against foodborne pathogens. Ir J Med Microbiol 17:230–242

44. Liao W, Badri W, Dumas E, Ghnimi S, Elaissari A, Saurel R, Gharsallaoui A (2021) Nanoencapsulation of essential oils as natural food antimicrobial agents: an overview. Appl Sci 11:5778
45. Grover S, Aggarwal P, Kumar A, Kaur S, Yadav R, Babbar N (2024) Utilizing citrus peel waste: a review of essential oil extraction, characterization, and food-industry potential. Biomass Conv Bioref 15:1–22
46. Mukurumbira A, Shellie R, Keast R, Palombo E, Jadhav S (2022) Encapsulation of essential oils and their application in antimicrobial active packaging. Food Control 136:108883
47. Ghosh J, Choudhury SR, Singh K, Gupta MM, Sharma D (2024) Biobased nanomaterials in nutraceuticals. In: Biobased nanomaterials: applications in biomedicine, food industry, agriculture, and environmental sustainability. Springer, Cham, pp 395–448
48. Karaca N, Demirci F (2024) In vitro cholinesterase, lipoxygenase inhibition evaluation of rosemary and coriander essential oil nanoemulsion and characterisation. J Herb Med 45:100874
49. Weisany W, Soufiania SP, Razmi J, Eshaghadadi AH (2024a) Nano-encapsulation of fenugreek and coriander essential oils using copper oxide NPs: novel approach for augmenting their effectiveness against Colletotrichum nymphaeae. Ind Crops Prod 219:119051
50. Souza VVMA, Almeida JM, Barbosa LN, Silva NCC (2022) Citral, carvacrol, eugenol and thymol: antimicrobial activity and its application in food. J Essent Oil Res 34:181–194
51. Peter S, Sotondoshe N, Aderibigbe BA (2024) Carvacrol and thymol hybrids: potential anticancer and antibacterial therapeutics. Molecules 29:2277
52. Ferreira VRF, Militani IA, de Almeida KJ, Lunguinho AdS, Saczk AA, Ionta M, da Silva GAlF, Felix FS, Nelson DL, Cardoso MdGa (2023) Antioxidant and cytotoxic activity of essential oils and their principal components: spectrophotometric, voltammetric, and theoretical investigation of the chelating effect of eugenol and carvacrol. ACS Food Sci Technol 3:350–360
53. Vaishnavi BC, Ananya SJ, Jitendra NY, Rajendra BS (2021) Strategies to improve stability of essential oils. J Pharm Sci Res 13:416–425
54. Arabpoor B, Yousefi S, Weisany W, Ghasemlou M (2021) Multifunctional coating composed of Eryngium campestre L. essential oil encapsulated in nano-chitosan to prolong the shelf-life of fresh cherry fruits. Food Hydrocoll 111:106394
55. Mushtaq A, Wani SM, Malik A, Gull A, Ramniwas S, Nayik GA, Ercisli S, Marc RA, Ullah R, Bari A (2023) Recent insights into Nanoemulsions: their preparation, properties and applications. Food Chemistry: X 18:100684
56. Kumar, S., Kumar, V., Sharma, R., Manhas, A., Chaudhary, S., Sharma, S., 2023. Edible Nanoemulsion coating materials: potential for fresh fruits and vegetables, nanotechnology horizons in food process engineering. Apple Academic Press, Palm Bay pp. 267–321
57. Sawale P, Patil P, Singh A, Xavier J, Kumar P, Dutta D (2024) Non-thermal techniques for microbiological safety, nutritional preservation, and enhanced efficiency in dairy processing. Funct Food Sci 4:180–203
58. Popescu P-A, Palade LM, Nicolae I-C, Popa EE, Miteluț AC, Drăghici MC, Matei F, Popa ME (2022) Chitosan-based edible coatings containing essential oils to preserve the shelf life and postharvest quality parameters of organic strawberries and apples during cold storage. Foods 11:3317
59. de Sousa DP, Damasceno ROS, Amorati R, Elshabrawy HA, de Castro RD, Bezerra DP, Nunes VRV, Gomes RC, Lima TC (2023) Essential oils: chemistry and pharmacological activities. Biomolecules 13:1144
60. Ozogul Y, Karsli GT, Durmuş M, Yazgan H, Oztop HM, McClements DJ, Ozogul F (2022) Recent developments in industrial applications of nanoemulsions. Adv Colloid Interf Sci 304:102685
61. Schneider G, Steinbach A, Putics Á, Solti-Hodován Á, Palkovics T (2023) Potential of essential oils in the control of listeria monocytogenes. Microorganisms 11:1364
62. Tian Y, Sun F, Wang Z, Yuan C, Wang Z, Guo Z, Zhou L (2024) Research progress on plant-based protein Pickering particles: stabilization mechanisms, preparation methods, and application prospects in the food industry. Food Chemistry: X 21:101066

63. Cheng Y, Cai X, Zhang X, Zhao Y, Song R, Xu Y, Gao H (2024) Applications in Pickering emulsions of enhancing preservation properties: current trends and future prospects in active food packaging coatings and films. Trends Food Sci Technol 151:104643
64. Baghi F, Ghnimi S, Dumas E, Chihib N-E, Gharsallaoui A (2023) Nanoemulsion-based multilayer films for ground beef preservation: antimicrobial activity and physicochemical properties. Molecules 28:4274
65. Pandey VK, Islam RU, Shams R, Dar AH (2022) A comprehensive review on the application of essential oils as bioactive compounds in Nano-emulsion based edible coatings of fruits and vegetables. Appl Food Res 2:100042
66. Fanzo J, McLaren R, Bellows A, Carducci B (2023) Challenges and opportunities for increasing the effectiveness of food reformulation and fortification to improve dietary and nutrition outcomes. Food Policy 119:102515
67. Rehman A, Qunyi T, Sharif HR, Korma SA, Karim A, Manzoor MF, Mehmood A, Iqbal MW, Raza H, Ali A (2021) Biopolymer based nanoemulsion delivery system: an effective approach to boost the antioxidant potential of essential oil in food products. Carbohydrate Polymer Technologies and Applications 2:100082
68. Kotta S, Aldawsari HM, Badr-Eldin SM, Alhakamy NA, Shaik RA (2021) Application of nanoemulsions to improve food system color, flavor, texture, and preservation. In: Nanoemulsions in food technology. CRC Press, Boca Raton, pp 205–228
69. Sneha K, Kumar A (2022) Nanoemulsions: techniques for the preparation and the recent advances in their food applications. Innov Food Sci Emerg Technol 76:102914
70. Preeti, Sambhakar S, Malik R, Bhatia S, Al Harrasi A, Rani C, Saharan R, Kumar S, Geeta, Sehrawat R (2023) Nanoemulsion: an emerging novel technology for improving the bioavailability of drugs. Scientifica 2023:6640103
71. Ashaolu TJ (2021) Nanoemulsions for health, food, and cosmetics: a review. Environ Chem Lett 19:3381–3395
72. Gharibzahedi SMT, Altintas Z (2022) Transglutaminase-induced free-fat yogurt gels supplemented with tarragon essential oil-loaded nanoemulsions: development, optimization, characterization, bioactivity, and storability. Gels 8:551
73. Yousefi S, Kavyanirad M, Aminifar M, Weisany W, Mousavi Khaneghah A (2022) Yogurt fortification by microencapsulation of beetroot extract (Beta vulgaris L.) using maltodextrin, gum arabic, and whey protein isolate. Food Sci Nutr 10:1875–1887
74. El-Wahab GMA, Shoala T, Amin BH, Masoud SA (2023) The potency of Syzygium aromaticum and Rosmarinus officinalis oils in the emulsion and nano-emulsion form to limit gray mould of strawberry fruits. Indian Phytopathol 76:853–865

Aromatic Plant-Based Nanoemulsions in Cosmetics

Elif Özlem Arslan Aydoğdu, Gülnihan Selim, Mine Koçyiğit, and Nihal Doğruöz Güngör

Abstract Aromatic plant-based nanoemulsions stand out as pharmaceutical delivery systems that offer natural and innovative solutions in the cosmetics field. This section primarily evaluates aromatic plants from a pharmaceutical botany perspective, examining the taxonomic classifications, botanical identification criteria, geographical distributions, and phytochemical contents of the main species used. The chemical structure of essential oils, consisting of key components (monoterpenes, sesquiterpenes, etc.), their chemotype differences, and biological effects are discussed. Furthermore, the antimicrobial mechanisms of essential oils and potential interactions related to resistance development are discussed. The pharmaceutical significance of nanoemulsion technology is evaluated through its ability to increase the bioavailability of bioactive substances, improve skin penetration, and provide targeted delivery. Additionally, microbiological stability, natural and synthetic preservative systems, GMP, and cosmetic quality standards in prepared herbal nanoemulsions are examined. The potential effects of aromatic nanoemulsions on skin microbiota, quality control, and pharmacopoeial compliance processes are discussed. From a clinical perspective, scientific evidence on anti-acne, antiaging, and anti-inflammatory effects is presented, and future perspectives of these systems as topical drug carriers are revealed.

Keywords Aromatic nanoemulsion · Cosmetic applications · Microbial stability · Natural preservative · Pharmacopeia standardization

E. Ö. Arslan Aydoğdu (✉) · N. Doğruöz Güngör
Department of Biology, Faculty of Science, Istanbul University, Istanbul, Turkey
e-mail: eoarslan@istanbul.edu.tr

G. Selim
Institute of Graduate Studies in Sciences, Istanbul University, Istanbul, Turkey

M. Koçyiğit
Department of Pharmaceutical Botany, Faculty of Pharmacy, Istanbul University, Istanbul, Turkey

A. Husen (ed.), *Sustainable Uses of Aromatic Plants in Nanotechnology*, Smart Nanomaterials Technology, https://doi.org/10.1007/978-981-95-7409-4_14

1 Introduction

Aromatic plants and their essential oils have been used for thousands of years in many areas, from medicine to food and cosmetics, and historical records confirm this extensive history of use. The use of aromatic plants, first used in traditional medicine and rituals, pioneered the use of herbal extracts and oils for cosmetic purposes [1].

The use of cosmetics in Egypt dates back to 5000 BC. Egyptians created cleansing creams (a mixture of animal/vegetable fat, lime, and perfume), red plant roots and soil-derived dyes for lip and cheek coloring, henna for nail and palm painting, and perfumes created from botanical ingredients such as myrrh, thyme, marjoram, chamomile, lavender, lily, mint, rosemary, cedar, rose, aloe, olive oil, sesame oil, and almond oil, for aesthetic and functional purposes. For the Sumerians, the use of perfumes and incense was also highly important, as they were seen as a way to connect with the gods. The most widely used and symbolic cosmetic of ancient Egypt was kohl, applied around the eyes and was green (malachite) in the early period and black (galena) in the dynastic period [2–4]. The *Ebers Papyrus*, a legacy of the Egyptians, mentions 700 plants, including garlic, onion, aloe, pomegranate, coriander, and juniper [5]. The use of cosmetics in the ancient Greek world emerged in the Bronze Age (3200–1100 BC). The ancient Greeks were influenced by Egypt in this regard [6]. Hippocrates (460–377 BC), who laid the foundations of Western medicine, used more than 400 plant species, especially aromatic medicinal plants, and studied disease development and treatment [7, 8].

In India, the use of herbs and spices as medicine is documented in the sacred books of Atharva and Rig Veda, as well as the Sushruta Samhita (1500 BC), as well as in Sanskrit texts dating back to approximately 1500 BC. Furthermore, Ayurveda, one of the oldest Indian treatment systems still practiced in India today, dates back to the second millennium BC. It utilizes many herbs, such as basil and turmeric [8, 9].

Traditional Chinese medicine is a treatment system that supports phytotherapy, with written sources dating back to 2700 BC. The Chinese Materia Medica, one of the earliest written sources, listed over 300 plants and formed the basis of herbal therapy. The Pen T' Sao Book describes 365 medicinal plants, some of which (rhubarb root, camphor, tea leaves, ginseng, cinnamon bark, and ephedra) are still used today. Chinese medicine emphasized the importance of using herbal mixtures in treatment and applied herbs as formulas rather than individually. During the Ming Dynasty (1368–1644), more than 60,000 herbal prescriptions were recorded, and Li Shizhen's pharmacopoeia, "Pen t'sao kang mu," described more than 1800 herbs and their properties [10].

In the twentieth century, synthetic and chemically modified ingredients became popular in cosmetics [6]. Today, as consumers become more aware, they are demanding cosmetic products with more natural ingredients, especially botanical ones, instead of synthetic ingredients that may have negative effects on health and the environment [11, 12]. Plant extracts, essential oils, and cold-pressed oils (olive oil, almond oil, coconut oil, etc.) are widely used in skin, beauty, and hair cosmetic formulations bearing natural and herbal labels [13, 14].

Nanoemulsions are ultrafine, stable liquid dispersions containing a submicron-sized oil phase and a water phase, along with a surfactant. The size of nanoemulsions (NEs) is generally on the nanometric scale, ranging from 1 to 100 nanometers [15–18]. While some sources indicate that the droplet size can range from 20 to 500 nm, the range of 10 to 200 nm is particularly emphasized for cosmetic applications [19, 20]. The physical appearance of nanoemulsions varies depending on their particle size. If they have a transparent or translucent appearance, the droplet diameter is around 20–50 nm. Because their particle size does not scatter light much, they can generally be produced as transparent solutions. The droplet diameter is less than 200 nm, usually in the range of 50–150 nm, with a translucent/beginning to opaque appearance [19, 21]. The creamy emulsion, however, refracts light differently, making it opaque (whitish, milky) and with a droplet size of 150–200 nm, which can reach 500 nm [22].

Nanotechnology has offered significant innovations for the pharmaceutical and cosmetics industries. Nanoemulsions typically have a monolayer of phospholipids surrounding a lipophilic inner core, making them ideal carriers for lipophilic active ingredients [18]. Nanoemulsions can encapsulate lipophilic and hydrophilic active ingredients, thus protecting them from physical and chemical degradation [23, 24]. They are considered excellent delivery systems, meaning they can effectively deliver lipophilic ingredients to the skin and prevent premature degradation due to biological reactions in the body. Furthermore, they offer advantages over traditional nanocarriers (niosomes, ethosomes, and liposomes) due to their high production costs and poor bioactive encapsulation performance [22, 25–27]. Preparation methods such as high-energy and low-energy techniques, formulation components such as oil phase, surfactant, and cosurfactant, and characterization parameters determine nanoemulsion performance. Therefore, intensive studies on production and characterization are ongoing [28].

2 Chemical Components of Essential Oils

Essential oils possess a wide range of biological activities, thanks to their complex structures composed of terpenoids, phenylpropanoids, and other small molecule compounds. However, properties such as volatility, low water solubility, and susceptibility to oxidation are limiting factors in terms of efficacy and stability in pharmaceutical and cosmetic applications. In this regard, nanotechnology offers innovative solutions for increasing the bioavailability and controlled release of essential oils. Systems such as nanoemulsions, liposomes, polymeric nanoparticles, and solid lipid nanoparticles both maintain the chemical stability of essential oils and serve as targeted delivery platforms. Therefore, the integration of nanotechnological approaches with the chemical properties of essential oils significantly expands the potential for medical and industrial uses of these natural products [29].

Volatile organic compounds (VOCs) constitute both floral fragrances and fruit aromas. The primary functions of these compounds include attracting pollinators and seed dispersers and acting as defense mechanisms against herbivores and pathogens. The chemical structure of VOCs is quite diverse and includes many different chemical groups, primarily monoterpenes, sesquiterpenes, benzenoids, phenylpropanoids, and fatty acid derivatives (Fig. 1).

The biosynthesis of volatile compounds occurs primarily through four main metabolic pathways:

- The MEP (2-C-methyl-D-erythritol-4-phosphate) pathway in plastids: It enables the production of monoterpenes and some sesquiterpenes.
- The MVA (mevalonic acid) pathway in the cytoplasm: It is responsible for the biosynthesis of some sesquiterpenes.
- The shikimate pathway: It provides precursor compounds for the biosynthesis of benzenoids and phenylpropanoids.
- Fatty acid and amino acid metabolism pathways: It plays a role in the production of aliphatic compounds and green leaf volatiles (GLVs).

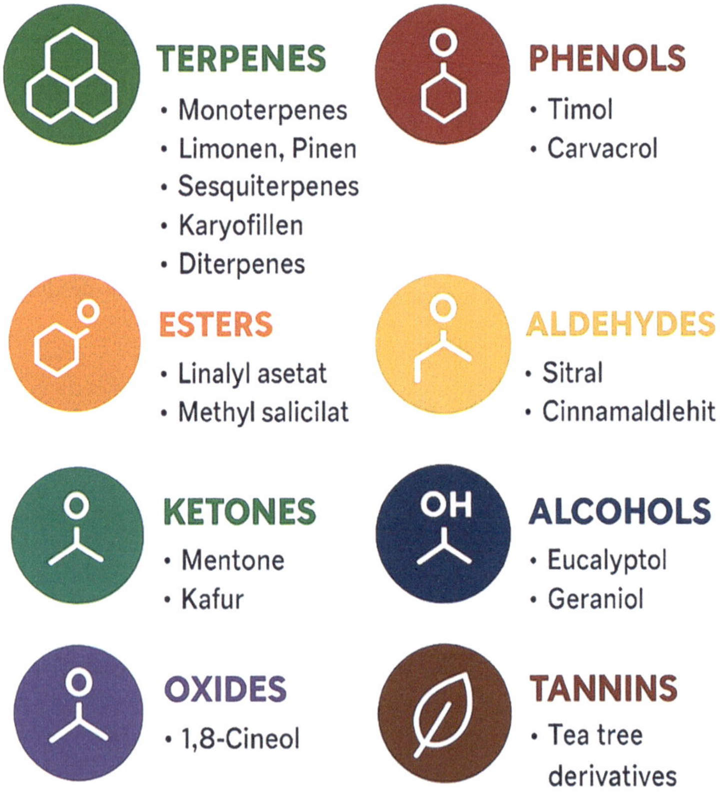

Fig. 1 Chemical components of essential oils

The production and release of VOCs are complexly regulated by various intrinsic and extrinsic factors, such as environmental factors, plant species, and genetic differences. Gene expression of key enzymes and transcription factors in the biosynthetic pathways plays a central role in this process [30]. Biosynthesis can also be affected by factors such as light, temperature, hormonal signals, and insect attacks. For example, phytohormones such as jasmonate and salicylic acid can activate plant defense mechanisms by increasing volatile compound production. Understanding the biosynthesis and regulation mechanisms of these compounds offers great potential for developing new strategies to improve the quality and yield of plant products [31].

2.1 Terpenes

Terpenes are critically important both as primary metabolites essential for plant growth and development (e.g., gibberellins, sterols) and as secondary metabolites that function in defense mechanisms. However, the volatility of their natural structures, their chemical instability, and their sensitivity to environmental conditions limit their direct use in agriculture and biopesticide applications. In this regard, nanotechnology plays a strategic role in preserving the biological functions of terpenes and enhancing their efficacy. Nanoformulations (nanoemulsions, nanocapsules, solid lipid nanoparticles, polymeric nanoparticles), in particular, increase the stability of terpenes against oxidation, enable controlled release, and enhance their bioavailability to target organisms (e.g., pathogens, herbivores). Furthermore, nanocarrier systems allow terpenes to exert their allelopathic or insecticidal effects at lower doses for longer periods, reducing environmental damage [32]. Increasing terpene production in plants through genetic engineering and supporting these metabolites with nanotechnological carriers stands out as a promising approach for the development of next-generation biopesticides in sustainable agriculture [33].

Monoterpenes are known to have a wide range of applications in medical and pharmaceutical fields. These compounds possess antioxidant, antibacterial, sedative, and anti-inflammatory properties. However, their potential toxic effects are also highlighted. The toxicity of monoterpenes varies depending on their chemical structure. Monoterpenes with double bonds and oxygen-containing functional groups can have cytotoxic, genotoxic, and mutagenic effects. For example, the most commonly used monoterpenes—pinenes, limonene, camphor, eucalyptol, menthol, and thymol—have potential side effects and toxicity when used in high doses or improperly. It has even been reported that camphor and eucalyptol can exhibit neurotoxic effects at high doses [34]. Although these compounds may have beneficial effects, they carry risks such as allergic reactions, skin irritation, and serious systemic side effects when taken in high doses [35].

2.2 Phenolics

Phenolic compounds are secondary aromatic metabolites that help plants adapt to changing environmental conditions. Environmental factors such as high light, cold, drought, and heavy metals increase the accumulation of phenolics to neutralize toxic effects in plants [36]. Phenolic compounds can be divided into two main groups: flavonoids and non-flavonoids. Flavonoids, the most diverse class of phenolics, include flavonols, flavanones, isoflavones, anthocyanins, and flavan-3-ols. Non-flavonoids are classified as phenolic acids, tannins, stilbenes, and lignans. Under normal conditions, flavonoids function as signaling molecules and UV protectors, while under stress, they protect against oxidative damage [37].

Abiotic stress factors (such as drought, salinity, heavy metals, and UV radiation) cause oxidative damage in plants by increasing the production of reactive oxygen species (ROS) [38, 39]. Phenolic compounds offer an important defense mechanism to prevent this damage. Drought triggers phenolic and flavonoid biosynthesis by modulating the phenylpropanoid pathway. This protects plants from the adverse effects of water deficit. Kaempferol and quercetin accumulation in tomatoes increases tolerance to drought conditions. High salinity causes osmotic disturbances and ROS production in plants. Phenolic compounds protect plants by scavenging these ROS through their antioxidant properties. Salt stress increases the production of various phenolic compounds by stimulating the phenylpropanoid biosynthesis pathway. Heavy metals lead to oxidative stress and toxicity in plants. Flavonoids play a critical role in counteracting this toxicity by chelating metals and reducing hydroxyl radicals. UV radiation leads to protein damage and DNA mutation. Plants overcome these harmful effects by increasing the production of phenolic compounds, which form a protective layer beneath epidermal cells. Flavonoids are potential ROS scavengers that protect plants from damage, thanks to their ability to absorb UV light [40, 41].

The potential use of essential oils as a natural alternative in the cosmetic and healthcare sectors is attracting the attention of researchers. A study examined the phytochemical structure and antioxidant and antimicrobial properties of the essential oils of *Thymus zygis* L. and *Thymus willdenowii* Boiss., confirming that their essential oils have distinct chemical profiles and that these profiles are associated with strong antioxidant and antimicrobial properties. These results support the potential use of the essential oils of these thyme species as novel and natural antiseptics in the pharmaceutical and food industries [42]. However, it is emphasized that further pharmacological and clinical research is needed to uncover the exact biological mechanisms of thyme. It has been stated that more detailed studies are needed, particularly on the effects of phytoncides on human health. Phytoncides are volatile organic compounds secreted by plants to protect themselves from viruses, bacteria, fungi, and protozoans. These compounds exhibit a variety of beneficial effects, including antimicrobial, anti-inflammatory, anti-stress, and analgesic [43].

2.3 Aldehydes

Aldehydes are a subgroup of volatile organic compounds (VOCs) found in various products, including fragrances, foods, and beverages. Aldehydes are found in personal care products such as perfumes, foods, beverages, cleaning products, and pharmaceuticals. While they can be derived naturally from plants, fruits, olive oil, and coffee, they can also arise from anthropogenic sources such as industrial waste, vehicle emissions, cigarette smoke, and high-temperature cooking methods. Based on their chemical structures, the main types of aldehydes are grouped as aliphatic aldehydes and aromatic aldehydes. Aliphatic aldehydes are formed by the thermal oxidation of fatty acids and include saturated aldehydes such as hexanal, heptanal, and decanal. These aldehydes can leach into mineral waters from packaging materials. Some aliphatic aldehydes, such as formaldehyde, acetaldehyde, and acrolein, are classified as carcinogenic by the IARC. Aromatic aldehydes are fragrance ingredients widely used in the perfume industry. Some, particularly those that create the scent of lilies, are frequently used in perfumery. Examples such as vanillin and anizaldehyde are also used as food additives due to their antimicrobial and antibiofilm activities.

Aldehydes can have adverse effects on human health due to their high reactivity. Short-term exposure can cause symptoms such as a burning sensation in the nose, eyes, and throat, cough, and skin irritation. Long-term exposure can lead to serious conditions such as asthma, cancer (especially lung cancer), neurodegenerative diseases, and other diseases related to aging. For example, formaldehyde is classified as a "Group 1" human carcinogen by the World Health Organization (WHO) and the International Agency for Research on Cancer (IARC). Some aldehydes, such as butylphenyl methylpropional (BMHCA), have been withdrawn from the market due to their potential toxicity and their allergenic, reproductive, and endocrine-disrupting properties. In conclusion, although aldehydes have beneficial biological activities in many areas, the potential toxic effects of these compounds should not be ignored, and more research and regulatory standards are needed, especially in cosmetics and medical products [44].

3 Phytochemical Properties of Essential Oils

In a study investigating the phytochemical structure and biological activities of essential oils (ESS) of six different aromatic medicinal plants (*Laurus nobilis* L., *Chamaemelum nobile* (L.) All., *Citrus × aurantium* L., *Pistacia lentiscus* L., *Cedrus atlantica* (Endl.) Manetti ex Carrière, and *Rosa × damascena* Herrm.) used in cosmetics, the essential oils of the plants were obtained by hydrodistillation, and the chemical compositions of the essential oils were analyzed by gas chromatography-mass spectrometry (GC-MS). According to the analysis results, the main component of each plant varies. While 1,8-cineole is the most abundant compound in

L. nobilis with 36.58%, angelical acetate is the most abundant compound with 41.79% in *C. nobile*. *C.* × *aurantium* is rich in linalool with 29.01%. *P. lentiscus* is dominated by 3-methylpentylangelate with 27.83%. *C. atlantica*: β-himachalene is the main component at 40.19%. *R. damascena*: n-nonadecane is rich at 44.89%. Antioxidant activity tests revealed that all essential oils examined had a high capacity to scavenge DPPH radicals. The oils with the highest antioxidant activity were those from *L. nobilis* (76.84%) and *P. lentiscus* (71.53%) [45].

A study conducted in Algeria investigated the chemical variability and chemotypes of essential oils (VFAs) from eleven different aromatic and medicinal plant species (including the genera *Thymus*, *Mentha*, *Salvia*, *Lavandula*, and *Eucalyptus*). The study aimed to identify the factors affecting the variability in the chemical composition of VFAs (vegetative cycle, differences between subspecies, environmental and genetic factors) and to delineate the limitations of the chemotype concept in this context. The profile in which individuals of the same species, even those with the same genome and phenotype, vary in terms of chemical composition and therefore biological activities, in other words, a profile with different biochemical properties, is defined as chemotype. The research has revealed that environmental factors such as climate, altitude, soil quality where the plant grows, and internal factors such as the vegetative cycle, significantly affect this chemical variability. Different chemotypes and chemical profiles were observed in the plants examined in the study. For example, *Thymbra capitata* (L.) Cav. (Syn.: *Thymus capitatus* (L.) Hoffmanns. & Link) collected from Western Algeria exhibited a single chemotype with the main component being carvacrol, while the three different subspecies of *Thymus munbyanus* Boiss. & Reut. (*T. munbyanus* subsp. *abylaeus* (Font Quer & Maire) Greuter & Burdet, *T. munbyanus* subsp. *ciliatus* (Desf.) Greuter & Burdet, *T. munbyanus* subsp. *coloratus* (Font Quer & Maire) Greuter & Burdet) each showed unique chemical profiles containing a different dominant compound. In the species *Salvia rosmarinus* Spenn., two different chemical profiles were detected: one rich in 1,8-cineole and the other rich in α-pinene, camphor, and camphene. It is also noted that the vegetative cycle of a plant can change the percentage of the dominant compound, leading to the emergence of different chemotypes at different times. For example, it was emphasized that the percentages of pulegone, eucalyptol, menthone, and neo-menthol in the composition of the UFAs of *Mentha pulegium* L. varied seasonally, indicating four distinct chemotypes. However, in the case of *Mentha* × *piperita* L., it was stated that the change in linalool and linalyl acetate ratios resulted from the vegetative cycle and should not be considered a new chemotype or phylogenetic variation. In conclusion, the article emphasizes the importance of traceability and standardization in the industrial use of UFAs, suggesting an approach to cultivate wild plants and screen their chemical compositions according to more specific standards for commercial use [46].

Another study examining the concept of chemotypes investigated the chemical diversity and chemotype classification of essential oils from *Artemisia herba-alba* Asso (ART) collected from the western and southern regions of Morocco. The study was based on the gas chromatography-mass spectrometry (GC-MS) analysis of 60 essential oil samples obtained from ten plants each from six different locations. The analyses identified 23 compounds, comprising 93.28% of the chemical composition of the essential oil. The major components were oxygenated monoterpenes (camphor, α-thujone, β-thujone, and eucalyptol) and hydrocarbon monoterpenes (camphene). Based on the chemical composition of the samples, statistical analyses revealed that the essential oils were divided into four main groups, predominantly determined by differences in the concentrations of camphor, α-thujone, β-thujone, eucalyptol, and camphene. In conclusion, the researchers emphasize that the variability in the chemical profile of *Artemisia herba-alba* essential oil is influenced by environmental factors such as climate, altitude, and soil type, as well as genetic differences between populations. The study highlights the potential of this species' essential oils not only for use in traditional medicine but also for the synthesis of sustainable nanomaterials for biomedical and environmental applications. In phytochemistry, chemotype is of great importance, particularly for the quality and efficacy of medicinal and aromatic plants. The therapeutic properties of a plant generally depend on the concentration of its main active compounds. Therefore, knowledge of the different chemotypes of a plant species is critical for the development of standardized herbal products and their appropriate therapeutic use [47].

4 Microbiological Properties of Essential Oils

Traditional chemical preservatives such as parabens can cause significant problems in the short term, such as skin irritation and allergies, and in the long term, they can promote cancer development. As a result of the proven dangers of synthetic ingredients and increased consumer awareness, there has been a growing trend toward using natural antimicrobial agents as alternatives to chemical preservatives in cosmetic products. Among the natural compounds used as preservatives, essential oils are highly preferred. Essential oils, which are also beneficial in the cosmetics industry due to their fragrance properties, are compounds that have anti-inflammatory, analgesic, antiseptic, and antimicrobial effects [48–50].

Essential oils can exert cytotoxic effects on various microorganisms through multiple mechanisms due to their complex chemical structures (Table 1). For example, phenolic monoterpenes such as thymol bind to the cell membrane, disrupting its integrity, increasing ion permeability, and ultimately causing cell death due to the disruption of the proton gradient. Disruption of membrane permeability can also lead to cell death by causing the release of vital metabolites from the cell or by disrupting the pH/ion balance. Additionally, these compounds cause enzyme inhibition because they can bind to sulfhydryl groups. This effect is particularly destructive on ATPase enzymes. Some compounds, such as aldehydes, bind to nucleic acids and suppress replication and transcription. Fatty acids that affect the synthesis or release of signaling molecules, such as acyl-homoserine lactones, block quorum-sensing and biofilm formation mechanisms. Furthermore, some volatile oils can cause irreversible damage to cells by increasing reactive oxygen species (ROS) in microorganisms through their effect on the oxidative stress response [60–63]. Essential oils,

Table 1 Antimicrobial effect mechanisms of essential oils, examples, and target microorganisms

Mechanism of action	Description	Example essential oils	Dominant components	Microorganisms	References
Structural disruption of the cell membrane	Disrupts membrane integrity, increases permeability	Oregano oil (*Origanum vulgare*)	Thymol, carvacrol	*Staphylococcus aureus*, *Escherichia coli*	Lombrea et al. [51]
Leakage of cell contents	ATP, proteins, and nucleic acids escape outside the cell	Cinnamon oil (*Cinnamomum verum*)	Cinnamaldehyde	*Listeria monocytogenes*, *Salmonella enterica*	Selma et al. [52], Özel et al. [53]
Enzyme and protein inhibition	Inhibits metabolic enzymes (ATPase, etc.)	Clove oil (*Syzygium aromaticum*)	Eugenol	*Bacillus subtilis*, *Pseudomonas aeruginosa*	Hekmatpanah et al. [54]
Effect on genetic material	Binds to DNA/RNA, suppressing replication and transcription	Sage oil (*Salvia officinalis*)	Phenolic monoterpenes	*E. coli*, *Candida albicans*	Bicak et al. [55]
Quorum-sensing and biofilm inhibition	Prevents biofilm formation by disrupting intercellular signaling	Tea tree oil (*Melaleuca alternifolia*); lemongrass oil (*Cymbopogon citratus*)	Terpinen-4-ol, geranial, neral	*Pseudomonas aeruginosa* (biofilm), *Staphylococcus epidermidis*	Prado et al. [56], Khosakueng et al. [57]
Disruption of metabolic pathways	Disrupts pH balance and energy metabolism	Rosemary oil (*Salvia rosmarinus*)	1,8-cineole, carvacrol	*S. aureus*, *Klebsiella pneumoniae*	Alaboudi et al. [58]
Oxidative stress induction	Causes cell damage by increasing ROS production	Lavender oil (*Lavandula angustifolia*), Eucalyptus oil (*Eucalyptus globulus*)	Linalool, 1,8-cineole	*Candida albicans*, *Aspergillus niger*	Hosseini and Heidari [59]

which can be obtained from almost every part of plants such as flowers, leaves, roots, and bark, can be extracted using various methods. Terpenes, which are chemically formed by the combination of isoprene units, phenylpropanoids synthesized from the shikimic acid pathway, and lipophilic compounds, including their oxygenated derivatives such as alcohols, aldehydes, ketones, esters, and phenols, constitute the main components of essential oils. This diversity forms the basis for the versatile use of essential oils in the pharmaceutical, cosmetics, and food industries [64, 65].

Microbial contamination of cosmetic products can lead to undesirable changes in the product's odor, color, viscosity, and performance, as well as cause adverse effects in users. Essential oils are important in cosmetics due to their chemical structure, pleasant odors, and bioactive potential. They exhibit antimicrobial activity across a broad spectrum, ranging from saprophytes such as *Bacillus* sp., *Micrococcus* sp., *Acinetobacter* sp., and *Aspergillus* sp. to human pathogens such as *Salmonella* sp. and *Candida* sp. [48, 66].

Numerous studies have been conducted on the antimicrobial properties of essential oils. The effectiveness of *Origanum onites* L. and *Thymus vulgaris* L. essential oils as antimicrobial emulsions containing collagen for cosmetic purposes has been proven on *Staphylococcus aureus*, *Escherichia coli*, *Enterococcus faecalis*, *Aspergillus fumigatus*, and *Candida albicans* [66]. *Cinnamomum verum* J.Presl (Syn., *C. zeylanicum* Blume) (Cinnamon), *Lavandula angustifolia* Mill. (Syn., *L. officinalis* Chaix) (Lavender), and *Melaleuca alternifolia* (Maiden & Betche) Cheel (tea tree) oils are also effective against various bacterial and fungal pathogen species [67]. *Artemisia afra* Jacq. ex Willd., *Pteronia incana* (Burm.f.) DC., *Lavandula angustifolia*, and *Salvia rosmarinus* (Rosemary) are sources of oils added to cream formulations to reduce microbial contamination [68]. Laboratory studies have also determined that different concentrations of *Citrus* × *aurantium* (orange), *Origanum vulgare* L. (oregano), *Origanum majorana* L. (marjoram), *Mentha* × *piperita* (peppermint), and *Helichrysum italicum* (roth) G.Don (immortelle) essential oils may be effective against fungal infections [67, 69]. The essential oil extract obtained from the leaves of *Ocimum gratissimum* L. (African Basil) has been proven to have antibacterial effects and also has antiseptic effects on minor wounds, boils, and pimples [70].

To enhance the effectiveness of essential oils and improve their stability, while also eliminating physicochemical limitations such as high volatility, low water solubility, and light-oxygen sensitivity, nanotechnology has begun to be utilized in cosmetics, as in many other industries [49]. One of the main problems solved by the use of nanocarriers is high volatility. Since essential oils are compounds that easily vaporize at low temperatures, it is difficult to maintain their stability in the product. Furthermore, as they are oil derivatives, they are hydrophobic in nature and do not distribute homogeneously in water-based products. Oxidation and phytodegradation also cause essential oils to lose their effectiveness rapidly [71].

Furthermore, the scent of essential oils can sometimes be potent and may not suit the nature of the product. When high concentrations are required to maintain effectiveness, scent is a crucial factor. Furthermore, high concentrations can trigger various reactions such as irritation and allergies. The use of nanocarriers to enable low concentrations plays a key role in solving this problem [72–74].

Volatile oils integrated with nanocarriers are more stable against physicochemical factors, have reduced hydrophobicity, enhanced antimicrobial effect and skin penetration, and increased bioactivity duration within the product through slow release. The concentration within the product is reduced, thereby decreasing sensory effects and the risk of irritation, and the advantages of the nanoparticles are bound to synergistically support the product positively, offering numerous benefits.

The main nanocarriers used in the application of essential oils in cosmetic products are nanoemulsions, nanogels, liposomes, phospholipid nanocapsules, and polymeric or biopolymeric nanocapsules [63, 75, 76].

Wattanasatcha et al. [77] examined the antimicrobial effect of free thymol, thymol loaded onto nanoparticles (ethylcellulose/methylcellulose coating), and methylparaben in cosmetic formulations such as lotions, gels, and creams. They found that encapsulated thymol was as practical a preservative as methylparaben, yielding promising results even when used at concentrations 12–25 times lower in all cosmetic formulations. Another study showed that urea-formaldehyde encapsulation of lemon essential oil increased stability [78]. Moraes-Lovison et al. [79] examined the antibacterial and physicochemical stability of nanoemulsions produced from *Origanum vulgare* essential oil and found that they increased the death of prostate cancer cells.

In a study published in 2024, keratin nanoparticles obtained from chicken feathers were combined with *Nigella sativa* L. extract, and the resulting nanoemulsion was found to have highly successful efficacy and release values. The study demonstrated that the product exhibited antibacterial effects against *S. aureus* and *E. coli* and highlighted its potential for cancer treatment [80]. Not only for antimicrobial effects but also for their nutritional properties, essential oils are being integrated with nanocarrier systems. Eugenol, obtained from cedarwood, rosemary, and cloves, is a good example of this. It is known that shampoos containing encapsulated eugenol nourish the scalp and are effective in promoting hair growth, strengthening, and repair. Furthermore, it has been reported that when combined with nanomaterials obtained from different emulsifying agents, such as beeswax, and used as a cream ingredient, its effect is enhanced and its application to the skin surface is facilitated [67, 81].

5 Pharmaceutical Importance of Nanoemulsion Technology

Nanoemulsions can increase the stability of cosmetic products and enhance the solubility of drugs. The nanometric dimensions of nanoemulsions contribute to increased bioavailability by providing properties such as large surface area, greater loading capacity, and improved solubility [82, 83]. Therefore, they hold an important place in the pharmaceutical field. The oil/water ratio is important for nanoemulsion effectiveness. Here, a surfactant or emulsifier is needed to provide stabilization and reduce the interfacial tension between the phases [82].

The skin is a structure that envelops the body and protects it from external damage (UV, microorganisms, dryness). The skin is composed of a multilayered structure consisting of the epidermis, dermis, hypodermis, and stratum corneum [20]. The multilayered structure of the skin also affects the absorption processes of the products used. Nanoemulsion structures can achieve significant cosmetic effects, such as minimizing water loss, increasing skin protection, and increasing the absorption of active ingredients in cosmetic products such as skin care products, moisturizers, and sunscreens [84, 85]. After products used in the cosmetics industry are topically applied, the absorption process begins. During the absorption process, the active compound of the applied substance must penetrate the skin, but factors such as the substance's molecular size and formula can affect absorption. Using this active ingredient in conjunction with a nanoemulsion can yield much better results [20].

Nanoemulsion structures are preferred in the cosmetics and pharmaceutical industries due to their various advantages (Fig. 2).

Essential oils are volatile substances derived from aromatic plants that generally have pleasant and aromatic odors and often have therapeutic effects. Essential oils are extracted in different ways depending on the plant from which they are obtained [82]. Aromatic plants, or their essential oils, have been used for many years as alternative therapeutic or preventative options. However, using only essential oils can lead to stability disadvantages. When used alone, essential oils can be affected by environmental factors such as light, oxygen, or temperature. Therefore, options such as nanoemulsions are important for increasing the stability of chemical compounds that may be effective for medical or cosmetic purposes [82].

6 Microbiological Stability and Preservative Systems in Plant-Based Nanoemulsions

Emulsifying essential oils in a biodegradable and biocompatible matrix can increase stability and promote controlled release, thus serving as a preservative. Essential oils are designated as generally recognized as safe (GRAS) by the FDA. Therefore, they are considered nontoxic and nonirritating and are frequently used in the food and cosmetics industries [71]. Decades of research have demonstrated that essential oil nanoemulsions have antimicrobial properties [88–92]. In addition to having antimicrobial effects, the fact that essential oils are transparent in structure suggests that they have the potential to be used in the cosmetics industry [93]. Essential oils derived from aromatic plants generally possess important bioactive properties. However, studies have shown that essential oils formulated with nanoemulsions exhibit better activity compared to essential oils alone. For example, a study by Badr et al. found increased antimicrobial properties in nanoemulsion formulations compared to crude lavender oils [94]. This difference is attributed to the lower stability of essential oils and their exposure to environmental factors. Essential oils formulated with nanoemulsions, on the other hand, offer advantages in the transportation of aromatic oils due to their greater stability [95]. Synthetic preservatives used in cosmetic products can cause allergies or irritation, prompting the search for more natural preservative options. Essential oils are among the most suitable materials for preserving dermocosmetic products. Essential oils from many aromatic plants, such as lavender, thyme, cinnamon, clove, sage, and tea tree, are known to have antimicrobial activity. Studies have shown that essential oils can be effective in

Fig. 2 Nanotechnological advantages in the cosmetics and pharmaceutical industries according to Chellapa et al. [86], Azmi et al. [84], and Preeti et al. [87]

preserving cosmetic products [49, 68, 96]. Nanoemulsion structures are good preservatives because of their antimicrobial properties and ability to reduce sensory effects. The antimicrobial activity of essential oil nanoemulsions against common pathogens such as *Listeria monocytogenes* [97], *Salmonella Typhimurium* [98], *Escherichia coli* [79], *Bacillus subtilis*, and *Staphylococcus aureus* [99] reveals their promise for use as natural preservatives [71].

7 Effects of Aromatic Plant-Based Nanoemulsions on Skin Microbiota

The skin is considered the largest organ in the human body. Like other microbiota in the body, the skin microbiota contains millions of microbial species (including bacteria, yeast, fungi, and viruses) and plays a major role in maintaining skin health and functions by acting as host protectors [100–102]. The ingredients of cosmetic products used for the skin can affect the skin microbiota. Green cosmetics and products derived from marine algae, which have become popular in cosmetic products in recent years, may indirectly affect the microbiota by providing positive effects such as moisturizing the skin, delaying aging, providing UV protection, and exfoliating the skin [103, 104]. For example, bioactive compounds derived from Spirulina species can both moisturize the skin and hair and slow down skin aging. Additionally, those derived from Chlorella species provide antiaging activity and UV protection [105].

Essential oils generally exhibit broad-spectrum antimicrobial activity, meaning they can affect both harmful and beneficial microorganisms. This has the potential to disrupt the delicate balance of the skin microbiota. Essential oils generally exert their antimicrobial effects by disrupting bacterial cell membrane integrity, causing potassium ion leakage, or affecting nucleic acids [106, 107]. Essential oils that target bacteria such as *Staphylococcus epidermidis* and *Cutibacterium acnes* (formerly *Propionibacterium acnes*), which are natural members of the skin microbiota and protectors of the host, can negatively impact the skin microbiota [108, 109].

Studies have shown that essential oils can be more potent when used in combination with synthetic antibiotics or other essential oils than when used alone. These combinations may reduce the risk of microbiota disruption while simultaneously increasing antimicrobial efficacy, and lower concentrations of natural and synthetic antimicrobials may be used. This use may reduce antimicrobial side effects and the development of resistance [110]. The interaction between nanoemulsions and the skin microbiota is of great importance for both skin health and product development strategies. Nanoemulsions can provide benefits or pose potential risks by targeting this microbiota balance.

Nanoemulsions offer an effective method for overcoming issues such as instability and degradation in plant-derived natural products. These systems protect active ingredients against physical and chemical degradation, increasing their bioavailability [111]. Plant-based bioactive compounds loaded into nanoemulsions have been tested for antibacterial, antifungal, and antiviral activities and have demonstrated in vitro antimicrobial activity against pathogens [112, 113].

It has been noted that nanotechnology-based products, when absorbed through the skin, can damage various organs, such as the skin, lungs, and brain, through the bloodstream. Additionally, some inorganic nanostructured materials (e.g., TiO_2 and ZnO) can be toxic to microorganisms by producing ROS [114, 115]. However, the toxicity of nanoparticles is also related to their size. Despite the benefits of nanoemulsions, careful research and regulatory oversight are essential regarding the potential toxicity of nanoparticles and their impact on the balance of the skin microbiota.

8 Quality Control, Standardization, and Pharmacopoeia Compliance

Pharmacopeias are official publications that define the quality requirements for natural and synthetic medicinal products essential for public health. One of the most important pharmacopoeias in this regard is the European Pharmacopoeia (EP). Essential oils are generally analyzed using parameters such as their image, relative density, refractive index, optical rotation, and acid value. For essential oils to be used for medical or commercial purposes, they must meet the criteria specified in the monographs. Furthermore, in addition to the aforementioned analyses, tests such as TLC and GC/MS are also relevant to the monographs. The aim here is to meet the criteria in the monographs and prevent public health hazards, which is the primary mission of the pharmacopoeia [116].

The European Pharmacopoeia defines essential oils as pure, defined plant raw materials obtained using methods such as hydrodistillation, steam distillation, and mechanical processes. The quality of essential oils is determined by organoleptic examination with sensory parameters and determination of physicochemical parameters such as specific gravity, refractive index, and optical rotation [49].

Comparable physicochemical characteristics may be assessed to support the quality control and standardization processes of plant-based nanoemulsions. These properties include particle size, polydispersity index, organoleptic properties, visual inspection, viscosity, acidity, zeta potential, surface morphology, and stability [83]. Properties such as color, clarity, and homogeneity of nanoemulsion droplets are determined through visual inspection. During visual inspection, the presence of adverse conditions such as sedimentation or phase separation is carefully examined to ensure quality standards [83]. Nanoemulsions are optically clear or translucent. Their small droplet size gives them this property and also makes them useful in cosmetic and pharmaceutical applications [82, 117]. Nanoemulsions are also subjected to organoleptic testing for properties such as taste, odor, and texture.

Especially in the food and cosmetics industries, situations such as bad odor or taste are undesirable [83].

Quality control, standardization, and pharmacopoeias that can be applied to essential oil nanoemulsions can be grouped as follows:

1. Sensory analysis: Odor, color, and aspect.
2. Physical analysis: Moisture content, specific gravity, optical rotation, refractive index, residue on evaporation, freezing or congelation point, and solubility in dilute alcohol.
3. Chemical analysis: Aldehyde index, acid index, ester index, ester index after acetylation, carbonyl index, and phenol index.
4. Instrumental techniques.
   (a) Spectrometric techniques: Ultraviolet and visible spectrophotometry, nuclear magnetic resonance, and mass spectrometry.
   (b) Seperation techniques: Thin layer chromatography, gas chromatography, high pressure chromatography.
   (c) Other techniques: Electronic noise, chemometrics [49].

The primary goal of these tests is to protect public health, but consumer satisfaction is also crucial. If product quality is not maintained and the product disturbs consumers in terms of taste, odor, or texture, this can lead to marketing problems [83]. Various patents have been filed today for the formulation of cosmetic products with nanoemulsions. For example, L'Oréal (Paris, France), a leading cosmetics brand, holds several patents based on nanoemulsions [82, 84]. This increases the potential for increased patents and increased use of nanoemulsion structures in the cosmetics sector.

9 Clinical and Therapeutic Effects: Scientific Evidence and Pharmaceutical Perspective

Nanoemulsions of plant-based aromatic compounds such as essential oils, polyphenols, flavonoids, and alkaloids enhance the fundamental benefits of these compounds, enabling their use at lower concentrations and ensuring long-term stability. Furthermore, these compounds, which exhibit increased skin penetration when applied as nanoemulsions, can be used safely and effectively in both dermocosmetic and pharmaceutical products. Aromatic plant-based nanoemulsions are innovative formulations that have gained significant attention in the cosmetics industry in recent years.

In recent years, with the increasing trend toward using natural products in the cosmetics industry, the role of aromatic plant-based nanoemulsions in dermocosmetic applications has significantly grown. Dermocosmetics are products that combine cosmetic and pharmaceutical properties; they contain biologically active ingredients to improve skin health and provide a youthful appearance. Plant extracts and essential oils are frequently preferred in these products. Botanical ingredients such as *Aloe vera*, lavender oil, chamomile extract, and green tea extract can offer antioxidant protection, anti-inflammatory properties, hydration, skin soothing, and antiaging effects. Furthermore, the application of plant-based aromatics with antimicrobial effects in nanoemulsion form means both a reduction in the chemical preservative content in product formulations and the production of new-generation alternatives to cosmetic products used against microbial skin problems [118, 119].

One of the most critical effects of aromatic nanoemulsions is their antioxidant and antiaging effects. Nanoformulations containing flavonoids, such as quercetin and luteolin, increase skin elasticity while reducing free radical damage and delaying the progression of photoaging. Thanks to nanoencapsulation, the slow release of flavonoids allows these effects to be beneficial for many hours after application [58, 67].

Not only flavonoids but plant extracts enriched with antioxidant compounds such as phenolics, tannins, stilbenes, terpenes, and steroids are essential for preventing oxidative stress caused by ROS. *Camellia sinensis* (L.) Kuntze (green tea) (Fig. 3), with its rich polyphenol content, is a good example of this. Additionally, *Phyllanthus niruri* L. flavonoids and terpenoids, *Vellozia squamata* Pohl proanthocyanidins, *Mauritia flexuosa* L.f. extract, and *Cymbopogon nardus* (L.) Rendle (citronella oil) are among the compounds that can be nanoemulsified with antioxidant properties. These plant extracts' nanoemulsions are among the sources that can contain not only antioxidant content but also various skin-nourishing by-products such as collagen, thereby enabling the production of richly formulated skin care products [120–124].

Another essential therapeutic effect is the anti-inflammatory effect. For this reason, aromatic nanoemulsions have pharmaceutical potential in the treatment of atopic dermatitis, irritant contact dermatitis, and minor skin lesions. For example, nanoemulsions containing aloe vera, chamomile (*Matricaria chamomilla* L.), and lavender oils suppress inflammation response mediators such as IL-6 and TNF-α, thereby supporting wound healing and epithelial tissue development [64, 76]. Chamomile oil (*M. chamomilla*) is rich in α-bisabolol and chamazulene, which inhibit the production of prostaglandins and leukotrienes, pro-inflammatory markers. Therefore, it soothes irritated and sensitive skin and is frequently included in formulations used to treat inflammation-related conditions such as dermatitis and eczema [125].

Curcuma longa L. (turmeric) essential oils have been reported to alleviate skin inflammation by modulating pro-inflammatory cytokines TNF-α, IL-6, and IL-1β, controlling the excessive production of oxidative markers, and repairing histopathological damage in a TPA-induced inflammation mouse model [126]. *Cymbopogon citratus* (DC.) Stapf (lemongrass) oil, which suppresses cyclooxygenase-2 while stimulating peroxisome proliferator-activated receptors α and γ, has a potent analgesic and anti-inflammatory effect [127]. Frankincense oil (*Boswellia sacra* Flück) contains boswellic acid, which inhibits the 5-lipoxygenase enzyme. It is also known to help reduce redness, swelling, and irritation in the inflamed area by decreasing leukotriene production [128].

Fig. 3 *Camellia sinensis* cultivation and harvest in northern Türkiye

One of the most common reasons for the use of aromatic compounds, particularly essential oils, in cosmetics is their antimicrobial effects. For example, when polyphenols such as gallic acid are loaded into nanoemulsion and nanoliposome systems, they exhibit both potent antimicrobial and anti-inflammatory effects against pathogens such as *S. aureus* and *C. albicans*. For this reason, such nanomaterials have proven their potential for the treatment of acne and dermatitis [129]. Sandalwood (*Santalum album* L.) oil also has anti-inflammatory and antimicrobial effects that are beneficial in the treatment of psoriasis, acne, and eczema. When Gri et al. [130] examined the potential use of five different nanoemulsions containing sandalwood oil in the treatment of burns, they demonstrated that the synthesized nanoemulsions enhanced skin restoration and regeneration capabilities and did not cause any irritating side effects on the skin.

Peppermint (*Mentha × piperita*), myrtle (*Myrtus communis* L.), oregano (*Origanum vulgare*), cinnamon (*Cinnamomum verum*), and clove (*Syzygium aromaticum* (L.) Merr. & L.M.Perry) can exhibit antiseptic, antibacterial, antifungal, antiviral, antiparasitic, or insecticidal activities. Thanks to these properties, they can be evaluated as cream ingredients for topical supplementation in diseases such as acne, fungal infection, and eczema, as well as forming the content of natural products produced to protect the skin against mosquitoes [52, 54, 63, 131]. Research on nanoemulsions of plant-based aromatics as active compounds in antiaging products has been increasing in recent years. In addition to their many effects, the antiaging properties of phytocosmetics make them preferable to synthetic alternatives in natural cosmetics [22]. For example, topical cream formulations containing geranium/calendula (*Calendula*) essential oil in ethanolic lipid vesicles have demonstrated deep skin penetration, photoprotective, and antioxidant effects [132].

Plant-based nanoemulsions containing bioactive compounds such as terpenes, phenolic compounds, and flavonoids can be used to alleviate or eliminate various signs of aging, such as wrinkles, hyperpigmentation, and loss of elasticity. ROS accumulation is one of the leading causes of skin aging and can be suppressed by antioxidant compounds such as flavonoids, phenolics, tannins, stilbenes, terpenes, and steroids. For example, *Rhodiola rosea* L. nanoemulsion, *Curcuma aromatica* Salisb. (Temulawak) extract nanoemulsion, *Agave sisalana* Perrine nanoemulsion, and *Vellozia squamata* extract nanoemulsion have the potential to be used in the cosmetics industry due to their antioxidant and antiaging effects [21, 133–135].

Additionally, plant-based nanoemulsions can support the skin by stimulating collagen and elastin synthesis in the skin or by inhibiting matrix metalloproteinases that break down collagen and elastin in the skin. The collagenase, elastase, and hyaluronidase inhibition effect of *Origanum vulgare* (thyme) essential oil nanoemulsion, the bioretinol effect of *Cullen corylifolium* (L.) Medik. (Syn.: *Psoralea corylifolia* L.) (Bakuchiol) nanoemulsion, and the anti-collagenase potential of *Premna odorata* Blanco leaf nanoemulsion are examples of this effect [83, 136, 137].

Nanoemulsions containing polyphenols and flavonoids can form a barrier that prevents the absorption of UV radiation in the skin. The *Melaleuca alternifolia* (tea tree) essential oil nanoemulsion is more effective than direct tea tree oil against photoaging, supports improvement in dermal echogenicity and collagen fibers, and, in addition to its antioxidant effect, reaches deeper skin layers with higher penetration through the stratum corneum, improving the morphology of the stratum granulosum [138].

Coriandrum sativum L. (coriander) essential oil nanoformulation also reduced UV-induced skin photoaging in vivo, causing significant decreases in MDA, AP-1, JNK, MMP-1, PEG-2, and COX-2 levels; it also exhibits anti-wrinkle effects by increasing skin collagen content and, through its antioxidant effects, significantly improves TGFβ, TGFβII, and SMAD3 protein expression [139]. Formulations containing rosemary and lavender oil nanoemulsions also neutralize UV-induced free radicals while reducing collagen degradation and increasing elasticity and are therefore considered serious antiaging agents [58, 67].

Nanotechnology-based products are used to enhance the delivery of nourishing and restorative ingredients to the hair and scalp in hair care products. Various nanocarriers such as liposomes, solid lipid nanoparticles (SLNs), transferosomes, etosomes, nanostructured lipid carriers (NLCs), fullerenes, carbon nanotubes, and cyclodextrins are used in hair and skin care products to provide advanced distribution of phytocomponents. In addition to their antioxidant, antiaging, anti-wrinkle, skin elasticity and firmness enhancing, moisturizing, trans-epidermal water loss reducing, dark spot lightening, anti-inflammatory, photoprotective, and wound healing effects, rosemary essential oil-loaded nanoemulsions also have significant effects on scalp health. Nanoparticles loaded with clotrimazole, formulated with natural oils such as rosemary or lavender oil, are also beneficial in treating scalp infections and conditions like dandruff due to their antifungal activity [81].

Similarly, tea tree oil nanoemulsions also contribute to scalp health by improving the skin barrier and morphological skin properties, providing moisturizing effects, reducing sebum levels, and exhibiting broad-spectrum antibacterial activity. Furthermore, tea tree oil in nanocapsule and nanoemulsion formulations has demonstrated antifungal activity in an onychomycosis model [140, 141]. Carrier systems are crucial for increasing the stability and efficacy of eucalyptus essential oil. It has been proven that the main component of the oil is 1,8-cineole and that this component is preserved for 90 days after nanoencapsulation. It is also known that chitosan films impregnated with eucalyptus oil nanoemulsion may be protective against bacterial problems on the scalp due to their antibacterial properties [142]. Nanoemulsification of peppermint oil, known for its antimicrobial properties, also enables stability and long-term use within the product. Studies have demonstrated that these emulsions exhibit significant antimicrobial activity against *S. aureus* and *E. coli* [143].

10 Conclusion and Future Perspectives

Aromatic plants and their oils have been used medicinally for thousands of years, as incense, perfume, and other purposes. Today, beyond medical use, they are widely used in mental health, fashion, beauty, and the cosmetics industry. The cosmetics industry, which has reached millions of dollars in value, has never been as popular as it is today. As in many other areas, the cosmetics industry will strive to move away from fossil-based ingredients and aim to offer its consumers natural, clean, vegan, sustainable, and highly effective ingredients. The rise in this field will continue in the coming years, with the discovery of many new plant extracts and oils. Further studies are needed on the potential use of aromatic plants in cosmetics, the composition of herbal extracts, their mechanisms of action, and optimum amounts. High-quality products will emerge, especially with the development of new isolation and extraction technologies. Aromatic plant-based nanoemulsions offer innovative approaches in the cosmetics industry, thus being of great importance for both skin health and product development strategies. Furthermore, developments in this area have great potential to increase the effectiveness of natural products and meet consumer demands. Aromatic plant-based nanoemulsions hold promise in the cosmetics industry, but some significant challenges must be overcome. Traditional overharvesting of aromatic plants can lead to deforestation and environmental degradation, as well as the extinction of numerous aromatic plant species. Therefore, the cultivation of these wild aromatic plant species, identified by experts, will ensure both consistent quality standards and the protection of plants and the environment through sustainable harvesting. When nanotechnology-based products are absorbed through the skin, they pass into the bloodstream and, if not properly used, can harm various organs. Depending on the inorganic nanostructured materials used, they can produce ROS and be toxic to microorganisms on the skin. Due to justified safety concerns regarding the use of nanomaterials in cosmeceuticals, restrictions and comprehensive safety testing before marketing are necessary. Clinical studies on cosmetic products are insufficient. Therefore, more stringent and streamlined legal guidelines are needed for the production of nanocosmeceuticals. The use of cosmetic products containing essential oils with antimicrobial properties can alter the skin's microbiota. If aromatic plant-based nanoemulsions are to be used in cosmetic products, a careful balance must be maintained to protect the microbiota. Several obstacles remain before transferring the successfully produced nanoemulsions produced in laboratory studies to the industrial scale. Chief among these obstacles is the development and standardization of environmentally friendly, low-energy, economical, and high-efficiency methods for this technology. Only when these challenges are addressed will these products become commercially available. Furthermore, by integrating artificial intelligence models into the nanoemulsion design process, very good results can be achieved in optimizing different combinations, eliminating the need for countless trial-and-error studies. In summary, aromatic plant-based nanoemulsions have the potential to enhance the efficacy and stability of natural ingredients, thus becoming a leading player in the cosmetics industry. However, to realize this potential, safety concerns must be addressed, regulatory deficiencies addressed, and interdisciplinary teams must be established to continue this scientific research. In conclusion, while aromatic plant-derived nanoemulsions currently offer significant advantages over traditional systems, their full potential in the cosmetics field is not yet fully realized. With further interdisciplinary studies on this topic, it is anticipated that nanoemulsion-based plant-derived cosmetics will hold a leading position in the cosmetics market in the coming years as safe, effective, and environmentally friendly alternatives.

Competing Interests The authors declare that they have no competing interest.

Declaration of Generative AI and AI-Assisted Technologies in the Writing Process During the preparation of this work, the book chapter author(s) have not taken any assistance from AI, and the book chapter author(s) reviewed and edited the content as needed and take(s) full responsibility for the content of the publication.

References

1. Giannenas I, Sidiropoulou E, Bonos E, Christaki E, Florou-Paneri P (2020) The history of herbs, medicinal and aromatic plants, and their extracts: past, current situation and future perspectives. In: Feed additives. Academic Press, New York, pp 1–18
2. Chaudhri SK, Jain NK (2009) History of cosmetics. Asian J Pharm 3:3
3. Pérez-Arantegui J (2021) Not only wall paintings—pigments for cosmetics. Archaeol Anthropol Sci 13:189
4. Scott DA (2016) A review of ancient egyptian pigments and cosmetics. Stud Conserv 61:185–202
5. Jones FA (1996) Herbs–useful plants. Their role in history and today. Eur J Gastroenterol Hepatol 8:1227–1231
6. McMullen RL, Acqua G (2023) History of natural ingredients in cosmetics. Cosmetics 10:71
7. Solomou AD, Martinos K, Skoufogianni E, Danalatos NG (2016) Medicinal and aromatic plants diversity in Greece and their future prospects: a review. Agric Sci 4:9–21
8. Sumner J (2000) The natural history of medicinal plants, p 235
9. Jaiswal YS, Williams LL (2017) A glimpse of ayurveda–the forgotten history and principles of indian traditional medicine. J Trad Comp Med 7:50–53
10. Lee KH (1999) Chinese materia medica: chemistry, pharmacology and applications by you-ping zhu (China– Netherlands medical and pharmaceutical Centre, Groningen, The Netherlands), vol 199. Harw Acad Publ, p 8
11. Mahesh SK, Fathima J, Veena VG (2019) Cosmetic potential of natural products: industrial applications. In: Natu. Bio. Acti. Comp. Volu. 2: chemistry, pharmacology and health care practices. Springer, Singapore, pp 215–250
12. Ribeiro AS, Estanqueiro M, Oliveira MB, Sousa Lobo JM (2015) Main benefits and applicability of plant extracts in skin care products. Cosmetics 2:48–65
13. Kanlayavattanakul M, Lourith N (2018) Skin hyperpigmentation treatment using herbs: a review of clinical evidences. J Cosmet Laser Ther 20:123–131
14. Mohammad IS, Naveed M, Ijaz S, Shumzaid M, Hassan S, Muhammad KS, Khan HMS (2018) Phytocosmeceutical formulation development, characterization and its in-vivo investigations. Biomed Pharmacother 107:806–817
15. Basudkar V, Gharat SA, Momin MM, Shringarpure M (2022) A review of anti-aging nanoformulations: recent developments in excipients for nanocosmeceuticals and regulatory guidelines. Critical Reviews™ in Therapeutic Drug Carrier Systems 39(3):45
16. Gupta V, Mohapatra S, Mishra H, Farooq U, Kumar K, Ansari MJ, Iqbal Z (2022) Nanotechnology in cosmetics and cosmeceuticals—a review of latest advancements. Gels 8:173
17. Souto EB, Cano A, Martins-Gomes C, Coutinho TE, Zielińska A, Silva AM (2022) Microemulsions and nanoemulsions in skin drug delivery. Bioengineering 9:158
18. Vaishampayan P, Rane MM (2022) Herbal nanocosmecuticals: a review on cosmeceutical innovation. J Cosmet Dermatol 21:5464–5483
19. Prado JCS, Prado GM, Aguiar FLL, Neves AM, do Nascimento JF, da Silva Abreu FOM, dos Santos Fontenelle RO (2024a) Nanoemulsions of plant-based bioactive compounds with antimicrobial applications: a review. Cienc Nat 46:e74325–e74325
20. Yukuyama MN, Ghisleni DDM, Pinto TDJA, Bou-Chacra NA (2016) Nanoemulsion: process selection and application in cosmetics–a review. Int J Cosmet Sci 38:13–24
21. Quintão FJ, Tavares RS, Vieira-Filho SA, Souza GH, Santos OD (2013) Hydroalcoholic extracts of vellozia squamata: study of its nanoemulsions for pharmaceutical or cosmetic applications. Rev Bras 23:101–107
22. Mohd-Setapar SH, John CP, Mohd-Nasir H, Azim MM, Ahmad A, Alshammari MB (2022) Application of nanotechnology incorporated with natural ingredients in natural cosmetics. Cosmetics 9:110

23. Padule K, Shinde S, Chitlange S, Giram P, Nagore D (2022) The advancement of herbal-based nanomedicine for hair. Cosmetics 9:118
24. Sethi M, Rana R, Sambhakar S, Chourasia MK (2024) Nanocosmeceuticals: trends and recent advancements in self care. AAPS PharmSciTech 25:51
25. Aziz ZAA, Mohd-Nasir H, Ahmad A, Mohd Setapar SH, Peng WL, Chuo SC, Mohamad Ibrahim MN (2019) Role of nanotechnology for design and development of cosmeceutical: application in makeup and skin care. Front Chem 7:739
26. Kaul S, Gulati N, Verma D, Mukherjee S, Nagaich U (2018) Role of nanotechnology in cosmeceuticals: a review of recent advances. J Pharm 2018:3420204
27. Lohani A, Verma A, Joshi H, Yadav N, Karki N (2014) Nanotechnology-based cosmeceuticals. ISRN Dermatol 2014:843687
28. Chauhan M, Mazumder R, Rani A, Mishra R, Pal RS (2024) Preparations, applications, patents, and marketed formulations of nanoemulsions-a comprehensive review. Pharm Nanotechnol
29. Ali MM, Hashim N, Aziz SA, Lasekan O (2020) Exploring the chemical composition, emerging applications, potential uses, and health benefits of durian: a review. Food Control 113:107189
30. Aboshi T, Musya S, Sato H, Ohta T (2020) Changes of volatile flavor compounds of watermelon juice by heat treatment. Biosci Biotechnol Biochem 84:2157–2159
31. Mostafa S, Wang Y, Zeng W, Jin B (2022) Floral scents and fruit aromas: functions, compositions, biosynthesis, and regulation. Front Plant Sci 13:860157
32. Karnwal A, Jassim AY, Mohammed AA, Sharma V, Al-Tawaha ARMS, Sivanesan I (2024) Nanotechnology for healthcare: plant-derived nanoparticles in disease treatment and regenerative medicine. Pharmaceuticals 17(12):1711
33. Ninkuu V, Zhang L, Yan J, Fu Z, Yang T, Zeng H (2021) Biochemistry of terpenes and recent advances in plant protection. Int J Mol Sci 22(11):5710
34. Yu J (2025) Chemical composition of essential oils and their potential applications in post-harvest storage of cereal grains. Molecules 30(3):683
35. Wojtunik-Kulesza KA (2022) Toxicity of selected monoterpenes and essential oils rich in these compounds. Molecules 27(5):1716
36. Qaderi MM, Martel AB, Strugnell CA (2023) Environmental factors regulate plant secondary metabolites. Plants 12(3):447
37. Pratyusha S (2022) Phenolic compounds in the plant development and defense. In: Plant stress physiology: perspectives in agriculture. Intechopen, London, p 125
38. Husen A (2021a) Harsh environment and plant resilience, Springer International Publishing AG, Cham pp 1–23 https://doi.org/10.1007/978-3-030-65912-7_1
39. Husen A (2021b) Plant performance under environmental stress. Springer International Publishing AG, Cham pp. 1–15 https://doi.org/10.1007/978-3-030-78521-5_1
40. Alami MM, Guo S, Mei Z, Yang G, Wang X (2024) Environmental factors on secondary metabolism in medicinal plants: exploring accelerating factors. Med Plant Biol 3(1):1
41. Kumar K, Debnath P, Singh S, Kumar N (2023) An overview of plant phenolics and their involvement in abiotic stress tolerance. Stress 3(3):570–585
42. Radi FZ, Bouhrim M, Mechchate H, Al-Zahrani M, Qurtam AA, Aleissa AM, Zair T (2021) Phytochemical analysis, antimicrobial and antioxidant properties of *Thymus zygis* L. and Thymus willdenowii Boiss essential oils. Plants 11(1):15
43. Thangaleela S, Sivamaruthi BS, Kesika P, Bharathi M, Kunaviktikul W, Klunklin A, Chaiyasut C (2022) Essential oils, phytoncides, aromachology, and aromatherapy—a review. Appl Sci 12(9):4495
44. Catalano A, Mariconda A, D'Amato A, Iacopetta D, Ceramella J, Marra M, Longo P (2024) Aldehydes: what we should know about them. Organ 5(4):395–342
45. Ailli A, Handaq N, Touijer H, Gourich AA, Drioiche A, Zibouh K, Zair T (2023) Phytochemistry and biological activities of essential oils from six aromatic medicinal plants with cosmetic properties. Antibiotics 12(4):721

46. Benomari FZ, Sarazin M, Chaib D, Pichette A, Boumghar H, Boumghar Y, Djabou N (2023) Chemical variability and chemotype concept of essential oils from Algerian wild plants. Molecules 28(11):4439
47. Chaari S, El Arnabi FZ, Benali A, Gaboun F, Lagnaoui N, Abdelmoumen H, Bakhy K (2025) Chemical diversity and chemotype classification of Artemisia herba-alba essential oils from western and southern Morocco. Biochem Syst Ecol 123:105106
48. Dreger M, Wielgus K (2013) Application of essential oils as natural cosmetic preservatives. Herb Pol 59:87–99
49. Guzmán E, Lucia A (2021) Essential oils and their individual components in cosmetic products. Cosmetics 8(4):114
50. Matwiejczuk N, Galicka A, Brzóska MM (2020) Review of the safety of application of cosmetic products containing parabens. J Appl Toxicol 40:176–210
51. Lombrea A, Antal D, Ardelean F, Avram S, Pavel IZ, Vlaia L, Danciu C (2020) A recent insight regarding the phytochemistry and bioactivity of *Origanum vulgare* L. essential oil. Int J Mol Sci 21:9653
52. Selma WB, Ferjeni S, Farouk A, Marzouk M, Boukadida J (2025) Antimicrobial activity of *Cinnamomum zeylanicum* essential oil against colistin-resistant gram-negative bacteria. Int J Environ Health Res 35:169–181
53. Özel Y, Çavuş İ, Ünlü G, Ünlü M, Özbilgin A (2024) Investigation of the antitrichomonal activity of cinnamaldehyde, carvacrol and thymol and synergy with metronidazole. Turkiye Parazitol Derg 48:92–100
54. Hekmatpanah A, Sharifzadeh A, Shokri H, Nikaein D (2022) Efficacy of *Syzygium aromaticum* essential oil on the growth and enzymatic activity of pathogenic *Candida albicans* strains. Curr Med Mycol 8:12–18
55. Bicak B, Gok B, Kecel-Gunduz S, Budama-Kilinc Y, Kocyigit M, Inan HA, Kartal M (2025) Structural analyses (spectroscopic and in silico), DNA-binding and in vitro studies of two popular sage species (*Salvia officinalis* and *Salvia fruticosa*). J Essent Oil Bear Plants 28:207–223
56. Prado GM, Prado JCS, Aguiar FLLD, Barbosa FCB, Vale JPD, Martins MR, Fontenelle RO (2024b) Antifungal, molecular docking and cytotoxic effect of the essential oil of *Cymbopogon citratus* (DC) Stapf. and *Cymbopogon nardus* (L.) Rendle against *Candida albicans*. An Acad Bras Cienc 96:e20230309
57. Khosakueng M, Taweechaisupapong S, Boonyanugomol W, Prapatpong P, Wongkaewkhiaw S, Kanthawong S (2024) *Cymbopogon citratus* L. essential oil as a potential anti-biofilm agent active against antibiotic-resistant bacteria isolated from chronic rhinosinusitis patients. Biofouling 40:26–39
58. Alaboudi KA, Aziz IM, Almosa AA, Farrag MA, Abalkhail T, Alshalan RM, Almuqrin AM (2025) In vitro and in silico pharmacological effects of *Rosmarinus officinalis* leaf methanolic extracts and essential oils. Sci Rep 15:10699
59. Hosseini R, Heidari M (2025) Impact of drought stress on biochemical and molecular responses in lavender (*Lavandula angustifolia* mill.): effects on essential oil composition and antibacterial activity. Front Plant Sci 16:1506660
60. Hou T, Sana SS, Li H, Xing Y, Nanda A, Netala VR, Zhang Z (2022) Essential oils and their antibacterial, antifungal and antioxidant activity applications: a review. Food Biosci 47:101716
61. Meenu M, Padhan B, Patel M, Patel R, Xu B (2023) Antibacterial activity of essential oils from different parts of plants against *salmonella* and *listeria* spp. Food Chem 404:134723
62. Swamy MK, Akhtar MS, Sinniah UR (2016) antimicrobial properties of plant essential oils against human pathogens and their mode of action: an updated review. Evid Based Complement Alternat Med 2016(1):3012462
63. Yap PSX, Yusoff K, Lim SHE, Chong CM, Lai KS (2021) Membrane disruption properties of essential oils—a double-edged sword? Processes 9:595

64. Cai J, Yan R, Shi J, Chen J, Long M, Wu W, Kuca K (2022) Antifungal and mycotoxin detoxification ability of essential oils: a review. Phytother Res 36:62–72
65. de Sousa DP, Damasceno ROS, Amorati R, Elshabrawy HA, de Castro RD, Bezerra DP, Lima TC (2023) Essential oils: chemistry and pharmacological activities. Biomolecules 13:1144
66. Yorgancioglu A, Bayramoglu EE (2013) Production of cosmetic purpose collagen containing antimicrobial emulsion with certain essential oils. Ind Crop Prod 44:378–382
67. Cunha C, Ribeiro HM, Rodrigues M, Araujo AR (2022) Essential oils used in dermocosmetics: review about its biological activities. J Cosmet Dermatol 21:513–529
68. Muyima NYO, Zulu G, Bhengu T, Popplewell D (2002) The potential application of some novel essential oils as natural cosmetic preservatives in an aqueous cream formulation. Flavour Fragr J 17(4):258–266
69. Nardoni S, Mugnaini L, Pistelli L, Leonardi M, Sanna V, Perrucci S, Pisseri F, Mancianti F (2014) Clinical and mycological evaluation of an herbal antifungal formulation in canine *Malassezia* dermatitis. J Mycol Med 24:234–240
70. Orafidiya LO, Oyedele AO, Shittu AO, Elujoba AA (2001) The formulation of an effective topical antibacterial product containing *Ocimum gratissimum* leaf essential oil. Int J Pharm 224:177–183
71. Maurya A, Singh VK, Das S, Prasad J, Kedia A, Upadhyay N, Dwivedy AK (2021) Essential oil nanoemulsion as eco-friendly and safe preservative: bioefficacy against microbial food deterioration and toxin secretion, mode of action, and future opportunities. Front Microbiol 12:751062
72. Bhavaniramya S, Vishnupriya S, Al-Aboody MS, Vijayakumar R, Baskaran D (2019) Role of essential oils in food safety: antimicrobial and antioxidant applications. Grain Oil Sci Technol 2:49–55
73. Das S, Singh VK, Dwivedy AK, Chaudhari AK, Dubey NK (2021) Eugenol loaded chitosan nanoemulsion for food protection and inhibition of aflatoxin B1 synthesizing genes based on molecular docking. Carbohydr Polym 255:117339
74. Lugani Y, Sooch BS, Singh P, Kumar S (2021) Nanobiotechnology applications in food sector and future innovations. In: Ray RC (ed) Microbial biotechnology in food and health. Academic Press, Cambridge, pp 197–225
75. Soni M, Yadav A, Maurya A, Das S, Dubey NK, Dwivedy AK (2023) Advances in designing essential oil nanoformulations: an integrative approach to mathematical modeling with potential application in food preservation. Foods 12:4017
76. Hasanin MS, Ayoob FA, Hashem AH, Emam M (2025) Synthesis of chitosan-based nanoemulsions and their characterization and antifungal activity toward fungi causing mucormycosis. Sci Rep 15:1–17
77. Wattanasatcha A, Rengpipat S, Wanichwecharungruang S (2012) Thymol nanospheres as an effective anti-bacterial agent. Int J Pharm 434:360–365
78. Park SJ, Shin YS, Lee JR (2001) Preparation and characterization of microcapsules containing lemon oil. J Colloid Interface Sci 241:502–508
79. Moraes-Lovison M, Marostegan LF, Peres MS, Menezes IF, Ghiraldi M, Rodrigues RA, Pinho SC (2017) Nanoemulsions encapsulating oregano essential oil: production, stability, antibacterial activity and incorporation in chicken pâté. Lwt 77:233–240
80. Antonysamy L, Devasahayam LRM, Muthukrishnan S, Anbazhagan S, Pushpam MA (2024) Optimization and characterisation of water-based keratin nanoparticles with *Nigella sativa* seed extract. Bioinformation 20:957
81. Uronnachi E, Atuegwu C, Umeyor C, Nwakile C, Obasi J, Ikeotuonye C, Attama A (2022) Formulation and evaluation of hair growth enhancing effects of oleogels made from rosemary and cedar wood oils. Sci Afr 16:e01223
82. Barradas TN, de Holanda e Silva KG (2021) Nanoemulsions of essential oils to improve solubility, stability and permeability: a review. Environ Chem Lett 19(2):1153–1171
83. Iskandar B, Liu TW, Mei HC, Kuo IC, Surboyo MDC, Lin HM, Lee CK (2024) Herbal nanoemulsions in cosmetic science: a comprehensive review of design, preparation, formulation, and characterization. J Food Drug Anal 32(4):428

84. Azmi NAN, Elgharbawy AA, Motlagh SR, Samsudin N, Salleh HM (2019) Nanoemulsions: factory for food, pharmaceutical and cosmetics. Processes 7(9):617
85. Mansour HM, Rhee YS, Wu X (2009) Nanomedicine in pulmonary delivery. Int J Nanomedicine 4:299–319
86. Chellapa P, Ariffin FD, Eid AM, Almahgoubi AA, Mohamed AT, Issa YS, Elmarzugi NA (2016) Nanoemulsion for cosmetic application. Europ J Biomed Pharm Sci 3(7):8–11
87. Preeti Sambhakar S, Malik R, Bhatia S, Al Harrasi A, Rani C, Sehrawat R (2023) Nanoemulsion: an emerging novel technology for improving the bioavailability of drugs. Scientifica 2023(1):6640103
88. Luo Y, Zhang Y, Pan K, Critzer F, Davidson PM, Zhong Q (2014) Self-emulsification of alkaline-dissolved clove bud oil by whey protein, gum arabic, lecithin, and their combinations. J Agric Food Chem 62(19):4417–4424
89. Mostaghimi M, Majdinasab M, Hosseini SMH (2022) Characterization of alginate hydrogel beads loaded with thyme and clove essential oils nanoemulsions. J Polym Environ 30(4):1647–1661
90. Salvia-Trujillo L, Rojas-Graü A, Soliva-Fortuny R, Martín-Belloso O (2015) Physicochemical characterization and antimicrobial activity of food-grade emulsions and nanoemulsions incorporating essential oils. Food Hydrocoll 43:547–556
91. Shehabeldine AM, Doghish AS, El-Dakroury WA, Hassanin MM, Al-Askar AA, AbdElgawad H, Hashem AH (2023) Antimicrobial, antibiofilm, and anticancer activities of syzygium aromaticum essential oil nanoemulsion. Molecules 28(15):5812
92. Sun R, Li Y, Su R, Cai X, Kong Y, Jiang T, Shi C (2025) Antibacterial effect of ultrasound combined with Litsea cubeba essential oil nanoemulsion on salmonella typhimurium in kiwifruit juice. Int J Food Microbiol 426:110898
93. Donsì F, Ferrari G (2016) Essential oil nanoemulsions as antimicrobial agents in food. J Biotechnol 233:106–120
94. Badr MM, Badawy ME, Taktak NE (2021) Characterization, antimicrobial activity, and antioxidant activity of the nanoemulsions of Lavandula spica essential oil and its main monoterpenes. J Drug Deliv Sci Technol 65:102732
95. Alam A, Ansari MJ, Alqarni MH, Salkini MA, Raish M (2023) Antioxidant, antibacterial, and anticancer activity of ultrasonic nanoemulsion of Cinnamomum cassia L. essential oil. Plants 12(4):834
96. Mangena T, Muyima NYO (1999) Comparative evaluation of the antimicrobial activities of essential oils of Artemisia afra, Pteronia incana and Rosmarinus officinalis on selected bacteria and yeast strains. Lett Appl Microbiol 28(4):291–296
97. Paudel SK, Bhargava K, Kotturi H (2019) Antimicrobial activity of cinnamon oil nanoemulsion against Listeria monocytogenes and Salmonella spp. on melons. Lwt 111:682–687
98. Hadian M, Rajaei A, Mohsenifar A, Tabatabaei M (2017) Encapsulation of Rosmarinus officinalis essential oils in chitosan-benzoic acid nanogel with enhanced antibacterial activity in beef cutlet against salmonella typhimurium during refrigerated storage. LWT 84:394–401
99. Zhang S, Zhang M, Fang Z, Liu Y (2017) Preparation and characterization of blended cloves/cinnamon essential oil nanoemulsions. LWT 75:316–322
100. Fournière M, Latire T, Souak D, Feuilloley MG, Bedoux G (2020) Staphylococcus epidermidis and Cutibacterium acnes: two major sentinels of skin microbiota and the influence of cosmetics. Microorganisms 8(11):1752
101. Imam MW, Luqman S (2024) Unveiling the mechanism of essential oil action against skin pathogens: from ancient wisdom to modern science. Arch Microbiol 206:347
102. Melnyk N, Vlasova I, Skowrońska W, Bazylko A, Piwowarski JP, Granica S (2022) Current knowledge on interactions of plant materials traditionally used in skin diseases in Poland and Ukraine with human skin microbiota. Int J Mol Sci 23:9644
103. Abrar S, Anjum A, Younas A, Abbas S, Kiran S, Iqbal S, Naveed HQ (2025) Sustainable cosmeceuticals based on natural colorants from plants. In: Sustainable cosmeceuticals. Springer Nature, Cham, pp 79–94

104. Mehmood T, Nawaz S, Nadeem F, Ahmed S (2025) Environment-friendly cosmetics from microbial sources. In: Sustainable cosmeceuticals: integrating natural and biotechnological innovations. Springer, Cham, pp 117–138
105. Zovia I, Ijaz A, Munir A, Mumtaz A, Arshad M, Ahmad A, Javed S (2025) Cosmeceuticals from marine sources: algal products as green cosmetics. In: Sustainable cosmeceuticals: integrating natural and biotechnological innovations. Springer, Cham, pp 139–176
106. Abdelhamed FM, Abdeltawab NF, ElRakaiby MT, Shamma RN, Moneib NA (2022) Antibacterial and anti-inflammatory activities of thymus vulgaris essential oil nanoemulsion on acne vulgaris. Microorganisms 10:1874
107. Carson CF, Mee BJ, Riley TV (2002) Mechanism of action of melaleuca alternifolia (tea tree) oil on staphylococcus aureus determined by time-kill, lysis, leakage, and salt tolerance assays and electron microscopy. Antimicrob Agents Chemother 46:1914–1920
108. Ganesan NG, Miastkowska MA, Pulit-Prociak J, Dey P, Rangarajan V (2023) Formulation of a stable biocosmetic nanoemulsion using a bacillus lipopeptide as the green-emulsifier for skin-care applications. J Dispers Sci Technol 44:2045–2057
109. Qadir A, Ullah SNMN, Gupta DK, Khan N (2022) Phytoconstituents-loaded nanomedicines for the management of acne. J Cosmet Dermatol 21:3240–3255
110. Orchard A, van Vuuren S (2017) Commercial essential oils as potential antimicrobials to treat skin diseases. J Evid Based Complement Altern Med 2017:4517971
111. Chaudhri N, Soni GC, Prajapati SK (2015) Nanotechnology: an advance tool for nano-cosmetics preparation. Int J Phar Res Rev 4:28–40
112. Goyal N, Jerold F (2023) Biocosmetics: technological advances and future outlook. Environ Sci Pollut Res Int 30:25148–25169
113. Manke A, Wang L, Rojanasakul Y (2013) Mechanisms of nanoparticle-induced oxidative stress and toxicity. Biomed Res Int 2013:942916
114. Leung YH, Xu X, Ma AP, Liu F, Ng AM, Shen Z, Gethings LA, Guo MY, Djurišić AB, Lee PK, Lee HK (2016) Toxicity of ZnO and TiO_2 to Escherichia coli cells. Scientific Reports 6(1):35243
115. Bano F (2025) Green-synthesized titanium dioxide nanoparticles inhibit and eradicate the biofilms of pathogenic bacteria through intracellular ROS production. Microbiology Research 16(2):48
116. Barak TH, Bölükbaş E, Bardakci H (2023) Evaluation of marketed rosemary essential oils (Rosmarinus officinalis L.) in terms of European pharmacopoeia 10.0 criteria. Turk Journal Pharm Sci 20(4):253
117. Dasgupta N, Ranjan S, Gandhi M (2019) Nanoemulsion ingredients and components. Environ Chem Lett 17(2):917–928
118. Hassanabadi N, Mahdavi Meymand Z, Ashrafzadeh A, Sharififar F (2024) Antioxidant and cytotoxicity activity of a nanoemulsion from Satureja kermanica (Lamiaceae). Ann Pharm Fr 82:645
119. Wang D, Bao A, Yuan Y, Wang Y, Li L, Song G (2024) Whey protein isolate-stabilized gardenia fruit oil nanoemulsions: ultrasonic preparation, characterization and applications in nutraceuticals delivery. Ind Crops Prod 212:118345
120. Kowalczyk T, Merecz-Sadowska A, Ghorbanpour M, Szemraj J, Piekarski J, Bijak M, Sitarek P (2023) Enhanced natural strength: Lamiaceae essential oils and nanotechnology in in vitro and in vivo medical research. Int J Mol Sci 24(20):15279
121. Olivero-Verbel J, Quintero-Rincón P, Caballero-Gallardo K (2024) Aromatic plants as cosmeceuticals: benefits and applications for skin health. Planta 260(6):132
122. Quintero-Rincón P, Pino-Benítez N, Galeano E, Rojo-Uribe C, Mesa-Arango AC, Flórez-Acosta OA (2023) Sloanea chocoana and S. Pittieriana (Elaeocarpaceae): chemical and biological studies of ethanolic extracts and skincare properties. Plants 12(23):3953

123. Sharma S, Chakkaravarthi S, Bhattacharya B (2023) Enhancement of oxidative stability of soybean oil via nano-emulsification of eggplant peel extract: process development and application. Food Chem 402:134249
124. Uchida VH, de Araújo Padilha CE, Rios NS, dos Santos ES (2023) Enzymatic inhibition of α-amylase and encapsulation of bioactive compounds by nanoemulsion from pulp extract Terminalia catappa Linn fruit. Results Chem 5:100736
125. Sah A, Naseef PP, Kuruniyan MS, Jain GK, Zakir F, Aggarwal G (2022) A comprehensive study of therapeutic applications of chamomile. Pharmaceuticals 15(10):1284
126. Kumar A, Agarwal K, Singh M, Saxena A, Yadav P, Maurya AK, Bawankule DU (2018) Essential oil from waste leaves of Curcuma longa L. alleviates skin inflammation. Inflammopharmacology 26(5):1245–1255
127. Pongkitwitoon B, Putalun W, Triwitayakorn K, Kitisripanya T, Kanchanapoom T, Boonsnongcheep P (2024) Anti-inflammatory activity of verbascoside-and isoverbascoside-rich Lamiales medicinal plants. Heliyon 10(1):e23644
128. Nischang V, Witt FM, Börner F, Gomez M, Jordan PM, Werz O (2024) Frankincense preparation promotes formation of inflammation-resolving lipid mediators by manipulating lipoxygenases in human innate immune cells. Front Pharmacol 14:1332628
129. Elkhalifa ME, Ashraf M, Ahmed A, Usman A, Hamdoon AA, Elawad MA, Ayaz M (2024) Polyphenols and their nanoformulations as potential antibiofilm agents against multidrug-resistant pathogens. Future Microbiol 19(3):255–279
130. Giri VP, Pandey S, Shukla P, Gupta SC, Srivastava M, Rao CV, Mishra A (2024) Facile fabrication of sandalwood oil-based nanoemulsion to intensify the fatty acid composition in burned and rough skin. ACS Omega 9(6):6305–6315
131. Cecchini ME, Paoloni C, Campra N, Picco N, Grosso MC, Perez MS, Bellingeri R (2021) Nanoemulsion of Minthostachys verticillata essential oil: in vitro evaluation of its antibacterial activity. Heliyon 7(1):e05896
132. Javed S, Mangla B, Salawi A, Sultan MH, Almoshari Y, Ahsan W (2024) Essential oils as dermocosmetic agents, their mechanism of action and nanolipidic formulations for maximized skincare. Cosmetics 11(6):210
133. Barreto SMAG, Maia MS, Benicá AM, de Assis HRBS, Leite-Silva VR, da Rocha-Filho PA, Ferrari M (2017) Evaluation of in vitro and in vivo safety of the by-product of Agave sisalana as a new cosmetic raw material: development and clinical evaluation of a nanoemulsion to improve skin moisturizing. Ind Crop Prod 108:470–479
134. Chen YS, Liou HC, Chan CF (2013) Tyrosinase inhibitory effect and antioxidative activities of fermented and ethanol extracts of Rhodiola rosea and Lonicera japonica. Sci World J 2013:612739
135. Wuttikul K, Sainakham M (2022) In vitro bioactivities and preparation of nanoemulsion from coconut oil loaded Curcuma aromatica extracts for cosmeceutical delivery systems. Saudi J Biol Sci 29(12):103435
136. Altyar AE, Ashour ML, Youssef FS (2020) Premna odorata: seasonal metabolic variation in the essential oil composition of its leaf and verification of its anti-ageing potential via in vitro assays and molecular modelling. Biomolecules 10:879
137. Laothaweerungsawat N, Sirithunyalug J, Chaiyana W (2020) Chemical compositions and anti-skin-ageing activities of Origanum vulgare L. essential oil trop Mediterr region. Molecules 25:1101
138. Infante VH, Campos PMM, Darvin M, Lohan S, Schleusener J, Schanzer S, Meinke M (2023) Cosmetic formulations with Melaleuca alternifolia essential oil for the improvement of photoaged skin: a double-blind, randomized, placebo-controlled clinical study. Photochem Photobiol 99(1):176–183
139. Salem MA, Manaa EG, Osama N, Aborehab NM, Ragab MF, Haggag YA, Hamdan DI (2022) Coriander (Coriandrum sativum L.) essential oil and oil-loaded nano-formulations as an anti-aging potentiality via TGFβ/SMAD pathway. Sci Rep 12(1):6578

140. Khorsand F, Hashemi SJ, Mahmoudi S, Kamali Sarvestani H, Amani A (2025) Nanoemulsion-based tea tree oil: a new frontier in enhancing clotrimazole antifungal activity. Mol Biotechnol:1–8
141. Su R, Fan W, Yu Q, Dong X, Qi J, Zhu Q, Lu Y (2017) Size-dependent penetration of nanoemulsions into epidermis and hair follicles: implications for transdermal delivery and immunization. Oncotarget 8(24):38214
142. de Godoi SN, Quatrin PM, Sagrillo MR, Nascimento K, Wagner R, Klein B, Ourique AF (2017) Evaluation of stability and in vitro security of nanoemulsions containing Eucalyptus globulus oil. Biomed Res Int 2017:2723418
143. Drais HK (2024) Development and evaluation of essential oils nanoemulgel as human skin sanitizer using novel method. Turk J Pharm Sci 21(5):456

Aromatic Plant-Based Nanoemulsions and Their Uses in Pharmaceuticals

Ali Umar, Shifa Shaffique, Muhammad Saleem Khan, Anis Ali Shah, and Muhammad Aaqil Khan

Abstract Aromatic plant-based nanoemulsions have emerged as promising nanocarrier systems for enhancing the therapeutic efficacy of essential oils and phytoconstituents. Their nanoscale size, high surface area, and amphiphilic composition overcome key limitations of conventional formulations, including poor solubility, instability, volatility, and low bioavailability. By enabling improved dispersibility, controlled release, and protection of labile compounds, nanoemulsions significantly augment pharmacological activities such as antimicrobial, antioxidant, anti-inflammatory, and anticancer effects. In addition, their adaptability for multiple routes of administration oral, dermal, nasal, and ocular broadens clinical utility, while alignment with green chemistry principles supports sustainable pharmaceutical development. Nevertheless, formulation stability, large-scale production, regulatory approval, and clinical validation remain critical challenges. Advances in biocompatible surfactants, scalable low-energy processes, and stimuli-responsive delivery systems are expected to overcome these limitations. Overall, aromatic plant-based nanoemulsions represent a transformative strategy that bridges traditional herbal medicine with modern nanotechnology for safe, effective, and sustainable therapeutic applications.

Ali Umar and Muhammad Saleem Khan contributed equally with all other contributors.

A. Umar · M. S. Khan
Department of Zoology, Faculty of Life Sciences, University of Okara, Okara, Pakistan

S. Shaffique (✉)
Department of Applied Biosciences, Kyungpook National University, Daegu, Republic of Korea
e-mail: shifa.2021@knu.ac.kr

A. A. Shah
Department of Botany, University of Education, Lahore, Pakistan

M. A. Khan
International Joint Research Laboratory for Global Change Ecology, School of Life Sciences, Henan University, Kaifeng, Henan, China

A. Husen (ed.), *Sustainable Uses of Aromatic Plants in Nanotechnology*, Smart Nanomaterials Technology, https://doi.org/10.1007/978-981-95-7409-4_15

Keywords Aromatic plants · Nanoemulsions · Essential oils · Drug delivery · Bioavailability · Sustainable pharmaceuticals

1 Introduction

The aromatic plants are rich in essential oils and secondary metabolites (terpenoids, flavonoids, and phenolics) and have long been known in terms of their therapeutic properties used in traditional and modern medicine. The bioactive compounds exhibit a wide variety of pharmacological properties, such as antimicrobial, antioxidant, anti-inflammatory, and anticancer properties, which opens up their usefulness in pharmacological applications [1–7]. They, however, have limited clinical delivery due to a lack of aqueous solubility, environmental instability, quick metabolism, and poor bioavailability [8]. Nanotechnology presents a way to resolve these problems: it can change plant-derived bioactives to improve weaknesses in delivery, stability, and efficacy. Nanoemulsions constitute one of the most exciting nanocarrier systems that can be used to encapsulate hydrophobic molecules, to enhance solubility, prolong circulation time, and/or release them in a controlled manner [9, 10]. Submicron-sized emulsions, i.e., with droplet diameters between 20 and 200 nm, have good thermodynamic and physicochemical properties and, therefore, distinct advantages over conventional emulsions [11] (Fig. 1).

There has been a surge in the formulation of nanoemulsions in recent years by using aromatic plant extracts and essential oils. These systems not only stabilize volatile and unstable compounds but also increase their bioavailability and pharmacological activity and offer a new clinical intervention possibility [12, 13]. The nanoemulsion systems of aromatic plant-based bioactives are, thus, a good move toward incorporating traditional medicine and novel drug delivery systems.

This chapter entails the design, description, and applications of plant-based nanoemulsions of aromatics and the concerns that are involved in this. It highlights the recent developments in the area, evaluates its implications in the area of pharmaceutical sciences, and makes several usage prospects in the field in the future.

2 Aromatic Plants as a Source of Bioactive Compounds

Another feature of the aromatic plants is also mainly the presence of volatile oils that contain aromatic terpenoids, phenylpropanoids, and other secondary metabolites: they are the ones that render them their special odor, as well as the excessive medicinal action. Others such as *Mentha* (mint), *Ocimum* (basil), *Lavandula* (lavender), *Rosmarinus* (rosemary), and *Cymbopogon* (lemongrass) could have been offered as ethnomedicine; antimicrobial agents, digestive, pain-killing agents, and sedatives. Then, the mentioned activities have been confirmed through pharmacological methods and aligned with essential oils and polyphenolic compounds [6, 7].

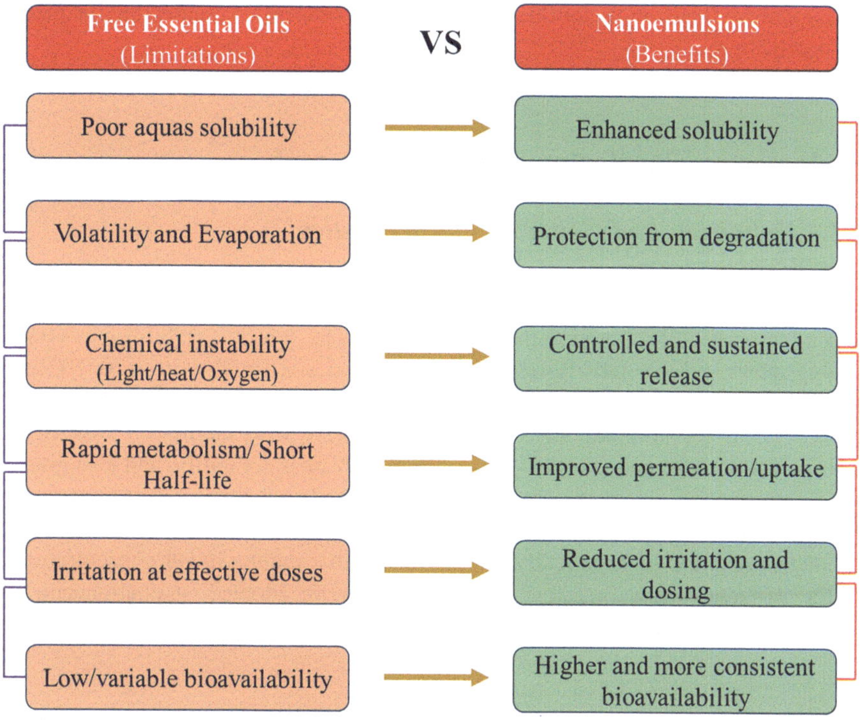

Fig. 1 Conceptual schematic illustrating the limitations of free essential oils (low solubility, instability, volatility, poor bioavailability) and how nanoemulsions overcome these drawbacks through improved dispersibility, stability, controlled release, and enhanced bioavailability

The odors of the herbal plants are generally complicated blends of active constituents, e.g., eugenol, linalool, thymol, carvacrol, citral, and menthol. These compounds also show prominent antimicrobial, antioxidant, anticancer, and anti-inflammatory properties, but their direct use in the pharmaceutical industry is highly restricted by physicochemical instability, photosensitiveness, volatility, and poor solubility in aqueous solutions. [13, 14]. As an example, citral of *Cymbopogon citratus* is very suitable in terms of its high antimicrobial effect, but it is readily destroyed by light and heat, whereas menthol of *Mentha piperita* has cooling and analgesic properties but has poor water solubility [15].

One of the effective ways of overcoming such challenges is the encapsulation of these bioactives in nanoemulsions. Nanoemulsions protect volatile compounds against the effects of degradation and help them distribute in aqueous conditions to enhance their stability and change their pharmacokinetic characteristics. This will not just conserve the natural therapeutic activity of the essential oils but will also enhance their application in the existing drug delivery formulations [12, 16].

Aromatic plants are therefore an unmatched source of bioactive compounds with high potential for drug enhancement. Their functionalization and integration into nanocarrier systems and nanoemulsions, in particular, allow introducing the traditional medicine and scientifically developed therapeutic systems into the medical practice.

3 Nanoemulsion Technology: Fundamentals

3.1 Definition and General Characteristics

Nanoemulsions are progressive colloidal suspensions where one liquid is dispersed in another unreceptive liquid in the nanoform of drops, which is stabilized by emulsifying agents and co-emulsifying agents. The drop size is usually within 20–200 nm, and this gives them unique characteristics as compared to other conventional emulsions [11]. As the dropping sizes are smaller than the wavelength of visible light, their nanoscale size increases the surface area, enhances solubility, and also grants optical transparency. In comparison to the microemulsions that are thermodynamically stable but require the use of relatively high concentrations of the surfactants, the nanoemulsions are kinetically stable systems that require lower concentrations of the surfactants and therefore are safer and more adopted in pharmaceutical uses [10].

3.2 Types of Nanoemulsions

There are basically three broad categories of nanoemulsions in terms of their structural character. The most frequent pharmaceutical formulation made using nanoemulsions is in oil-in-water (O/W) emulsions that allow the dispersion of hydrophobic aromatic essential oils (such as thymol, menthol, and citral) into water systems [7]. W/O nanoemulsions are less used in the context of oral drug delivery but may be useful in dermal formulations involving the delivery of hydrophilic bioactives that are suspended in an oil phase. Bicontinuous nanoemulsions present a hybrid system with the coexistence of oil and water phases, and they can effectively solubilize hydrophilic and lipophilic bioactives at the same time [17]. The nanoemulsions used depend on the bioactive substance's physicochemical characteristics and targeted route of delivery.

3.3 Preparation Methods

Because of the critical need, the formation of nanoemulsions necessitates perfect control of the droplet size and stability. The high-energy techniques, including ultrasonication, high-pressure homogenization, and microfluidization, depend on the

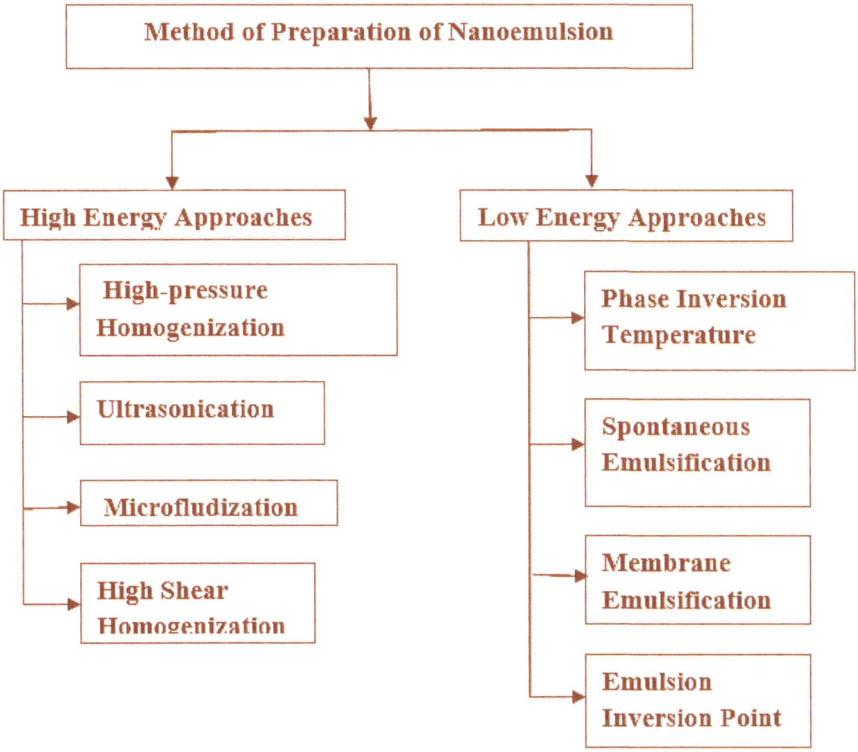

Fig. 2 Preparation methods of nanoemulsions, classified into high-energy (mechanical disruption) and low-energy (physicochemical self-assembly) approaches

mechanical forces to mechanically disintegrate the droplets to nanoscale sizes. These procedures are routinely used in pharmaceutical research because they are reproducible and can produce very stable dispersions, but special devices are needed, as well as the expenditure of energy [10]. Low-energy formulations, including spontaneous emulsification, phase inversion temperature (PIT), and self-emulsifying drug delivery systems (SEDDS), are based on the self-packing ability of surfactants and oils to spontaneously form nanodroplets without extensive mechanical treatment. The methods are environmentally friendly, scalable, and cheap to scale but require proper optimization of the surfactant systems [17] (Fig. 2).

3.4 *Function of Formulation Ingredients*

The functionality of a nanoemulsion is significantly determined by what it is composed of. Surfactants lower the interfacial tension as well as stabilize the droplets, with nonionic types of them, namely, Tween 20, Tween 80, and natural emulsifiers (including lecithin), being preferred due to their biocompatibility and low toxicity

[16]. Cosurfactants such as alcohols and glycols increase flexibility at the oil/water interface and also make the droplets smaller. Carrier oils are usually medium-chain triglycerides or vegetable oils and serve the dual purpose of solubilizer as well as having potential therapeutic value as derived substances of aromatic plants, in this case, *Mentha* or *Cymbopogon* [12]. In this case, the essential oil lines the operational and treatment aspects in unity as the bioactive agent and oil phase.

3.5 Physicochemical Characteristics

Several physicochemical factors control the performance and stability of nanoemulsions. The size of the droplet and polydispersity index (PDI) should be uniform; a good system will provide droplets less than 200 nm as well as a PDI less than 0.3. Surface charge (zeta potential) is also essential to stability; formulations with absolute values higher than plus or minus 30 mV will not aggregate much or at all, and steric stabilization becomes possible using polymeric surfactants [14]. Transparency of formulation will tend to reflect particle size, with the smaller particles making clear dispersions. They affect the viscosity and rheology and release profile, and the spreadability, particularly in topical formulations [15].

3.6 Mechanisms Behind the Bioavailability Increase

Nanoemulsions improve the bioavailability of bioactive flavonoids in plants in several ways. Their nanometer size enhances their surface area and dissolution, which is essential to poorly water-soluble compounds. Encapsulation can also give protection to the labile and volatile molecules during oxidative, thermal, and photodegradation processes. The small drops have good interaction with biomembranes and facilitate epithelial absorption. In most instances, the absorption will be through lymphatic flow, not passing through the liver first-pass, hence enhancing circulatory bioavailability [8, 13]. Membrane permeability may also be altered by the surfactants in the formulation to trigger cellular uptake, which adds to therapeutic efficacy.

3.7 Pharmaceutical Relevance

Nanoemulsions have pharmaceutical significance that was experienced in several studies. The process of encapsulating oils enhances their stability and augments their pharmacologic efficacy, as well as reduces the toxicity that is likely to occur in large quantities of unadulterated oils. To illustrate, both thymol and carvacrol are potent antimicrobials applicable as such, but in the form of nanoemulsions, the irritation generated by the free forms of these agents can be considerably lessened without affecting the therapeutic activity [7]. Also, peppermint and lemongrass oils have better antimicrobial and antioxidant activity when encapsulated in nanoemulsions than when used in their free form [12]. Their versatility allows nanoemulsions to be currently investigated in oral, dermal, ocular, nasal, and parenteral drug delivery, and therefore, they are one of the most diverse delivery systems in the modern drug research quest.

4 Formulation of Aromatic Plant-Based Nanoemulsions

4.1 Selection of Aromatic Plant Extracts and Essential Oils

The preparation of aromatic plant-based nanoemulsions starts by ensuring the choice of the bioactive oil or extract is carefully made. Essential oils obtained from *Mentha piperita* (peppermint), *Cymbopogon citratus* (lemongrass), *Ocimum basilicum* (basil), *Rosmarinus officinalis* (rosemary), or lavender (*Lavandula angustifolia*) are used in large quantities because they contain terpenoids, phenolics, and flavonoids [6, 7]. These compounds have strong antimicrobial, antioxidant, anti-inflammatory, and anticancer activity. They are, however, hampered in their direct use by hydrophobicity, volatility, and lack of resistance to oxidative degradation. When they are introduced into the nanoemulsion structures, they could be effective in providing better solubility, stability, and bioavailability [12].

4.2 Oil Phase Selection

Formulation of nanoemulsion is founded on oil. It is useful as a solubilizing agent and as a pharmacological agent when the essential oils are to be administered directly. One of them is that lemongrass essential oil contains a lot of citral and has strong activity in the form of an antimicrobial, whereas peppermint essential oil is rich in menthol and known because of its analgesic and cooling effect [15]. Other edible oils use medium-chain triglycerides (MCTs) or other oils to enhance the stability and release kinetics of the plant essential oils [14]. This blending might not eliminate the volatility of the ingredients, but not the therapeutic action of the volatile ingredients.

4.3 Choice of Surfactants-Cosurfactants

Surfactants play an important role in minimizing the interfacial tension and stabilizing nanodroplets. Pharmaceutical formulations are usually done using biconononic surfactants, which are the least toxic and include Tween 20, Tween 80, and lecithin

[16]. Included cosurfactants such as ethanol, propylene glycol, or polyethylene glycol enhance interfacial flexibility and allow smaller droplet sizes and greater stability [12]. Pseudo-ternary phase diagrams are frequently used to define the surfactant and cosurfactant system to be used, in terms of the surfactant/cosurfactant ratio and cosurfactant-to-surfactant molar ratio, to ensure a region of nanoemulsion formation. The hydrophilic-lipophilic balance (HLB) is also a critical parameter: oil-in-water nanoemulsions are usually made using surfactants that have an HLB value of 8–18 [8].

4.4 Optimization of Formulation Parameter

The concentration of surfactants, the ratio of oil to water, and processing parameters can be varied to obtain the desired properties in the systematic variation of nanoemulsion formulations. These properties are droplet size, polydispersity index (PDI), zeta potential, and encapsulation efficiency, which are established through measuring parameters. Droplet sizes smaller than 100 nm and a low distribution (PDI < 0.3) are preferable in the case of plant-based formulations of aromatic formulas, which is necessary to reach high penetration and stability [13]. The nanoresultants will be electrostatically stable at a potential of 30 mV and above, and though the values immersed, about 20 mV can still provide sufficient stability in the event of steric hindrance caused by surfactants [14].

4.5 Stability Considerations

Stability is one of the most crucial parameters of the applicability of pharmaceuticals. Essential oils are readily oxidized and evaporated and thus may lose their therapeutic value. Light, oxygen, and heating are encased by the nanoemulsions to guarantee that the encapsulation remains stable (B Faster stability studies at different storage conditions, e.g., temperature changes, pH changes, and ionic strength changes, are the studies most often performed to make predictions regarding the shelf life). Together with this, some antioxidants such as vitamin E or ascorbic acid may also be included in the formula in an attempt to further stabilize its loss.

4.6 Formulations in the Literature

A variety of works showed the successful development of aromatic plant-based nanoemulsions. Nanoemulsions of peppermint oil were prepared by high-pressure homogenization with droplet sizes of less than 80 nm and high antimicrobial efficacy against foodborne pathogens [12]. Nanoemulsions of lemongrass oil made

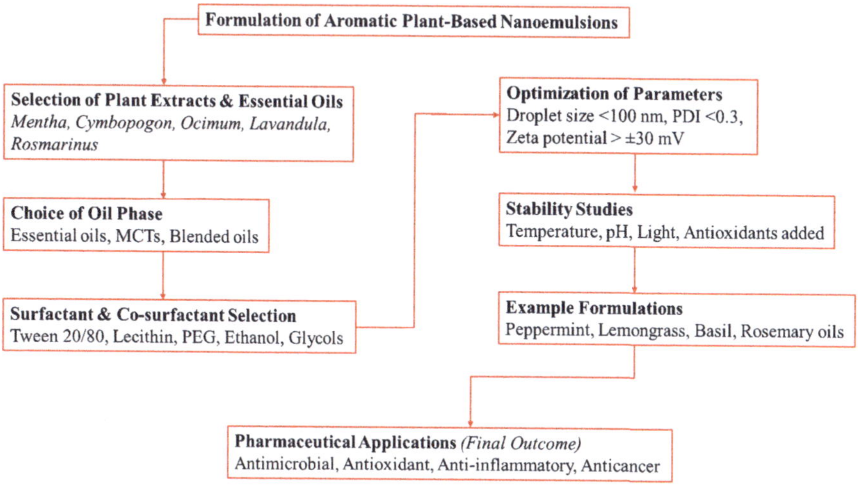

Fig. 3 Stepwise schematic of the formulation process for aromatic plant-based nanoemulsions, from raw material selection through oil and surfactant choice, optimization, and stability studies, to final pharmaceutical applications

using spontaneous emulsification not only showed enhanced solubility but also prolonged antimicrobial activity compared to the free oil [13]. Equally, rosemary oil nanoemulsions have been said to improve antioxidant powers in nutraceutical and pharmaceutical products [6]. These examples confirm the rate and flexibility of nanoemulsion systems in boosting the efficiency of plant-sourced actives (Fig. 3).

5 Characterization of Aromatic Plant-Based Nanoemulsions

Characterization of aromatic plant-based nanoemulsions is essential to ensure their pharmaceutical applicability, as it directly influences stability, bioavailability, and therapeutic potential. Size and dispersion of the droplets are the two most important parameters to be measured, as an ideal size of less than 200 nm and a polydispersity index (PDI) of less than 0.3 are required for homogeneous and stable. Morphological verification with the help of transmission electron microscopy (TEM) or scanning electron microscopy (SEM) normally gives spherical droplets. Data on the surface charge and stability is also obtained by the further analysis of the zeta potential, with a value exceeding 30 mV indicating that the particles are repelled electrostatically and will not aggregate even in the presence of the latter is produced under the assistance of biocompatible surfactants such as Tween 80 or lecithin [14, 16]. The efficiency of encapsulation (EE) and the drug loading capacity are also important parameters this way because the effective formulations usually possess an envelope of aromatic plant bioactives above 80 percent to ensure that volatile compounds (if citral, thymol, or menthol) are not lost and destroyed by the nanodroplets [7].

Stability testing in different conditions of pH, temperature, and light is a general practice, and the accelerated testing under ICH direction gives some clue of prediction of shelf life as well as the stability of the product. Inclusion of natural antioxidants such as vitamin E has been found to increase the oxidative stability of the essential oil-loaded nanoemulsions considerably [9, 13]. Biphasic profiles, which can be characterized by exposure to a burst and the slow release in vitro resulting in the subsequent high release and sustained release, are common phenomena in in vitro release studies, generally carried out using dialysis bag diffusion or using the Franz diffusion method [8]. Permeation studies show improvement in transdermal and mucosal penetration of essential oils using complementary permeation compared to free essential oils. Lastly, in every case, biological characterization with antimicrobial, antioxidant, and cytotoxicity studies was found to assess the enhanced bioactivity and inhibited toxicity of encapsulated oils compared to unformulated oils, thus validating the pharmaceutical significance of aromatic flora-sourced nanoemulsions [6, 12].

6 Pharmacological Applications of Aromatic Plant-Based Nanoemulsions

Synthesis of aromatic plant-based nanoemulsions has opened more opportunities in the pharmaceutical industry, most especially due to the success of addressing the intrinsic problems of the essential oils, such as poor solubility, instability, and fast metabolism attributes. Nanoemulsions may be an alternative to increase the potential therapeutic value of bioactive plant constituents by protecting against degradation, increasing dispersibility, and improving pharmacokinetic profiles. Besides having a higher interaction with the biological membranes due to their nanoscale droplet size, they can penetrate the biological barrier, such as the barrier across epithelia and other biological barriers, with higher efficiency, thanks to the fact that their droplet size is so small. The pharmacological uses of these systems, in particular, antimicrobial, antioxidant, anti-inflammatory, anticancer, and neuroprotective activity, are discussed in detail.

6.1 Antimicrobial Applications

Aromatic plant essential oils are characterized by longstanding antimicrobial activity, but their volatility and lack of stability have proved to be impediments to their clinical applicability. Nanoemulsions improve their properties by maintaining a defined release, allowing their performance to be more effective and using lower concentrations, extending their duration of activity [7, 13]. As an example, the most significant bioirritants in thyme and oregano oils, thymol and carvacrol, have strong activity against gram-positive and gram-negative bacteria. Nanoemulsions enhance the activity of chitosan against *Escherichia coli*, *Salmonella enterica*, and *Staphylococcus aureus*, with the probable reason being the better dispersion in aqueous systems and a closer interaction with the microbial membranes [6]. The main mechanism is the disruption of cell membranes, showing the leakage of their own contents and the blocking of important enzyme processes.

In antifungals, lemongrass oil nanoemulsions with a high percentage of citral are potent inhibitors of *Candida albicans* and *Aspergillus niger*. In some cases, the administration of free oils may need high concentrations to achieve a therapeutic threshold, but the nanoemulsions will have better therapeutic activity, even at low doses, thus lowering the risk of toxicity [12]. Moreover, peppermint oil nanoemulsions have also been demonstrated to inhibit the formation of biofilms, an aspect that allows the establishment of diverse infections, which can result in drug resistance, and thus suggests that it may be useful in chronic infection management. Significantly, such systems can represent a way to solve the global antimicrobial resistance crisis, because they provide natural, multitargeted therapeutic options to conventional antibiotics (Fig. 4).

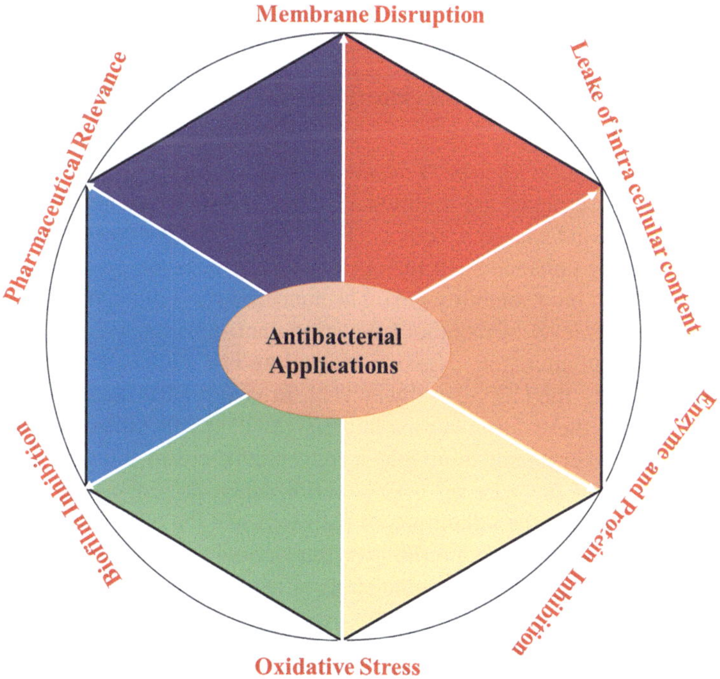

Fig. 4 Radial schematic of antibacterial mechanisms of aromatic plant-based nanoemulsions, highlighting membrane disruption, leakage, enzyme inhibition, oxidative stress, biofilm inhibition, and overall pharmaceutical relevance

6.2 Antioxidant Properties

One of the most appreciated pharmacological properties of the aromatic plant-derived compounds is their antioxidant potential. Eugenol, rosmarinic acid, linalool, and citral, as well as other phenolic compounds, can quench reactive oxygen species (ROS), which are the key mediators of aging, cardiovascular disease, and neurodegenerative diseases (B Their clinical application is, however, hampered by a lack of stability to heat, light, and oxygen. The antioxidant activity of these compounds is not only maintained by hydrophobic encapsulation in nanoemulsions, but the encapsulation also extends their antioxidant activity [7].

Radical-scavenging activity of rosemary, basil, and lavender oil nanoemulsions has been significantly improved relative to the free oils. The additional surface area of nanodroplets also gives the hydrophobic antioxidants more potential to be dispersed in aqueous media and thus more resources to interact with free radicals in the biological system [14]. An example is that rosemary oil nanoemulsions have been documented to diminish lipid peroxidation and enhance antioxidant enzyme activity in models of oxidative stress very significantly. Potential applications to prevent or alleviate diabetes, atherosclerosis, and cancer, among others are proposed by this upgrade, given that the primary cause of these ailments is due to oxidative stress.

6.3 Anti-Inflammatory and Analgesic Effects

Aromatic plants: *Mentha piperita*, *Lavandula angustifolia*, *Ocimum basilicum*, as folk medicine. These plants are traditionally used in folk medicine with regard to their antiphlogistic and analgesic activity. The use of nanoemulsions of their essential oils can greatly enhance such therapeutic effects because essential oils have better solubility and bioavailability [15]. The nanoemulsions of peppermint oil containing a significant level of menthol have strong anti-inflammatory properties, as they suppress pro-inflammatory cytokines, including TNF-alpha and IL-6, and have the ability to modify the COX-2 pathway [12]. In vivo models have demonstrated that topical use of these formulations has a significant anti-inflammatory effect, reducing edema and erythema more than a conventional cream containing free oils.

Nanoemulsions of lavender oil have superior analgesic properties, supposedly because of synergies on the inflammatory pathways and the nervous system. Their tiny droplet size permits a deeper infiltration into dermal tissues and hence local and systemic analgesia. Similarly, nanoemulsions of basil oil were found to inhibit the NF-kB pathway, as a result of which the expression of pro-inflammatory factors decreased and tissue recovery was increased [7]. Taken together, these data indicate that aromatic nanoemulsions may find applications in the treatment of diseases involving joint or muscle pain and skin inflammation, where current drugs cause adverse side effects (Fig. 5).

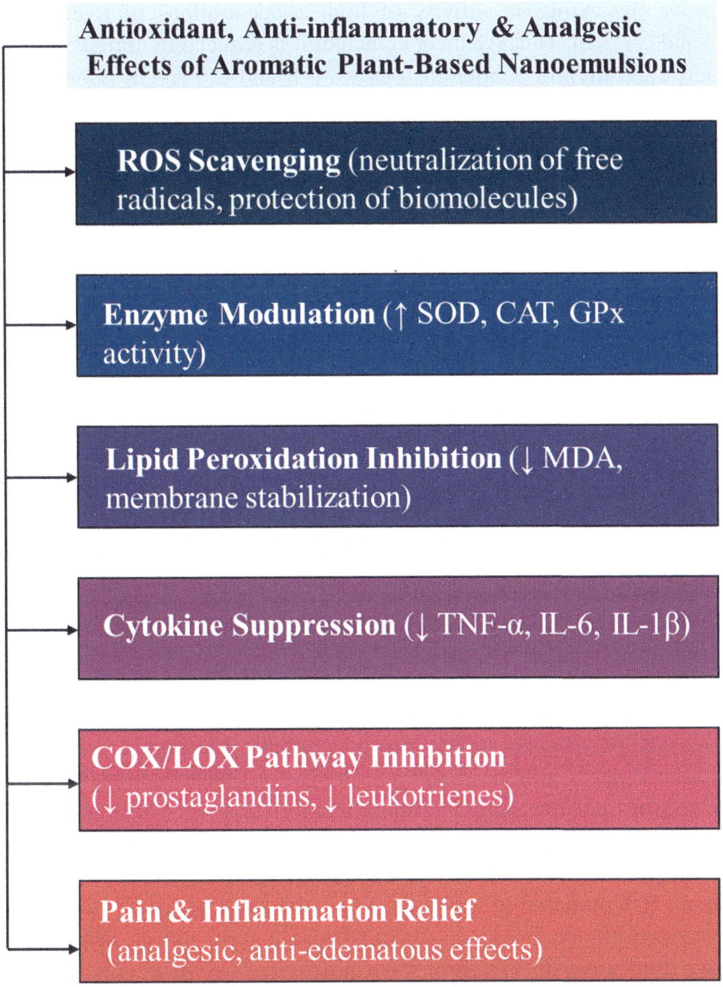

Fig. 5 Radial schematic representation of antioxidant, anti-inflammatory, and analgesic effects of aromatic plant-based nanoemulsions, showing ROS scavenging, enzyme modulation, inhibition of lipid peroxidation, suppression of cytokines, COX/LOX pathway inhibition, and resulting relief of pain and inflammation

6.4 Apoptotic (Anticancer) Potential

Aromatic essential oils have potent bioactive compounds demonstrated to have anti-cancer effects, and though many drugs are in the process of being developed, they are currently hampered by hydrophobicity and unstable behavior. Nanoemulsions address these issues in that they have sustained phytoconstituent release and target delivery. One example is eugenol in the clove oil that causes apoptosis on cancer cells by producing ROS and triggering caspase-driven pathways. Nanoemulsions also improve the cytotoxic activity of high concentrations of it against breast (MCF-7) and cervical (HeLa) cancer cells, but it is reduced in normal cells [6].

Carvacrol and thymol nanoemulsions were found to prevent tumor growth by inhibiting angiogenesis and regulating matrix metalloproteinases (MMPs), which are essential in metastasis [7]. Basil oil nanoemulsions have also been indicated to be effective against lung and prostate cancers since they caused cell cycle arrest in the G2/M stage. Besides, the nanodelivery mechanism promotes deeper infiltration and uptake in the tumor, which locally promotes elevated levels of the cancer drug without cardiac exposure. The properties of nanoemulsions lend them the potential of being complementary or alternatives to conventional chemotherapies.

6.5 Neuroprotective and Cardioprotective Uses

The aromatic essential oils of aromatic plants are progressively becoming seen to have neuroprotective effects, especially in the treatment of neurodegenerative illnesses like Alzheimer's and Parkinson's. The nanoemulsions allow bioactive agents to penetrate the blood-brain spinal cord, a major obstacle for the delivery of drugs to the central nervous system [12]. The nanoemulsions of rosemary oil that contain a lot of carnosic acid and rosmarinic acid have been observed to protect oxidative-stress-induced neuronal apoptosis, whereas lavender oil-based nanoemulsions lowered anxiety and enhanced memory in the animal models. These insights present nanoemulsions as the prospects of natural neurotherapeutic transport.

In cardiovascular health, basil and lemongrass oils have demonstrated potential and have shown lipid-lowering and antioxidant effects when formulated in nanoemulsions. These formulations positively affect atherosclerosis and hypertension by lowering the production of oxidative stress and improving the conditions of lipid metabolism [7]. Compared to synthetic drugs, the capacity to stabilize and deliver cardioprotective phytoconstituents in bioavailable form makes nanoemulsions a prospective drug alternative in chronic cardiovascular treatment (Fig. 6).

6.6 Broader Pharmaceutical Applications

The variety of all applications of pharmaceutical products using aromatic plant-based nanoemulsions is virtually unlimited. They can be successfully formulated in oral supplements, transdermal delivery systems, nasal sprays, ocular preparations, and injectable forms. An example is the nasal administration of peppermint oil nanoemulsion, whose penetration was enhanced into the mucosa, and ocular administration of basil oil nanoemulsions that increase corneal cells' antioxidant defense [8]. Their flexibility about a variety of routes of administration improves patient adherence to medications and increases the clinical utility of these drugs.

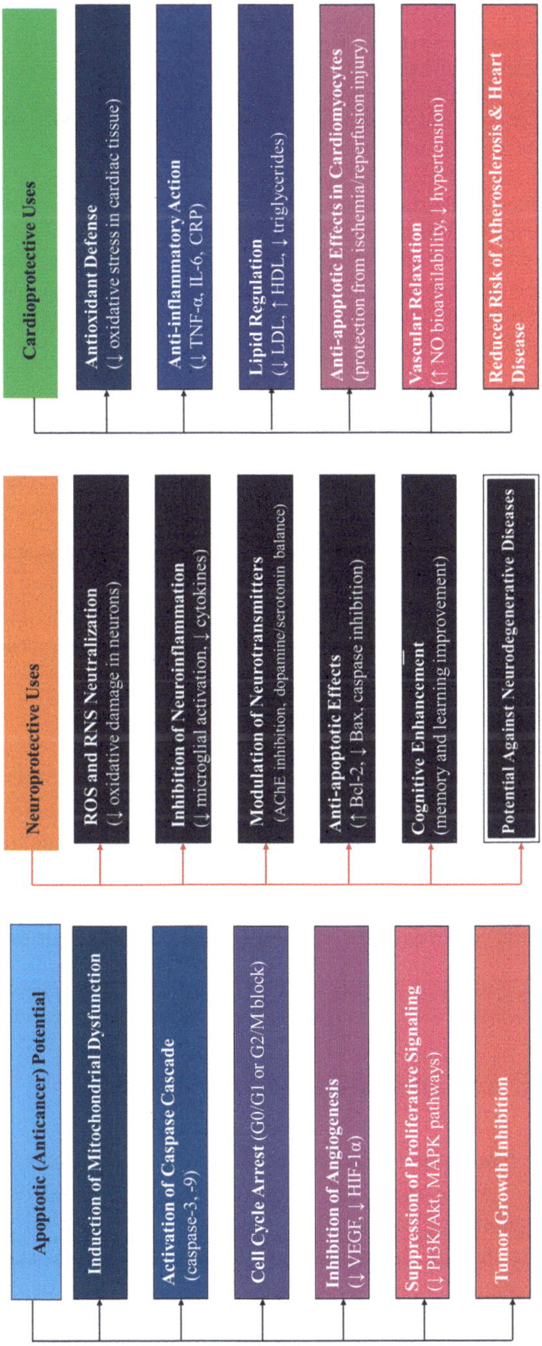

Fig. 6 Stepwise schematic representation of apoptotic (anticancer), neuroprotective, and cardioprotective uses of aromatic plant-based nanoemulsions, illustrating their key molecular mechanisms and therapeutic outcomes

Table 1 Pharmacological applications and mechanisms of aromatic plant-based nanoemulsions

Application	Strictly verified findings	Mechanism of Action
Antimicrobial	EO nanoemulsions show enhanced antibacterial and antifungal activity; improved dispersibility and controlled delivery; effective against *Listeria monocytogenes* and other pathogens	Disruption of microbial cell membranes; leakage of cytoplasmic content; inhibition of essential enzymes; prevention of biofilm formation
Antioxidant	EO nanoemulsions/nanoemulgels enhance stability and radical-scavenging potential of phenolic compounds; higher antioxidant activity than free oils	Direct scavenging of free radicals (DPPH, ABTS); reduction of lipid peroxidation; enhancement of endogenous antioxidant enzymes
Anti-inflammatory and analgesic	Nanoemulsions enhance the penetration and efficacy of EO with anti-inflammatory and analgesic properties	Downregulation of pro-inflammatory cytokines (TNF-α, IL-6); inhibition of NF-κB and COX-2 pathways; modulation of nociceptor activity
Anticancer	Preclinical studies report that EO nanoemulsions induce apoptosis, inhibit proliferation, and show anticancer potential	Induction of ROS-mediated apoptosis; caspase activation; inhibition of angiogenesis and metastasis (downregulation of MMPs, VEGF)
Topical/dermal delivery	EO nanoemulsions and nanoemulgels improve skin penetration, stability, and prolonged therapeutic effects	Enhanced transdermal permeation due to nanoscale droplet size and surfactant-mediated barrier disruption; controlled release prolongs activity

Moreover, the environmentally friendly properties of the bioactives of vegetable origin, along with the low intensity of biocompatible surfactants, coincide with the global tendency of humanitarian caring about the environment and safe pharmaceutical products. Not only are such formulations useful in their therapeutic value, but they are also rising to fulfill the consumer interest of using natural over synthetic drugs. The aromatic plant-based nanoemulsions have, therefore, considerable potential as therapeutic and preventive heterogeneous multifunctional healthcare platforms (Table 1).

7 Advantages Over Conventional Formulations

The use of aromatic plant-based nanoemulsions in pharmaceuticals has multiple and unique benefits over the traditional formulations of essential oils and plant extracts. The main benefits of nanoemulsions when compared to micronized systems concern the fully developed phenomena of nanoscale droplet size, enhanced surface area, enhancement of stability, and control of antirelease. These factors provide an even better therapeutic capability and a wider clinical safety net.

7.1 Better Solubility and Bioavailability

The only main drawback of essential oils and other plant-derived biological actives is that of low aqueous solubility, limiting absorption, and lowering systemic availability. Conventional-based formulations are usually characterized by a slow rate of bio-dispersion and a fast metabolic rate, which lead to subtherapeutic target site concentrations. The nanoemulsions circumvent this barrier by changing the droplet size to less than 200 nm and hence increasing solubility and dispersibility in water [8]. Tracing back to low surface tension, which makes the large interfacial area of nanodroplets easy to dissolve, the surfactants enhance the crossing of the mucosal and dermal barriers. The result of this is higher oral bioavailability and better delivery via nasal, ocular, and dermal means [12].

7.2 Enhanced Stability of Volatile Compounds

Essential oils are of high volatility and degrade easily in environmental factors, including heat, light, and oxygen. Conventional emulsions and solutions are ineffective at preserving these compounds, thus causing rapid activity loss and a reduced shelf life. Nanoemulsions offer encapsulation, protecting the delicate compounds against external stresses and therefore maintaining their integrity and extending their shelf lives. The surfactants and antioxidants used to formulate are also biocompatible, providing added protection, as this helps to maintain the pharmacological action during storage and treatment.

7.3 Controlled and Sustained Release

Contrary to the typical chemistry that controls the release of bioactives in rapid and uncontrolled ways, nanoemulsions are capable of controlled and sustained release. The biphasic release profile, where the burst release is followed by a longer-acting release onto the droplet core, results in therapeutic concentrations and efficacy over extended periods of time [7]. This decreases the dosing frequency, increases patient compliance, and decreases the risk of side effects due to changes in plasma concentrations. In topical products, a controlled release also enhances the length of the therapeutic effect on the skin surface or within the deeper innermost dermal layers.

7.4 Less Toxicity and More Safety

Although essential oils have beneficial properties as therapeutics, their lack of restraint is related to irritation, cell killing, or probable in vivo poisonings in high doses. Nanoemulsions mitigate these risks since they enhance delivery and allow the same effect with the use of less concentration. Bioactives are encapsulated, which eliminates the direct contact of tissues with subjecting irritant compounds, in turn, mitigating local adverse responses [14]. In addition, the choice of food grade to use as a surfactant and natural oils will also aid in the safety of nanoemulsions and therefore can be used in sensitive delivery methods like ocular or nasal.

7.5 Multi-Route and Targeted Delivery

Nanoemulsions are also very adaptable and can be employed in various routes of administration, which include oral, transdermal, ocular, nasal, and parenteral administration. They can penetrate biological barriers, such as the stratum corneum and gastrointestinal mucosa, which provides the possibility to deliver them in a site-specific form or systemically [15]. In addition, they can be passively targeted (e.g., tumor or foci of inflammation) because of the increased permeability and retention (EPR) sites owing to their nanoscale. This advantage manipulates the possible easy-target therapeutic results with no convoluted changes.

7.6 Green and Sustainability Pharmaceutical Compatibility

The other interesting strength is that plant-based nanoemulsions are in the direction of environmentally friendly chemistry and green pharmaceutical design. Conventional synthetic drugs are usually contaminated in the environment as they are being made and when they are discarded. They are prepared using biodegradable, renewable, and biocompatible substances, contrary to aromatic and plant-based nanoemulsions of aromatic composition. Low energy is the most environmentally friendly way of preparation, such as spontaneous emulsification [17]. The aspects make nanoemulsions appropriate as natural alternatives in pharmaceutical innovations.

8 Challenges and Future Perspectives

Although aromatic plant-based nanoemulsions significantly hold promise as therapeutic and pharmaceutical remedies, several challenges should be overcome before they can deliver on this potential. Such challenges are mainly associated with formulation stability, large-scale manufacturing, and acceptance of regulatory and translation of preclinical results into clinically proven products. Nevertheless, constant developments of nanotechnology, formulation science, and regulations are promising in their future inclusion in everyday medicine.

8.1 Formulation and Stability Problems

Even though nanoemulsions are characterized by greater kinetic stability in comparison to conventional emulsions, they cannot be called thermodynamically stable systems. During prolonged storage, they can also experience phase separation, Ostwald ripening, or alternation in drop size distribution [8]. It takes a lot of stability to encapsulate highly volatile essential oils. Moreover, physicochemical properties are highly dependent on surfactants and cosurfactants, and the toxicity aspects of synthetic surfactants are elusive. The emphasis to be made in future research should be the development of robust, environmentally friendly, and biocompatible systems of surfactants that guarantee stability and safety.

8.2 Upscaling and Industry Production

Although numerous nanoemulsion formulations have been successfully scaled up to laboratory scale, there is still a challenge to effectively transfer these processes to an industrial scale. High-energy emulsification processes, e.g., high-pressure homogenization or ultrasonication, are not economical at large production scales [17]. On one hand, low-energy techniques frequently have a hard time maintaining reproducibility and batch to batch consistency. Novel methods of manufacturing that are scalable, energy-efficient, and cost-effective are needed to encourage their adoption in the pharmaceutical setting.

8.3 Regulatory and Safety Issues

The issue of incorporating the nanoemulsions in food, cosmetics, and medicine is an issue that presents some regulatory concerns. Regulatory bodies like the FDA and EMA do not yet have guidelines that are specific to nanoemulsion-based herbal products, and thus, it is not clear how such products may be approved [12]. Also, the traditional stability of many surfactants and excipients utilized in nanoemulsions is GRAS, but the long-term toxicological effects of chronic exposure to the nanocarriers are not well known. A key factor impacting clinical translation will be how well safety concerns are addressed via panels of systematic toxicology studies and how internationally harmonized guidelines on regulatory requirements can be developed.

8.4 Translating to the Clinic and Validation of Efficacy

Available research on aromatic plant-based nanoemulsions is most often used at the preclinical stage and is mostly limited to in vitro testing or small animal models. Clinical trials of these systems critically lack well-designed architectures that can confirm their therapeutic efficacy, their potential to boost bioavailability, and how safe they are in humans [6]. Also, reproducibility is hampered by the variability of essential oil composition as a direct result of differences in plant source geographic origin and extraction process. Raw materials will have to be standardized and the testing protocol harmonized to gain acceptance in clinics.

8.5 Future Directions

To create a multifunctional nanoemulsion design, further research should be so. Essential oils' combinations with other nanocarriers (liposomes or solid lipid nanoparticles) can be a potential supplement to the effect and positively affect the bioavailability. New site destination of site-specific drug delivery can be offered by a breakthrough in stimuli-responsive nano-emulsions can release their bioactives at locations due to variations in pH, temperature, or enzyme destination [7]. In addition, weighting nanoemulsions with new ideas such as personalized medicine, artificial intelligence with optimization of formulation, and the use of omics to identify biomarkers, the capability in precision healthcare will significantly rise with this combination.

9 Conclusion

Aromatic plant-based nanoemulsions have become an innovative system used to increase the therapeutic potential of essential oils and bioactive phytoconstituents, which has a better solubility, stability, bioavailability, and regulated release than traditional formulations. Their broad and diverse pharmacological effects, such as antimicrobial, antioxidant, anti-inflammatory, and anticancer effects, coupled with the fact that they are compatible with oral, dermal, nasal, and ocular delivery, show their widespread clinical potential. Although formulation stability, large-scale production, regulatory acceptance, and clinical validation still are major challenges, ongoing innovations in nanotechnology, green chemistry, and personalized medicine modalities are bound to eliminate these challenges. Therefore, the nanoemulsions are a game changer and a long-term solution that can facilitate the classic utilization of plant-based medicine alongside contemporary pharmaceutical advancements.

Competing Interests The authors declare that they have no competing interest.

Declaration of Generative AI and AI-Assisted Technologies in the Writing Process During the preparation of this work, the book chapter author(s) have not taken any assistance from AI, and the book chapter author(s) reviewed and edited the content as needed and take(s) full responsibility for the content of the publication.

References

1. Husen A (2024a) Antimalarial medicinal plants. Taylor & Francis Group, LLC, New York. https://doi.org/10.1201/9781003378396
2. Husen A (2024b) Medicinal spices and condiment crops. Taylor & Francis Group, LLC, New York. https://doi.org/10.1201/9781003387046
3. Husen A (2023a) Antidiabetic medicinal plants and herbal treatments. CRC Press, Boca Raton, FL, pp 33487–32742. https://doi.org/10.1201/b23347
4. Husen A (2022) Herbs, shrubs and trees of potential medicinal benefits. CRC Press, Boca Raton, FL. https://doi.org/10.1201/9781003205067
5. Husen A (2023b) Exploring poisonous plants: medicinal values, toxicity response and other potential uses. CRC Press, Boca Raton, FL, pp 33487–32742. https://doi.org/10.1201/b23017
6. Jamir Y, Singh S, Verma R (2024) Plant-based essential oil encapsulated in nanoemulsions: formulation, characterization, and applications. Environ Chem Lett 22:1585–1602. https://doi.org/10.1007/s10311-023-01626-1
7. Prado JC, Prado G, Linhares L, Neves A, Nascimento J, Abreu F, Fontenelle R (2024) Nanoemulsions of plant-based bioactive compounds with antimicrobial applications: a review. Cienc Nat 46:e74325. https://doi.org/10.5902/2179460X74325
8. McClements DJ (2021) Nanoemulsion-based technologies for delivering natural antimicrobials in foods. Front Sustain Food Syst 5:643208. https://doi.org/10.3389/fsufs.2021.643208
9. Bolgen UMG, Demirci Kayiran S, Ozogul Y, Ozogul F (2025) Essential oil-based nanoemulsions with current knowledge: formulation, characterization, and applications in food and pharmaceuticals. Ind Crop Prod 233:121411. https://doi.org/10.1016/j.indcrop.2025.121411
10. Soni M, Patel K, Shah D (2023) Advances in designing essential oil nanoformulations: encapsulation strategies, stability, and applications. Foods 12(21):4017. https://doi.org/10.3390/foods12214017
11. McClements DJ (2011) Edible Nanoemulsions: fabrication, properties, and functional performance. Royal Society of Chemistry. https://doi.org/10.1039/9781849733116
12. Iskandar B, Alawiyah N, Rahayu ES (2024) Herbal nanoemulsions in cosmetic science. J Cosmet Dermatol 23(5):1421–1434. https://doi.org/10.1111/jocd.16121

13. Maurya A, Singh AK, Yadav M (2021) Essential oil nanoemulsion as eco-friendly and safe antibacterial agents: inhibition of listeria monocytogenes in food. Front Microbiol 12:783161. https://doi.org/10.3389/fmicb.2021.783161
14. Yap XF, Tan WN, Ong CS (2024) Plant essential oil nanoemulgel as a cosmeceutical: formulation, characterization and applications. Cosmetics 11(4):116. https://doi.org/10.3390/cosmetics11040116
15. Chaurasiya CK, Ganju E (2021) Herbal nanoemulsion in topical drug delivery and skin applications. J Res Pharm Sci 12(2):88–96
16. Shirole A, Shete D, Ghawate V, Pund A (2023) A comprehensive review on: herbal nanoemulsion formulation, characterization, and its application. Int J Pharm Pharmaceut Res 27(3):1–15
17. Cholakova D, Denkov ND, Tcholakova S (2022) Self-emulsification in chemical and pharmaceutical technologies. arXiv preprint. https://doi.org/10.48550/arXiv.2202.07502

Aromatic Plant-Based Nanoemulsions as Insect Repellents

Alolika Sarkar, Dusan Losic, and Peetam Mandal

Abstract The use of aromatic plant-derived essential oils as natural insect repellents has gained renewed attention as safe and eco-friendly alternatives to synthetic chemicals. Despite their proven efficacy, limitations such as high volatility, instability, and poor solubility have hindered their practical utility. Nanoemulsion technology has emerged as a transformative strategy to overcome these barriers, offering nanoscale droplet dispersion, enhanced stability, prolonged release, and improved dermal penetration of bioactive compounds. This chapter provides a comprehensive review of aromatic plant-based nanoemulsions, with a focus on their formulation, physicochemical characterization, mechanisms of insect-repellent action, safety considerations, and commercialization prospects. Specific attention is given to the encapsulation of essential oils such as citronella, neem, basil, eucalyptus, peppermint, clove, and lemongrass, highlighting their potential to deliver long-lasting and broad-spectrum repellency. Toxicological and environmental assessments underscore the eco-safety and biodegradability of these systems compared with synthetic repellents. Furthermore, challenges of scalability, reproducibility, and regulatory compliance are discussed in the context of industrial applications. Finally, future research perspectives, including smart nanoemulsions, synergistic oil blends, and integration into integrated pest management (IPM), are outlined. Collectively, this chapter emphasizes the role of medicinal plant-based nanoemulsions as innovative, sustainable solutions for global vector control.

Keywords Aromatic medicinal plants · Essential oils · Nanoemulsion · Natural insect repellents · Encapsulation

---

A. Sarkar
Department of Chemistry, Krish Biotech Research Private Limited, OECD-GLP Test Facility, Kalyani, India

D. Losic
School of Chemical Engineering, Faculty of Sciences, Engineering and Technology, The University of Adelaide, Adelaide, SA, Australia

P. Mandal (✉)
Department of Chemistry, National Institute of Technology, Raipur, Chhattisgarh, India

A. Husen (ed.), *Sustainable Uses of Aromatic Plants in Nanotechnology*, Smart Nanomaterials Technology, https://doi.org/10.1007/978-981-95-7409-4_16

1 Introduction

Insects have coexisted with humans, animals, and plants since the earliest stages of prehistoric evolution, playing indispensable roles in ecosystem balance while simultaneously posing significant threats to health, agriculture, and economic sustainability. The relationship between humans and insects is therefore paradoxical: while many species contribute to pollination, soil fertility, and biodiversity maintenance, others act as vectors of diseases and destructive pests of crops. This duality underlines the pressing need for effective insect management strategies that minimize harm without disturbing ecological equilibrium. Among these strategies, insect repellents occupy a pivotal role in both public health and agricultural domain. They provide a frontline defense against vector-borne diseases, safeguard food production systems, and contribute to the overall security of human livelihoods.

The significance of insect repellents in public health cannot be overstated since a large proportion of the global disease burden is attributable to insect vectors, with mosquitoes, ticks, sandflies, and tsetse flies among the most important culprits. Mosquitoes alone are responsible for transmitting malaria, dengue, chikungunya, Rift Valley fever, filariasis, yellow fever, West Nile virus, Japanese encephalitis, and the Zika virus [1]. Malaria continues to afflict millions of people annually, with the World Health Organization (WHO) estimating more than 240 million cases and over half a million deaths each year, disproportionately affecting children and pregnant women in endemic regions [2]. Dengue, another mosquito-borne illness, has emerged as one of the most rapidly spreading vector-borne diseases, with explosive outbreaks occurring in urban and semi-urban areas worldwide. Ticks are vectors of Lyme disease, babesiosis, and tick-borne encephalitis, while sandflies transmit leishmaniasis, a neglected tropical disease with significant morbidity. Repellents are indispensable in such scenarios because they act at the level of individual protection, reducing contact between humans and vectors and thereby disrupting pathogen transmission cycles. In areas where vector control infrastructure is weak, where healthcare access is limited, or during humanitarian emergencies, repellents often serve as the most practical and immediately deployable preventive measure. Their utility is not restricted to endemic regions; global travel and climate change are extending the range of many vector species, making repellents increasingly relevant worldwide.

In addition to their role in public health, insect repellents are equally vital in agriculture. Global agriculture suffers losses of gargantuan magnitude due to insect pests, with conservative estimates indicating that 20–40% of crop production is lost annually [3]. These losses translate into reduced farmer incomes, food insecurity, and economic instability, particularly in developing countries where agriculture is the backbone of livelihoods. Insects cause damage in two major ways: through direct feeding on crops and through the transmission of plant pathogens. Sap-sucking insects like aphids and whiteflies, for example, not only reduce plant vigor but also act as efficient vectors of plant viruses. Fruit flies devastate horticultural products, limiting both domestic consumption and export potential. The economic burden of such infestations is further compounded by the costs of pest management and the loss of market access due to strict residue standards in international trade. In this context, repellents function as preventive tools that reduce insect colonization, feeding, and oviposition on crops, thereby minimizing reliance on broad-spectrum pesticides. However, there remain very few experimental validations of repellent action in agriculture. One small greenhouse study demonstrated reduced pest reinfestation following essential oil application, while field trials in California using encapsulated essential oils suggested a deterrent effect long after contact toxicity had waned [4]. Their integration into sustainable agricultural systems helps protect yields while reducing chemical inputs, aligning with global calls for more eco-friendly food production systems (Fig. 1).

While synthetic repellents and insecticides have dominated the landscape for much of the twentieth and early twenty-first centuries, their limitations have become increasingly evident. Among synthetic repellents, N,N-diethyl-meta-toluamide (DEET) has long been considered the benchmark for effectiveness. First developed during World War II, DEET remains the most widely used repellent to this day. Its efficacy against a broad spectrum of biting insects is undeniable, and its relatively long duration of action has made it the gold standard for personal protection [5]. However, concerns over toxicity, environmental impact, and resistance development have eroded its universal acceptance. Reports of adverse effects such as skin irritation, neurotoxicity, and in rare cases, seizures, especially among children, have contributed to public apprehension regarding DEET use. Although regulatory agencies generally consider DEET safe when used as directed, the perception of

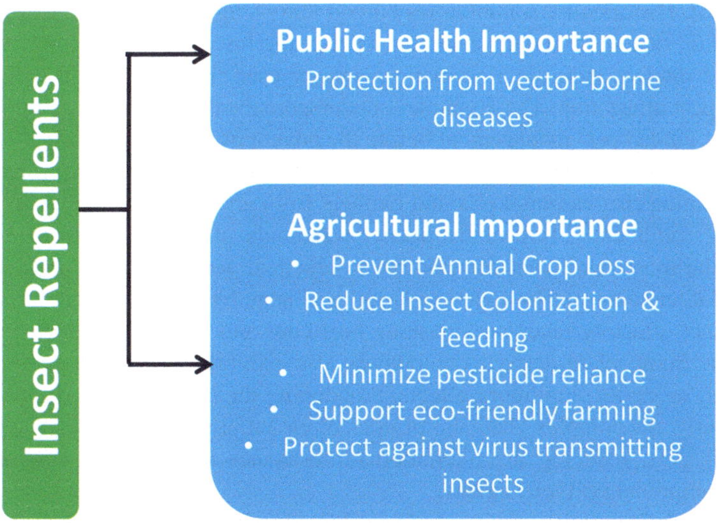

Fig. 1 Multifaceted role of insect repellents in safeguarding public health and agriculture by reducing vector-borne disease transmission and minimizing crop losses

chemical hazard remains a deterrent for many consumers, particularly those seeking "green" and natural alternatives.

Resistance to synthetic repellents and insecticides represents another critical challenge. Continuous exposure of insect populations to the same chemical agents exerts strong selective pressure, enabling resistant phenotypes to proliferate. Documented cases of mosquito resistance to pyrethroids and DEET-like compounds highlight the urgency of diversifying our arsenal of repellents. Resistance not only diminishes the efficacy of existing tools but also threatens global disease control programs that rely heavily on chemical interventions.

Equally concerning are the environmental consequences associated with synthetic repellents and pesticides. Their persistence in soil and water systems, coupled with volatilization and runoff, contributes to contamination of natural resources. Nontarget organisms, including beneficial insects such as pollinators and natural enemies of pests, are often adversely affected. Aquatic ecosystems are particularly vulnerable, as even low concentrations of chemical residues can disrupt fish, amphibians, and aquatic invertebrates. At a broader scale, the synthesis, application, and disposal of synthetic chemicals contribute to ecological imbalance and environmental degradation. The tightening of regulatory frameworks in many countries, along with growing consumer preference for safe, sustainable, and eco-friendly products, has accelerated the search for alternatives to synthetic repellents.

It is in this context that aromatic plants emerge as promising candidates for insect management. Humans have long recognized the insect-repellent properties of aromatic plants, embedding them into cultural and traditional practices across civilizations. Plant species such as citronella (*Cymbopogon* spp.), neem (*Azadirachta indica*), clove (*Syzygium aromaticum*), eucalyptus (*Eucalyptus* spp.), basil (*Ocimum* spp.), and mint (*Mentha* spp.) have been extensively used in various forms to deter insects [6]. Historically, communities burned leaves, applied crushed plant materials on skin, or used aromatic smoke to repel biting insects. In agriculture, farmers practiced intercropping with aromatic plants or used them in storage systems to prevent pest infestation. These practices underscore the deep ethnobotanical knowledge that guided pest management before the advent of synthetic chemicals.

Modern scientific investigations have validated many of these traditional uses, demonstrating that essential oils and extracts from aromatic plants contain potent bioactive compounds. Terpenoids, phenylpropanoids, alkaloids, and flavonoids are among the most significant contributors to repellent activity. Compounds such as citronellal, eugenol, linalool, geraniol, and limonene have been shown to interfere with insect olfactory receptors, masking host cues and reducing biting or feeding behavior. Some plant-derived compounds also exhibit insecticidal, oviposition-deterrent, or growth-inhibitory effects, broadening their utility beyond repellency. Importantly, the complex chemical profiles of essential oils act on multiple targets within insects, reducing the likelihood of resistance development compared to single-compound synthetic agents.

Despite their promise, plant-based repellents face practical challenges that limit their widespread adoption. Essential oils are volatile, unstable, and prone to degradation when exposed to light, heat, or oxygen. Their poor solubility in water complicates formulation and delivery, while their rapid evaporation shortens the duration of protection. These limitations have necessitated advanced formulation technologies to enhance stability, bioavailability, and efficacy of plant-derived repellents. This is where nanotechnology, and specifically nanoemulsion systems, offers a transformative solution.

Nanoemulsions are colloidal dispersions of oil and water stabilized by surfactants, with droplet sizes typically ranging from 20 to 200 nanometers. At this nanoscale, the physical and functional properties of emulsions differ significantly from conventional formulations. Nanoemulsions exhibit high kinetic stability, enhanced solubilization of hydrophobic compounds, and increased surface area for interaction. When applied to plant-based repellents, these attributes translate into several advantages. Volatile essential oil components are encapsulated within nanoscale droplets, protecting them from environmental degradation and reducing evaporation losses. This encapsulation improves shelf stability and prolongs the duration of repellent action. The small droplet size also facilitates better penetration and adherence to the skin or plant surfaces, ensuring more uniform coverage and improved bioefficacy. Moreover, nanoemulsions can be engineered to provide controlled and sustained release of active compounds, reducing the need for frequent reapplication and enhancing user convenience.

Beyond personal protection, nanoemulsified plant-based repellents offer immense potential in agriculture. When applied as foliar sprays, soil drenches, or seed coatings, these formulations provide targeted deterrence of insect pests while minimizing nontarget effects and residue accumulation. Their biodegradable nature aligns with the principles of sustainable agriculture and integrated pest management, supporting the global agenda of reducing chemical pesticide dependency. Furthermore, by enhancing the efficiency of active compounds, nanoemulsions reduce the amount of plant material required, promoting resource conservation and eco-friendliness.

The integration of aromatic plant extracts with nanoemulsion technology represents a convergence of traditional knowledge and modern science. It acknowledges the efficacy of natural bioactive compounds that humans have used for centuries while employing cutting-edge nanotechnology to overcome their practical limitations. Recent research has demonstrated the superior performance of nanoemulsions prepared from essential oils of citronella, lemongrass, clove, eucalyptus, and others, showing prolonged repellency and improved stability compared to conventional formulations. Such findings reinforce the potential of nanoemulsion-based repellents as next-generation solutions for both public health and agriculture.

2 Phytochemistry of Aromatic Plants with Insect-Repellent Properties

Aromatic medicinal plants such as *Ocimum*, *Mentha*, *Thymus*, *Origanum*, *Cymbopogon*, *Eucalyptus*, *Melaleuca*, and clove (*Syzygium aromaticum*) have long been used in traditional medicine for protection against insects. Their repellency arises from essential oils (EOs), which are complex mixtures dominated by oxygenated terpenoids and phenylpropanoids. Recent research confirms that EO activity is determined not only by the identity of major constituents but also by their ratios, volatility, and synergistic effects [7, 8].

The essential oils of these medicinal plants comprise a wide range of volatile compounds. Monoterpenes (C10), such as limonene and α-pinene, together with their oxygenated derivatives like linalool, citronellal, menthol, and 1,8-cineole, are abundant components. Sesquiterpenes (C15), including β-caryophyllene and nerolidol, also contribute significantly, while phenylpropanoids such as eugenol and anethole add further activity. Other oxygenated compounds, including aldehydes like citronellal and citral, ketones such as carvone, and esters like linalyl acetate, diversify the chemical profile. These groups differ in volatility, polarity, and receptor-binding capacity, which explains the variations in repellency strength and persistence observed across different oils [9].

Within this spectrum, several compounds stand out for their specific activities. Linalool, abundant in *Ocimum* and *Lavandula*, consistently repels mosquitoes, and recent nanoemulsion studies have confirmed its larvicidal efficacy [7]. Eugenol, the principal active in clove oil, strongly deters mosquitoes; a 2025 study further demonstrated that eugenyl acetate lacked similar activity, emphasizing the importance of the free phenolic OH group [10]. Thymol and carvacrol, phenolic monoterpenes derived from *Thymus* and *Origanum*, show broad repellent effects against both mosquitoes and ticks, and their stability is enhanced through encapsulation methods [11]. Citronellal, the signature compound of *Cymbopogon*, has been widely tested in human trials, with recent mechanistic studies linking its action to mosquito TRPA1 ion channels, although high temperatures were found to reduce its efficacy [12]. Similarly, compounds from *Mentha* species, such as menthol and carvone, display dual activities: carvone-rich oils deter oviposition, while menthol combines repellency with a soothing dermal effect [13].

Overall, phytochemistry underscores that oxygenated terpenoids and phenylpropanoids are the primary drivers of repellency in medicinal aromatic plants. Their limitations, including volatility, instability, and short protection time, highlight the need for nanoemulsion-based delivery systems to achieve sustained efficacy. Chemotype selection, such as linalool-rich *Ocimum* or carvone-rich *Mentha*, provides opportunities for tailored repellent properties. Moreover, insights from structure-activity relationships (SAR) and quantitative structure-activity relationship (QSAR) analyses support rational design strategies. Taken together, the phytochemistry of medicinal aromatic plants offers both the molecular foundation and the strategic roadmap for developing effective nanoemulsion-based insect repellents, bridging traditional use with contemporary formulation technology.

3 Conventional Plant-Based Insect Repellents: Challenges and Limitations

For centuries, aromatic plants have been employed in traditional medicine and household practices to ward off insects. Crude formulations such as burning leaves of *Ocimum*, *Cymbopogon*, and *Eucalyptus*, applying oils extracted from mint or thyme, and using clove or neem pastes are widely documented in ethnobotanical records [14, 15]. These practices provided communities with accessible and eco-friendly means of personal protection, particularly in regions where synthetic insecticides were unavailable. Despite their long history, conventional plant-based repellents face significant limitations when evaluated against modern public health and agricultural requirements.

One major issue is stability. Essential oils (EOs), the primary active agents of aromatic plants, are highly volatile and chemically unstable under environmental conditions such as heat, light, and oxygen exposure [16]. Constituents such as citronellal, linalool, and eugenol readily oxidize, leading to reduced efficacy and sometimes the formation of allergenic by-products. This instability contributes to inconsistent protection levels, making crude EO formulations unreliable in field applications.

Closely tied to stability is the challenge of volatility. The same chemical property that gives EOs their fragrance also causes rapid evaporation from skin or environmental surfaces. For instance, citronella oil—long used as a traditional mosquito repellent—typically provides less than 30–60 min of effective protection before reapplication is needed, compared with several hours of protection provided by synthetic repellents such as DEET or picaridin [17, 18]. This short duration of activity is a critical barrier for practical use in disease-endemic settings where extended coverage is essential.

Poor solubility is another limitation. Most EO constituents are lipophilic and display limited solubility in aqueous media, restricting their formulation into sprays, lotions, or gels without the use of surfactants or solvents. In crude preparations, phase separation often occurs, leading to uneven distribution of active compounds and variable repellent performance [19]. Furthermore, attempts to enhance solubility using organic solvents can compromise safety and skin tolerability.

In addition, some EO constituents cause skin irritation and sensitization. Eugenol, citral, and limonene are well-known contact allergens when applied in concentrated form [20]. This restricts their direct topical use, particularly for children and individuals with sensitive skin. Traditional crude applications—such as undiluted oils or plant pastes—can therefore pose dermatological risks, undermining their suitability for widespread adoption.

When compared to synthetic standards, plant-based repellents clearly show a shorter duration of protection. DEET, for instance, can provide 4–8 h of reliable coverage against *Aedes aegypti*, while picaridin offers comparable protection with lower odor and skin irritation potential [21]. In contrast, most plant oils require frequent reapplication, which reduces user compliance and practicality in real-world scenarios.

Taken together, these limitations—instability, volatility, poor solubility, dermal irritation, and short protective duration—underscore why traditional plant-based repellents, while safe and culturally significant, are often considered inadequate as stand-alone solutions for vector management. These challenges also highlight the urgent need for modern formulation strategies such as nanoemulsions, which can stabilize volatile actives, enhance skin penetration, prolong release, and ultimately bridge the gap between ethnobotanical wisdom and contemporary insect-repellent technology. The contrasting differences between conventional plant-based repellents and synthetic repellents are provided below (Table 1):

Table 1 Comparative overview of conventional plant-based and synthetic insect repellents across key parameters

Parameter	Conventional plant-based repellents (EOs, crude extracts)	Synthetic repellents (DEET, picaridin)	Key references
Traditional use	Used in crude forms (burning leaves, oils, pastes) for centuries; strong ethnobotanical base	Developed in the twentieth century for military and civilian use	[22, 23]
Stability	Chemically unstable; prone to oxidation and photodegradation (citronellal, linalool, eugenol)	High chemical stability; shelf life >2 years	[24, 25]
Volatility	Highly volatile → rapid evaporation, reduced efficacy in hot/humid climates	Lower volatility → gradual release maintains long protection	[22, 26]
Skin effects	Risk of irritation, sensitization (eugenol, citral, thymol); safety depends on dilution and formulation	Generally well tolerated; DEET sometimes causes mild irritation; picaridin has better skin tolerability	[27, 28]
Solubility/ formulation	Poor aqueous solubility; unstable in sprays/lotions without surfactants; prone to phase separation	Readily formulated into lotions, gels, and sprays with consistent performance	[24, 29]
Duration of protection	Short (30–90 min typical; up to ~2–6 h for lemon eucalyptus oil). Requires frequent reapplication	Long (4–8 h, sometimes longer depending on formulation)	[17, 30]
User compliance	Low, due to need for frequent reapplication and odor intensity	High, due to longer protection and milder odor (esp. picaridin)	[17, 31]
Environmental impact	Biodegradable, low persistence, safer ecotoxicological profile	DEET can contaminate waterways; picaridin considered less persistent	[32, 33]

4 Fundamentals of Nanoemulsion Technology and the Key Concepts

Nanoemulsions are colloidal dispersions of two immiscible liquids, typically oil and water, stabilized by surfactants, with droplet diameters usually ranging between 10 and 500 nm. Unlike microemulsions, which are thermodynamically stable, nanoemulsions are only kinetically stable, meaning that they will eventually destabilize but may persist for extended durations under appropriate formulation conditions. Based on the nature of the dispersed and continuous phases, nanoemulsions are generally classified into three structural types. Oil-in-water (O/W) systems consist of oil droplets dispersed within a continuous aqueous phase, making them the most common form in pharmaceutical, cosmetic, and insect-repellent applications. Water-in-oil (W/O) systems, on the other hand, contain water droplets dispersed in a continuous oil phase, which makes them useful for encapsulating hydrophilic compounds in nonpolar environments. A third category comprises bi-continuous emulsions, in which both oil and water form interpenetrating continuous phases, or more complex multiple emulsions such as W/O/W or O/W/O, which allow staged release of active ingredients. These structural forms are determined by formulation composition, surfactant selection, and processing method [34, 35].

Droplet size plays a critical role in nanoemulsion performance. Smaller droplets, typically within the 10–200 nm range, enhance stability by reducing gravitational separation phenomena such as creaming or sedimentation while also providing a higher surface area for interfacial interactions [36]. A narrow droplet size distribution, reflected in a low polydispersity index (PDI), is essential for stability, since broader distributions facilitate coalescence or Ostwald ripening, both of which promote destabilization. Equally important is the zeta potential (ζ), which reflects the electrical charge at the droplet surface. Values above ±30 mV confer moderate stability, while magnitudes exceeding ±40 mV impart strong electrostatic repulsion that effectively prevents aggregation [37]. Other parameters, including temperature, pH, surfactant type and concentration, and viscosity, also influence overall stability and release characteristics.

The role of surfactants and cosurfactants is central to maintaining stability, as they lower interfacial tension and provide steric or electrostatic stabilization of droplets [38]. While synthetic surfactants such as Tweens and Spans are widely employed, there is growing interest in natural, biodegradable alternatives for sustainable and safe use, particularly in food, cosmetic, and medicinal applications. Among these, saponins extracted from sources such as Quillaja bark have shown strong emulsifying capacity, producing droplets of approximately 20 nm with zeta potentials below −40 mV [39]. Sucrose esters, derived from sucrose and fatty acids, offer both biodegradability and versatility due to their tunable hydrophilic-lipophilic balance [40]. Biosurfactants also hold promise, as they often exhibit lower critical micelle concentrations than synthetic surfactants, leading to improved efficiency. In selecting surfactants, key considerations include biocompatibility, low required concentrations, long-term stability, and interfacial activity appropriate to the intended O/W or W/O system.

Preparation techniques for nanoemulsions are broadly classified into high-energy and low-energy methods. High-energy methods employ intense mechanical forces to break down droplet size. Among them, high-pressure homogenization exposes emulsions to pressures exceeding 30–120 MPa to generate very fine droplets, with effectiveness depending on the energy input, temperature, and formulation composition [41]. Microfluidization follows a similar principle but often achieves more uniform distributions with fewer passes. Ultrasonication employs high-amplitude sound waves to produce cavitation, disrupting droplets, while specialized equipment such as jet dispersers or high-shear mixers are also employed for nanoscale emulsification [42]. These methods are effective in generating very small, uniformly distributed droplets with comparatively low surfactant requirements, which is particularly advantageous for food-grade or sensory-sensitive formulations. However, the intense mechanical forces can generate localized heating that risks degrading heat-sensitive bioactive compounds. Incorporating cooling strategies during processing is therefore critical to preserve functional stability.

In contrast, low-energy methods rely on intrinsic physicochemical interactions and phase behavior rather than mechanical force. Spontaneous emulsification, also referred to as self-emulsification, occurs when system components self-assemble into nanoemulsions under specific mixing or dilution conditions. Phase inversion techniques, based on changes in temperature or composition, flip the system from W/O to O/W or vice versa at inversion points, producing fine droplets. Additional low-energy approaches include D-phase emulsification, membrane emulsification, microemulsion dilution, and solvent displacement [43]. These methods are energy-efficient and require less equipment, reducing production costs [44]. However, their reliance on synthetic surfactants has sometimes limited their suitability for food or natural-product-focused applications, though increasing use of natural emulsifiers is helping overcome this limitation [45]. Consequently, interest in "greener" self-emulsifying systems such as self-emulsifying drug delivery systems (SEDDS) is rising, as they combine reduced energy consumption with eco-friendly formulations.

5 Formulation of Aromatic Plant-Based Nanoemulsions

The formulation of nanoemulsions containing aromatic plant-derived essential oils represents a transformative approach in the development of safer, sustainable, and highly effective insect-repellent systems. Unlike conventional repellents, which frequently rely on synthetic actives such as DEET or picaridin, nanoemulsions harness bioactive terpenoids, phenylpropanoids, and other phytochemicals extracted from medicinal plants. These volatile compounds, while exhibiting potent insect-repellent activity, are chemically unstable and susceptible to degradation by light, oxygen, and temperature. Nanoemulsion technology provides a means to encapsulate, stabilize, and control the release of these bioactives, thereby enhancing their functional efficacy.

This section discusses in detail the encapsulation of essential oils from prominent medicinal aromatic plants, the optimization of surfactant-to-oil ratios, the influence of formulation parameters on stability and release kinetics, and the design of final dosage forms such as gels, sprays, and creams that are suitable for topical application against insect vectors.

5.1 Encapsulation of Essential Oils from Medicinal Plants

The encapsulation of essential oils within nanoemulsions fulfills several critical functions: it reduces volatility, shields labile components from oxidative degradation, minimizes the risk of skin irritation, and prolongs repellency duration. A number of medicinal plants traditionally used in ethnomedicine have emerged as prime candidates for nanoemulsion-based insect repellents, offering both enhanced efficacy and improved user acceptability.

Citronella (*Cymbopogon winterianus*, *C. nardus*) represents one of the most widely studied examples. Its oil, rich in citronellal, citronellol, and geraniol, has long been valued for mosquito repellency. Nanoemulsification improves its retention on skin surfaces and reduces rapid evaporation, thereby extending the duration of protection [22]. Neem (*Azadirachta indica*) oil, which contains azadirachtin and other limonoids with insect growth-regulating and repellent activities, poses formulation challenges due to its high viscosity and poor miscibility in water. Encapsulation within nanoemulsions enables dispersion into aqueous systems and facilitates incorporation into user-friendly sprays and gels [46].

Eucalyptus (*Eucalyptus globulus*, *E. citriodora*) oils, dominated by 1,8-cineole and citronellal, also benefit from nanoemulsification, which enhances dermal penetration and provides sustained release, leading to improved repellency compared with crude formulations [47]. Similarly, peppermint (*Mentha piperita*) oil, characterized by menthol and menthone, is known for its insect-repellent and cooling properties. Encapsulation reduces dermal irritation while imparting a refreshing sensory effect in topical applications [48]. Clove (*Syzygium aromaticum*) oil, rich in the phenylpropanoid eugenol, is highly effective but prone to oxidative degradation. Nanoemulsion encapsulation stabilizes eugenol, thereby enhancing both efficacy and shelf life in skin-applied products [49].

Other examples include lemongrass (*Cymbopogon citratus*), whose citral-rich oil demonstrates excellent repellency but suffers from volatility and sensitization issues. Nanoemulsion delivery mitigates these concerns by improving bioavailability and reducing irritancy [50]. Basil (*Ocimum basilicum*, *O. sanctum*) oils, which contain linalool and eugenol, are similarly optimized through nanoemulsion encapsulation, enabling controlled release for both immediate and sustained protection [51].

Taken together, these case studies illustrate the broad utility of nanoemulsion systems in enhancing the safety, stability, and functional performance of essential oils. By improving formulation characteristics and user acceptability, nanoemulsions provide a scientifically robust pathway for reducing reliance on synthetic repellents and advancing sustainable, plant-based alternatives.

5.2 Optimization of Surfactant-to-Oil Ratios

The successful preparation of nanoemulsions largely depends on the surfactant-to-oil ratio (SOR). This parameter governs droplet size, physical stability, and encapsulation efficiency. In the case of aromatic medicinal plant oils, optimization is particularly challenging because of their chemical heterogeneity and high volatility.

Nonionic surfactants, especially those with established safety profiles such as Tween 20, Tween 80, and Span 80, are commonly used. They are frequently combined with cosurfactants like ethanol, propylene glycol, or polyethylene glycol to reduce interfacial tension. For example, nanoemulsions of neem oil stabilized with Tween 80 have demonstrated average droplet sizes below 50 nm, high stability against creaming, and improved bioactivity in topical formulations [52].

Optimization studies typically employ pseudo-ternary phase diagrams, which map out the regions of nanoemulsion formation across varying SORs. Ratios between 1:1 and 4:1 (surfactant-oil) are often required to maintain stability, although the ideal ratio varies with oil polarity, surfactant HLB value, and intended route of application [53]. Higher surfactant concentrations may ensure smaller droplet sizes but can raise concerns of dermal irritation; hence, the balance between efficacy and biocompatibility is crucial. For insect-repellent formulations, the optimization process must also consider sensory acceptability, such as non-greasiness and ease of spreadability, which are influenced by the surfactant system.

5.3 Influence of Formulation Parameters on Stability and Release Kinetics

The stability and release behavior of nanoemulsions are strongly influenced by key formulation parameters, including droplet size distribution, zeta potential, surfactant characteristics, viscosity, and environmental resilience. Careful optimization of these factors is essential to ensure that medicinal plant-based nanoemulsions are not only physically stable but also functionally effective as insect repellents.

Droplet size plays a central role in determining both kinetic stability and release kinetics. Smaller droplets, typically below 100 nm in diameter, provide enhanced stability against coalescence and offer a larger interfacial area for controlled release. Dynamic light scattering (DLS) is frequently employed to monitor size distribution, and experimental studies on citronella and peppermint oils have shown that reductions in droplet size are directly correlated with prolonged repellency on human skin.

Equally important is the contribution of zeta potential to colloidal stability. High absolute zeta potential values, generally greater than ±30 mV, impart electrostatic repulsion that minimizes droplet aggregation. In systems stabilized by neutral, nonionic surfactants, steric stabilization also becomes significant. For example, basil oil nanoemulsions formulated with Tween 20 exhibited zeta potentials of approximately −25 mV, which were sufficient to maintain physical stability under conditions relevant to topical application.

Rheological properties further influence performance, as the incorporation of viscosity-modifying agents such as Carbopol or xanthan gum can transform liquid nanoemulsions into semisolid gels. This modification improves user compliance, reduces volatilization of active oils, and prolongs their release, thereby enhancing overall functional performance. Similarly, release kinetics are shaped by the encapsulation process, which governs both immediate burst release and sustained release phases. Clove oil nanoemulsions, for instance, have been shown to exhibit biphasic release, with an initial rapid liberation of surface-associated eugenol followed by gradual diffusion of the encapsulated fraction. Such controlled release profiles are vital for achieving long-lasting repellency with minimal need for reapplication.

Finally, the environmental stability of nanoemulsions is a critical determinant of their practical applicability. Exposure to temperature fluctuations, pH variation, and light can compromise stability, but well-optimized formulations demonstrate resilience under these stresses. Neem and lemongrass oil nanoemulsions, for example, retained stability up to 40 °C, highlighting their suitability for deployment in tropical and resource-limited settings.

Taken together, these findings emphasize that the functional efficacy of nanoemulsions is closely tied to formulation design. By systematically optimizing droplet size, charge, rheology, and release behavior, while ensuring stability under environmental stresses, medicinal plant-based nanoemulsions can be advanced into reliable and sustainable insect-repellent products.

5.4 Dosage Forms: Nanoemulsion Gels, Sprays, and Creams

For the successful translation of nanoemulsion research from laboratory development to real-world application, incorporation into suitable dosage forms is essential. The choice between gels, sprays, or creams is influenced by multiple factors, including the intended site of application, the preferences of the end -user, and the specific insect vector being targeted. These formulations not only facilitate user-friendly administration but also enhance the stability and functional performance of nanoemulsions under practical conditions.

Nanoemulsion gels represent one of the most widely studied dosage forms. By incorporating appropriate gelling agents such as Carbopol or hydroxypropyl methylcellulose, liquid nanoemulsions can be converted into stable semisolid systems.

This transformation offers several advantages, including improved ease of topical application, reduced volatility of the active essential oils, and enhanced adherence to the skin surface. Empirical studies have demonstrated that citronella and basil oil nanoemulsion gels can extend repellency for up to 8 h, significantly surpassing the efficacy of their crude oil counterparts [47]. In addition, their nonsticky texture contributes to higher user compliance and comfort during use.

Spray-based nanoemulsions offer another promising dosage form, available in both aerosol and pump formats. Their fine droplet dispersion ensures uniform surface coverage, thereby creating an effective protective barrier on skin or clothing. Successful examples include eucalyptus and lemongrass oil nanoemulsions formulated as sprays, in which the addition of natural humectants prevents excessive skin dryness and maintains comfort during prolonged application [54]. This delivery system combines convenience with efficiency, making sprays highly attractive for consumer adoption.

Nanoemulsion creams and lotions constitute a third category of formulations, particularly suitable for daily topical use. Oil-in-water nanoemulsion bases can be thickened to yield creams or lotions that balance repellency with additional dermocosmetic benefits. Neem oil nanoemulsion creams, for example, provide not only insect-repellent activity but also antimicrobial properties, broadening their utility as multifunctional products for skincare and protection [55].

Across all dosage forms, several critical considerations determine their success. Parameters such as skin compatibility, spreadability, and long-term storage stability strongly influence both consumer acceptance and regulatory approval. The incorporation of natural surfactants and biocompatible excipients further enhances the safety profile of these formulations, which is particularly important for sensitive populations such as children and individuals with delicate skin.

6 Mechanisms of Insect-Repellent Action

The use of medicinal plant-derived aromatic oils and their nanoemulsion formulations as insect repellents is underpinned by a range of complex mechanisms that operate at the molecular, physiological, and behavioral levels. These mechanisms not only highlight the biological sophistication of essential oils but also explain their effectiveness against different insect vectors of medical relevance. Among the most prominent processes are olfactory interference and odor masking, neurotoxic interactions with insect sensory receptors, and the disruption of host-seeking and feeding behaviors. The following sections outline these mechanisms in detail, with emphasis on studies involving nanoemulsified formulations of citronella, neem, eucalyptus, peppermint, clove, lemongrass, and basil oils.

6.1 Olfactory Interference and Odor Masking

Nanoemulsified essential oils such as citronella, lemongrass, and basil generate a competing olfactory layer that disrupts an insect's ability to detect human-emitted signals, effectively cloaking the host. This masking effect is dramatically prolonged when oils are formulated as nanoemulsions, due to sustained release of volatile terpenoids (e.g., citral, linalool, geraniol). Studies demonstrate that citronella nanoemulsions extend mosquito protection times compared to conventional formulations by modulating release kinetics—optimized by surfactant and glycerol content—to suppress droplet evaporation and maintain effective vapor concentration [56]. The altered odor plume prevents recognition by the insect's olfactory receptors, reducing attraction and approach behaviors—a critical first line of defense in preventing vector contact.

6.2 Neurotoxic Effects on Insect Sensory Receptors

Certain aromatic compounds in essential oils, such as eugenol (from clove) and 1,8-cineole (from eucalyptus), exert neurotoxic effects on insects by targeting receptor proteins and ion channels exclusive to invertebrates, such as GABA-gated chloride and octopaminergic receptors. While evidence from molecular electrophysiology is emerging, behavioral studies indicate these compounds disrupt normal sensory neuron function, impair host odor detection, and cause disorientation. In nanoemulsion form, enhanced penetration increases exposure at receptor sites—even at lower concentrations—improving efficacy. For instance, in grain weevils, clove terpenoids like eugenol and caryophyllene oxide have been shown to elicit reduced mobility and decreased respiration, suggesting interference with neural signaling pathways [57]. Though not all studies focus specifically on mosquitoes, the receptor-targeting mode of action is widely observed across arthropod pests.

6.3 Behavioral Repellence and Disruption of Host-Seeking Activity

Beyond receptor-level interference, medicinal plant nanoemulsions affect insect behavior by forming spatial repellents that deter landing, biting, and host approach. Nanoemulsions release repellents into the surrounding air long enough to create an effective vapor barrier, leading to reduced contact. Case evidence includes citronella and lemongrass nanoemulsions providing over 4 hours of protection in vivo (versus significantly less for crude oil) [58]. Additionally, nanogel formulations of aromatic oils (e.g., *Acroptilon repens* essential oil) achieved protection times

(~310 min) notably longer than DEET in tests against *Anopheles stephensi*, underscoring sustained repellence mediated by controlled release in behavioral assays.

6.4 Case Studies with Medicinal Plant-Based Nanoemulsions

See Table 2.

A growing body of case studies highlights the efficacy of medicinal plant-based nanoemulsions against diverse vectors of medical importance. Among mosquitoes, nanoemulsion formulations of citronella and eucalyptus oils have been extensively investigated, with results consistently demonstrating enhanced repellency duration compared to their crude oil counterparts. Controlled trials reported that citronella nanoemulsion sprays maintained over 90% repellency against *Aedes* mosquitoes for up to 6 h, whereas conventional emulsions offered protection for only two to 3 h.

Ticks represent another vector for which medicinal plant nanoemulsions have proven effective. Neem oil nanoemulsions significantly reduced tick attachment rates and blood-feeding success, with additional studies showing that combinations of neem with peppermint or clove oils in nanoemulsions exert synergistic effects, further limiting host-parasite interactions. Similarly, investigations involving flies have demonstrated promising outcomes. Lemongrass and basil nanoemulsions reduced landing rates and feeding attempts of houseflies and stable flies, effects that were attributed both to olfactory disruption and to the irritant properties of citral and methyl chavicol.

Beyond these primary vectors, nanoemulsion formulations of medicinal plant essential oils are increasingly being evaluated against other medically relevant arthropods such as sandflies, which transmit leishmaniasis, and lice. Across these diverse studies, a consistent pattern emerges: nanoemulsion encapsulation enhances efficacy, prolongs duration of action, and improves skin tolerability relative to crude essential oils.

Table 2 Reported applications of medicinal plant-based nanoemulsions in insect-repellent formulations and their key outcomes

Insect vector	Medicinal plant oil	Formulation type	Key result	Key references
Aedes aegypti	Citronella, basil, vetiver	Nanoemulsion (NE)	~4.7 h protection, enhanced stability	[59]
Aedes, *Anopheles*, *Culex* spp.	Nutmeg oil	Nanoemulsion (NT levels)	Improved repellency over oil alone, stable, low surfactant <10%	[60]
Sitophilus granarius (weevil)	Clove, cinnamon oils	Terpenoid bioassay	Reduced respiration, repellency, and toxic effects	[57]
Aedes aegypti, others	Citronella/basil/vetiver mix	Nanoemulsion (NE)	Small droplet (150–160 nm), high zeta stability, prolonged release, longer repellency	[59]

The mechanisms underlying this enhanced repellency are multifaceted. Aromatic plant-based nanoemulsions act through olfactory masking, neurotoxic interactions with sensory receptors, and behavioral interference. By improving the volatility, bioavailability, and stability of essential oils, nanoemulsion technology markedly strengthens their repellent performance across a range of insect vectors. Collectively, the available evidence underscores the promise of medicinal plant-derived nanoemulsions as sustainable, safe, and effective alternatives to synthetic repellents such as DEET, with the added advantage of reduced environmental toxicity.

7 Characterization and Evaluation of Nanoemulsions

The development of aromatic plant-based nanoemulsions for insect-repellent applications necessitates thorough characterization and rigorous performance evaluation. This section reviews key assessment domains including physicochemical profiling, stability testing, in vitro release dynamics, and bioassays—with all evidence drawn from studies involving medicinal plant essential oils.

7.1 Physicochemical Analysis

Stability assessment of nanoemulsion systems encompasses the evaluation of physicochemical integrity under diverse stress conditions, including variations in temperature, pH, ionic strength, and light exposure. While numerous investigations have reported stability under ambient conditions, more comprehensive evaluations addressing accelerated aging, pH variability, and photostability remain comparatively limited. As highlighted in an extensive review, most essential oil nanoemulsion studies focus primarily on droplet size and polydispersity index (PDI) under standard storage conditions, with far fewer extending to environmental stress testing [61].

Experimental findings with neem oil nanoemulsions illustrate the potential for robust stability. Such formulations retained kinetic stability over several weeks without evident phase separation or creaming, and droplet diameters increased only marginally during the observation period [62]. Similarly, seed oil nanoemulsions derived from tropical plants demonstrated stability across a pH range of 4.3 to 6.5 while maintaining zeta potential values between −32 and − 58 mV. These findings suggest that electrostatic repulsion provides a strong stabilizing mechanism under moderate environmental fluctuations.

Additional evidence comes from studies employing low-energy preparation methods. For example, nanoemulsions formulated with *Protium heptaphyllum* resin oil exhibited excellent kinetic stability and resistance to sedimentation, even in the absence of thermal or photostability testing [63]. Such results reinforce the promise of medicinal plant-based nanoemulsions but also underscore the gaps in systematic evaluation of stability under more demanding environmental conditions.

In summary, while current literature confirms favorable stability profiles for a wide range of medicinal plant nanoemulsions, the absence of consistent stress-testing protocols highlights a critical need for more rigorous assessments. Greater emphasis on environmental stability studies will be essential to support the practical deployment and regulatory acceptance of these formulations in real-world applications.

7.2 Stability Studies

Stability assessment of nanoemulsion systems encompasses the evaluation of physicochemical integrity under diverse stress conditions, including variations in temperature, pH, ionic strength, and light exposure. While numerous investigations have reported stability under ambient conditions, more comprehensive evaluations addressing accelerated aging, pH variability, and photostability remain comparatively limited. As highlighted in an extensive review, most essential oil nanoemulsion studies focus primarily on droplet size and polydispersity index (PDI) under standard storage conditions, with far fewer extending to environmental stress testing [61].

Experimental findings with neem oil nanoemulsions illustrate the potential for robust stability. Such formulations retained kinetic stability over several weeks without evident phase separation or creaming, and droplet diameters increased only marginally during the observation period [62]. Similarly, seed oil nanoemulsions derived from tropical plants demonstrated stability across a pH range of 4.3 to 6.5 while maintaining zeta potential values between −32 and − 58 mV. These findings suggest that electrostatic repulsion provides a strong stabilizing mechanism under moderate environmental fluctuations.

Additional evidence comes from studies employing low-energy preparation methods. For example, nanoemulsions formulated with *Protium heptaphyllum* resin oil exhibited excellent kinetic stability and resistance to sedimentation, even in the absence of thermal or photostability testing [63]. Such results reinforce the promise of medicinal plant-based nanoemulsions but also underscore the gaps in systematic evaluation of stability under more demanding environmental conditions.

In summary, while current literature confirms favorable stability profiles for a wide range of medicinal plant nanoemulsions, the absence of consistent stress-testing protocols highlights a critical need for more rigorous assessments. Greater emphasis on environmental stability studies will be essential to support the practical deployment and regulatory acceptance of these formulations in real-world applications.

7.3 *In Vitro Release Studies*

The understanding of how essential oils release from nanoemulsions is critical for achieving optimal repellency duration. Controlled release patterns are inferred from physicochemical data and indirect bioactivity assays in most studies; explicit in vitro diffusion tests (e.g., Franz cells) are uncommon in the insect-repellent literature. However, in repellency trials, the extended protection observed with nanoemulsions strongly suggests that release is sustained compared to crude oils [64]. For instance, *Mentha piperita* and *Eucalyptus globulus* nanoemulsions provided 4.96 ± 0.21 h and 6.06 ± 0.20 h protection, respectively, in field tests—evidence of prolonged release enabled by the nanoemulsion matrix. To robustly quantify release kinetics, future work should incorporate standardized in vitro release assays across formulations.

7.4 *Bioassays for Insect Repellency*

The functional efficacy of repellents is typically assessed through established bioassays such as the arm-in-cage method, cone tests, and controlled field trials. These assays are most informative when nanoemulsion formulations are evaluated alongside their corresponding crude oils and benchmark standards such as DEET, thereby enabling direct performance comparisons.

Arm-in-cage methodologies have been widely adopted to determine complete protection times (CPTs) for medicinal essential oils. For instance, Sutthanont et al. [65] demonstrated that ten essential oils, including basil, peppermint, and petitgrain, provided between 120 and 360 minutes of protection against *Culex quinquefasciatus*, with petitgrain affording 270 min of protection against *Aedes aegypti*. Building on these findings, comparative studies have highlighted the benefits of nanoemulsion delivery systems. A *Mentha piperita* nanoemulsion achieved nearly 5 h of protection, a significant improvement over the crude oil, though still somewhat inferior to DEET. In contrast, a *Eucalyptus*-based nanoemulsion was found to statistically match the performance of DEET under similar conditions [64].

Neem oil has also been investigated, with nanoemulsified formulations substantially enhancing efficacy. Specifically, protection against *Anopheles stephensi* increased from the typically shorter durations observed with crude oil to a CPT of 240 ± 39 min, approaching DEET's benchmark of 360 ± 24 min. Field-based trials further underscore the promise of these systems. For example, a eucalyptus oil microemulsion at 15% w/w achieved a protection time of 170.7 ± 26 min, a performance that was comparable to DEET in controlled laboratory settings [66].

Collectively, these findings confirm that medicinal plant nanoemulsions can provide extended and reliable repellency, narrowing the efficacy gap with synthetic standards. Their ability to enhance the protective capacity of crude oils while

Table 3 Summary of key characterization and evaluation parameters of medicinal plant-based nanoemulsions developed for insect-repellent applications

Assessment area	Key parameters	Medicinal plant examples
Physicochemical analysis	Particle size, PDI, viscosity, interfacial tension	Neem, *Caesalpinia* seed oil nanoemulsions
Stability studies	Thermal, pH, ionic strength, light	Neem, tropical seed oil, *Protium heptaphyllum*
In vitro release	Release kinetics (inferred via protection times)	*Mentha piperita*, *Eucalyptus globulus*
Bioassay evaluation	Arm-in-cage, field trials, CPT	Petitgrain, basil, neem, eucalyptus, peppermint NE

minimizing safety and environmental concerns positions them as credible candidates for next-generation insect-repellent technologies (Table 3).

Rigorous characterization and evaluation of medicinal plant-based nanoemulsions underpin their translation into effective insect-repellent formulations. Standard characterization metrics—particle size, PDI, viscosity, and interfacial tension—provide a foundation for understanding delivery dynamics. Stability under diverse conditions ensures shelf life and field applicability. Although in vitro release testing remains underrepresented, extended protection in human assays offers promising surrogate evidence. Finally, bioassays, particularly arm-in-cage and field trials, validate functional performance and highlight the tangible advantages of nanoemulsified essential oils over crude oil formulations. For future research, implementing standardized in vitro release models, expanded accelerated stability testing, and deeper interrogation of structure-activity relationships (e.g., droplet size vs. protection time) will strengthen the evidence base. The convergence of precise physicochemical control and validated repellency research positions medicinal plant nanoemulsions as viable, eco-friendly alternatives to synthetic repellents.

8 Toxicological and Environmental Considerations

The growing interest in aromatic plant-based nanoemulsions for insect-repellent applications stems not only from their efficacy but also from their promise of improved ecological and human safety. This section systematically explores four critical dimensions: biodegradability and eco-safety, cytotoxicity and dermal safety, impacts on ecological accumulation and ecotoxicity, and a comparative perspective with synthetic repellents.

8.1 Biodegradability and Eco-Safety of Plant-Based Nanoemulsions

A fundamental advantage of plant-derived nanoemulsions lies in their bio-based composition, which typically affords enhanced biodegradability and reduced environmental persistence compared to synthetic alternatives. Studies underscore that bio-based nanoemulsion systems, particularly those derived from botanical essential oils, demonstrate lower ecotoxicological footprints, and their nanoscale structure improves functional performance without compromising eco-safety [67]. Moreover, emerging ecotoxicological assessments indicate that essential oil-based nanoemulsions (e.g., clove, ginger, henna oils) degrade more rapidly in environmental matrices and exhibit minimal residues, thereby minimizing long-term ecological hazards [68]. This eco-safety profile positions medicinal plant nanoemulsions as crucial tools for sustainable repellent solutions, especially in ecosystems where environmental accumulation of synthetic agents is a significant concern.

8.2 Cytotoxicity and Dermal Safety Assessments

Safety assessment for human use is central to product development. Contemporary studies evaluate cytotoxicity (e.g., via keratinocyte assays) and dermal irritation/sensitization for nanoemulsified essential oils. Albeit direct data on specific oils remain limited, research on similar biocompatible nanoemulsion systems frequently indicates negligible cytotoxicity and high dermal tolerability, particularly when nonionic or lecithin-based surfactants are utilized. These platforms typically avoid harsh chemical residues and maintain safety even after prolonged exposure [69]. Additionally, studies emphasize the importance of surface capping or coatings to reduce nanoparticle reactivity while retaining biological activity—a principle applicable in smoothing nanoemulsion-host interactions [70]. Rigorous in vitro and in vivo dermal safety testing tailored to each essential oil system is still needed, but extant evidence supports a general favorable safety margin for plant-based nanoemulsion repellents.

8.3 Risk of Nanoparticle Accumulation and Ecotoxicity

While nanoscale dimensions enhance bioefficacy, they also raise concerns about potential bioaccumulation, interspecies transfer, and nano-toxicity in nontarget organisms. Although plant-based emulsions typically use biodegradable constituents, their long-term environmental behavior remains understudied. By analogy with broader nanomaterial literature, nanoparticle size, surface chemistry, and coating influence environmental fate and toxicity, with uncapped or reactive surfaces posing higher hazards [71]. Hence, to mitigate such risks, saturating nanoemulsion surfaces with benign surfactants or biocompatible polymers—analogous to nanoparticle capping strategies—could reduce ecological persistence and ensure safer degradation profiles.

8.4 Comparison with Synthetic Repellents

Synthetic repellents such as N,N-diethyl-meta-toluamide (DEET) continue to dominate the market owing to their proven efficacy; however, their use is not without significant drawbacks. Concerns have been raised regarding potential neurotoxicity, skin sensitization, and long-term environmental persistence. Moreover, DEET residues have been detected in water bodies and shown to exhibit moderate toxicity toward aquatic organisms, including algae and fish. Although the compound is biodegradable under aerobic conditions, its persistence in anaerobic environments raises the possibility of bioaccumulation and associated ecological risks.

By contrast, plant-based nanoemulsions represent a promising alternative that can achieve protection times comparable to those of synthetic repellents. For example, certain *Eucalyptus*-derived nanoemulsions have demonstrated performance on par with DEET while minimizing risks to both human health and the environment [72]. Medicinal plant oils such as citronella and basil are generally recognized as safe by regulatory bodies, and their nanoemulsified formulations typically display minimal environmental footprints and low toxicity profiles. Although isolated cases of skin irritation have been reported, the overall safety record of these botanical compounds remains favorable compared to their synthetic counterparts.

The advantages of nanoemulsified medicinal plant repellents extend beyond individual safety to encompass broader ecological considerations. Their biodegradability, reduced human toxicity, and limited environmental impact render them particularly attractive in ecologically sensitive regions or health-critical settings. Nevertheless, achieving widespread adoption will depend on continued efforts to ensure environmental and dermatological safety. Standardized skin compatibility testing, long-term ecotoxicological evaluations, and formulation strategies that emphasize degradable and inert carrier systems are essential steps toward establishing these products as credible alternatives.

Through rigorous scientific development and careful regulatory validation, nanoemulsions based on medicinal plant oils have the potential to emerge as sustainable, high-performance repellents that not only rival but also surpass synthetic agents such as DEET in terms of safety, sustainability, and societal acceptance.

9 Industrial Applications and Commercialization Prospects

Medicinal plant-based nanoemulsions hold increasing promise in the insect-repellent market, reflecting rising consumer demand, advances in scalable manufacturing, evolving intellectual property landscapes, and shifting commercial offerings. This section examines market growth trends, challenges in scale-up, patent activity, and emerging product trajectories.

9.1 Growing Market Demand for Natural Insect Repellents

In recent years, the global market for natural insect repellents has experienced robust growth, reflecting heightened consumer interest in safer, plant-based alternatives to synthetic formulations. The global natural insect-repellent market was estimated at USD 2.79 billion in 2024 and is projected to reach USD 4.37 billion by 2030, exhibiting a CAGR of 8.0% from 2025 to 2030 [73]. Another forecast underscores this upward trajectory, predicting growth from USD 3.1 billion in 2025 to USD 6.5 billion by 2035, at a CAGR of 7.7% [74]. In the USA specifically, the natural insect-repellent market was valued at USD 726 million in 2024, with a projected increase to USD 1.13 billion by 2030, growing at approximately 7.9% annually [75]. These trends emphasize increasing opportunities for medicinal plant-based nanoemulsions, driven by performance improvements—such as prolonged release and reduced volatility—and strong eco-friendly appeal. A few noteworthy examples of commercialized encapsulated products already exist, including terpene-based fungicides developed by Eden Research (UK) and plant-oil-based biopesticides from Vegalab (Switzerland). These illustrate the translational potential of encapsulation technologies from laboratory innovation to agricultural markets.

9.2 Challenges in Large-Scale Manufacture and Commercial Scale-Up

Translating nanoemulsion technology from laboratory proof of concept to industrial-scale production entails several complex challenges that extend beyond mere technical replication. One critical limitation lies in the cost and energy requirements of scale-up. Processes such as high-pressure homogenization, ultrasonication, and other high-energy methods that are feasible at the laboratory level become both economically and operationally burdensome when applied to larger production volumes [76].

Equally important is the issue of reproducibility and batch-to-batch consistency. Achieving uniform droplet size distribution, maintaining a narrow polydispersity index (PDI), and ensuring interfacial stability across large-scale batches necessitate stringent process control and advanced analytical monitoring systems [77]. Without such oversight, variability in product performance can undermine both regulatory compliance and market acceptance.

Furthermore, equipment design exerts a significant influence on scale-up outcomes. The scaling process is not linear, and challenges such as uneven heating, mixing inefficiencies, and altered shear profiles frequently compromise product quality. These limitations demand specialized engineering solutions, including tailored equipment design and extensive process validation [78].

Industry stakeholders have begun to address these obstacles. For instance, Delphi Scientific, LLC has refined critical process parameters and optimized equipment design to enhance reproducibility, reduce costs, and improve the overall performance of large-scale nanoemulsion manufacturing systems. Such innovations provide a model for bridging the gap between laboratory success and commercial viability.

9.3 Patent Landscape and Intellectual Property Trends

Patent filings reflect growing innovation in medicinal plant-based nanoemulsion repellents. A citronella oil nanoemulsion patent (CN102166314A) outlines a stable, low-viscosity O/W nano-formulation with mosquito-repellent, antibacterial, and mood-enhancing properties while also emphasizing scalable and low-toxicity production [79]. Earlier general patents, such as WO2010061354A2, propose aqueous formulations combining multiple essential oils, tinctures, and emulsifiers, though these are not specifically designed as nanoemulsions [80]. Despite these developments, formal patent coverage focused specifically on medicinal plant nanoemulsions for insect-repellent use remains limited, suggesting considerable opportunity for innovators to secure tailored intellectual property.

9.4 Commercial Products and Emerging Industry Trends

Botanical-based repellents have increasingly transitioned from laboratory research to the commercial marketplace, reflecting both consumer demand and regulatory validation. One notable example is oil of lemon eucalyptus (PMD), a compound derived from medicinal plant sources that has obtained registration from the US Environmental Protection Agency (EPA). Its widespread use highlights the growing regulatory acceptance of botanically derived repellents as viable alternatives to synthetic formulations.

In parallel, several commercial brands, including Murphy's Naturals and Wondercide, have successfully positioned themselves in the market by formulating natural repellents that incorporate plant-derived oils such as peppermint, lemongrass, and cedar. These products emphasize eco-friendly branding and consumer safety; however, they generally do not rely on advanced delivery technologies such as nanoemulsions [75].

At the industry level, the global mosquito repellent market was valued at approximately USD 4.59 billion in 2025 and is projected to reach USD 6.58 billion by 2030, with a compound annual growth rate (CAGR) of nearly 7.5%. This sustained growth trajectory highlights the considerable opportunity for premium, next-generation delivery systems, particularly nanoemulsions, to capture a greater share of the expanding market [81].

Key trends include convergence between cosmetic and functional use cases, consumer preference for natural ingredients with enhanced aesthetics or sensory profiles, and an expectation for multifunctional product formats such as lotions or sprays delivering both fragrance and protection. The commercialization of aromatic medicinal plant-based nanoemulsion repellents aligns with growing consumer demand for natural, sustainable, and effective products. While significant challenges in scale-up, reproducibility, regulatory compliance, and intellectual property positioning remain, the potential market—backed by robust growth projections—offers substantial opportunity for innovators.

Navigating this path effectively will involve process optimization, sound regulatory strategies, strategic patenting, and market differentiation through nanoemulsion benefits (e.g., extended duration, skin safety, eco-friendliness). With continued interdisciplinary collaboration across formulation science, manufacturing engineering, and regulatory frameworks, the next generation of botanical repellents stands poised to redefine both consumer expectations and industry standards.

10 Future Directions and Research Perspectives

The advancement of aromatic plant-based nanoemulsions as insect repellents is entering a transformative stage, where cross-disciplinary innovations can shape both scientific progress and practical applications. Although significant achievements have been made in demonstrating efficacy, several areas remain underexplored. The following subsections outline the key research perspectives for the coming years.

10.1 Integration with Other Nanocarriers

One promising future direction involves the integration of nanoemulsions with other nanocarriers such as liposomes, polymeric nanoparticles, and nanogels. This strategy can improve stability, enhance controlled release, and provide dual protective mechanisms by combining the advantages of multiple delivery platforms [82]. For example, liposomes loaded with essential oils like clove or eucalyptus demonstrate deeper dermal penetration, while nanoemulsions enable faster dispersion and absorption. Together, hybrid systems could generate next-generation repellents offering extended duration and reduced volatility of active molecules.

10.2 Smart and Stimuli-Responsive Nanoemulsions

Another forward-looking approach is the development of smart, stimuli-responsive nanoemulsions. These systems release essential oils in response to specific environmental or physiological cues such as temperature, pH, or enzymatic activity [83]. For instance, thermo-responsive nanoemulsions could prevent premature oil evaporation in tropical climates, while pH-sensitive formulations may activate upon contact with the slightly acidic skin surface, ensuring efficacy only at the time of use [84]. Such smart systems can minimize wastage and improve the longevity of repellence. Translationally, smart nanoemulsions for skin benefit from the broader nanomedicine literature, which details triggered release strategies and biocompatible polymers to minimize irritation while enhancing residence at the stratum corneum—principles that can be directly applied to botanical repellents to achieve longer protection with lower reapplication frequency [84].

10.3 Synergistic Combinations of Essential Oils

Another priority is systematic exploration of synergy among chemotyped essential oils and their key constituents (e.g., citronellal, eugenol, thymol, geraniol), where complementary modes of repellence (olfactory masking, irritancy) or insecticidal sublethal stress (e.g., neuroenzymatic disruption) can be balanced. Earlier broad screens across 60 essential oils identified several lead candidates and suggested that nanoemulsification can prolong activity by moderating volatility, thereby creating a favorable backdrop for combining botanicals [85]. Recent work underscores that mixtures may yield additive or synergistic effects but can also be antagonistic, necessitating rigorous design of experiments to optimize ratios and avoid counterproductive combinations [26]. For example, binary mixtures of essential oils or their principal components have shown enhanced insecticidal performance against storage pests, illustrating principles that could be translated to repellent end points with mosquitoes and ticks when formulated via nanoemulsions for controlled evaporation [86]. Likewise, a 2024–2025 study on EO combinations with inorganic nanostructures (e.g., zeolite nanoparticles) indicated synergistic activity and distinct physical modes of action, hinting at hybrid botanical-inorganic designs worth vetting for personal protection products so long as dermal safety is demonstrated [87].

Complementary to mixture studies, narrative and scoping reviews continue to consolidate evidence that aromatic-plant essential oils possess meaningful repellent profiles against *Aedes* spp., supporting rational selection of pairs/triads for synergy testing within nanoemulsions [88]. At the product concept level, nanostructured lipid carriers (NLCs) and other hybrid lipidic systems with essential oils have demonstrated prolonged repellency compared with conventional emulsions, providing an additional lever for synergy by combining carrier effects with multi-oil blending [89].

10.4 Role in Integrated Pest Management (IPM) Strategies

Within IPM frameworks, medicinal-plant nanoemulsion repellents are best positioned as behavior-modifying components that reduce human-vector contact in combination with physical measures (e.g., nets) and environmental management, thereby decreasing reliance on synthetic chemistries and delaying resistance. Reviews on botanical interventions emphasize their compatibility with ecosystem health and potential to fit into community-level programs, provided that efficacy and duration are reliably demonstrated under field conditions [88]. A growing body of literature documents EO efficacy against mosquitoes and ticks, two priority vectors for human health, and calls for standardized bioassays to enable evidence-based inclusion in IPM [90]. Methodological harmonization is advancing as well: recent analyses critique laboratory repellency methods and recommend more predictive protocols bridging lab, semi-field, and field trials—guidance that will be crucial for regulatory and programmatic adoption of nanoemulsified botanicals [91].

From a technology standpoint, biopolymeric and lipid-based nanostructures tailored for mosquito repellents are now surveyed through a translational lens, highlighting barriers to clinical/consumer translation (scale-up, reproducibility, safety dossiers) and proposing strategies for carrier selection and process design, which is essential for IPM procurement and deployment [92]. Newer reports on ecological repellent preparations using natural polymers with essential oils also align with IPM values by emphasizing biodegradability and low ecotoxicity, and these could be adapted into controlled-release sprays or wearable in tandem with nanoemulsions [93]. Finally, the broader evidence synthesis on EO-based repellents for public health continues to expand, with recent human-vector studies evaluating clove, cinnamon, and other medicinal-plant oils in head-to-head repellency assays, thus enriching the evidence base for programmatic decisions [10].

Hence, sustainability will remain a core focus in the future development of nanoemulsions. Researchers are exploring solvent-free preparation methods, biodegradable surfactants, and renewable excipients to reduce the environmental footprint of production. Coupling green nanotechnology with medicinal plant oils not only meets consumer demands for eco-conscious products but also aligns with global regulatory and environmental standards.

11 Conclusion

The exploration of aromatic plant-based nanoemulsions as insect repellents underscores a rapidly advancing field at the convergence of nanotechnology, medicinal plant sciences, and public health. This chapter has addressed the multifaceted aspects of these systems, ranging from formulation fundamentals and encapsulation strategies for essential oils to characterization, biological efficacy, and environmental safety. Collectively, these discussions highlight the promise of nanoemulsion-based repellents in mitigating vector-borne diseases while supporting ecological sustainability. Yet their adoption in large-scale agriculture remains limited, with only preliminary greenhouse and field trials suggesting potential utility, and a handful of encapsulated products having reached commercialization. Bridging this gap between innovation and practice will be crucial for future impact.

A central conclusion is the ability of nanoemulsions to overcome the inherent limitations of essential oils. While oils such as citronella, neem, peppermint, and clove are established repellents, their volatility, instability, and poor aqueous solubility limit practical use. Nanoemulsions, through nanoscale droplet size, large surface area, and kinetic stability, enhance dispersion, prolong release, and improve dermal penetration. Consequently, they not only preserve but also amplify repellent efficacy in real-world applications.

Equally important is the systematic approach to formulation and optimization. The choice of surfactants, oil-to-surfactant ratios, and emulsification techniques critically determines droplet size, stability, and biological performance. Coupled with physicochemical characterization and biological assays—such as zeta potential, in vitro release studies, and standardized repellency bioassays—these methods ensure reliable translation from laboratory to product development.

Toxicological and ecological considerations further reinforce the relevance of plant-based nanoemulsions. Their biodegradability, reduced risk of nanoparticle accumulation, and dermal safety distinguish them from synthetic repellents. These qualities make them particularly suited for communities requiring frequent, safe, and eco-friendly protection against vectors such as *Aedes* and *Anopheles* mosquitoes.

From a translational perspective, rising consumer demand for natural, sustainable repellents aligns with the opportunities nanoemulsions provide. Although scalability, reproducibility, and regulatory approval remain challenges, innovations in processing and green nanotechnology are steadily addressing these barriers. Integration into holistic integrated pest management frameworks further enhances their applicability.

Looking ahead, future research is expected to focus on hybrid carriers, stimuli-responsive systems, synergistic oil blends, and personalized repellents, ensuring that these technologies evolve in step with ecological and public health needs. Ultimately, aromatic plant-based nanoemulsions embody the fusion of traditional medicinal resources with nanoscience innovation, offering sustainable solutions for global vector control and human well-being.

Competing Interests The authors declare that they have no competing interest.

Declaration of Generative AI and AI-Assisted Technologies in the Writing Process During the preparation of this work, the book chapter author(s) have not taken any assistance from AI, and the book chapter author(s) reviewed and edited the content as needed and take(s) full responsibility for the content of the publication.

References

1. Lee H, Halverson S, Ezinwa N (2018) Mosquito-borne diseases. Prim Care 45(3):393–407
2. Venkatesan P (2024) The 2023 WHO world malaria report. Lancet Microbe 5(3):e214
3. Ma C-S, Wang B-X, Wang X-J, Lin Q-C, Zhang W, Yang X-F, van Baaren J, Bebber DP, Eigenbrode SD, Zalucki MP (2025) Crop pest responses to global changes in climate and land management. Nat Rev Earth Environ 6(4):264–283
4. da Camara CA, Akhtar Y, Isman MB, Seffrin RC, Born FS (2015) Repellent activity of essential oils from two species of citrus against Tetranychus urticae in the laboratory and greenhouse. Crop Prot 74:110–115
5. Haleem ZM, Yadav S, Cushion ML, Tanner RJ, Carek PJ, Mainous AG III (2020) Exposure to n, n-diethyl-meta-toluamide insect repellent and human health markers: population based estimates from the national health and nutrition examination survey. Am J Trop Med Hygiene 103(2):812
6. Mandal P, Saha M (2021) Challenges and perspectives of plant-derived green solvents. In: Green sustainable process for chemical and environmental engineering and science. Elsevier, Amsterdam, pp 157–170
7. Ragavendran K, Selvakumaran J, Muthukanagavel M, Alharbi NS, Thiruvengadam M, Mutheeswaran S, Ignacimuthu S, Ganesan P (2024) Chemical composition and mosquitocidal properties of essential oil from indian indigenous plants *Ocimum tenuiflorum l.* And *Ocimum americanum l.* Against three vector mosquitoes. Exp Parasitol 258:108709
8. Wang H, Ren J, Zhou S, Duan Y, Zhu C, Chen C, Liu Z, Zheng Q, Xiang S, Xie Z (2024) Molecular regulation of oil gland development and biosynthesis of essential oils in *citrus* spp. Science 383(6683):659–666
9. Ben Miri Y (2025) Essential oils: chemical composition and diverse biological activities: A comprehensive review. Nat Prod Commun 20(1):1934578X241311790
10. Lopez AD, Whyms S, Luker HA, Galvan CJ, Holguin FO, Hansen IA (2025) Repellency of essential oils and plant-derived compounds against *Aedes aegypti* mosquitoes. Insects 16(1):51
11. Tabanca N, Cloonan KR, Nesterkina M, Gill MA, Montgomery WS, Kravchenko I, Kendra PE (2024) Behavioral and electrophysiological responses of the male medfly, *Ceratitis capitata*, to thymol and carvacrol ethers. Pest Manag Sci 81:7314
12. Park Y, Piermarini PM (2025) Heat activation desensitizes *Aedes aegypti* transient receptor potential ankyrin 1 (AaTRPA1) to chemical agonists that repel mosquitoes. Pestic Biochem Physiol 209:106326
13. Abbas MG, Binyameen M, Azeem M, Majeed S, Sarwar ZM, Nazir A, Sharif MMI, Parveen A, Mozūratis R (2025) Chemical analysis, repellent, larvicidal, and oviposition deterrent activities of plant essential oils against *Aedes aegypti*, *Anopheles gambiae*, and *Culex quinquefasciatus*. Front Insect Sci 5:1582669
14. Isman MB (2020) Botanical insecticides in the twenty-first century—fulfilling their promise? Annu Rev Entomol 65(1):233–249
15. Nerio LS, Olivero-Verbel J, Stashenko E (2010) Repellent activity of essential oils: A review. Bioresour Technol 101(1):372–378

16. Pavela R, Benelli G (2016) Essential oils as ecofriendly biopesticides? Challenges and constraints. Trends Plant Sci 21(12):1000–1007
17. Maia MF, Moore SJ (2011) Plant-based insect repellents: A review of their efficacy, development and testing. Malar J 10(Suppl 1):S11
18. Sharma R, Rao R, Kumar S, Mahant S, Khatkar S (2019) Therapeutic potential of citronella essential oil: A review. Curr Drug Discov Technol 16(4):330–339
19. Anton N, Benoit J-P, Saulnier P (2008) Design and production of nanoparticles formulated from nano-emulsion templates—a review. J Control Release 128(3):185–199
20. Na M, Ritacco G, O'Brien D, Lavelle M, Api AM, Basketter D (2021) Fragrance skin sensitization evaluation and human testing: 30-year experience. Dermatitis 32(5):339–352
21. Debboun M, Frances SP, Strickman D (2014) Insect repellents handbook. CRC Press, Boca Raton
22. Sarmah K, Anbalagan T, Marimuthu M, Mariappan P, Angappan S, Vaithiyanathan S (2025) Innovative formulation strategies for botanical-and essential oil-based insecticides. J Pest Sci 98(1):1–30
23. Raghuvanshi TS, Singh PP, Kohar N, Prakash B (2023) Essential oils: from traditional to modern-day applications with special reference to medicinal and aromatic plants in India. In: Plant essential oils: from traditional to modern-day application. Springer, Cham, pp 1–26
24. Ayllón-Gutiérrez R, Díaz-Rubio L, Montaño-Soto M, Haro-Vázquez M d P, Córdova-Guerrero I (2024) Applications of plant essential oils in pest control and their encapsulation for controlled release: a review. Agriculture (Basel) 14(10):1766
25. Pavela R, Pavoni L, Bonacucina G, Cespi M, Cappellacci L, Petrelli R, Spinozzi E, Aguzzi C, Zeppa L, Ubaldi M (2021) Encapsulation of *Carlina acaulis* essential oil and carlina oxide to develop long-lasting mosquito larvicides: microemulsions versus nanoemulsions. J Pest Sci 94(3):899–915
26. Farina P, Conti B (2024) Liabilities of essential oils as insect repellents. Curr Opin Environ Sci Health 40:100564
27. Hasler T, Fehr J, Held U, Schlagenhauf P (2019) Use of repellents by travellers: A randomised, quantitative analysis of applied dosage and an evaluation of knowledge, attitudes and practices (kap). Travel Med Infect Dis 28:27–33
28. Lalko J, Api A (2006) Investigation of the dermal sensitization potential of various essential oils in the local lymph node assay. Food Chem Toxicol 44(5):739–746
29. Kaspute G, Ivaskiene T, Ramanavicius A, Ramanavicius S, Prentice U (2025) Terpenes and essential oils in pharmaceutics: applications as therapeutic agents and penetration enhancers with advanced delivery systems for improved stability and bioavailability. Pharmaceutics 17(6):793
30. Matthews M (2025) Are essential oils enough to keep mosquitoes away? Real Simple. https://www.realsimple.com/are-essential-oils-enough-to-keep-mosquitoes-away-11762409
31. Feuser ZP, Colonetti T, Grande AJ, Uggioni MLR, Roever L, da Rosa MI (2020) Efficacy of the deet, ir3535, and picaridin topical use against *Aedes aegypti*: systematic review. Infect Dis Clin Pract 28(6):327–341
32. Arena JS, Rossetti MR, Salvo A (2025) Overall repellent effect of plant essential oils on insect choice behaviour: a meta-analysis. J Pest Sci 98:1–11
33. Nelson PN, Mulder WH (2024) Thermodynamic and kinetic models for the photolytic and hydrolytic degradation of picaridin: A computational and theoretical study. Comput Theor Chem 1238:114706
34. Chevalier RC, Gomes A, Cunha RL (2022) Role of aqueous phase composition and hydrophilic emulsifier type on the stability of w/o/w emulsions. Food Res Int 156:111123
35. Sun G, Wang L, Dong Z, Zhang Y, Yang Y, Hu M, Fang H (2025) The current status, hotspots, and development trends of nanoemulsions: A comprehensive bibliometric review. Int J Nanomedicine 20:2937–2968
36. Elsewedy HS (2025) Insights of nanoemulsion as a drug delivery system: an overview of current trends and applications. Ind J Pharm Edu Res 59(2):472–492

37. Alhamdany AT, Saeed AM, Alaayedi M (2021) Nanoemulsion and solid nanoemulsion for improving oral delivery of a breast cancer drug: formulation, evaluation, and a comparison study. Saudi Pharm J 29(11):1278–1288
38. Che Marzuki NH, Wahab RA, Abdul Hamid M (2019) An overview of nanoemulsion: concepts of development and cosmeceutical applications. Biotechnol Biotechnol Equip 33(1):779–797
39. Schreiner TB, Santamaria-Echart A, Ribeiro A, Peres AM, Dias MM, Pinho SP, Barreiro MF (2020) Formulation and optimization of nanoemulsions using the natural surfactant saponin from quillaja bark. Molecules 25(7):1538
40. Cholakova D, Tcholakova S (2024) Sucrose ester surfactants: current understanding and emerging perspectives. Curr Opin Colloid Interface Sci 73:101832
41. Jacob S, Kather FS, Boddu SH, Shah J, Nair AB (2024) Innovations in nanoemulsion technology: enhancing drug delivery for oral, parenteral, and ophthalmic applications. Pharmaceutics 16(10):1333
42. Jasmina H, Džana O, Alisa E, Edina V, Ognjenka R (2017) Preparation of nanoemulsions by high-energy and lowenergy emulsification methods. In: CMBEBIH 2017: proceedings of the international conference on medical and biological engineering 2017. Springer, pp 317–322
43. Cholakova D, Vinarov Z, Tcholakova S, Denkov ND (2022) Self-emulsification in chemical and pharmaceutical technologies. Curr Opin Colloid Interface Sci 59:101576
44. Safaya M, Rotliwala YC (2020) Nanoemulsions: a review on low energy formulation methods, characterization, applications and optimization technique. Mater Today Proc 27:454–459
45. Komaiko JS, McClements DJ (2016) Formation of food-grade nanoemulsions using low-energy preparation methods: A review of available methods. Compr Rev Food Sci Food Saf 15(2):331–352
46. Anjali C, Sharma Y, Mukherjee A, Chandrasekaran N (2012) Neem oil (*Azadirachta indica*) nanoemulsion—a potent larvicidal agent against culex quinquefasciatus. Pest Manag Sci 68(2):158–163
47. Subramanian K (2025) Green herbicide revolution: development and characterization of *Eucalyptus globulus* essential oil-loaded nanoemulsion for sustainable weed control. BioNanoScience 15(3):1–15
48. Heydari M, Amirjani A, Bagheri M, Sharifian I, Sabahi Q (2020) Eco-friendly pesticide based on peppermint oil nanoemulsion: preparation, physicochemical properties, and its aphicidal activity against cotton aphid. Environ Sci Pollut Res 27(6):6667–6679
49. Ikawati S, Himawan T, Abadi AL, Tarno H (2021) Toxicity nanoinsecticide based on clove essential oil against *Tribolium castaneum* (herbst). J Pestic Sci 46(2):222–228
50. Abdullah NNA, Kamari A, Wong STS, Yusof N, Othman H, Hussein MZ, Hargreaves JS (2024) Lemongrass essential oil nanoemulsion formulations based on tragacanth gum and guar gum as durable anti-mosquito fabric finishing agents. Polym Bull 81(18):16903–16933
51. Bisma HAT, Adeel S, Hassan A, Bilal U, Mia R (2024) 17 insect repellent role. In: Advances in renewable natural materials for textile sustainability. CRC Press, Boca Raton, p 313
52. Iskandar B, Liu T-W, Mei H-C, Kuo I-C, Surboyo MDC, Lin H-M, Lee C-K (2024) Herbal nanoemulsions in cosmetic science: A comprehensive review of design, preparation, formulation, and characterization. J Food Drug Anal 32(4):428
53. Mahdi ES, Sakeena MH, Abdulkarim MF, Abdullah GZ, Sattar MA, Noor AM (2011) Effect of surfactant and surfactant blends on pseudoternary phase diagram behavior of newly synthesized palm kernel oil esters. Drug Des Devel Ther 5:311–323
54. Gago C, Artiga-Artigas M, Antunes M, Faleiro M, Miguel M, Martín-Belloso O (2019) Effectiveness of nanoemulsions of clove and lemongrass essential oils and their major components against *Escherichia coli* and *Botrytis cinerea*. J Food Sci Technol 56(5):2721–2736
55. Rinaldi F, Hanieh PN, Longhi C, Carradori S, Secci D, Zengin G, Ammendolia MG, Mattia E, Del Favero E, Marianecci C (2017) Neem oil nanoemulsions: characterisation and antioxidant activity. J Enzyme Inhib Med Chem 32(1):1265–1273
56. Sakulku U, Nuchuchua O, Uawongyart N, Puttipipatkhachorn S, Soottitantawat A, Ruktanonchai U (2009) Characterization and mosquito repellent activity of citronella oil nanoemulsion. Int J Pharm 372(1–2):105–111

57. Plata-Rueda A, Campos JM, da Silva Rolim G, Martínez LC, Dos Santos MH, Fernandes FL, Serrão JE, Zanuncio JC (2018) Terpenoid constituents of cinnamon and clove essential oils cause toxic effects and behavior repellency response on granary weevil, *Sitophilus granarius*. Ecotoxicol Environ Saf 156:263–270
58. Zeni V, Canale A, Lucchi A, Bruschi F, Benelli G (2019) Green micro-and nanoemulsions for managing parasites, vectors and pests. Green Synthesis of Nanomaterials 9:184
59. Pavoni L, Pavela R, Cespi M, Bonacucina G, Maggi F, Zeni V, Canale A, Lucchi A, Bruschi F, Benelli G (2019) Green micro-and nanoemulsions for managing parasites, vectors and pests. Nano 9(9):1285
60. Mohd Narawi M, Chiu HI, Yong YK, Mohamad Zain NN, Ramachandran MR, Tham CL, Samsurrijal SF, Lim V (2020) Biocompatible nutmeg oil-loaded nanoemulsion as phyto-repellent. Front Pharmacol 11:214
61. Pavoni L, Perinelli DR, Bonacucina G, Cespi M, Palmieri GF (2020) An overview of micro- and nanoemulsions as vehicles for essential oils: formulation, preparation and stability. Nano 10(1):135
62. Taheri M, Iman M, Khoobdel M, Dehghan O, Zarei SM (2020) Nanoemulsion formulation of cold pressed neem oil as a repellent against *Anopheles Stephensi*
63. Faustino CG, de Medeiros FA, Galardo AKR, Rodrigues ABL, da Costa ALP, Martins RL, Brandão LB, Santos LL, de Medeiros MAA, de Castro CP (2021) Biocidal activity of a nano-emulsion containing essential oil from *Protium heptaphyllum* resin against *Aedes aegypti* (diptera: Culicidae). Molecules 26(21):6439
64. Mohammadi R, Khoobdel M, Talebi AA, Negahban M, Norani M, Moradi M, Dehghan O (2020) In vivo evaluation of the repellency effects of nanoemulsion of *Mentha piperita* and *Eucalyptus globulus* essential oils against mosquitoes. Open Biotechnol J 14(1):145
65. Sutthanont N, Sudsawang M, Phanpoowong T, Sriwichai P, Ruangsittichai J, Rotejanaprasert C, Srisawat R (2022) Effectiveness of herbal essential oils as single and combined repellents against *Aedes aegypti*, *Anopheles dirus* and *Culex quinquefasciatus* (diptera: Culicidae). Insects 13(7):658
66. Navayan A, Moghimipour E, Khodayar MJ, Vazirianzadeh B, Siahpoosh A, Valizadeh M, Mansourzadeh Z (2017) Evaluation of the mosquito repellent activity of nano-sized micro-emulsion of *Eucalyptus globulus* essential oil against culicinae. Jundishapur J Nat Pharm Prod 12(4):e55626
67. Mishra P, Balaji A, Mukherjee A, Chandrasekaran N (2019) Bio-based nanoemulsions: an eco-safe approach towards the eco-toxicity problem. In: Handbook of ecomaterials. Springer, Cham, pp 1985–2006
68. Mahran HA (2022) Using nanoemulsions of the essential oils of a selection of medicinal plants from jazan, Saudi Arabia, as a green larvicidal against *Culex pipiens*. PLoS One 17(5):e0267150
69. Hoelscher F, Cardoso PB, Candiotto G, Guindani C, Feuser P, Araújo PH, Sayer C (2021) In vitro degradation and cytotoxicity response of biobased nanoparticles prepared by thiol-ene polymerization in miniemulsion. J Polym Environ 29(11):3668–3678
70. Bhatt R, Goyal A, Kachhwaha S, Kothari S (2021). Capping of nanoparticles: an alternative approach for reducing nanoparticle toxicity in plants. arXiv preprint arXiv:2111111355
71. Buzea C, Pacheco II, Robbie K (2007) Nanomaterials and nanoparticles: sources and toxicity. Biointerphases 2(4):MR17–MR71
72. Mohammadi R, Khoobdel M, Negahban M, Khani S (2019) Nanoemulsified *Mentha piperita* and *Eucalyptus globulus* oils exhibit enhanced repellent activities against *Anopheles stephensi*. Asian Pac J Trop Med 12(11):520–527
73. Research G. V (2024). Natural insect repellent market size, share & trends analysis report by type, by distribution channel, by region, and segment forecasts, 2024–2030. https://www.grandviewresearch.com/industry-analysis/natural-insect-repellent-market-report. Accessed 30 Aug 2025
74. Insights FM (2025) Natural insect repellent market size & trends 2025–2035. Future Market Insights. https://www.futuremarketinsights.com/reports/natural-ingredient-insect-repellent-market?utm_source=chatgpt.com. Accessed 30 Aug 2025

75. Research G. V (2025) U.S. Natural insect repellent market size, share & trends analysis report by product (sprays/aerosols, creams/lotions, essential oils, candle, liquid vaporizers & diffusers), and segment forecasts, 2025–2030. Grand view research. https://www.grandviewresearch.com/industry-analysis/us-natural-insect-repellent-market-report. Accessed 30 Aug 2025
76. Patel DM, Patel NN, Patel JK (2021) Nanomedicine scale-up technologies: feasibilities and challenges. In: Emerging technologies for nanoparticle manufacturing. Springer, Cham, pp 511–539
77. Sammasagi SS, Sutar KP, Hooli S (2025) Scale-up and quality control challenges in the industrial manufacturing of nanoformulations: current trends and future perspectives. IJSAT-Int J Sci Technol 16(2):1
78. VandenBerg MA, Dong X, Smith WC, Tian G, Stephens O, O'Connor TF, Xu X (2025) Learning from the future: towards continuous manufacturing of nanomaterials. AAPS Open 11(1):7
79. Wang W, Li L (2011) Nano emulsion of essential oil and its preparation method. Patent Number: CN102166314A. Application Number: CN201110068884.1. Zhejiang Gongshang University, China. https://patents.google.com/patent/CN102166314A/en?oq=CN102166314A
80. Elraz H (2010) Insect-repellent formulations
81. Intelligence M (2024) Mosquito repellent market—growth, trends, covid-19 impact, and forecasts (2024–2029). https://www.mordorintelligence.com/industry-reports/mosquito-repellent-market. Accessed 30 Aug 2025
82. Chatzidaki MD, Mitsou E (2025) Advancements in nanoemulsion-based drug delivery across different administration routes. Pharmaceutics 17(3):337
83. Ge X-h, Mo L, Yu A, Tian C, Wang X, Yang C, Qiu T (2022) Stimuli-responsive emulsions: recent advances and potential applications. Chin J Chem Eng 41:193–209
84. Huang H, Li Y, Huang C, Gao J (2024) A review of stimuli-responsive microemulsions: responsive mechanism, physicochemical property, and application. J Mol Liq 401:124666
85. Wu W, Yang Y, Feng Y, Ren X, Li Y, Li W, Huang J, Kong L, Chen X, Lin Z (2022) Study of the repellent activity of 60 essential oils and their main constituents against aedes albopictus, and nano-formulation development. Insects 13(12):1077
86. Et-Tazy L, Lamiri A, Krimi BS, Errati H, Hashem A, Avila-Quezada GD, Abd-Allah EF, Satrani B, Essahli M, Satia L (2025) Exploring synergistic insecticidal effects of binary mixtures of major compounds from six essential oils against *Callosobruchus maculatus*. Sci Rep 15(1):15180
87. El-Bakry AM, Youssef HF, Abdelmaksoud NM, Abdel-Aziz NF, Sammour EA (2025) Synergistic insecticidal effects of zinc-loaded zeolite nanoparticles combined with essential oils against *Callosobruchus maculatus*. Sci Rep 15(1):1–17
88. Corzo-Gómez JC, Espinosa-Juárez JV, Ovando-Zambrano JC, Briones-Aranda A, Cruz-Salomón A, Esquinca-Avilés HA (2024) A review of botanical extracts with repellent and insecticidal activity and their suitability for managing mosquito-borne disease risk in Mexico. Pathogens 13(9):737
89. Kechagia A, Dimaki VD, Mourelatou E, Avgoustakis K, Lamari FN, Hatziantoniou S (2024) Enhanced stability and prolonged insect-repellent action of essential oil-loaded nanostructured lipid carriers. Appl Sci 14(23):11309
90. Le Mauff A, Norris EJ, Li AY, Swale DR (2024) Repellent activity of essential oils to the lone star tick, *Amblyomma americanum*. Parasit Vectors 17(1):202
91. Luker HA (2024) A critical review of current laboratory methods used to evaluate mosquito repellents. Front Insect Sci 4:1320138
92. Nwagwu C, Onugwu A, Echezona A, Uzondu S, Agbo C, Kenechukwu F, Ogbonna J, Ugorji L, Nwobi L, Nwobi O (2024) Biopolymeric and lipid-based nanotechnological strategies for the design and development of novel mosquito repellent systems: recent advances. Nanoscale Adv 6(19):4751–4780
93. Owczarek M, Wiśniewska-Wrona M, Bartosik K, Buczek A, Sikora M, Piekarska K, Cichacz P, Śniarowski P, Mrozińska Z, Kudzin MH (2024) Ecological repellent preparations based on natural polymers with the addition of essential oils acting on ticks. Insects 15(12):931

Various Nano-Priming for Higher Performance of Aromatic Plants' Growth and Development

Srishti Jagota, Shashank Shekhar Pandey, Anju Sharma, Rajinder Kaur, and Arpit Sand

Abstract The nano-priming technique aids in increasing productivity and improving the quality of essential oils in aromatic plants. This new approach leverages the specific properties of NPs to stimulate beneficial plant responses, showing promising potential for enhancing growth, development, yield, and secondary metabolite production. Moreover, the use of NPs can also aid in reducing abiotic stresses, further supporting the overall growth and productivity of aromatic plants. The dual benefit of increasing growth while alleviating stress makes nano-priming a vital tool in sustainable agriculture for aromatic plant cultivation. Incorporating nano-priming techniques into aromatic plant cultivation not only boosts growth but also increases resilience against environmental challenges, thus aligning with sustainable agricultural practices. Nano-seed priming has become a groundbreaking agricultural biotechnology technique for enhancing seed vigor, improving germination rates, and strengthening crop resilience under diverse environmental stresses. Drawing insights from many past studies reveals the multifaceted applications and mechanisms of nano-priming in modern sustainable agriculture. The use of engineered nanomaterials (ENMs), like silicon dioxide (SiO_2), titanium dioxide (TiO_2), zinc oxide (ZnO), and carbon-based nanostructures, has been shown to improve physiological and biochemical processes during the initial stages of plant growth. These NPs act as elicitors that initiate metabolic reprogramming in seeds, leading to faster germination, improved seedling growth, and enhanced resistance to abiotic stresses, including salinity, drought, and temperature extremes.

Keywords Nano-priming · NPs · Aromatic plants · Nanotechnology

S. Jagota
Department of Sciences (Chemistry), Manav Rachna University, Faridabad, Haryana, India

Department of Forensic Science, Lloyd Institute of Forensic Science, Greater Noida, India

S. S. Pandey · A. Sharma · A. Sand (✉)
Department of Sciences (Chemistry), Manav Rachna University, Faridabad, Haryana, India

R. Kaur
Department of Plant Sciences, University of Idaho, Mosow, ID, USA

A. Husen (ed.), *Sustainable Uses of Aromatic Plants in Nanotechnology*, Smart Nanomaterials Technology, https://doi.org/10.1007/978-981-95-7409-4_17

Abbreviations

ABA	Abscisic acid
APX	Ascorbate peroxidase
CAT	Catalase
CNTs	Carbon nanotubes
CQDs	Carbon quantum dots
ENMs	Engineered nanomaterials
FeO	Ferrous oxide or iron (II) oxide
GA	Gibberellin or gibberellic acid
GO	Graphene oxide
MgO	Magnesium oxide
NPs	Nanoparticles
nZVI	Zero-valent iron
PEG	Polyethylene glycol
PGBR	Plant growth-promoting rhizobacteria
POX	Peroxidase
ROS	Reactive oxygen species
Se	Selenium
SiO2	Silicon oxide
SOD	Superoxide dismutase
TCM	Traditional Chinese medicine
TiO_2	Titanium oxide
ZnO	Zinc oxide

1 Introduction

One method that is employed in agriculture is nano-priming. A pre-sowing seed treatment method called nano-priming uses NPs to prime seeds to increase agricultural productivity, stress tolerance, seed germination, and seedling vigor. It combines the ideas of controlled hydration or seed priming, with nanotechnology to produce a novel and effective method of promoting plant growth. This technique involves the controlled application of ENMs such as SiO_2, ZnO, TiO_2, or carbon-based NPs to seeds before sowing. These nanomaterials can penetrate seed coats and activate internal biochemical pathways, accelerating germination and improving early seedling development. In the context of aromatic plants, such as basil, lemongrass, and mint, nano-priming plays a pivotal role in ensuring uniform germination and stress resilience, which are critical for maintaining both yield and quality.

Nano-priming techniques are broadly divided into dry nano-priming (application of dry NPs) and wet nano-priming (NPs dispersed in water or solutions). Some high-tech techniques also include coated nano-priming and bio-nano-priming, where NPs are combined with biological agents or polymers to enhance efficacy.

Many factors influence the efficiency of nano-priming, including NP type, size, concentration, exposure time, seed type, and environmental conditions like temperature and humidity. Observing and optimizing these parameters is essential for achieving consistent and desired outcomes. The application of nano-priming in aromatic plants promises enhanced productivity, better resource use, and sustainable cultivation under diverse agroclimatic conditions.

1.1 Classification of Nano-Priming Techniques

Nano-priming may be divided into three categories depending on the type of supporting NPs and the type of treatment the seeds undergo. "Chemical nano-priming" includes the application of metal or metal oxide NPs such as silver, ZnO, TiO_2, or iron oxide. These NPs enhance the activity of particular enzymes, help in stress alleviation, and promote nutrient uptake. For instance, ZnO NPs in basil increase the chlorophyll content and essential oil yield, while FeO_4NPs enhance the root systems of mint. "Physical nano-priming" involves nonchemical techniques such as the use of magnetic fields, ultrasounds, and lights, often in conjunction with NPs, to improve the mobility of the seed coat, water uptake, and energy metabolism [1]. Exposure of lemongrass seedlings to magnetic NPs and rosemary has improved seedling vigor and antioxidant levels, respectively, when combined with UV-activated TiO_2. "Biological nano-priming" applies green-synthesized, biocompatible, eco-friendly, and nontoxic NPs obtained from the extracts of plants or through the activity of microbes. These, in addition to stimulating beneficial microbial populations in the rhizosphere, also help to stimulate the activity of beneficial microbes. For example, green-synthesized silver NPs promote the oil yield and growth rate in citronella, while biogenic NPs made from fungi stimulate the growth of plants [2].

1.2 Nano-Primers and Their Working

Nano-priming is an intelligent and innovative method of seed treatment that, unlike traditional seed priming, employs the multifunctional avenues of nanotechnology. In nano-priming, seeds are exposed to controlled amounts of nanomaterials: ZnO, SiO_2, TiO_2, and other NPs, under controlled moisture, temperature, and time. This treatment takes place before sowing and is designed to excite numerous physiological and biochemical functions within the seed without the seed actually germinating [3]. The unique position of NPs allows them to interact within the seed directly modulating certain activities like gene expression, hormonal signaling, and digestion of antioxidant super macromolecules and acting as carriers of nano-energizers to the embryonic tissues.

1.3 Mechanism of Action

Nano-primers undergo mechanisms of penetration but also activation, ROS signaling, ion transport, and expression of certain genes. The NPs penetrate the seed coats at the points of origin through pores. The enzymatic activities are triggered, like amylase and protease [4]. The defense pathway defends and gets simulated with certain levels of metal oxidizing as ROS levels. The presence of some of the ions provides a metal base and improves nutrient availability. Then, these modulate genes related to growth and stress resistance. This mechanism is followed by all in the same way. These show the following functions, like improved germination under adverse conditions and enhanced yield [5]. They can be used on other specific crops like rice, maize, legumes, wheat, and some other aromatic plants. They are the best eco-friendly alternatives to chemical seed treatments.

1.4 Aromatic Plants

Aromatic plants are plants that produce and store essential oils, volatile, aromatic compounds used in flavoring, fragrance, cosmetics, traditional medicine, and aromatherapy. These plants often belong to families like Lamiaceae (mint family), Apiaceae, Rutaceae, Poaceae, and Asteraceae, contain essential oil glands in leaves, flowers, stems, or roots, and produce secondary metabolites like terpenes, phenolics, and alkaloids [6]. Medicinal and aromatic plants are known for their strong fragrance, which comes from volatile oils. They are generally well adapted to hot, dry, or semiarid climates. Common examples include lemongrass (*Cymbopogon citratus*), which has antimicrobial properties and is widely used in culinary and cosmetic products; basil (*Ocimum basilicum*), valued for its essential oils, flavoring, and medicinal benefits; peppermint (*Mentha piperita*), used in pharmaceuticals, candies, and aromatherapy; rosemary (*Rosmarinus officinalis*), a popular culinary herb with antioxidant properties and a reputation as a memory aid; turmeric (*Curcuma longa*), rich in curcumin and known for its medicinal and anti-inflammatory effects; and eucalyptus (*Eucalyptus globulus*), commonly used for respiratory treatments and essential oil extraction. These plants have diverse applications. In the pharmaceutical sector, they are recognized for their antimicrobial, antifungal, antioxidant, and anti-inflammatory properties. In cosmetics, they are key ingredients in soaps, lotions, perfumes, and hair oils. In the food industry, they serve as spices and flavoring agents, while in agriculture, they are used as natural pesticides and soil enhancers. Aromatherapy makes use of their oils for stress relief and mood enhancement, and in perfumery, they are valued for fragrance extraction in perfumes and deodorants. Globally, major producers include India, China, Brazil, and Mediterranean countries. Beyond their economic value, the cultivation of medicinal and aromatic plants also supports rural livelihoods and sustains small-scale industries. They are also used in Ayurveda, Unani, and traditional Chinese medicine.

1.5 Application on Aromatic Plants

When applied to aromatic plants such as basil, lemongrass, lavender, mint, fennel, and rosemary, nano-priming shows remarkable potential to improve crop performance. These plants are cultivated primarily for their essential oils, which are highly valued in the pharmaceutical, cosmetic, food, and aromatherapy industries. However, their productivity and oil quality are often vulnerable to environmental and physiological stresses [7]. It also enhances tolerance to drought, salinity, and pest pressure while stimulating secondary metabolite production, including terpenes and phenolics, which are vital for the aroma, flavor, and medicinal value of these plants. The process typically begins with preparing a priming solution in which seeds are soaked in a NP suspension commonly containing ZnO, FeO, SiO_2, TiO_2, or chitosan NPs. During the activation phase, these NPs initiate metabolic activity, triggering enzyme activation and strengthening antioxidant defenses. In the final dry-back phase, seeds are returned to their original moisture content, ready for sowing, and exhibit faster, more robust growth in the field [8]. Mechanistically, nano-priming benefits aromatic plants by activating enzymes such as α-amylase, which accelerates starch hydrolysis to release energy for germination; enhancing antioxidant capacity to reduce oxidative stress; improving nutrient uptake efficiency, especially for micronutrients like zinc and iron; regulating genes responsible for aroma-linked metabolites; and stimulating terpenoid biosynthesis to increase the concentration and quality of essential oils.

Several NPs have been tested for their specific effects in aromatic plant nano-priming. ZnO NPs, for example, have been shown to enhance germination rates, chlorophyll levels, and zinc uptake in basil and mint. In lavender and lemongrass, SiO_2 NPs enhance antioxidant defense and drought tolerance. According to Kanwal et al. [9], chitosan NPs increase oil yield in coriander and fennel while also promoting immunity. Iron oxide NPs enhance root growth and biomass production in lemongrass, while TiO_2 NPs boost photosynthesis and essential oil content in rosemary. These advantages are supported by experimental evidence: Chitosan NP priming enhanced menthol yield and resistance to biotic stress in Japanese mint; ZnO NP priming accelerated germination and leaf area expansion in basil, resulting in a higher eugenol content in the essential oil; and nano-priming with FeO and ZnO NPs enhanced root length, chlorophyll concentration, and citral content in lemongrass [10].

1.6 Application of Nano-Primers in Agriculture

The ability of nano-priming to improve germination, growth, and resilience in a range of crops has led to its widespread use in agriculture. It has been especially successful in increasing the amount of essential oil and biomass yield in aromatic and therapeutic plants like mint, lemongrass, and basil. Even in the face of stressors like drought or salinity, nano-priming has demonstrated the ability to raise germination rates and yields in cereal crops like rice and wheat. Following nano-priming treatments, vegetables and legumes also have improved disease resistance and nutrient uptake [11]. Making a suspension of NPs, such as 50 ppm of ZnO NPs, is usually the first step in a streamlined nano-priming procedure. Under carefully monitored conditions, seeds are soaked in this solution for eight to twelve hours. Once the soaking phase is complete, the seeds are dried back to their original moisture content before sowing. This pretreatment ensures that seeds germinate faster and more uniformly, resulting in healthier and more vigorous plants in the field.

1.7 Factors Influencing Growth and Development

Many variables underlie the effectiveness of nano-priming. The type of particle used causes different physiological changes and levels of stress adaptation. Size and surface area are important because smaller particles with more surface area are likely to interact better with the seed surface as well as with the cell walls. Also, the number of NPs in solution must be evaluated carefully because certain doses can enhance growth, while large doses can inhibit growth and induce oxidative stress. Also, timing, or how long the seeds are allowed to be preconditioned, impacts the efficiency, NP uptake, and subsequent biochemical activation in the seed [12]. Different types of seeds and sensitivities to particular species are also fundamental and observable differences in crop specificity greatly differ. Environmental conditions (temperature, humidity, and light) during the nano-priming and the metabolic status after priming will have implications for being absorbed along with metabolic activity during the period of time before germination. Different modes of application also determine how the NPs will interact with the seed, such as dry coating, wet suspension, or encapsulation, and will also impact potential efficacy [13]. The standard process generally includes preconditioning the seeds by partially moistening them with an NP solution, allowing the NPs to either penetrate the seed coat and/or enter the embryo through preexisting pores and canals facilitated by the movement of water. Following the relatively short period of immersion and/or hydration, the NPs would be conducting multiple biochemical mechanisms that are anticipated to be activated to influence GA entry and distribution and define hydration thresholds. This would result in budding, typically after a period of 3–7 h.

Some of the factors provide good water uptake with metabolic activation, thanks to nano-priming. Nano-priming techniques also show improvement in the root and shoot growth. They also induce protective enzymes and osmolytes to improve stress tolerance. They are also used to deliver micronutrients like Fe, Cu, and Zn. Because of the use of nano-primers, the need for the use of agrochemicals has been reduced and also enhanced performance. Some of the nano-primers are ZnO, SiO_2, and TiO_2 [3]. They also participate in different functions like germination and growth. Some are used to improve drought and salt stress tolerance. Some NPs are used to promote photosynthesis and the synthesis of chlorophyll with the supply of iron. Chitosan NPs are used as a growth promoter.

2 Types of NPs in Nano-Priming

Nano-priming, a seed treatment approach that merges traditional priming methods with nanotechnology, utilizes NPs of various types for seed germination, stress condition resistance, and promoting early plant growth [14] (Fig. 1). Some types of commonly used NPs include silver NPs among metal-based particle categories used because of their antimicrobial properties and germination ability with reducing seed-borne pathogens [1] and ZnO NPs that promote vigor measurement as plant seeds germinate and can have noticeable effects on seedling enzyme activities [15]. Iron oxide NPs improve nutrients and photosynthesis [16], while cupric NPs provide benefits of regulating oxidative stress potential and pathogen resistance. SiO_2 NPs provide benefits in plant resistance [17] to drought and salinity stress [18]. Carbon-based nanomaterials such as CNTs and GO penetrate seed coats and have been proven to improve water and nutrient uptake by the seed [19]. Chitosan NPs can also provide the characteristic of biocompatibility since they are biopolymer based and can be used for controlled release of growth-promoting substances or pesticides [20]. The type of NP depends on what type of outcome is desired, such as either increased germination, combined disease resistance, or abiotic stress tolerance, allowing for possible application in sustainable agriculture and increasing the tools to utilize in an agricultural ecosystem or crop assessment [21].

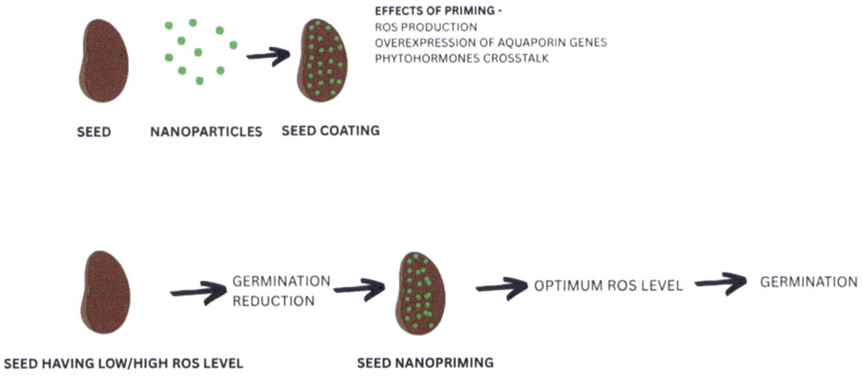

Fig. 1 Effects of seed nano-priming during germination and on ROS levels

2.1 Metal and Metal Oxide NPs

Metals and metal oxide NPs are preferred for nano-priming because they have the properties and traits that can influence plant growth and induction seed germination [12, 22, 23].

Metal, specifically silver, gold, and copper, prevents infection in seeds and nullifies the issue of poor seedling establishment. Seed germination is induced by silver NPs because of decontamination; however, copper induces both oxidation and stress tolerance [17].

Metal oxide NPs, ZnO, Fe_3O_4, TiO_2, and MgO, are used primarily for nutrient delivery.

ZnO induces activity of enzymes related to germination, while Fe_3O_4 can lead to improved nutrition by iron availability and help in chlorophyll biosynthesis. TiO_2 increases photosynthesis and imbibition [24], and MgO also increases chlorophyll biosynthesis activity.

Maximum concentrations of NPs can both induce seed vigor, stress tolerance, and seedling emergence; however, this action is dose dependent, and excess is phytotoxic [2]. This must be optimized for agricultural applications carefully.

2.2 Carbon-Based NPs

Carbon NPs are important in nano-priming, as they stimulate seed germination, nutrient absorption, and stress tolerance in plants [14]. CNTs, GO, fullerenes, and CQDs are the most frequently researched carbon NPs in nano-priming. To improve germination and root development and facilitate the absorption of nutrients and water, CNTs pierce seed coverings. To boost seedling vigor and stress tolerance, GO will use its large surface area for nutrient adsorption and delivery [25]. Fullerenes can reduce reactive oxygen intermediates and aid in plant metabolism, while CQDs can be used for real-time monitoring and the significance of nutrient delivery in seeds and plants based on size and photoluminescence, as well as to enhance growth.

2.3 Biopolymeric-Based NPs

To enhance seed performance and plant development at the seedling stage of establishment, biopolymeric-based NPs—green, biodegradable, organic polymer-based nanomaterials—are often incorporated in seed nano-priming. They may be less toxic and more available or have a more controlled release priority than inorganic NPs. They may also act as nutrient carriers, hormonal/growth regulators, or protection agents. Biopolymeric NPs that tend to be often used are cellulose, starch, alginate, and chitosan. Chitosan NPs are derived from chitin, which is known for having strong film-forming, antimicrobial, and bio-stimulant capabilities. Incorporation of chitosan NPs into nano-priming methods has been shown to stimulate seed germination and enhance certain mechanisms related to defense-related enzyme activity and pathogen protection [26].

Alginate NPs are derived from brown algae and are frequently used for encapsulating micronutrients or plant growth regulators and provide a slow and targeted release of compounds as seedlings develop. Starch and cellulose NPs are also widely obtained from plant biomass and provide biodegradability and enhancement of water retention by seeds, as well as delivery of nutrients and physical protection during the seed nano-priming process [27].

3 Mechanism of Nano-Priming on Aromatic Plant Physiology

When seeds of aromatic plants are nano-primed, a variety of physiological and molecular pathways are stimulated, which ensure rapid germination and enhance secondary metabolite production and stress tolerance. To enhance water uptake to facilitate imbibition, NPs, particularly metal oxide NPs like CeO_2, ZnO, Ag, or SiO_2, first engage with the coat of the seed and form nanopores. In the meantime, enhance the levels of aquaporin gene expression [28]. The controlled release of ROS, particularly H_2O_2, following water absorption is the primary trigger for dormancy release and metabolism initiation, and antioxidant enzymes (SOD, POD, CAT, APX) are then used to maintain ROS balance [29]. At the same time, α-amylase activation by GA promotes rapid starch hydrolysis, which will also provide energy to support embryonic development and, especially, the vitality of seedlings [30]. Nano-priming results in inconsistently altering the transcriptome and molecular variation in plants' biochemistry.

To increase gene expression involved in first or second-tertiary metabolites, which include multiple biosynthetic pathways, including phenylpropanoid and flavonoid metabolism, nano-priming generally involves interfering with signaling cascades linked to ROS and phytohormone cross talk with ethylene, SA, and JA (do Espirito Santo [31]). In terms of promoting glandular trichomes and facilitating metabolite flux toward terpenes and phenolics, this control especially enhances the accumulation of essential oils in aromatic plants, allowing production and quality-wise improvement. The effectiveness of nano-priming in enhancing the yield and resilience of aromatic crops arises from that interaction, which encompasses improving water uptake, ROS signaling, enzyme-mediated activation, hormonal adjustment, and secondary metabolism regulation [32].

3.1 Enhanced Germination and Seedling Vigor

Nano-priming is an exceptionally successful priming technique to promote germination and seedling vigor. The practice of priming seeds with NPs is called "nano-priming," and it has proven to be particularly effective with aromatic and medicinal plants. Recent research has shown that NPs stimulate aquaporin protein expression to promote imbibition and germination while also creating nanopores in the seed's outer coat to allow for faster water uptake [30]. At the same time, nano-priming elicits the production of ROS within regulated time windows described as "oxidative windows." Also, ROS acts by signaling molecules to trigger hormonal pathways and trigger antioxidant defenses, for example, GA-mediated activation of α-amylase, which facilitates the hydrolysis of starch into soluble sugars that can support early growth of the embryo [33]. SiO_2 NP priming (ideal at ~10 ppm) considerably increased root, shoot, and seedling biomass (in the range of 283%, 169%, as well as 220%, respectively) and almost doubled the percentage of germination and rate in species such as *Stevia rebaudiana* [33]. Enhanced α-amylase activity, greater sucrose buildup, as well as increased antioxidant enzyme activities, such as CAT and POX, which shield seedlings from oxidative stress, were the main drivers of these changes [34]. In addition, it has been shown that ZnO NPs stimulate water uptake, amylase activity, and antioxidant enzyme activity (i.e., SOD, CAT, and POXs); these treated plants had decreased levels of lipid peroxidation and increased levels of phenolics and flavonoids, important metabolites in the physiology of aromatic plants [35]. On a macro-level, studies highlight improvements in surface reactivity in conjunction with increased electron exchange in the plant tissues from nano-priming that influence hormonal balances, stimulate secondary metabolite pathways (including flavonoids and phenolics), increase stress tolerance, and therefore improve growth and yield (do Espirito Santo [31]). These results have also been validated in numerous crop species, in which nano-priming was able to induce both transcriptional and metabolic reprogramming, promote cell wall loosening, enhance mobilization of reserves, and coordinate germination under ideal and stress conditions [35]. The combined effects of these processes on medicinal and aromatic plants enhance secondary metabolite production and aromatic quality while increasing seedling vigor and establishment and resistance to environmental stresses (do Espirito Santo [31]).

3.2 Regulation of Phytohormones and Enzyme Activities

Applying nano-priming of seeds with NPs changes enzyme activities and phytohormonal processes while improving physiological characteristics in aromatic plants. This method also affects substantial hormonal balances, such as those of GA and ABA, both of which are significant regulators of seed dormancy and germination. For instance, it has been demonstrated how the seeds were affected by changes to

GA and ABA synthesis through priming seeds with NPs made of FeO, ZnO, and Se oxide (SeO); this caused positive effects on seed vigor and germination rates [17]. Nano-priming equally influences both enzymatic activity and phytohormonal control. It stimulated soluble sugar concentration, which promoted seedling growth through α-amylase efficiency, the main enzyme regulating starch hydrolysis. The relationship between stimulation of enzyme activity and GA signaling pathways contains an indicative synergistic relationship between NPs and phytohormones in regulating metabolic activity at germination [36].

ROS can act as signaling molecules, inducing antioxidant defense systems, and are produced in fixed amounts from nano-priming (this in effect represents production of antioxidants that act synergistically in reducing oxidative stress and safeguarding the cell, i.e., APX, CAT, and SOD). In addition, encasim (expression of aquaporin genes) refers to increased water uptake, which is the most important factor for seed germination and early seedling growth [30].

Such physiological adaptations provide an advantage to aromatic plant growth and possibly enhance metabolites associated with aroma and flavor, which are also more tolerant to environmental stresses. If nothing else, we can define nano-priming as a flexible agrotechnique for growing aromatic plants, given that it allows the modulation of hormonal and enzymatic pathways.

3.3 NP-Mediated Stress Tolerance Mechanisms

NP-aided stress tolerance through nano-priming will also evoke physiological resilience of aromatic plants by activating numerous biochemical and molecular pathways. The NP influence on the nano-priming approach is when it ruptures the outer seed coat or root system to elicit a sugar-like signaling cascade involving numerous phytohormones that regulate germination, growth, and stress tolerance, including GAs, SA, ABA, and JA [1].

More notably, the activity of POD, CAT, and SOD antioxidant enzymes through NP application can be stimulated to minimize the hazards arising from ROS accumulation during biological stressors, including salinity, heavy metals, and drought [37]. The vegetative and aromatic exploration of *Ocimum basilicum* and *Mentha* spp. is reliant on the ability to improve membrane integrity and maintain chlorophyll concentrations and efficiency of photosynthetic processes, which are modifiers of and can reduce oxidative damage. Among other improvements, nano-priming stimulates nutrient uptake through increasing root exudation behaviors and activating ion transporters, to stimulate nitrogen and phosphorus digestion [38].

NPs mediate the molecular modulation of stress-responsive genes, ultimately causing an upregulation of osmolytes, such as proline, and secondary metabolites, leading to the production of aromatic compounds while enhancing stress tolerance [39].

In the end, NP-mediated nano-priming is an integrated process whereby aromatic plants suppress their enzymatic defenses, are metabolically primed, and germinate vigorously and therefore are able to survive and yield under stress.

4 Impact of Nano-Priming on Growth and Yield of Aromatic Plants

Nano-priming, which is the application of NPs to seeds before seeding, is a new best practice that can enhance the growth and productivity of aromatic plants. It generally improves overall plant resilience under stress, in part increasing seed germination and seedling vigor. For instance, nano-priming with ZnO NPs greatly improved water uptake characteristics, enzymatic and antioxidant defenses, establishment, and growth of seedlings [34].

Nano-priming has benefits, regulating genes related to growth and stress response and enhancing biomass acquisition, photosynthetic performance/symbiotic relationships, and increasing chlorophyll concentrations [40]. Also, nano-priming can improve nutrient efficiency and eliminate the reliance on chemical fertilizers or herbicides, creating sustainable agricultural standards [41]. Nano-priming has been shown to increase the production of secondary metabolites in aromatic plants, which have important value from an oils and health perspective, which can produce greater yield and quality, even production, and more synchronized germination of the seeds, which is also important in maximizing production [42].

4.1 Root and Shoot Biomass Enhancement

Regarding aromatic and related plant systems, NPs, generally referred to as nano-priming pretreatment, can greatly increase root and shoot biomass. For example, after fourteen days of growth, the low concentrations of nZVI NPs (~20 mg/L) primed seeds of rice (a chemically aromatic varietal) had improved seedling vigor as a result of increased biomass and root and shoot length. Treating plants with TiO_2 NPs around 50–200 mg/L improved both the shoot and root fresh biomass in coriander (another aromatic species), although larger doses presented concerns for oxidative stress [32]. In both aromatic and similar plants, as well as in *Brassica nigra* and *Trigonella foenum-graecum*, green nano-priming with magnesium sulfide NPs also showed increased root and shoot lengths and earlier germination [32].

Low to slightly elevated concentrations of NPs consistently enhanced water absorption, cell division, and overall metabolic activity in all of these trials, which resulted in biomass accumulation. This amounts to cautioning to balance concentrations so that phytotoxicity would not occur.

4.2 Improved Essential Oil Biosynthesis

Nano-priming (the use of NPs applied to seeds) can stimulate secondary metabolism via regulated oxidative stress and improved nutrient delivery and confirms higher rates of essential oil production in aromatic plants. The use of SiO_2 and TiO_2

NPs in *Ocimum tenuiflorum* (holy basil) improved essential oil production due to modification of the trichomes of peltate glands that produce important secondary metabolites such as cis-β-elemene and eugenol [43]. Because ROS-induced stimulation of biosynthetic pathways, as well as better micronutrient management, results in higher levels involving valuable secondary metabolites, such as essential oils, in medicinal and aromatic plants treated with NPs, nano-priming is a promising method for increasing both the quantity and quality of the bioactive compounds [32]. Additionally, mechanistic reviews show nano-priming increases germination and stress resilience by generating moderate ROS (as signaling signals) to increase secondary metabolites and essential oils [30].

4.3 Nano-Priming Effects on Flowering and Biomass Accumulation

The use of NPs for seed pre-sowing, or "nano-priming," has emerged as a valuable way to improve biomass accumulation and flower traits in plants, particularly in aromatic species. In studies with aromatic herbs (lemon balm) with SiO_2 NPs, and with *Pseudomonas fluorescens* or *P. putida* combined with SiO_2 NPs, the following increases occurred in biomass, leaf water content, photosynthetic-derived pigments, and essential oil yield—revealing that the use of nanomaterials as an approach to nano-priming may act synergistically to improve growth and general biochemical properties. Research with *Melissa officinalis* (a medicinal herb, lemon balm) documented that cold plasma priming with ZnO NPs and Se NPs enhanced stem, root, leaf sizes, and overall biomass yields, which would be an indication of increased vegetative growth that may result in an earlier or greater flowering abundance [44]. Additionally, with nZVI seed priming, there was obvious seedling vigor noted in aromatic rice (an aromatic rice cultivar, *Oryza sativa*), increased shoot, root biomass, and pigment levels and formed a superior baseline for increased later stage development, such as flowering [5]. Therefore, as a whole, we may note these increases in biomass, plant vigor, and physiological resiliency. Lenda means to potentially benefit flowering initiation and duration; however, the author has an understanding that professionals may not have a direct knowledge base that demonstrates nano-priming advances flowering duration compared to non-nano-primed flowering aromatic plants. Nano-priming likely helps whatever aromatic plants reach reproductive maturity and could potentially lead to increased flowering through supporting plant health and bootstrapping early growth.

5 Case Studies and Recent Advances

Evidence for nano-priming has only grown with studies showing positive impacts of nano-priming on seed germination, stress tolerance, and crop yield. For example, a study on ZnO NPs for nano-priming wheat shows improvement in the germination percentage, chlorophyll content, and activity of antioxidant enzymes under salt stress. Similarly, an experiment on chitosan NPs on maize seeds shows the positive impact of drought resistance, i.e., root growth and water intake. The effect of single- and multiwalled CNTs on tomato seeds was even more striking, as they showed greater germination rates and biomass due to greater water and nutrient intake. Similarly, studies on rice crops used FeO NPs and showed enhanced seedling vigor and greater iron content in rice seedlings in an effort to reduce micronutrient deficiency.

Additional innovation has employed stimuli-responsive nanocarriers as a delivery vehicle for nutrient uptake and/or plant growth regulators and their application at different levels according to environmental signals perceived by plants. Toxicity is not being explored directly, but there have been reports of environmental formulations that utilize green synthesis to produce NPs from raw materials (e.g., fungus or plants) to decrease their toxicity [45].

5.1 Nano-Priming in Basil, Mint, and Lemongrass

Nano-priming of lemongrass, mint, and basil has been discovered to be an effective alternative to seed germination, homogeneity, and early-stage seedling establishment, especially under stressful environmental conditions. These medicinally and aromatically used plants possess hard- or slow-germinating seeds and are therefore best suited to be utilized for nano-priming treatment.

In basil (*Ocimum basilicum*), it has been found through studies that nano-priming with silver NPs or ZnO improves the germination percentage, root and shoot growth elongation, and essential oil biosynthesis by increasing enzymatic as well as metabolic processes. In mint (*Mentha* spp.), likewise, nano-priming with chitosan NPs has been found to improve germination under drought stress and help in the induction of higher antioxidant activity as well as essential oil content during the early vegetative phase. Biocompatibility and antimicrobial activity of chitosan also help in preventing infection by fungi in seeds.

For lemongrass (*Cymbopogon citratus*), nano-priming with SiO_2 NPs or iron oxide has been shown to enhance seedling vigor, root elongation, and abiotic stress tolerances such as salinity or nutrient stress. Carbon NP priming (such as GO) has also been shown to be effective in the enhancement of the yield of secondary metabolites in such medicinally valuable crops.

5.2 Field-Level Trails and Commercial Potential

Field trials and commercial feasibility of nano-priming are also attracting significant attention as researchers are outscaling their laboratory experiments to field scale to determine real-world viability in diverse agroclimatic settings. Certain field trials on wheat, maize, rice, and tomato crops have established that nano-priming with metal oxide NPs like ZnO, Fe_3O_4, and TiO_2 yields greater germination percentage, improved seedling establishment, improved nutrient use efficiency, and improved yield under normal and stress conditions ([46]), such as drought and salinity. These field trials confirm the fact that the advantages realized under controlled conditions can be scaled.

In medicinal and aromatic crops like basil, mint, and lemongrass, seeds have been nano-primed with chitosan, Ag, or SiO_2 NPs, which have been reported to enhance biomass, oil yield, and resistance against pests and diseases in small-scale field experiments. The usage of chitosan, a biopolymeric NP with the additional benefits of environmental safety and compatibility with organic farming, is of special interest as it increases its commercial attractiveness.

The commercial potential of nano-priming is that it can attain improved seed performance with a reduced environmental impact. It requires low levels of NPs, which is cost-effective and scalable. Seed treatment technology firms are now targeting nano-coatings and nano-priming formulations as value-added products. But regulatory barriers, NP toxicity, and the formulation of validated protocols need to be resolved before large-scale commercialization [47]. Nevertheless, with ongoing progress and favorable trial results, nano-priming is a future sustainable agriculture and precision seed technology contender.

5.3 Synergistic Effect of Nano-Bioformulation

The nano-bioformulation synergy of nano-priming is attributed to improved performance generated by the synergistic interaction of nanoparticles with biological factors like PGPR biofertilizers or biostimulants [48]. The NPs offer controlled release, enhance the surface area, and facilitate targeted delivery, whereas the biological factor offers ecological benefit, inducing seed germination, plant growth, nutrient uptake, and resistance to biotic and abiotic stress. For example, chitosan NPs can increase the viability and survival of desirable microbes (or enzymes) in the rhizosphere, so there is an improved opportunity for colonization of the root system and increased solubilization of nutrients. Likewise, ZnO or iron oxide NPs and biofertilizers provide a source of plant nutrition and, at the same time, trigger plant hormones that will also lead to an environmentally mediated proliferation induction of PGPR. The nano-bio formulation-injected seedlings will increase the seedling vigor (plant growth in the early phase), photosynthetic activity, and antioxidant defense mechanisms compared to soil, potting media, or hydroponics-grown seedlings exposed to abiotic stress conditions, like drought or salinity.

This is supported by a number of studies that nano-biopriming is more effective in early stages of plant growth than both factors in isolation. The use of nano-bioproducts reduces the chemical and fertilizer requirements, enhances sustainable agriculture, and enhances soil health and yield.

6 Challenges in Using Nano-Primers

Nevertheless, there are some issues and concerns with nano-priming despite its potential. According to Kesraoui et al. [49], excessive concentrations of NPs can have phytotoxic effects, harm seed tissues, or prevent growth. The efficacy and environmental sustainability of NPs in soils and water systems are a source of concern, as they may disrupt microbial communities or bioaccumulate in food chains. There may also be inconsistencies due to a lack of standardized procedures on NP size, concentration, exposure times, and mode of application. Additionally, nano-priming technologies may be hindered by expense and scalability in large or resource-limited farming systems [5]. To enable safe and long-term application of nano-priming to maintain aromatic plants, these challenges must be overcome by developing safe and environmentally adjustable nanomaterials, proportionate dosing methodologies, and regulatory frameworks. Additional challenges include variation in doses of NPs used, duration of exposure, and type of NP; low NP doses do not enhance the production of secondary metabolites, including essential oils; NP cost and stability limitations may prevent on-farm adoption at scale and lack of knowledge on the long-term effect of plant aroma and oil quality.

6.1 Impact of Nano-Primers on Aromatic Plants

The effects of aromatic plants are evident at various developmental stages of growth. On germination, it means enhanced performances of enzymes including amylase and CAT impact on faster and uniform germination and, most importantly, the greater vigor index and longer shoots and roots for seedling growth. Plants produce more leaves, with more chlorophyll, and can deal with photosynthesis better when growing vegetatively. Furthermore, when used to fortify roots, nano-priming increases nutrient uptake. During the flowering phase, plants often also can produce more in florescence and flower earlier [50]. In general, yield of essential oils is higher, and volatile compounds are of superior quality. Associated with this, nano-primed plants possess enhanced tolerance to stressors, including oxidative damage, salinity, and drought.

Because of these advantages, nano-priming is a viable and sustainable method of enhancing the phytochemical diversity, growth, and development of aromatic plants, particularly in settings that are prone to stress.

In addition to their fragrance, aromatic plants are essential to industry and agriculture for their ecological and financial advantages. They are the main suppliers of essential oils used in food flavorings, aromatherapy, cosmetics, and medications. For example, India is a major producer of essential oils from plants like vetiver, citronella, and palmarosa [51]. These crops are perfect for small and marginal farmers because they are high value and require less land. Mint cultivation in particular demonstrates how intercropping aromatic plants with conventional crops can boost total income and enhance land-use efficiency, as seen with mint cultivation in Uttar Pradesh and Punjab or geranium production in the hills of Tamil Nadu [52]. Ecological and agricultural balance also aid in crop diversification, soil health maintenance, and biodiversity. Another benefit of aromatic plants is their sustainability; many species are inherently more resistant to pests and diseases and require fewer agrochemicals. Some, such as basil, neem, and citronella, support organic and environmentally friendly farming methods by acting as natural pest repellents. These plants are essential to traditional medical systems like Ayurveda and Unani, and they have important medicinal and therapeutic qualities beyond their economic worth, such as antimicrobial, antioxidant, and anti-inflammatory effects [53].

6.2 Limitations of NPs on Aromatic Plants

By lessening the stress of monocropping and improving agroecological practices this approach promotes long-term soil health and ecosystem resilience. Although traditional seed priming, which uses water or simple chemical solutions as a pre-sowing treatment, has been used for a long time to increase seed germination and early seedling growth, it has several drawbacks when used on aromatic plants like lavender, mint, basil, and lemongrass. These techniques mostly concentrate on increasing vigor and germination rate; they have little effect on improving metabolites, especially essential oil biosynthesis, which is a crucial characteristic of aromatic plants. Water-based priming does not increase the quantity or quality of essential oils because it does not effectively stimulate secondary metabolic pathways. Loss of water-soluble substances, such as essential oil precursors, could harm seed membranes, especially in fragile seeds like thyme or basil [54].

The benefits of such methods are often short term, limited to the early growth stages, without providing long-term epigenetic or molecular priming effects that could enhance oil content, aroma, or yield. Furthermore, priming with basic solutions like water, salts, or PEG lacks precision, offering no targeted delivery of nutrients, hormones, or stimulants to specific seed tissues, and is ineffective in activating glandular trichomes where essential oils are synthesized. It also fails to efficiently trigger secondary metabolic pathways responsible for the synthesis of terpenoids and phenolics, which are vital for aromatic plant quality. Traditional methods exclude the use of modern biostimulants, nanocarriers, or micronutrient enhancers, making them insufficient to meet the growing agricultural demand for improved aroma, biomass, and oil composition in aromatic crops.

6.3 Challenges and Safety Concerns with Future Trends

In contrast, nano-priming, where nanomaterials are used for seed enhancement, holds significant potential for improving crop performance but comes with its own set of challenges, safety concerns, and future considerations. Various key challenges include the lack of standardized protocols for NP size, shape, concentration, priming duration, and drying time, resulting in variable responses across plant species; inconsistent efficacy in different aromatic and medicinal crops; difficulties in scaling up from laboratory to field applications due to cost and mechanization constraints; and the problem of NPs aggregation or degradation, which reduces bioavailability.

Safety concerns involve potential seed and plant toxicity from excessive NP doses, which can damage seed membranes, DNA, and enzymes, such as the case where silver NPs may inhibit root elongation if overdosed, along with environmental risks from NP leaching into soil and water, disrupting microbial flora. Nutrient cycling and potentially cause bioaccumulation in the food chain. There is also the risk of human and animal exposure through inhalation, ingestion, or skin contact, when handling NP-treated seeds, compounded by a lack of regulatory frameworks, standardized toxicology evaluations, environmental labeling, and certification in most countries [55].

6.4 Toxic Effects

The toxicological profile of nanomaterials shows that they can penetrate cell membranes and organelles, generating ROS that cause lipid peroxidation, DNA fragmentation, and mitochondrial dysfunction, while some can directly interact with DNA, leading to mutations, chromosomal aberrations, and cell division inhibition. In plants, high doses can cause root growth inhibition, chlorophyll degradation, and impaired nutrient uptake, while in animals, certain NPs can cross the blood-brain barrier and cause developmental defects. NPs are also mobile in soil and water, with their fate influenced by pH, ionic strength, soil organic matter, and surface coatings, and they may persist in the environment, undergo transformations like aggregation, oxidation, or dissolution, and disrupt beneficial soil microbes and enzymatic activity, thus affecting nutrient cycling. In aquatic systems, NPs can enter via runoff, be absorbed by plankton or fish, and undergo biomagnification [56]. The limitations of nano-priming in agriculture, especially for aromatic plants, include risks of nanotoxicity that can harm sensitive species:

Lack of crop-specific protocols leading to inconsistent results, environmental accumulation affecting soil health, limited understanding of molecular mechanisms influencing secondary metabolite pathways, high costs, and limited availability of safe NPs [57].

Regulatory and safety challenges and potential unintended impacts on essential oil yield or composition. Despite these concerns, the future of sustainable nano-priming in aromatic crops shows promising trends, such as the use of green nano-materials derived from plants for eco-safety, the development of smart nano-formulations for controlled release of nutrients or hormones, and integration of micronutrients and phytohormones into priming systems.

Adoption of IoT and precision agriculture tools for real-time monitoring, improved biosafety and toxicity testing, and multifunctional nanocarriers capable of promoting growth, enhancing disease resistance, and improving aroma yield.

7 Future Scope

The future scope involves developing crop-specific nano-priming protocols to optimize essential oil yield, focusing on green and biodegradable nanomaterials to minimize environmental risks, enhancing nutritional and medicinal properties of aromatic plants, conducting large-scale field validations for commercialization, and establishing comprehensive regulatory frameworks to ensure safe adoption by farmers worldwide. Tailored nano-priming techniques can be optimized for different aromatic plant species (e.g., mint, basil, lemongrass) to enhance uniformity and maximize essential oil yield [58]. Future research will focus on the synthesis of eco-friendly, plant-based, or biopolymer-coated NPs to reduce environmental risks and ensure sustainable application. Combining nano-priming with precision agriculture tools, sensors, and data analytics could enable real-time monitoring and optimization of plant growth conditions. Genomics, transcriptomics, proteomics, and metabolomics will help decode the complex signaling and metabolic changes induced by NPs during seed priming [59]. Large-scale field trials under diverse agroclimatic conditions are needed to translate lab-scale benefits into real-world agricultural practices. Developing international guidelines for safe NP use in agriculture will be crucial to gain farmer trust and encourage adoption across regions.

8 Conclusion

Nano-priming is a promising technology for enhancing germination, yield, and essential oil content in aromatic crops. However, toxicity risks, regulatory gaps, and environmental uncertainties must be addressed. The future lies in green, safe, and targeted nano-formulations integrated with sustainable agriculture practices. Here is a well-rounded conclusion based on the whole study and the current scenario in the field of nano-priming, particularly concerning aromatic plants and agriculture.

In the present era of climate uncertainty and increasing global demand for sustainable agriculture, nano-priming has emerged as a transformative technique with the potential to revolutionize crop production, particularly in high-value aromatic plants. Modeling nano-priming focuses on controlled application of engineered nanomaterials on pre-germinating seeds to improve seed germination, enhance seedling growth, and improve resistance to environmental stresses such as water stress, salinity stress, and temperature extremes. This is particularly important for aromatic crops; the growth and quality of aromatic crops/plants is important as the essential oil yield and economic value of the crop directly relate to consistency in growth and quality. However, nano-priming benefits remain largely aspirational; at this time, it is an emerging technology. The challenges related to a lack of species-specific methodology, an incomplete understanding of chronic effects of nanomaterials in nature, and unclear regulations inhibiting widespread agricultural implementation are not insurmountable; however, they do represent obstacles to agricultural and health-wise growth in areas of promising agricultural technology. The cross-disciplinary considerations - ranging from NP interactions with the environment and toxicity to cost-effective methods for large-scale implementation (at least scalable and >5 kg) - are essential to ensure effectiveness, foster growth, and facilitate collaboration with stakeholders interested in nano-priming and clean agricultural technologies. As the scientific community continues to explore eco-friendly and biodegradable nanomaterials, integration of multi-omics tools, AI-based real-time modeling, and field-validation studies will be pivotal in establishing nano-priming as a reliable, safe, and scalable seed enhancement technology and practice.

Competing Interests The authors declare that they have no competing interest.

Declaration of Generative AI and AI-Assisted Technologies in the Writing Process During the preparation of this work, the book chapter author(s) have not taken any assistance from AI, and the book chapter author(s) reviewed and edited the content as needed and take(s) full responsibility for the content of the publication.

References

1. Mahakham W, Sarmah AK, Maensiri S, Theerakulpisut P (2017) Nano-priming technology for enhancing germination and starch metabolism of aged rice seeds using phytosynthesized silver NPs. Sci Rep 7:8263
2. El-Saadony MT, Saad AM, Soliman SM, Salem HM, Desoky ESM, Babalghith AO, AbuQamar SF (2022) Role of NPs in enhancing crop tolerance to abiotic stress: a comprehensive review. Front Plant Sci 13:946717
3. Singh A, Sengar RS, Sharma R, Rajput P, Singh AK (2021) Nano-priming technology for sustainable agriculture. Biogeosyst Tech 8:79–92
4. Rai-Kalal P, Tomar RS, Jajoo A (2021) Seed nano-priming by SiO_2 oxide improves drought stress alleviation potential in wheat plants. Funct Plant Biol 48:905–915
5. Bakkali F, Averbeck S, Averbeck D, Idaomar M (2008) Biological effects of essential oils: a review. Food Chem Toxicol 46:446–475
6. Burt S (2004) Essential oils: their antibacterial properties and potential applications in foods: a review. Int J Food Microbiol 94:223–253
7. Servin A, White JC (2016) Nanotechnology in agriculture: next steps for understanding engineered NP exposure and risk. Nano Impact 1:9–12

8. Farooq M, Wahid A, Basra SM, Rehman A, Siddique KH (2019) Improving crop resistance to abiotic stresses through seed invigoration. In: Handbook of plant and crop stress, 4th edn. CRC Press, pp 773–792
9. Kanwal N, Hanif MA, Khan MM, Ansari TM, Khalil-ur-Rehman (2016) Effect of micronutrients on vegetative growth and essential oil contents of Ocimum sanctum. J Essent Oil Bear Pl19:980–988
10. Jisha KC, Vijayakumari K, Puthur JT (2013) Seed priming for abiotic stress tolerance: an overview. Acta Physiol Plant 35:1381–1396
11. Chaithanya G, Rao D (2023) Nano-priming: an innovative tool for seed quality enhancement for sustainable agriculture. Int J Agric Invent 8:170–181
12. Dimkpa CO, Bindraban PS (2017) Nanofertilizers: new products for the industry? Jasmonic Acid Agric Food Chem 66:6462–6473
13. Jain A, Ranjan S, Dasgupta N, Ramalingam C (2018) Nanomaterials in food and agriculture: an overview on their safety concerns and regulatory issues. Crit Rev Food Sci Nutr 58:297–317
14. Abbasi Khalaki M, Moameri M, Asri Lajayer B, Astatkie T (2021) Influence of nano-priming on seed germination and plant growth of forage and medicinal plants. Plant Growth Regul 93:13–28
15. Dimkpa CO, Singh U, Bindraban PS, Elmer WH, Gardea-Torresdey JL, White JC (2019) ZnO NPs alleviate drought-induced alterations in sorghum performance, nutrient acquisition, and grain fortification. Sci Total Environ 688:926–934
16. Rui M, Ma C, Hao Y, Guo J, Rui Y, Tang X, Zhu S (2016) Iron oxide NPs as a potential iron fertilizer for peanut (Arachis hypogaea). Front Plant Sci 7:815
17. Lee JH, Kasote DM (2024) Nano-priming for inducing salinity tolerance, disease resistance, yield attributes, and alleviating heavy metal toxicity in plants. Plants 13:446
18. Ochoa-Chaparro EH, Patiño-Cruz JJ, Anchondo-Páez JC, Pérez-Álvarez S, Chávez-Mendoza C, Castruita-Esparza LU, Sánchez E (2025) Seed nano-priming with ZnO and SiO_2 enhances germination, seedling vigor, and antioxidant defense under drought stress. Plants 14:1726
19. Begum P, Fugetsu B (2012) Phytotoxicity of multiwalled CNTs on red spinach and role of ascorbic acid. Toxicol Lett 211:S199
20. Abd-Elsalam KA, Hashim AF, Ahmed FK, Thomas S (eds) (2024) Biopolymeric NPs for agricultural applications. Springer, Cham
21. Çiğ F, Özek R, Erman M, Soysal S, Uçar Ö, Erden Z, Efe M (2025) Nano-priming as seed priming technology for sustainable agriculture. In: Nanotechnology for agriculture. Springer, pp 35–57
22. Sharma A, Sharma MMM, Sharma P, Husen A (2025) Effect of silver nanoparticles on seed germination and seedling growth. In: Husen A (ed) Plant response to silver nanoparticles. Smart nanomaterials technology. Springer, Singapore, pp 123–136., ISBN: 978-981-97-7354-1. https://doi.org/10.1007/978-981-97-7352-7_8
23. Sharma MMM, Kapoor D, Loyal A, Kumar R, Sharma P, Husen A (2024) Effect of gold nanoparticles on seed germination, plant growth, and plant protection. In: Husen A (ed) Plant response to gold nanoparticles. Smart nanomaterials technology. Springer, Singapore, pp 155–175., ISBN: 978-981-97-7173-8. https://doi.org/10.1007/978-981-97-7174-5_8
24. Nair R, Varghese SH, Nair BG, Maekawa T, Yoshida Y, Kumar DS (2010) Nanoparticulate material delivery to plants. Plant Sci 179:154–163
25. Ganjavi AS, Oraei M, Gohari G, Akbari A, Faramarzi A (2021) Glycine betaine functionalized GO as a new engineering NP lessens salt stress impacts in sweet basil (Ocimum basilicum L.). Plant Physiol Biochem 162:14–26
26. Tamindžić G, Azizbekian S, Miljaković D, Ignjatov M, Nikolić Z, Budakov D, Grahovac M (2024) Assessment of various nano-primings for boosting pea germination and early growth in both optimal and drought-stressed environments. Plants 13:1547
27. Meel S, Saharan BS (2024) Enhancing crop resilience towards drought: by integrating nanotechnology, microbiomes, and growth-promoting rhizobacteria. Discov Agric 2:112

28. Imtiaz H, Shiraz M, Mir AR, Siddiqui H, Hayat S (2023) Nano-priming techniques for plant physio-biochemistry and stress tolerance. J Plant Growth Regul 42:6870–6890
29. Khan MN, Li Y, Fu C, Hu J, Chen L, Yan J, Li Z (2022) CeO2 NPs seed priming increases SA level and ROS scavenging ability to improve rapeseed salt tolerance. Glob Chall 6:2200025
30. Nile SH, Thiruvengadam M, Wang Y, Samynathan R, Shariati MA, Rebezov M, Kai G (2022) Nano-priming as emerging seed priming technology for sustainable agriculture: recent developments and future perspectives. J Nanobiotechnology 20:254
31. Pereira AES, Oliveira HC, Fraceto LF, Santaella C (2021) Nanotechnology potential in seed priming for sustainable agriculture. Nano 11:267
32. Kralova K, Jampilek J (2021) Responses of medicinal and aromatic plants to engineered NPs. Appl Sci 11:1813
33. Hasanaklou NT, Mohagheghi V, Hasanaklou HT, Ma'mani L, Malekmohammadi M, Moradi F, Dalvand Y (2023) Seed nano-priming using SiO_2 NPs: effects on seed germination and physiological properties of Stevia Rebaudiana Bertoni. Chem Biol Technol Agric 10:96
34. Geremew A, Stovall L, Woldesenbet S, Ma X, Carson L (2025) Nano-priming with ZnO: a novel approach to enhance germination and antioxidant systems in amaranth. Front Plant Sci 16:1599192
35. Salam A, Afridi MS, Javed MA, Saleem A, Hafeez A, Khan AR, Gan Y (2022) Nano-priming against abiotic stress: a way forward towards sustainable agriculture. Sustainability 14:14880
36. Chandrasekaran U, Luo X, Wang Q, Shu K (2020) The germination of nanoprimed seeds? Front Plant Sci 11:832
37. Tripathi DK, Singh S, Singh VP, Prasad SM, Chauhan DK, Dubey NK (2016) SiO_2 NPs more efficiently alleviate arsenate toxicity than SiO_2 in maize cultivar and hybrid differing in arsenate tolerance. Front Environ Sci 4:46
38. Lateef A, Nazir R, Jamil N, Alam S, Shah R, Khan MN, Saleem M (2016) Synthesis and characterization of zeolite nano-composite: an environment-friendly slow-release fertilizer. Microporous Mesoporous Mater 232:174–183
39. Afzal S, Sharma D, Singh NK (2021) Eco-friendly synthesis of phytochemical-capped iron oxide NPs as nano-priming agent for boosting seed germination in rice (*Oryza sativa* L.). Environ Sci Pollut Res 28:40275–40287
40. Yang L (2024) Nano-priming boosts seed vigor: deeper insights into the effect mechanism. Plant Physiol Biochem 214:108895
41. Kumar B (2022) Nano-priming in sustainable agriculture: recent advances, emerging challenges and future prospective. In: New and future developments in microbial biotechnology and bioengineering: sustainable agriculture: revisiting green chemicals. Elsevier, pp 339–365
42. Aggarwal S (2025) Green nano-priming: a comparative analysis of chemical and bioinspired approaches for sustainable agriculture. Seed Biol 4:e001
43. Bhat UH (2024) Enhancing growth, vitality, and aromatic richness: unveiling the dual magic of SiO_2 dioxide and TiO_2 NPs in *Ocimum tenuiflorum* L. Front Plant Sci 15:1335965
44. Shelar A (2021) Sustainable agriculture through multidisciplinary seed nano-priming: prospects of opportunities and challenges. Cells 10:2428
45. Zaman W, Khalil AAK, Amin A, Ali S (2025) Nano fabrication techniques for enhancing plant-microbe interactions in sustainable agriculture. Nano 15:1086
46. Kumar R, Dadhich A, Dhiman M, Sharma L, Sharma MM (2024) Stimulatory effect of ZnO NPs as a nanofertilizer in seed priming of pearl millet (Pennisetum glaucum) and their bioactivity studies. S Afr J Bot 165:30–38
47. Sharma B, Tiwari S, Kumawat KC, Cardinale M (2023) Nano-biofertilizers as bio-emerging strategies for sustainable agriculture development: potentiality and their limitations. Sci Total Environ 860:160476
48. Verma KK, Joshi A, Song XP, Singh S, Kumari A, Arora J, Li YR (2024) Synergistic interactions of NPs and plant growth-promoting rhizobacteria enhancing soil-plant systems: a multigenerational perspective. Front Plant Sci 15:1376214

49. Kesraoui S, Andrés MF, Berrocal-Lobo M, Soudani S, Gonzalez-Coloma A (2022) Direct and indirect effects of essential oils for sustainable crop protection. Plants 11:2144
50. Pandey G, Tripathi S, Bajpai S, Kamboj M (2023) Approaches, challenges, and prospects of nanotechnology for sustainable agriculture. In: Agricultural and environmental nanotechnology: novel technologies and their ecological impact, pp 83–103
51. Rizwan M (2019) ZnO NPs improved the growth and stress tolerance in plants. Environ Sci Pollut Res
52. MS A, Sridharan K, Puthur JT, Dhankher OP (2021) Priming with nanoscale materials for boosting abiotic stress tolerance in crop plants. *J Agric Food Chem 69*:10017–10035
53. Ghosh S, Thongmee S, Kumar A (eds) (2022) Agricultural nanobiotechnology: biogenic NPs, nanofertilizers and nanoscale biocontrol agents. Woodhead Publishing
54. Aslani F, Bagheri S, Muhd Julkapli N, Juraimi AS, Hashemi FSG, Baghdadi A (2014) Effects of engineered nanomaterials on plant growth: an overview. Sci World J 2014:641759
55. Adhikari T (2016) ZnO NPs asseed priming agent to enhance growth and productivity in maize. J Plant Nutr 39:614–628
56. Guha T, Ravikumar KVG, Mukherjee A, Mukherjee A, Kundu R (2018) Nano-priming with zero valent iron (nZVI) enhances germination and growth in aromatic rice cultivar (Oryza sativa cv. Gobindabhog L.). Plant Physiol Biochem 127:403–413
57. Kah M (2018) Nano-pesticides: state of knowledge, environmental fate, and exposure modeling. Crit Rev Environ Sci Technol 48:447–468
58. Lubbe A, Verpoorte R (2011) Cultivation of medicinal and aromatic plants for specialty industrial materials. Ind Crop Prod 34:785–801
59. Rico CM (2011) Interaction of NPs with edible plants and their possible implications in the food chain. J Agric Food Chem 59:3485–3498

Effects of Carbon-Based Nanoparticles on Aromatic Plants

Shrejal Tiwari, Anant, Neeraj Negi, Pankaj Sharma, and Azamal Husen

Abstract Carbon-based nanomaterials (CBNMs) have gained significant attention across multiple scientific disciplines through remarkable mechanical and physicochemical characteristics including high surface area, stability, and biocompatibility. Since their discovery in the mid-1980s, these materials, including carbon nanotubes, graphene oxide, fullerenes, carbon quantum dots, and nanofibers, have found wide application in agriculture as growth stimulants, nanofertilizers, nanopesticides, and carriers of biomolecules, in aromatic plants such as *Ocimum sanctum* (holy basil), *Ocimum basilicum* (sweet basil), *Mentha arvensis* (mint), and *Lavandula angustifolia* (lavender). By promoting oxidative and hormonal signaling pathways, low concentrations of carbon nanoparticles (CNPs) have been shown to enhance germination, biomass accumulation, and phytochemical content through modifications to photosynthesis, nutrient uptake, and stress-response pathways, thereby stimulating oxidative and hormonal signaling pathways. For instance, graphene oxide treatments in basil and holy basil increased the content of essential oils and phenolic compounds, highlighting their role in modulating secondary metabolism. However, excessive or uncontrolled application may induce cytotoxicity, oxidative stress, and genetic alterations, leading to growth inhibition and metabolic disruption. Therefore, even though CBNMs are a promising new development in

S. Tiwari · N. Negi
Department of Floriculture and Landscaping, College of Horticulture and Forestry, Rani Lakshmi Bai Central Agricultural University, Jhansi, India

Anant
Department of Plant Breeding and Genetics, College of Agriculture, Rani Lakshmi Bai Central Agricultural University, Jhansi, India

P. Sharma (✉)
University of Illinois Chicago, Chicago, IL, USA
e-mail: sharmap@uic.edu

A. Husen
Wolaita Sodo University, Wolaita, Ethiopia

Department of Biotechnology, Graphic Era (Deemed to be University), Dehradun, Uttarakhand, India

A. Husen (ed.), *Sustainable Uses of Aromatic Plants in Nanotechnology*, Smart Nanomaterials Technology, https://doi.org/10.1007/978-981-95-7409-4_18

aromatic plant biotechnology that could help with issues like abiotic stress, decreased fertility, and climate variability, their safe and long-term integration requires careful optimization of dosage, morphology, and application method optimization.

Keywords Carbon-based nanoparticles · Aromatic plants · Essential oil · Secondary metabolites

Abbreviations

CBNMs	Carbon-based nanomaterials
CNPs	Carbon-based nanoparticles
CBNPs	Carbon-based nanoparticles
CNTs	Carbon nanotubes
GO	Graphene oxide
CQDs	Carbon quantum dots
MWCNTs	Multiwalled carbon nanotubes
MWCNT-COOH	MWCNTs functionalized with carboxyl group
f-MWCNT	MWCNTs functionalized with phenylalanine
SWCNTs	Single-walled carbon nanotubes
ROS	Reactive oxygen species
PAL	Phenylalanine ammonia-lyase
TAT	Tyrosine aminotransferase
HPPR	4-hydroxyphenylpyruvate reductase
SOD	Superoxide dismutase
GP	Guaiacol peroxidase
APX	Ascorbate peroxidase
CAT	Catalase
TPC	Total phenolic content
TFC	Total flavonoid content

1 Introduction

Magic carbon of carbon-based nanomaterials has been extensively used in various scientific disciplines because of their extraordinary qualities (physical, chemical, and mechanical) [1–3]. The most stable atom, carbon, with balanced thermodynamic properties, has given rise to different structures with just a change in reaction conditions, such as carbon nanotubes, fullerenes, graphene sheets, nanocones, horns, dots, beads, and nanofibers [4–6]. These nanomaterials are differentiated from others by their high surface area, chemical stability, and biocompatibility. In agriculture, these are used to promote plant growth, enhance nutrient uptake in plants, and boost resistance against stressors [7–12]. The unique physicochemical properties enable them to interact with plant tissue at the cellular and subcellular levels, potentially influencing the metabolic pathways involved in the production of secondary metabolites [13].

Aromatic plants, such as *Lavandula angustifolia* (lavender), *Ocimum sanctum* (holy basil), *Mentha arvensis* (mint), and *Satureja hortensis* (savory), have long been recognized for their ability to produce essential oils and volatile compounds that are widely utilized in perfumery, traditional medicine, cosmetics, and the food industry. These secondary metabolites, predominantly terpenes, phenolics, flavonoids, and alkaloids, serve ecological functions, including plant defense, pollinator attraction, and stress response regulation. Due to their high market value and pharmacological potential, aromatic plants are extensively cultivated across diverse agroecological zones.

However, the cultivation of aromatic crops faces considerable challenges: reduced soil fertility, erratic rainfall, and increasing incidences of abiotic stress such as drought, salinity, and temperature extremes. As a result, agricultural and biotechnological strategies are being explored to enhance the productivity and metabolic output of these plants. Among them, the use of nanotechnology, specifically CBNPs, has emerged as a promising approach to improving plant growth and stimulating secondary metabolism.

Many medicinal and aromatic plants, whether whole or in extract form, have been used as food or in the treatment of various diseases and disorders in humans and animals due to their complex phytochemical components and associated pharmacological activity. Aromatic plants such as *Mentha arvensis* (mint), *Ocimum basilicum* (basil), and *Lavandula angustifolia* (lavender) are important due to their commercial value for the essential oils used in pharmaceuticals, cosmetics, perfumery, and the food industry. These plants contain two important secondary metabolites that are rich in bioactive compounds, such as terpenoids, phenolics, and flavonoids [14], which contribute to their immense importance. Enhancing the content of these by applying nanoparticles or carbon nanomaterials holds high potential for farmers in the commercial market.

Despite the overall goal of using nanomaterials in plants being to enhance agricultural productivity, some research studies have raised serious concerns about the perceived negative impact of NPs on human health and the environment. Unique characters of the NMs and enumerated their fruitful uses as pesticides, seed sprout and growth enhancers, and carriers of phytohormones, herbicides, DNA, and fertilizers to the cells of plants [15]. As CBNP is smaller than other growth regulators, it has better surface coverage, which allows it to penetrate through the stomatal opening or root system and localize in cell walls and chloroplasts. This can positively as well as negatively affect plant growth and functioning [16–20]. For example, they have been observed to enhance yield, regulate plant functioning, improve nutrient uptake, and provide protection against various biotic stresses such as diseases and insect attacks, as well as abiotic stresses. At the same time, they may also exhibit toxicity to plants, leading to cytotoxic effects and genetic modifications. Moreover, studies indicate that CNPs can induce mild oxidative stress, which may serve as a trigger for enhanced secondary metabolite biosynthesis as part of the plant's defense response [21]. For example, *Ocimum sanctum* treated with nanoparticle showed an increase in essential oil content and phenolic compounds, possibly due to the modulation of stress signaling pathways [22, 23].

2 Effect of Carbon-Based Nanoparticles (CBNPs) on Aromatic Plants

Some of the many nanomaterials currently developed to be applied to agriculture include plant germination and growth stimulants, novel soil and water cleaning methods, nanopesticides, and nanofertilizers that can be used in lesser quantities and still deliver improvements to food production and quality [24, 25]. Recent studies, however, have demonstrated that certain nanoparticles can harm plants through mechanisms such as phytotoxicity and delayed plant seedling germination [26, 27]. However, others can enhance plant development by activating cellular signaling pathways [28]. The physical-chemical characteristics of the nanoparticles, such as their size and zeta potential, dictate the biological reactions [29].

As mentioned earlier, carbon nanoparticles include materials like carbon nanotubes, graphene oxide (sp2-C atoms in a checkered six-crystal lattice), and carbon quantum dots. Fullerene (20-polyhedron facet symmetry, allotrope) also has a significant and complex impact on aromatic plants [6]. Aromatic plants are highly valued for their essential oils and bioactive compounds. At the same time, the application of these nanoparticles affects not only the growth and development but also the production of secondary metabolites. When introduced in controlled, low concentrations, CNPs have been found to enhance seed germination, root and shoot growth, and overall biomass production. This is largely attributed to their ability to improve water and nutrient uptake, increase chlorophyll content, and stimulate photosynthesis. Some nanoparticles may even interact with plant hormones, such as auxins, promoting cell elongation and division, as illustrated in Fig. 1. Most nanoparticles have been shown to adversely affect crop growth and development at low doses, as suggested by Zhao et al. [30]. At the same time, it shows a positive impact at high concentrations. These changes may differ in morphology, dosage, covering, and composition of the CNP used. However, the toxicity of these may lead to root growth inhibition, formation of RO species, and a rise in lipid peroxidation, as observed in some other plant species.

3 Influence on Growth and Morphological Traits

When applied thoughtfully, CBNPs have the remarkable capacity to invigorate the growth and vitality of plants by supporting numerous physical traits, inner biochemical processes, and other traits (Fig. 1). These tiny particles are adept at slipping through seed coats, allowing seeds to absorb water more effectively, a process that leads to improved germination and robust root growth [31]. When used in small amounts, carbon-based nanoparticles encourage plants to grow by boosting cell

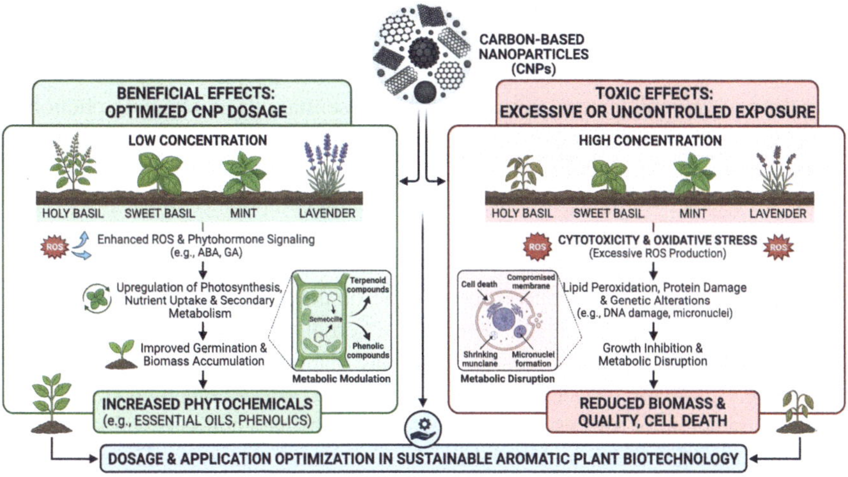

Fig. 1 Effects of carbon-based nanoparticles (CNPs) on aromatic plants. This figure explained the mechanistic overview of the effects of CNPs, such as graphene oxide, on aromatic plants (holy basil, sweet basil, mint, and lavender). At low concentrations, CNPs enhance oxidative and hormonal signaling pathways, leading to improved germination, biomass accumulation, and increased phytochemical content (e.g., essential oils, phenolic compounds) by modulating photosynthesis, nutrient uptake, and secondary metabolism. However, excessive or uncontrolled exposure may result in cytotoxicity, oxidative stress, genetic alterations, growth inhibition, and metabolic disruption, emphasizing the need for dosage and application optimization in sustainable aromatic plant biotechnology

division, which helps roots and shoots develop more vigorously. This helps the plant absorb more nutrients from the soil and also encourages the growth of bigger, healthier leaves with deeper green color, giving the plant everything it needs to capture sunlight better, photosynthesize more effectively, and flourish in its surroundings [32, 33].

In addition to the visible growth effects, carbon-based nanoparticles work behind the scenes to help plants operate more smoothly at the cellular level. They encourage the plant to produce more chlorophyll, making leaves better at capturing sunlight, and improve the flow of carbon dioxide into the leaves' essential ingredients for photosynthesis. By boosting the activity of Rubisco, a key enzyme that turns carbon dioxide into the sugar plants need, CBNPs help plants harness sunlight more effectively so they can grow stronger and healthier.But their influence isn't limited to growth alone. CBNPs can also gently stress the plant by generating reactive oxygen species (ROS), which, when kept in check, act as important messengers to kick-start protective enzymes. These enzymes are part of the plant's antioxidant systems (including SOD and CAT), helping to keep cellular damage at bay while also activating pathways that create beneficial compounds like phenolics and flavonoids, as suggested in a study done by Halawani et al. [34] and Hasanuzzaman et al. [35]. In aromatic plants, carbon-based nanoparticles play a special role by boosting the production of valuable natural compounds like essential oils and healing phenolics, which enhance both the plant's fragrance and its medicinal benefits. They do this by turning on important regulatory proteins (known as transcription factors like WRKY and MYB) and activating genes that create aromatic compounds and terpenes. This leads to a richer buildup of natural chemicals such as eugenol, linalool, menthol, and rosmarinic acid. As a result, plants treated with CBNPs often have stronger, more vibrant scents and enhanced therapeutic qualities, highlighting the exciting potential of these nanoparticles in farming and herbal medicine [36, 37].

Research has revealed that CDs can penetrate plant cells and migrate with water and minerals along the roots to the stems and leaves. They are taken up via intercellular and extracellular spaces and through plasmodesmata and cell walls. Thence, they go through the cortex and into the xylem via the plastid route. The steady and unambiguous fluorescent emissions of CDs represent a reliable means of plant tracking. Plant absorption of CDs can be demonstrated using methods such as Raman spectroscopy measurement, fluorescence imaging, and transmission electron microscopy. For example, it has been shown that nanoparticles significantly increase vegetative growth in *Lavandula angustifolia*, indicating improved photosynthesis and resource allocation [38].

More than 90 percent of crop biomass is attributed to effective photosynthesis and serves as the primary material source for plant development, making biomass accumulation crucial to crop growth and production. One of the major physiological functions of CDs is to increase plant photosynthesis. The two energy conversion processes constituting photosynthesis include light energy to electric energy conversion and electric energy to chemical energy conversion, all of which involve light absorption, electron transfer, photophosphorylation, and carbon assimilation, among other critical reaction steps. As a result of the differing nature and content of surface groups, and differences in the oxygen/nitrogen content of carbon nuclei, CDs may also display light absorption into the visible light range (though they tend to be very strong absorbers in the ultraviolet region (200–400 nm). The longwave absorptions for CDs lie within 500–800 nm, which converts UV light into visible light that plants can utilize.

Seed germination is a crucial part of plant growth, and healthy seed germination can lead to the development of more nutritious plants. Treading of seeds with a CD aqueous solution has been found to increase the seed germination of both wheat and rice. The absorption and assimilation of water and nutrients affect seed germination, crop growth and development, and yield formation. One of the reasons why CDs promote seed germination and enhance seed vitality is the ability to penetrate the hard seed coat, promote water retention and absorption, and trigger seed germination and seed growth. The surface hydrophilic groups (-OH and -COOH) of CDs influence seed germination, root growth, seed moisture content, and seedling length. The hydrophilic moieties on the surface of CDs provide a large number of binding sites for water molecules, which the plant imbibes. Adequate imbibition of water stimulates plant germination and accelerates seedling growth. In addition to serving as water adsorption sites, CDs enhance the expression of seed aquaporins, activate aquaporins, and lower the pH of the rhizosphere. This helps in water and nutrient absorption, microbial environment in the rhizosphere, seed germination, and growth.

It is also at this point that crops take up nutrients as they take up water. Additionally, these hydroxyl and carboxyl groups confer the ability of CDs to adsorb a range of nutrient ions, such as K^+, Ca^{2+}, Mg^{2+}, Cu^{2+}, Zn^{2+}, Mn^{2+}, and Fe^{3+}, which are essential for crop growth. They bind to the hydrophilic groups on the surface of the CDs through hydrogen bonding and electrostatic forces, accumulating on the surface. As the CDs enter the plant, the concentration of nutrient ions increases along with the CDs, which is why the nutrients are released slowly and steadily through the xylem. We can see that the nutrient ions can enter the plant using CDs, as the content of nutrient ions in *Arabidopsis* subjected to CDs is increased compared to the control. When coriander was sprayed with 40 mg/L of CDs, the content of K, Ca, Mg, P, Mn, and Fe increased by 64.3%, 21.0%, 26.2%, 12.8%, 56.0%, and 125%, respectively. When the plants were dipped in 0.02 mg mL $^{-1}$ of CDs, the content of N, P, and K in the plants increased by 4.4, 10.8, and 16.5, respectively. Nevertheless, the expression of genes associated with aquaporin membrane proteins involved in gas, nutrients, and water exchange is enhanced following treatment with CDs, leading to improved absorption of water and mineral ions and promoting crop growth. To illustrate this, experiments done by Tan et al. [39] estimated the interaction of MWCNTs and rice suspension cells and discovered that they were producing and accumulating reactive oxygen species (ROS), which reduced cell growth and eventually resulted in cell death. Application of MWCNTs in suspension cell culture media reduced *Arabidopsis* cell viability, dry weight, chlorophyll content, and superoxide dismutase (SOD) activity, according to Lin et al. [40]. Recent studies by Tiwari et al. [41] demonstrated that seedlings of *Zea mays* that were grown in agar gel in the presence of different concentrations of pristine MWCNTs exhibited improved growth and water uptake.

4 Uptake and Translocation of Carbon Nanoparticles

Clarifying the uptake and translocation pathways of CNPs is essential to comprehending their effects on aromatic plants. Root absorption is the main way that CNPs enter the soil after application. According to studies, nanoparticles can enter the xylem for systemic distribution after passing through the apoplastic and symplastic pathways that penetrate the rhizodermal and cortical tissues of roots. Bypassing the conventional size-exclusion limits of the root apoplast, carbon nanotubes (CNTs), graphene oxide (GO), and carbon quantum dots (CQDs) can enter vascular tissues and move to aerial parts because of their nanoscale dimensions [19, 20, 42].

Furthermore, it has been demonstrated that CNTs function as nano-channels, facilitating the movement of ions and water across root membranes, thereby enhancing the effectiveness of water uptake and serving as carriers of nutrients or signaling molecules simultaneously [43]. Because the distribution of nanoparticles through xylem flow directly affects the leaf tissues where essential oil biosynthesis takes place, root-mediated uptake is especially crucial for aromatic plants like *Mentha* and *Ocimum*.

Another popular technique for introducing CNPs into plants is foliar application. Nanoparticles can enter the leaf surface through cuticular nanopores or stomatal apertures when sprayed or suspended and then make their way to the mesophyll tissues. Because it avoids the root-soil interface, foliar uptake is thought to be effective in minimizing nanoparticle losses from microbial interactions or soil adsorption [44]. Once within the mesophyll, nanoparticles can interact with the mitochondria, enter chloroplasts, or accumulate in intercellular spaces, all of which can impact vital physiological functions such as respiration, photosynthesis, and the biosynthesis of secondary metabolites. Due to the localized stimulation of stress-related signaling pathways, the foliar application of CNPs has been associated with increased oil yield and improved phenolic content in aromatic plants, where leaf tissues are the primary sites for essential oil accumulation in glandular trichomes [45].

A third promising tactic to enhance plant performance from the outset of development is seed priming with nanoparticles. It has been observed that priming seeds with CNPs during imbibition accelerates water uptake, promotes the emergence of radicles, and activates key enzymes for early metabolism and germination [46]. Through their interaction with embryonic tissues and penetration of seed coats, nanoparticles alter hormonal balance and initiate early defense responses. For example, it has been demonstrated that CNT-mediated priming increases the activity of proteases and α-amylase, which improves seedling vigor and mobilizes more seed reserves [43]. Nanomaterial seed priming has been linked to improved establishment, increased biomass accumulation, and higher essential oil content in mature plants of aromatic species such as *Ocimum basilicum* and *Mentha arvensis* [47]. The potential of nanoparticle seed priming as a sustainable agronomic technique for high-value aromatic crops is highlighted by its capacity to affect long-term plant physiology.

Depending on the growth stage, environmental factors, and intended phytochemical results, each route may be optimized and offer unique physiological benefits. However, it's also crucial to recognize that several variables, including nanoparticle size, surface charge, chemical functionalization, and the inherent anatomical characteristics of the target plant species, influence the extent to which CNPs are absorbed and subsequently distributed within plant tissues.

5 Influence on the Production of Secondary Metabolites, Such as Essential Oil Production and Composition

5.1 Effects on Essential Oil Profile and Secondary Metabolites

For more than ten years, CNPs have been utilized to enhance the secondary metabolite profiles, essential oil yield, and antioxidant properties of various medicinal plants. Secondary metabolites are abundant in medicinal plants. Under field-grown conditions, yields and contents are often relatively low. Therefore, CNT-augmented in vitro cultures may be a useful technique for raising total secondary metabolite production. Businesses that deal with pharmaceuticals and drugs in particular may find this to be profitable. The studies have focused on a wide array of medicinal and aromatic plants, such as *Satureja khuzestanica* (Khuzistan summer savory), *Simmondsia chinensis* (jojoba), *Catharanthus roseus* (periwinkle), *Ocimum basilicum* (sweet basil), *Salvia nemorosa* (woodland sage), *Salvia verticillata* (lilac sage), *Thymus daenensis* (denaian thyme), *Dracocephalum moldavica* (Moldavian balm), *Stevia rebaudiana* (stevia), *Nigella sativa* (black caraway), *Mentha piperita* (peppermint), *Melissa officinalis* (lemon balm), *Calotropis procera* (milkweed), and *Isatis tinctoria* (woad). CNPs are quite versatile, e.g., CNT (single and multiwalled), along with their functionalized and pristine forms, carbon dots, and graphene, and their elicitation effects have been described as follows.

5.2 Changes in Essential Oil Chemical Composition

Satureja khuzestanica (Khuzistan summer savory), a member of the Lamiaceae family, is a medicinal plant native to Iran, recognized for its analgesic and antioxidant properties. The antioxidative properties of the plant are primarily due to a phenylpropanoid compound, rosmarinic acid. Ghorbanpour and Hadian [48] incorporated MWCNTs into the in vitro *Satureja* leaf segment cultures. They observed that while the highest phenolics, flavonoid content, and rosmarinic acid were observed at 100 $\mu g\ ml^{-1}$, an elicitation in caffeic acid content was observed with 250 $\mu g\ ml^{-1}$ concentrations of MWCNTs, respectively. In another study [49], MWCNTs were used in combination with methyl jasmonate in the *Satureja* nodal segments. The rosmarinic acid content was found to be 13 times higher than that of the field-grown control plants.

Simmondsia chinensis (jojoba) is a commercially important plant due to its pharmaceutical and cosmetic value in the market. Under in vitro conditions, Gaafar et al. [50] demonstrated that at the lowest dose (0.002 g/l) of MWCNTs in the fortified media, the content of secondary metabolites (total phenolics, flavonoids, and tannins) was found to be highest in the seedlings compared to other MWCNT treatments.

Catharanthus roseus (rose periwinkle) is popularly known for its anticancerous properties. Particularly, vincristine and vinblastine have been identified as key compounds for the manufacture of several pharmaceutical drugs. These compounds are chemical derivatives of terpenoid indole alkaloids. McGehee [51] reported that while the total alkaloid content in all CBN treatments was twice that of untreated plants, the yield of vincristine and vinblastine was enhanced under MWCNT- and graphene-treated callus, under both light and dark conditions, respectively. Ghasempour [52] has also reported an elicited alkaloid production response in MWCNT-treated calluses of *Catharanthus*.

Salvia (sage) is one of the major pharmaceutical herbs, which serves as a potential source of phenolic acids (rosmarinic acid, caffeic acid, and ferulic acid). It has been used for the treatment of various diseases like Alzheimer's, dementia, and diabetes. Two particular species of *Salvia*, namely, *Salvia nemorosa* (woodland sage) and *Salvia verticillata* (lilac sage), have been studied for the evaluation of the effect of CNTs on secondary metabolite production. Rahmani et al. [53] have demonstrated that MWCNTs were sprayed on *S. verticillata* plants, and a concentration of 50–1000 mg l^{-1} was effective in enhancing the RA (rosmarinic acid) content, which was four times that of the control plants. On the other hand, Rahmani and Radjabian [54] used MWCNTs in combination with methyl jasmonate in the form of an aqueous spray on *S. verticillata* leaves. A significant increase in levels of jasmonic acid and salicylic acid was observed 8 h after the exposure. Furthermore, MWCNTs led to an enhanced profile of rosmarinic acid (23.63 ± 2.49 mg g^{-1} DW), salvianolic acid A (0.80 ± 0.04 mg g^{-1} DW), salvianolic acid B (2.03 ± 0.14 mg g^{-1} DW), and caffeic acid (0.96 ± 0.04 mg g^{-1} DW). Heydari et al. [55] have incorporated MWCNT-COOH in the cell suspension cultures of *S. nemorosa* and concluded that rosmarinic acid, salvianolic acid, cinnamic acid, and ferulic acid content of callus treated with 100 mg/l MWCNT-COOH were 13, 14.2, 20, and 3 times higher than wild *S. nemorosa* plant at flowering stage. Hir et al. [56] used CNTs (80 mg/l) in combination with Kinetin (4 mg/l) to increase carotenoids, phenols, and flavonoids in in vitro cultures of *S. nemarosa*.

The Lamiaceae family includes the medicinal herb thyme (*Thymus daenensis*). The plant, which is native to Iran, has anti-inflammatory, carminative, and tonic properties. The elicitation effects of MWCNTs on *Thymus* secondary metabolite production have been thoroughly investigated by Samadi et al. [57–60]. They looked at how MWCNTs affected *T. daenensis* seed cultures grown in vitro [60]. The highest concentrations of TPC (6.70 ± 0.06 mg GAE g^{-1} DW) and TFC (8.19 ± 0.01 mg QUE g^{-1} DW) were found in media fortified with 250 mg ml^{-1} MWCNTs.

Another study examined the effect of single-walled carbon nanotubes (SWCNTs) on *Thymus* callus culture [59]. The impact of single and multiwalled carbon nanotubes on in vitro seed and callus cultures was once more investigated in 2023. The highest levels of TPC and TFC were found in cultures treated with MWCNTs at 62.5 μg ml^{-1} [57]. However, callus cultures treated with MWCNTs demonstrated enhanced biosynthesis of secondary metabolites, including rosmarinic acid (25.25 mg g^{-1} DW), ferulic acid (1.97 mg g^{-1} DW), caffeic acid (1.02 mg g^{-1} DW), catechin (1.99 mg g^{-1} DW), vanillin (0.91 mg g^{-1} DW), and hesperidin (2.83 mg g^{-1} DW) [58].

Moldavian balm (*Dracocephalum moldavica*) is a medicinal and aromatic herb used for its antirheumatic, antioxidant, antitumor, and antimutagenic properties, as demonstrated by the essential oil and extract. Proline-functionalized graphene oxide nanoparticles, "GO–Pro NPs" (Pro), were used in conjunction with salt stress in plants [61]. In both salinity levels (50 and 100 mm salt stress), the majority of the dominant components (geranial, z-citral, geraniol, and nerol) showed increased values under 50 mg l^{-1} Pro treatment. On the other hand, under both salinity conditions, 100 mg l^{-1} pro application increased the geranial and z citral values while not affecting the nerol value.

Lemon balm (*Melissa officinalis*) is a medicinal herb from the Lamiaceae family, used for the treatment of insomnia and anxiety, as it acts as a natural antidepressant. Several polyphenolic compounds contribute to its therapeutic properties. Carboxylic-group functionalized MWCNTs were sprayed, and Ebrahimi et al. [62] reported that a 100 mg l^{-1} spray on two-month-old plant shoots boosted total phenols, flavonoids, and phenolic acid levels. Further analysis showed an increase in active constituents like rosmarinic acid, salvianolic acid B, apigenin, and kaempferol.

Isatis tinctoria (European indigo) is one of the underutilized medicinal crops from the Brassicaceae family. Studies have suggested that it has inhibitory effects as an antiviral and antileukemic agent. Huang et al. [63] utilized carbon dots as treatments to influence the pharmaceutical properties. The concentration of active constituents, such as indigo and indirubin, was increased by 128% and 27%, respectively, compared to the control plants.

Calotropis (*Calotropis procera*) has also been shown to have an increased concentration of metabolites like camphor, linalool, p-cymene, germacrene D, citronellal, geranial, and citronellol by using MWCNTs (200 mg l^{-1}) in the in vitro *C. procera* seedlings [64].

6 Case Studies and Experimental Evidence

6.1 *Basil (***Ocimum basilicum***)*

Ocimum basilicum (sweet basil) is an aromatic and medicinal herb valued for several pharmaceutical properties, such as antibacterial, anti-inflammatory, and analgesic, to name a few. Several phenylpropanoids (eugenol, chavicol, linalool, geraniol) and phenolic compounds have been derived from the essential oil extracted from leaves. Research on *Ocimum* has primarily focused on functionalized carbon nanotubes. Functionalization has been proven to improve the dispersion of CNTs into the desired cells. Gohari et al. [65] have used MWCNT-COOH in the basil plants cultured hydroponically. An increase in the phenolic content was observed with a 50 mg/l treatment of the CNTs under 50 mM NaCl stress. MWCNTs have also been functionalized with phenylalanine (f-MWCNT), along with pristine (non-adulterated) forms and commercial forms, as different treatments in the callus cultures of *Ocimum*. According to Holghoomi et al. [66], the highest concentrations of phenolic compounds were found in 200 mg/l of f-MWCNT, 50 mg/l of pristine MWCNT, and 100 mg/l of MWCNT.

They have further reported an increase in total phenols, flavonoids, and anthocyanins in 200 mg/l of f-MWCNT [67]. Also, they have further elucidated different components affected by functionalized MWCNTs in different concentrations. The highest percentages of alpha-copaene, trans-alpha-bergamotene, alpha-guaiene, bicyclogermacrene, 1,10-di-epi-cubenol, and alpha-eudesmol compounds were observed at 150 mg/l of f-MWCNT, while the highest percentages of compounds 1,8-cineole and eugenol were observed at 100 mg/l of f-MWCNT. The compounds linalool, camphor, and anethole also showed their highest amounts in treatments of 200, 150, and 50 mg/l of carbon nanotube, respectively.

*6.2 Peppermint (***Mentha piperita***)*

The Lamiaceae family's *Mentha* (*Mentha piperita*) is a well-liked aromatic herb prized for its essential oil. Menthone, menthol, and menthyl acetate are the main ingredients of the essential oil. The effects of MWCNTs on *Mentha* under salt stress in greenhouse settings were examined by Fallah et al. [68]. The results indicated that the highest flavonoid content was found at 100 and 200 μg/ml MWCNTs combined with 200 mM NaCl, while the application of 200 mM MWCNTs levels increased the phenol content (41%) in plants treated with 200 mM NaCl in comparison to the control plants, as shown in Table 1.

6.3 Modulation of Biosynthetic Pathway

The molecular mechanism behind the elicitation effects of carbon nanoparticles has been revealed as a function of upregulation in certain genes involved in biosynthetic pathways. Relevant data on the expression profiles of biosynthetic genes in connection to the accumulation of rosmarinic acid in *Satureja khuzistanica* were presented by Fatemi et al. [49]. PAL, TAT, and HPPR all showed upregulated expression patterns at 100 mg/L MWCNTs, while downregulated levels were noted at 250 mg/L. In every MWCNT treatment, RAS was upregulated.

Moreover, upregulation in the transcription of the DAT (deacetylvindoline-4-O-acetyl transferase) gene, responsible for the biosynthesis of vincristine and vinblastine metabolites, resulted from MWCNT supplementation in *Catharanthus roseus* [52]. The transcriptomic sequence analysis by Huang et al. [63] in *Isatis tinctoria* has further elucidated that CDs promote the upregulation of genes involved in indole biosynthesis, including *ASA1* and *TSA1*. In the photosynthetic pathway (Ko00195), the expression levels of *PSAK* genes in photosystem I (PSI) and *PSBY* and *PSBP1* genes in PSII were significantly upregulated in the CD-treated group compared with the control group. Their upregulation indicates that CDs improve photosynthesis in *Isatis tinctoria* by improving its light capture efficiency and facilitating electron transfer between PSI and PSII. Similar findings were made by Asakereh et al. [69], who found that foliar spraying 600 ppm MWCNTs (multiwalled carbon nanotubes) for 48 h was effective in increasing the expression of genes related to the silymarin synthesis pathway, including PAL, CHI, CHS1, CHS2, CHS3, CHS1, SmTIP41,

Table 1 Positive effect of CNPs on aromatic plants

Plant name	Common name	Active compound	Carbon nanoparticle used	Results	Key references
Ocimum basilicum	Sweet basil	Eugenol, chavicol, linalool, geraniol	MWCNT-COOH	An increase in the phenolic content was observed with 50 mg l^{-1} treatment of the CNTs with 50 mM NaCl stress can mitigate salinity stress by boosting enzymatic antioxidant components (APX, CAT, and GP activity)	Gohari et al. [65]
			200 mg/l of f-MWCNT	Increase in total phenols, flavonoids, and anthocyanins	Holghoomi et al. [67]
			f-MWCNT, along with pristine	The highest levels of TPC were observed in 200 mg/l of f-MWCNT, 50 mg/l of pristine MWCNT, and 100 mg/l of MWCNT, respectively With improved PPO and POD activity	Holghoomi et al. [66]
Mentha piperita	Peppermint	Menthone, menthol, and menthyl acetate	MWCNTs on *Mentha* under salt stress	Highest flavonoid content and enhanced level of phenols with high SOD and peroxidase	Fallah et al. [68]
Satureja spp.	Summer savory	Phenylpropanoid compound, rosmarinic acid	MWCNTs	Increased flavonoid content and rosmarinic acid and the highest PPO and POD activities help in stress management	Ghorbanpour and Hadian [48]
			Used MWCNT with methyl jasmonate	13 times more rosmarinic acid content	[49]
Simmondsia chinensis	Jojoba	TPC, TFC, and tannins	Lowest dose (0.002 g/l) of MWCNTs	Improved the content of secondary metabolites	Gaafar et al. [50]
Nigella sativa	*Nigella*	Thymoquinone, carvacrol, α-pinene, β-pinene, and thymol	High concentration carbon nanotubes (in vitro) (5.0 to 25.0 mg/l)	DPPH antioxidant activity increased	Ambreen et al. [70]

(continued)

Table 1 (continued)

Plant name	Common name	Active compound	Carbon nanoparticle used	Results	Key references
Catharanthus roseus	Rose periwinkle	Vincristine and vinblastine, chemical derivatives of terpenoid indole alkaloids	MWCNTs and graphene-treated callus	Improved content of alkaloids	McGehee [51]
			MWCNT	Increased CAT, GP, and PAL activities	Ghasempour [52]
Salvia verticillata	Lilac sage	Rosmarinic acid, caffeic acid, and ferulic acid	MWCNTs sprayed at 50–1000 mg l^{-1}	Effective in enhancing the RA	Rahmani et al. [53]
			MWCNTs in combination with methyl jasmonate in the form of an aqueous spray	A significant increase in levels of jasmonic acid and salicylic acid, with improved RA, and salvianolic acid A and B	Rahmani and Radjabian [56]
S. nemorosa	Woodland sage	Salvianolic acid A and salvianolic acid B, cinnamic acid, and ferulic acid	CNTs (80 mg/l) in combination with kinetin (4 mg/l	Increase in carotenoids, phenols, and flavonoids	Hir et al. [54]
Thymus daenensis	Thyme	Thymol and carvacrol, para-cymene and linalool	MWCNTs	High conc of TPC and TFC with increased antioxidant activities (73.88 ± 0.47%) and maximum PAL activity	Samadi et al. [60]
			SWCNTs	Maximum amounts of TPC and TFC	Samadi et al. [57]
			MWCNT-treated callus cultures	Improved rosmarinic acid (25.25 mg/gDW), ferulic acid (1.97 mg/g DW), caffeic acid (1.02 mg/g DW	Samadi et al. [58]

Dracocephalum moldavica	Moldavian balm	Geranial, z-citral, geraniol, and nerol	Proline-functionalized GO nanoparticles, "GO-Pro NPs" (Pro), were used along with salt stress	Improved geranial, z-citral at 50 and 100 mg l^{-1}	Fatehi et al. [61]
Melissa officinalis	Lemon balm	Rosmarinic acid, salvianolic acid B, apigenin, and kaempferol	MWCNTs were sprayed on a two-month-old shoot	Boosted TPC, TFC, and phenolic acid levels	Ebrahimi et al. [62]
Calotropis procera	*Calotropis*	Camphor, linalool, p-cymene, germacrene D, citronellal, geranial	Using MWCNTs (200 mg l^{-1}) in the in vitro	Improved secondary metabolites	Razavizadeh et al. [64]

SmMON1, and SmNDHF. At 600 ppm, there was the greatest increase in CHS33 gene expression, 7.23 times greater than the control.

6.4 Effect on Antioxidant Enzyme Activity

Studies on *Satureja khuzestanica* have shown that it affects the highest PPO and POD activities on media treated with 500 μg ml^{-1} MWCNTs [48]. When compared to control cultures, Ghasempour [52] found that the application of MWCNT in periwinkle doubled the activities of catalase, peroxidase, and phenylalanine ammonia lyase. Gohari et al. [65] have also reported that MWCNTs-COOH can reduce the effects of salinity stress by increasing the activity of enzymatic antioxidant components in *Ocimum basilicum*, such as GP, CAT, and APX. In contrast, the highest PPO and POD activity was noted in basil at 200 mg/l of f-MWCNT and 50 mg/l of MWCNT application, respectively [66]. In *Thymus daenensis*, Samadi et al. [60] found that plants fortified with 250 mg ml^{-1} MWCNTs exhibited higher antioxidant activities (73.88 ± 0.47%) and maximum PAL activity (1.25 ± 0.08 mM cm/g FW). Once more, it has been proposed that thyme's biochemical parameters may be impacted by SWCNTs. Enzymes that aid in the biosynthesis of PAL were elevated by SWCNTs (50 μg ml^{-1}) [57–59]. High concentrations of carbon nanotubes (5.0–25.0 mg/l) were found to increase the antioxidant activity of DPPH in *Nigella sativa*. On the other hand, the PAL activity of adventitious roots decreased as CNT concentrations and pretreatment times increased [70]. According to Fallah et al. [68], under salinity stress, *Mentha piperita* plants treated with both MWCNTs and NaCl exhibited increased antioxidant enzyme activities, including SOD, CAT, and POD content.

6.5 Cytotoxic/Genotoxic Effects at Higher Doses

The adverse effects of carbon-based nanoparticles are often observed at higher doses. Oxidative damage and toxicity from CNTs have been reported in various medicinal plants, e.g., in basil [65], *Salvia* [53], *Stevia* [71], thyme [57, 58], lemon balm [62], and *Calotropis* [64], which is further summarized in Table 2.

Basil developed toxicity symptoms as a result of aggregations inside plant tissue brought on by high concentrations of MWCNTs-COOH (100 mg l^{-1}) [65]. Similarly, CNTs caused an excessive increase in ROS generation and oxidative damage in the treated *Salvia* leaves at higher concentrations (>100 mg l^{-1}). One of the main signs of toxicity was the reduction in photosynthetic capacity [53]. According to Ramezan et al. [71], increased MWCNT application in the culture medium combined with high light intensity (200 and 400 μmol m^{-2} s^{-1}) has been shown to cause damage to the promoter site and the sequence of the UGT76G1 gene in stevia, which is involved in the synthesis of SVglys (steviol glycosides).

Table 2 Negative effect of CNPs on aromatic plants

Plant name	NP used	Toxic effect observed	Key references
Basil	High concentration of MWCNTs-COOH (100 mg l^{-1}	Aggregations inside plant tissue	Gohari et al. [65]
Salvia	Higher concentrations (>100 mg l^{-1}), CNTs	Excessive increase in the ROS generation and oxidative damage with a decline in photosynthetic efficiency	Rahmani et al. [53]
Stevia	MWCNTs in the culture medium (200 and 400 μmol m^{-2} s^{-1})	Damage to the promoter site and the sequence of the *UGT76G1* gene, involved in the synthesis of SVglys (steviol glycosides)	Ramezan et al. [71]
Thymes	Higher concentrations of SWCNTs	Negative effects on morphological traits and oxidative damage caused by increased activity of PPO and POD	Samadi et al. [57, 58]
Melissa	MWCNTs 250 mg l^{-1} dosage	Caused cytoplasmic disruption in abaxial epidermal cells and altered the localization and ultrastructure of chloroplasts	Ebrahimi et al. [62]
Calotropis	A higher amount of MWCNT (400 mg l^{-1})	Reduction in chlorophyll content and increase in SOD and APX activities	Razavizadeh et al. [64]

Accordingly, culture media containing higher concentrations of SWCNTs had detrimental effects on the morphological characteristics of *T. daenensis* seedlings (such as shoot height, root elongation, and biomass accumulation) as well as the production of secondary metabolites as a result of cellular oxidative damage brought on by increased PPO and POD activity, which in turn produced more toxic oxidation products like quinone [57, 58].

Similarly, damaging properties of higher doses of MWCNTs were observed in *Melissa officinalis* [62], where a 250 mg l^{-1} dosage caused cytoplasmic disruption in abaxial epidermal cells and altered the localization and ultrastructure of chloroplasts. Still, they had no particular effects on secondary metabolite production. A recent in vitro study by Razavizadeh et al. [64] on *Calotropis* has also concluded that, at a higher amount of MWCNT (400 mg l^{-1}), a reduction in chlorophyll content and an increase in SOD and APX activities were observed. High concentrations of MWCNTs might cause depletion of iron in plants, thereby hindering chlorophyll production and increasing the activities of SOD and POD [72].

7 Future Research

Plants take up nanoparticles through various pathways, primarily via their roots and leaves. Nanoparticles first contact plant roots by adhering to the root surface, where root hairs and chemical secretions influence their accumulation and absorption [73].

They can enter root tissues through pores in the root epidermis or, in some cases, by creating new openings, moving inside through different mechanisms like endocytosis or binding to transport proteins. Once inside, nanoparticles travel throughout the plant via apoplastic and symplastic pathways, moving cell to cell or through the spaces outside cells. Foliar uptake also occurs mainly through stomata and cuticles on leaves, with nanoparticles crossing these barriers to reach internal tissues. Several factors affect nanoparticle uptake and transport in plants, including particle size, surface charge, chemical composition, and plant species. Smaller particles tend to move more easily, and positively charged particles often accumulate more due to attraction with negatively charged plant surfaces. Understanding these complex routes and interactions is essential to developing safe and effective agricultural applications of nanoparticles [74].

8 Conclusion

Carbon nanoparticles are a promising approach in the cultivation of aromatic plants. When applied in low doses of <100 ppm, they enhance photosynthetic efficiency, stimulate biosynthesis of essential oils and secondary metabolites, and activate vital physiological and biochemical processes in the plant cell. Further, the application of CNP triggers activities of antioxidant enzymes like SOD, POD, CAT, APX, and GP, which mitigate biotic and abiotic stress in plants, supporting plant resilience to environment. These combined effects highlight the potential of CNPs to revolutionize aromatic crop management by increasing yields while simultaneously improving the quality and durability of these commercially and medicinally important species. While CNPs can promote plant growth and improve metabolic activities, their use in agriculture requires careful management to protect the environment. Excessive or unregulated application of CNPs may cause harmful effects such as toxicity to plants, increased oxidative stress, and unforeseen environmental impacts. Therefore, it is critical to develop balanced strategies that maximize the benefits of CNPs while minimizing potential risks to plants, soil ecosystems, and human health. Thoughtful and responsible use of CNPs will enable their sustainable integration into aromatic plant cultivation, unlocking their full potential without compromising environmental safety.

Competing Interests The authors declare that they have no competing interest.

Declaration of generative AI and AI-assisted technologies in the writing process: During the preparation of this work, the book chapter author(s) have not taken any assistance from AI, and the book chapter author(s) reviewed and edited the content as needed and take(s) full responsibility for the content of the publication.

References

1. Bennett SW, Adeleye A, Ji Z, Keller AA (2013) Stability, metal leaching, photoactivity and toxicity in fresh water systems of commercial single wall carbon nanotubes. Water Res 47:4074–4085
2. Hurt RH, Monthioux M, Kane A (2006) Toxicology of carbon nanomaterials: status, trends and perspectives on the special issue. Carbon 44:1028–1033
3. Srivastava V, Gusain D, Sharma YC (2015) Critical review on the toxicity of some widely used engineered nanoparticles. Ind Eng Chem Res 54(24):6209–6233
4. Baptista FR, Belhout SA, Giordani S, Quinn SJ (2015) Recent developments in carbon nano material sensors. Chem Soc Rev 44:4433–4453
5. Baughman RH, Zakhidov AA, de Heer WA (2002) Carbon nanotubes—the route toward applications. Science 97:787–792
6. Wagay JA, Singh S, Raffi MM, Rahman QI, Husen A (2019) Effect of carbon-based nanomaterials on plant functioning and rhizosphere. In: Husen A, Iqbal M (eds) N*anomaterials* and plant potential. Springer International Publishing AG, Gewerbestrasse 11, 6330 Cham, pp 553–575. https://doi.org/10.1007/978-3-030-05569-1_22
7. Husen A (2024) Plant response to gold nanoparticles. Springer Nature Singapore Pte Ltd., 152 Beach Road, #21–01/04 Gateway East, Singapore 189721, Singapore
8. Husen A (2025a) Plant response to silver nanoparticles. Springer Nature Singapore Pte Ltd., 152 Beach Road, #21–01/04 Gateway East, Singapore 189721, Singapore
9. Husen A (2025b) Emerging carbon nanomaterials for sustainable agricultural practices (synthesis, plant growth, performance, production and protection). Springer Nature Singapore Pte Ltd., 152 Beach Road, #21–01/04 Gateway East, Singapore 189721, Singapore
10. Husen A (2025c) Nanobiosensors for agricultural and other related sectors. Springer Nature Singapore Pte Ltd., 152 Beach Road, #21–01/04 Gateway East, Singapore 189721, Singapore
11. Husen A (2025d) Nanobiotechnology for abiotic stress adaptation and mitigation in agricultural crops. Springer Nature Singapore Pte Ltd., 152 Beach Road, #21–01/04 Gateway East, Singapore 189721, Singapore
12. Husen A (2020) Carbon-based nanomaterials and their interactions with agricultural crops. In: Husen A, Jawaid M (eds) Nanomaterials for agriculture and forestry applications. Elsevier Inc., 50 Hampshire St., 5th Floor, Cambridge, MA 02139, USA, pp 199–218
13. Xu X, Wan J, Liu G, Lu C, Mao X, Wu J et al (2025) Physiological, transcriptomic and metabolomic analyses reveal the mechanism of CuO and silicon nanoparticles involved in *Polygonatum kingianum* response to root rot. Chem Biol Technol Agric 12(1):102
14. Garg S, Bhavya Surendran VS, Avanthi A (2024) Recent trends and applications of biochar and nanoparticles from plant biomass. In: Medicinal and aromatic plants: current research status, value-addition to their waste, and agro-industrial potential (Vol II). Springer Nature Switzerland, Cham, pp 203–216
15. Sahithi BVN, Lakshmaiah VV, Nagella P (2025) Nanomaterials for plant growth promotion. In: Plant nanotechnology fundamentals and methodologies. Springer Nature Switzerland, Cham, pp 217–239
16. Chaachouay N, Zidane L, Husen A (2025) Beneficial and adverse effects of silver nanoparticles on plant. In: Husen A (ed) Plant response to silver nanoparticles. Smart nanomaterials technology. Springer, Singapore. https://doi.org/10.1007/978-981-97-7352-7_2
17. Husen A, Siddiqi KS (2014a) Phytosynthesis of nanoparticles: concept, controversy and application. Nanoscale Res Lett 9(229):1–24. https://doi.org/10.1186/1556-276X-9-229
18. Husen A, Siddiqi KS (2014b) Carbon and fullerene nanomaterials in plant system. J Nanobiotechnol 12(16):1–10. https://doi.org/10.1186/1477-3155-12-16
19. Sharma MMM, Kapoor D, Loyal A, Kumar R, Sharma P, Husen A (2025b) Impact of carbon-based nanomaterials on the plant growth-promoting Rhizobacteria and sustainable agricultural crop plant production. In: Husen A (ed) Emerging carbon nanomaterials for sustainable

agricultural practices. Smart nanomaterials technology. Springer, Singapore. https://doi.org/10.1007/978-981-97-5104-4_16
20. Sharma P, Sharma MMM, Kapoor D, Loyal A, Husen A (2025a) Beneficial and adverse effects of silver nanoparticles on rhizosphere biology. In: Husen A (ed) Plant response to silver nanoparticles. Smart nanomaterials technology. Springer, Singapore. https://doi.org/10.1007/978-981-97-7352-7_3
21. Rico CM, Majumdar S, Duarte-Gardea M, Peralta-Videa JR, Gardea-Torresdey JL (2011) Interaction of nanoparticles with edible plants and their possible implications in the food chain. J Agric Food Chem 59:3485–3498
22. Singh S, Singh S, Khan MMA (2024a) Enhancing growth, vitality, and aromatic richness: unveiling the dual magic of silicon dioxide and titanium dioxide nanoparticles in. Novel Ways Methodol Improv Crop Yield
23. Bhat UH, Uddin M, Chishti AS, Singh S, Singh S, Khan MM, Mukarram M (2024) Enhancing growth, vitality, and aromatic richness: unveiling the dual magic of silicon dioxide and titanium dioxide nanoparticles in *Ocimum tenuiflorum* L. Frontiers in Plant Science. 15:1335965. https://doi.org/10.3389/fpls.2024.1335965
24. Do Espirito Santo Pereira A, Caixeta Oliveira H, Fernandes Fraceto L, Santaella C (2021) Nanotechnology potential in seed priming for sustainable agriculture. Nano 11(2):267
25. Yadav N, Garg VK, Chhillar AK, Rana JS (2023) Recent advances in nanotechnology for the improvement of conventional agricultural systems: a review. Plant Nano Biol 4:100032
26. Pelegrino MT, Kohatsu MY, Seabra AB, Monteiro LR, Gomes DG, Oliveira HC, Lange CN (2020) Effects of copper oxide nanoparticles on growth of lettuce (*Lactuca sativa* L.) seedlings and possible implications of nitric oxide in their antioxidative defense. Environ Monit Assess 192(4):232
27. Pourkhaloee A, Haghighi M, Saharkhiz MJ, Jouzi H, Doroodmand MM (2011) Carbon nanotubes can promote seed germination via seed coat penetration. Seed Technol 33:155–169
28. Acharya P, Jayaprakasha GK, Crosby KM, Jifon JL, Patil BS (2020) Nanoparticle-mediated seed priming improves germination, growth, yield, and quality of watermelons (*Citrullus lanatus*) at multi-locations in Texas. Sci Rep 10(1):5037
29. Pérez-de-Luque A (2017) Interaction of nanomaterials with plants: What do we need for real applications in agriculture? *Frontiers in Environmental Science, 5*, 12. https://doi.org/10.3389/fenvs.2017.00012
30. Zhao L, Lu L, Wang A, Zhang H, Huang M, Wu H et al (2020) Nano-biotechnology in agriculture: use of nanomaterials to promote plant growth and stress tolerance. J Agric Food Chem 68(7):1935–1947
31. Awasthi K, Gupta K, Srivastava R (2024) Synthesis, characterization, and impact of different carbon-based nanomaterials on gram (Cicer arietinum) plant growth and soil properties. Curr Agric Res J 12(3)
32. Chen N, Tian X, Yang M, Xu J, Tan T, Wang J (2024) Effect of carbon nanoparticles on the growth and photosynthetic property of *Ficus tikoua* Bur. plant. PeerJ 12:e17652
33. Sigala-Aguilar NA, López MG, Fernández-Luqueño F (2024) Carbon-based nanomaterials as inducers of biocompounds in plants: potential risks and perspectives. Plant Physiol Biochem 212:108753
34. Halawani RF, AbdElgawad H, Aloufi FA, Balkhyour MA, Zrig A, Hassan AH (2023) Synergistic effect of carbon nanoparticles with mild salinity for improving chemical composition and antioxidant activities of radish sprouts. Front Plant Sci 14:1158031
35. Hasanuzzaman M, Bhuyan MHMB, Zulfiqar F, Raza A, Mohsin SM, Mahmud JA, Fujita M, Fotopoulos V (2020) Reactive oxygen species and antioxidant defense in plants under abiotic stress: revisiting the crucial role of a universal defense regulator. Antioxidants (Basel) 9(8):681. https://doi.org/10.3390/antiox9080681
36. Pradhan J, Pramanik K, Jaiswal A, Kumari G, Prasad K, Jena C, Srivastava AK (2024) Biosynthesis of secondary metabolites in aromatic and medicinal plants in response to abiotic stresses: a review. J Exp Biol Agric Sci 12(3):318–334. https://doi.org/10.18006/2024.12(3).318.334

37. Sun Y, Alseekh S, Fernie AR (2023) Plant secondary metabolic responses to global climate change: a meta-analysis in medicinal and aromatic plants. Glob Chang Biol 29(2):477–504
38. Arslan MBAÇH (2023) Cytotoxic and genotoxic effects of copper nanoparticles synthesis and characterized from black grass (*Lavandula stoechas* L.) by green synthesis method on Rtg-2 cell line. ICOFAAS 2023, 148
39. Tan XM, Lin C, Fugetsu B (2009) Studies on the toxicity of multi-walled carbon nanotubes on suspension rice cells. Carbon 47(15):3479–3487
40. Lin C, Su Y, Takahiro M, Fugetsu B (2010) Multi-walled carbon nanotubes induce oxidative stress and vacuolar structure changes to Arabidopsis T87 suspension cells. Nano Biomed 2(2):170–181
41. Tiwari DK, Dasgupta-Schubert N, Villaseñor Cendejas LM, Villegas J, Carreto Montoya L, Borjas García SE (2014) Interfacing carbon nanotubes (CNT) with plants: enhancement of growth, water and ionic nutrient uptake in maize (Zea mays) and implications for nanoagriculture. Appl Nanosci 4(5):577–591
42. Zhu H, Han J, Xiao JQ, Jin Y (2019) Uptake, translocation, and accumulation of engineered nanomaterials in plants. Crit Rev Environ Sci Technol 49(22):1979–2018
43. Khodakovskaya MV, Dervishi E, Mahmood M, Xu Y, Li Z, Watanabe F, Biris AS (2012) Carbon nanotubes are able to penetrate plant seed coat and dramatically affect seed germination and plant growth. ACS Nano 3(10):3221–3227
44. Wang P, Lombi E, Zhao FJ, Kopittke PM (2016) Nanotechnology: a new opportunity in plant sciences. Trends Plant Sci 21(8):699–712
45. Tripathi DK et al (2017) Uptake, accumulation and toxicity of engineered nanoparticles in higher plants. Environ Sci Nano 4(2):423–439
46. Joshi A, Kaur S, Dhar P (2018) Nanoparticle-based seed priming: a review. J Plant Growth Regul 37(1):83–102
47. Sharma A et al (2020) Carbon quantum dots as novel elicitors for secondary metabolite production and stress tolerance in plants. Plant Physiol Biochem 146:47–56
48. Ghorbanpour M, Hadian J (2015) Multi-walled carbon nanotubes stimulate callus induction, secondary metabolites biosynthesis and antioxidant capacity in medicinal plant *Satureja khuzestanica* grown in vitro. Carbon 94:749–759
49. Fatemi F, Abdollahi MR, Mirzaie-Asl A, Dastan D, Garagounis C, Papadopoulou K (2019) Identification and expression profiling of rosmarinic acid biosynthetic genes from *Satureja khuzistanica* under carbon nanotubes and methyl jasmonate elicitation. Plant Cell Tissue Organ Cult 136(3):561–573
50. Gaafar AA, Taha RA, Abou-Baker NH, Shaaban EA, Salama ZA (2018) Evaluation of regeneration, active ingredients and antioxidant activities in jojoba tissue cultures as affected by carbon nanotubes. Biosci Res 15(3):2283–2392
51. McGehee DL, Alimohammadi M, Khodakovskaya MV (2019) Carbon-based nanomaterials as stimulators of production of pharmaceutically active alkaloids in cell culture of *Catharanthus roseus*. Nanotechnology 30(27):275102
52. Ghasempour M, Iranbakhsh A, Ebadi M, Oraghi Ardebili Z (2019) Multi-walled carbon nanotubes improved growth, anatomy, physiology, secondary metabolism, and callus performance in Catharanthus roseus: an in vitro study. 3 Biotech 9(11):404
53. Rahmani N, Radjabian T, Soltani BM (2020) Impacts of foliar exposure to multi-walled carbon nanotubes on physiological and molecular traits of *Salvia verticillata* L., as a medicinal plant. Plant Physiol Biochem 150:27–38
54. Rahmani N, Radjabian T (2024) Integrative effects of phytohormones in the phenolic acids production in Salvia verticillata L. under multi-walled carbon nanotubes and methyl jasmonate elicitation. *BMC Plant Biol 24*, 56 (2024). https://doi.org/10.1186/s12870-023-04719-5
55. Heydari HR, Chamani E, Esmaielpour B (2020) Cell line selection through gamma irradiation combined with multi-walled carbon nanotubes elicitation enhanced phenolic compounds accumulation in Salvia nemorosa cell culture. Plant Cell Tissue Organ Cult 142(2):353–367
56. Hir YP, Chamani E, Ahadzadeh M, Shaker S, Sanjbod RN (2024) Investigation on the Effects of Carbon Nanotubes and Kinetin on Growth and Regeneration of Salvia nemorosa under in vitro Conditions. *Journal of Horticultural Science 38*(2):353–403.

57. Samadi S, Saharkhiz MJ, Azizi M, Samiei L, Ghorbanpour M (2023a) Exposure to single-walled carbon nanotubes differentially affect in vitro germination, biochemical and antioxidant properties of Thymus daenensis celak. Seedlings. BMC Plant Biol 23(1):579
58. Samadi S, Saharkhiz MJ, Azizi M, Samiei L, Ghorbanpour M (2023b) Exploring potential of multi-walled carbon nanotubes to establish efficient callogenesis, elicitation of phenolic compounds and antioxidative activities in thyme plants (Thymus daenensis): an in vitro assay. S Afr J Bot 157:602–613
59. Samadi S, Saharkhiz MJ, Azizi M, Samiei L, Karami A, Ghorbanpour M (2021) Single-wall carbon nano tubes (SWCNTs) penetrate Thymus daenensis Celak. Plant cells and increase secondary metabolite accumulation in vitro. Ind Crop Prod 165:113424
60. Samadi S, Saharkhiz MJ, Azizi M, Samiei L, Ghorbanpour M (2020) Multi-walled carbon nanotubes stimulate growth, redox reactions and biosynthesis of antioxidant metabolites in Thymus daenensis celak. In vitro. Chemosphere 249:126069
61. Fatehi SF, Oraei M, Gohari G, Akbari A, Faramarzi A (2022) Proline-functionalized graphene oxide nanoparticles (GO–Pro NPs) mitigate salt-induced adverse effects on morpho-physiological traits and essential oils constituents in Moldavian Balm (*Dracocephalum moldavica* L.). J Plant Growth Regul 41(7):2818–2832
62. Ebrahimi S, Abrishamchi P, Radjabian T, Saboora A (2024) Elicitation with multi-walled carbon nanotubes improved growth and production of polyphenolic compounds in *Melissa officinalis*. Plant Biosyst—An International Journal Dealing with all Aspects of Plant Biology 158(5):1125–1135
63. Huang X, Wu J, Wang Y, Qin H, Fan R, Yang C et al (2025) The foliar application of carbon dots promoted the photosynthesis performance, growth, and pharmaceutical ingredient enrichment in *Isatis tinctoria* L. ACS Appl Mater Interfaces
64. Razavizadeh R, Adabavazeh F, Gholipour Z (2025) Application of multi-walled carbon nanotubes in *Calotropis procera* in vitro cultures for improved element uptake and enhanced production of monoterpenes and antioxidants. Plant Cell Tissue Organ Cult 161(1):1–15
65. Gohari G, Safai F, Panahirad S, Akbari A, Rasouli F, Dadpour MR, Fotopoulos V (2020) Modified multiwall carbon nanotubes display either phytotoxic or growth promoting and stress protecting activity in *Ocimum basilicum* L. in a concentration-dependent manner. Chemosphere 249:126171
66. Holghoomi R, Sarghein SH, Khara J, Hosseini B (2021) Effect of functionalized-carbon nanotube on growth indices in *ocimum basilicum* l. grown in vitro. Russ J Plant Physiol 68(5):958–972
67. Holghoomi R, Hosseini Sarghein S, Khara J, Hosseini B, Rahdar A, Kyzas GZ (2023) Foliar application of Phenylalanine functionalized multi-walled carbon nanotube improved the content of volatile compounds of basil grown in greenhouse. Environ Sci Pollut Res 30(31):77385–77407
68. Fallah R, Gerami M, Ramezani M (2023) Beneficial role of multi-walled carbon nanotubes on physiological and phytochemical responses of *Mentha piperita* L under salinity stress. J Essent Oil Bear Plants 26(2):323–342
69. Asakereh F, Etminan A, Azizinezhad R, Ghorbanpour M, Pour-Aboughadareh A (2025) Effect of multi-walled carbon nanotube exposure on the expression pattern of genes involved in silymarin synthesis pathway in Silybum Marianum L. BMC Plant Biol 25(1):1033
70. Ambreen, Khan MA, Raza A, Hussain T, Ali H (2023) Unleashing the potential of carbon nanotubes for production of medicinal metabolites in *Nigella sativa* adventitious roots. Plant Cell Tissue Organ Cult 155(2):581–601
71. Ramezan D, Farrokhzad Y, Mokhtassi-Bidgoli A, Rasouli-Alamuti M (2023) Multi-walled carbon nanotubes interact with light intensity to affect morpho-biochemical, nutrient uptake, DNA damage, and secondary metabolism of Stevia rebaudiana. Environ Sci Pollut Res 30(13):36915–36927
72. Li P, Xia Y, Song K, Liu D (2024) The impact of nanomaterials on photosynthesis and antioxidant mechanisms in gramineae plants: research progress and future prospects. Plants 13:984
73. Wang X, Xie H, Wang P, Yin H (2023) Nanoparticles in plants: uptake, transport and physiological activity in leaf and root. Materials 16(8):3097
74. Bhati VS, Singh PK, Parmar YJ, Rokins PG (2025) Foliar uptake and translocation of nanoparticles in the plants: a review. J Plant Nutr 1–38. https://doi.org/10.1080/01904167.2025.2538638

Effects of Metal-Based Nanoparticles (Au, Ag, Zn, and Cu) on Aromatic Plants

Ayeesha Khatoon, Sazia Siddiqui, and Aisha Kamal

Abstract Aromatic and medicinal plants are rich reservoirs of essential oils and bioactive phytochemicals with wide applications in the food, cosmetic, and pharmaceutical sectors. Recent advances in nanotechnology highlight the potential of metal-based nanoparticles (NPs) particularly Au, Ag, Zn, and Cu to modulate growth, stress tolerance, and secondary metabolite production in these species. Their nanoscale size, high surface reactivity, and tunable physicochemical properties enable unique interactions with plant tissues, influencing nutrient uptake, photosynthetic efficiency, antioxidant defenses, and phytochemical biosynthesis. At optimal concentrations, NPs act as bio-stimulants, enhancing enzymatic and nonenzymatic antioxidant activity, improving essential oil yield, and enriching compounds such as phenolics, flavonoids, and terpenoids, thereby elevating both therapeutic and commercial value. However, their effects are strongly dose- and species dependent. While low to moderate NP levels stimulate plant defense mechanisms and metabolic pathways, excessive accumulation can induce oxidative stress, DNA damage, impaired photosynthesis, and growth inhibition, underscoring the importance of defining threshold concentrations. This chapter synthesizes current evidence on the uptake, transport, and mechanistic actions of Au, Ag, Zn, and Cu NPs in aromatic plants, highlighting their dual role as stimulants or stressors. It emphasizes the need for optimized application strategies and biosafety assessments to harness nanotechnology for sustainable cultivation and enhanced medicinal value of aromatic crops.

Keywords Aromatic plants · Metal-based nanoparticles · Essential oils · Phytochemical modulation · Antioxidant activity · Phytotoxicity thresholds

---

A. Khatoon · S. Siddiqui
Department of Biosciences, Integral University, Lucknow, India

A. Kamal (✉)
Department of Bioengineering, Integral University, Lucknow, India

A. Husen (ed.), *Sustainable Uses of Aromatic Plants in Nanotechnology*, Smart Nanomaterials Technology, https://doi.org/10.1007/978-981-95-7409-4_19

1 Introduction

Aromatic and medicinal plants are valued sources of essential oils and bioactive phytochemicals used across food, cosmetic, and pharmaceutical sectors [1–9]. Harnessing nanotechnology in this context offers new levers to modulate plant growth, stress tolerance, and secondary metabolite biosynthesis central to aroma and therapeutic quality [10–13]. Metal-based nanoparticles particularly Au, Ag, ZnO, and Cu/CuO exhibit size-dependent physicochemical behavior (1–100 nm) such as high surface-area-to-volume ratios and tunable surface chemistry, which underlie their distinctive interactions with plant tissues [14]. Increasingly, these NPs are synthesized via "green" routes using plant extracts whose phytoconstituents act as reducing and capping agents, improving dispersion and biocompatibility attributes especially pertinent for agricultural use with aromatic crops [15–17]. Applied judiciously, metal NPs can enhance germination, photosynthetic efficiency, antioxidant enzyme activities, and essential oil yield and quality in species such as basil, mint, rosemary, and lavender [13, 18]. At the same time, their benefits are tightly dose- and context dependent. NP chemistry, concentration, particle size/charge, coating, plant genotype, and environmental factors (pH, temperature, soil properties) govern uptake, translocation, and downstream responses; at supraoptimal levels, metal NPs can trigger oxidative stress, membrane injury, and growth inhibition with implications for plant health and rhizosphere ecology [19]. These dual potentials underscore the need for well-characterized materials, optimized dosing, and environmentally responsible application strategies [20, 21]. This chapter explores current evidence on Au, Ag, ZnO, and Cu/CuO NPs in aromatic plants. It examines (a) key physicochemical features shaping plant interactions; (b) uptake and transport mechanisms; (c) impacts on growth, physiology, and essential oil composition; and (d) modulation of antioxidant and enzymatic systems; and (e) toxicity thresholds and comparative species responses. Collectively, the goal is to clarify when metal NPs act as stimulants versus stressors and to outline research and practical considerations for precision, sustainable cultivation of aromatic crops.

2 Physicochemical Properties and Synthesis of Metal Nanoparticles

The interaction of metal nanoparticles (NPs) with aromatic plants is primarily influenced by their physicochemical properties and synthesis pathways, which collectively determine their stability, uptake, translocation, and subsequent physiological and biochemical responses [14–16]. These NPs typically range in size from 1 to

100 nm, and their nanoscale dimensions confer a high surface-area-to-volume ratio and unique surface phenomena absent in bulk materials. Smaller particles generally penetrate root tissues more efficiently and move through apoplastic and symplastic pathways, while shape variations such as spherical, rodlike, or cubic forms influence dissolution behavior and bioactivity [13, 17, 20]. Surface charge, commonly expressed as zeta potential, governs colloidal stability and adhesion to root surfaces, with positively charged NPs often displaying enhanced interaction with negatively charged cell wall components [18, 22]. Another critical factor is solubility and ion release, particularly in metal oxide NPs such as ZnO and CuO, which dissolve partially in aqueous environments, releasing Zn^{2+} and Cu^{2+} ions that can function as micronutrients at optimal concentrations but cause toxicity through oxidative stress and enzyme inhibition when excessive [19, 23]. Crystalline structure and redox potential further influence NP reactivity; for instance, Au NPs exhibit inert behavior with minimal toxicity, whereas Cu NPs, despite their strong antimicrobial activity, may induce oxidative imbalance at higher doses [24].

Surface functionalization also plays a decisive role; green synthesis methods employing plant extracts introduce phytochemical capping agents such as flavonoids, polyphenols, and terpenoids, which improve dispersion, stability, and biocompatibility compared to synthetic surfactants, thereby reducing ecological risk and enhancing plant compatibility [15, 16, 25]. Evidence suggests that when applied in appropriate concentrations, ZnO and Ag NPs enhance plant growth, chlorophyll content, essential oil yield, and antioxidant enzyme activities in species such as *Ocimum basilicum* and *Mentha arvensis*, whereas overapplication can lead to oxidative stress, membrane damage, lipid peroxidation, and reduced photosynthesis. Stress responses triggered by NP exposure can also upregulate secondary metabolite synthesis, including phenolics, terpenoids, and flavonoids, which contribute to the therapeutic value of aromatic plants [18, 26–28].

The synthesis approach strongly dictates NP properties: chemical methods such as reduction and precipitation allow size control but involve hazardous reagents, whereas green synthesis utilizing aromatic plant extracts offers an eco-friendly alternative by integrating phytochemicals as reducing and capping agents, thereby producing functionalized NPs suitable for agricultural applications [14, 15, 17]. Physical methods like laser ablation provide high-purity products but are energy intensive and less practical for large-scale use [13]. Accurate characterization through techniques such as TEM and SEM for morphology, XRD for crystallinity, DLS for size and zeta potential, and FTIR for functional group analysis is essential for correlating NP properties with biological outcomes [15]. Overall, the impact of metal NPs on aromatic plants is a function of their size, morphology, surface chemistry, dissolution behavior, and synthesis route, underscoring the importance of controlled production, appropriate dosing, and rigorous characterization to harness their agricultural potential while minimizing adverse effects [14, 16, 24].

3 Uptake, Translocation, and Interaction Mechanisms in Aromatic Plants

Metal-based nanoparticles (NPs) such as Au, Ag, Zn, and Cu are commonly introduced into plant systems through diverse application techniques, including foliar spraying, soil amendment, and hydroponic systems (Fig. 1). Regardless of the delivery route, the main entry points for NPs include stomata, the cuticle, root epidermis, and damaged tissues, which act as gateways for internalization. At the root-soil interface, NPs undergo physicochemical interactions with root exudates and rhizospheric microbial populations, which influence their solubility, stability, and bioavailability.

Plant uptake of NPs occurs through both passive and active mechanisms. Passive uptake takes place via apoplastic channels, where NPs cross cell walls and intercellular spaces without penetrating the plasma membrane, an especially important route for larger particles or aggregates. In contrast, active uptake involves processes such as endocytosis, aquaporins, or ion transporters and depends on factors such as particle size, surface charge, and surface coating, as well as plant species [29]. Once internalized, NPs can follow the symplastic pathway, moving through the cytoplasm and plasmodesmata. For efficient intracellular mobility, however, particles often undergo size reduction or dissolution into ionic forms [30]. Long-distance transport is primarily achieved through the xylem, facilitating systemic distribution of NPs across plant tissues.

Research demonstrates that Ag and Zn NPs, owing to their higher solubility and reactivity, frequently release ions that mimic essential nutrients such as Zn^{2+} or

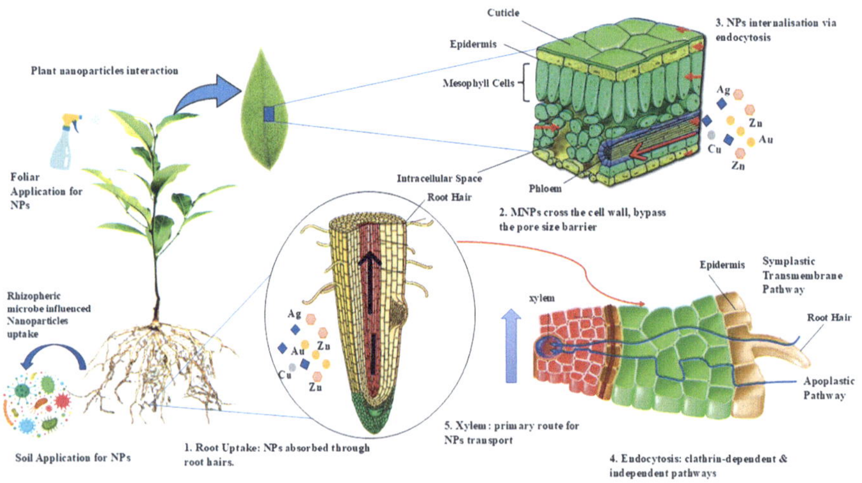

Fig. 1 Mechanisms of nanoparticle uptake and transport in medicinal plants through foliar and root pathways, cellular internalization, and vascular translocation

Cu^{2+}, thereby exploiting natural metal transport pathways [31]. In aromatic plants, species-specific accumulation patterns are evident. For example, *Ocimum basilicum* and *Pelargonium graveolens* preferentially retain Au and Cu NPs in roots, followed by xylem-mediated translocation to aerial tissues, where they are either stored in vacuoles or converted into less reactive forms [13]. Similarly, in *Mentha arvensis*, CuO and ZnO NPs are absorbed by roots and transported to shoots, stimulating monoterpene and antioxidant enzyme production [13, 32]. In *Lavandula angustifolia*, foliar-applied Au and Ag NPs penetrate the stomata and cuticle, eventually localizing in secretory tissues and modulating the synthesis of aromatic metabolites such as linalool and linalyl acetate [33]. In *Rosmarinus officinalis*, ZnO and Cu NPs absorbed via root epidermal cells accumulate in leaves, altering essential oil composition and antioxidant activity, likely through reactive oxygen species (ROS)-mediated signaling pathways [13, 34]. Moreover, the surface chemistry of NPs plays a critical role in internalization and distribution. Organic acid or polysaccharide coatings enhance stability and biocompatibility, thereby improving uptake [35].

Once inside plant tissues, NPs may undergo biotransformation processes such as chelation, redox reactions, or enzymatic degradation, which reduce toxicity and modulate mobility [36]. The abundance of secondary metabolites in aromatic plants also influences NP absorption. These compounds can act as ligands or reducing agents, enhancing NP uptake and accumulation. For instance, *Matricaria chamomilla* and *Salvia officinalis*, rich in phenolics and essential oils, show enhanced NP accumulation in trichomes and secretory tissues, enabling targeted enrichment that drives secondary metabolite biosynthesis [13].

NP uptake is further influenced by physiological stress responses. For example, Ag NPs often generate ROS, which trigger secondary metabolic pathways leading to enhanced production of flavonoids and phenolic compounds with defensive and therapeutic roles, as reported in *Hypericum perforatum* [37].

Such ROS-mediated elicitation forms the basis of nanopriming strategies, which aim to stimulate biosynthesis of pharmacologically important molecules without impairing plant health. At the molecular level, nanoparticle exposure can cause transcriptional reprogramming in roots, increasing the expression of metal transporter families such as ZIP and NRAMP, thereby facilitating selective NP entry [38]. Notably, NPs synthesized via green methods using plant extracts tend to exhibit greater biocompatibility, higher absorption efficiency, and lower cytotoxicity, making them a sustainable option for medicinal plant cultivation [38].

These findings indicate that metal-based NPs are absorbed by plants through intricate and multifaceted mechanisms, including passive diffusion, active transport, and endocytosis. Their subsequent movement through vascular tissues is dictated by NP properties, plant physiology, and environmental conditions. In medicinal and aromatic plants, this intricate interplay not only influences growth and metabolism but also offers opportunities to strategically enhance the biosynthesis of valuable phytochemicals [13].

4 Impact on Growth, Physiological Traits, and Essential Oil Composition

The introduction of nanotechnology in agriculture has opened up exciting new possibilities for improving the growth, resilience, and bioactivity of medicinal and aromatic plants. Metal NPs especially those made of Au, Ag, Zn, and Cu are increasingly utilized as nano-fertilizers, antimicrobial agents, and modulators of abiotic stress in plant systems. These NPs have both stimulatory and inhibitory effects on medicinal and aromatic plants based upon their concentration, size, and method of implementation [39].

4.1 Gold Nanoparticles

Au NPs are generally less phytotoxic than other metal nanoparticles and primarily influence metabolomic pathways rather than causing major changes in plant morphology. They enhance the expression of genes associated with the phenylpropanoid pathway, leading to elevated levels of phenolics, terpenoids, and flavonoids in medicinal and aromatic plants. In *Ocimum sanctum*, Au NPs increased the accumulation of rosmarinic acid and eugenol, both key bioactive compounds with therapeutic relevance [40]. These effects are attributed to the ability of Au NPs to interact with plant signaling networks at transcriptomic and proteomic levels, thereby modulating secondary metabolite biosynthesis [39].

4.2 Silver Nanoparticles

Ag NPs are well recognized for their antimicrobial properties and their ability to influence plant microbe interactions in the rhizosphere, thereby modifying nutrient uptake and assimilation. At low concentrations, Ag NPs have been shown to promote plant growth and enhance essential oil production. For example, in *Cymbopogon citratus*, Ag NPs improved plant height, leaf number, and essential oil yield [41]. Similarly, in *Nigella sativa* and *Matricaria chamomilla*, they stimulated root and shoot elongation as well as volatile oil biosynthesis [42]. Their growth-promoting effects are often linked to the regulation of hormonal signaling pathways, particularly gibberellins and cytokinins, which govern cell division and elongation.

In *Pelargonium graveolens*, Ag NPs significantly increased levels of geraniol and citronellol, accompanied by improvements in leaf morphology, trichome density, and aroma profile, thereby enhancing both aesthetic and therapeutic value [43]. Likewise, in vitro exposure to Ag NPs enhanced artemisinin production by 3.9-fold in *Artemisia annua* hairy roots [44], while treatment with 2 mg/kg Ag NPs improved both growth and diosgenin content in fenugreek (*Trigonella foenum-graecum*) [45].

However, excessive accumulation of Ag NPs in soil may negatively impact beneficial microbial communities and disturb plant microbe symbiosis, highlighting the need for careful regulation of application strategies.

4.3 Zinc Nanoparticles

ZnO NPs are among the most widely studied nanomaterials in plant science due to their essential roles in chlorophyll biosynthesis, auxin metabolism, and enzyme activation. They have been reported to enhance seed germination, photosynthetic efficiency, and antioxidant enzyme activities in several aromatic and medicinal species, including *Ocimum basilicum*, *Salvia officinalis*, and *Mentha arvensis* [13]. Their nanoformulation improves Zn bioavailability, which supports metabolic processes without inducing phytotoxicity when applied at appropriate concentrations [46].

Application of ZnO NPs, either as a foliar spray or soil additive, has been shown to enrich essential oil profiles by increasing compounds such as gamma-terpinene and p-cymene, particularly in *Satureja hortensis* [47]. In *M. arvensis*, ZnO NPs significantly boosted essential oil production, while in *O. basilicum*, they enhanced catalase and peroxidase activities, strengthened the antioxidant defense system, and improved photosystem II efficiency, critical for stress tolerance and biomass yield [18]. Furthermore, ZnO NPs conferred drought tolerance by promoting osmolyte accumulation and improving water retention capacity. Collectively, these physiological and biochemical improvements lead to greater biomass and secondary metabolite accumulation, positioning ZnO NPs as a sustainable tool for improving the productivity and therapeutic potential of aromatic plants [48].

4.4 Copper Nanoparticles

Cu NPs play a dual role in medicinal and aromatic plants, acting both as an essential micronutrient and as a redox-active agent capable of generating ROS. At low concentrations, they enhance shoot and root biomass, stimulate metabolite synthesis, and improve overall plant vigor in species such as *Rosmarinus officinalis*, *Mentha piperita*, and *Lavandula angustifolia* [13]. These benefits are largely attributed to the increased bioavailability of Cu in nanoparticle form, which activates enzymatic and nonenzymatic antioxidant defense systems. For instance, in *L. angustifolia*, Cu NPs promoted the accumulation of flavonoids and phenolic compounds, most likely through ROS-mediated signaling that triggers secondary metabolite formation. Similarly, in *M. piperita* and *R. officinalis*, nanomolar doses of Cu NPs improved biomass production and key physiological functions [49]. Their antimicrobial activity, effective against phytopathogens, further underscores their agronomic value [50].

In *Coriandrum sativum*, Cu NPs not only enhanced plant height, leaf area, and chlorophyll content but also increased linalool concentration, with the improvements linked to overexpression of Cu-dependent enzymes such as polyphenol oxidase, which strengthen phenolic pathways [51]. However, when applied at excessive concentrations, Cu NPs can lead to ROS accumulation beyond the plant's detoxification capacity, causing oxidative stress that impairs vital enzyme activity and reduces photosynthetic efficiency. Thus, optimizing Cu NP dosage is critical to maximizing their growth-promoting and metabolic benefits while minimizing phytotoxic risks [51].

5 Antioxidant and Enzymatic Activity Modulation by Nanoparticles

The interaction of metal-based nanoparticles (NPs) has emerged as a promising strategy for modulating antioxidant and enzymatic activity, thereby offering new avenues for the cultivation and improvement of aromatic plants. Owing to their rich phytochemical profile and abundance of secondary metabolites, aromatic species exhibit distinctive responses when exposed to NPs such as Au, Ag, Zn, and Cu. These NPs influence plant redox homeostasis by altering reactive oxygen species (ROS) concentrations, activating antioxidant defense mechanisms, and modulating key enzymatic pathways linked to stress tolerance and metabolic regulation [27].

The plant antioxidant system operates through two major components: enzymatic and nonenzymatic defenses. Enzymatic antioxidants such as superoxide dismutase (SOD), catalase (CAT), ascorbate peroxidase (APX), guaiacol peroxidase (GPX), and glutathione reductase (GR) function synergistically to detoxify ROS and preserve redox homeostasis. In parallel, nonenzymatic antioxidants such as phenolics, flavonoids, ascorbic acid, carotenoids, and essential oils contribute to ROS scavenging while enhancing the medicinal and aromatic properties of plants [52]. Metal-based NPs can influence both categories either directly through biochemical interactions or indirectly via gene regulation. For instance, ZnO NPs enhanced SOD and CAT activities in *Ocimum basilicum*, lowering lipid peroxidation and safeguarding membrane integrity [53], while Ag NPs elevated APX and GPX activity in *Mentha piperita*, concomitant with higher phenolic content and antioxidant potential [54].

The underlying modulation mechanism is multifaceted. NPs can penetrate plant tissues via root uptake or foliar absorption and subsequently localize in cellular organelles such as chloroplasts and mitochondria, which are primary sites of ROS generation. Due to their high surface reactivity, NPs interact with electron transport chains, leading to mild oxidative signaling that triggers overexpression of antioxidant enzyme genes [55]. For example, Au NPs enhanced SOD and CAT gene expression in *Lavandula angustifolia*, thereby improving resilience to abiotic stress [56]. However, excessive NP accumulation may disturb redox homeostasis. Elevated

CuO NP concentrations in *Rosmarinus officinalis* reduced CAT activity and increased malondialdehyde (MDA) content, signaling oxidative stress and cellular injury [57].

Beyond enzymatic defenses, NPs influence secondary metabolism, particularly pathways that generate antioxidant-rich compounds. ZnO and CuO NPs upregulated phenylalanine ammonia-lyase (PAL), a pivotal enzyme in the phenylpropanoid pathway, thereby enhancing phenolic biosynthesis [58]. This contributes not only to stronger plant defense but also to improved essential oil quality and pharmacological potential. For instance, ZnO NPs elevated phenolic and flavonoid concentrations in *Salvia officinalis*, leading to greater DPPH radical scavenging activity [13].

The antioxidant and enzymatic responses of aromatic plants to NPs depend on multiple factors, including particle type, size, concentration, surface coating, and duration of exposure. Generally, smaller NPs with higher surface reactivity elicit stronger effects at lower doses, whereas larger particles require higher concentrations for comparable outcomes. In species such as basil, mint, lavender, and rosemary, optimized NP treatments enhanced antioxidant activity as measured by assays like FRAP (ferric reducing antioxidant power), ABTS radical cation decolorization, and ORAC (oxygen radical absorbance capacity) [59].

From an agricultural and pharmaceutical perspective, carefully regulated NP application can biofortify aromatic plants with elevated antioxidant properties, thereby strengthening their adaptability to environmental stressors and improving their medicinal efficacy. Striking a balance between beneficial stimulation and toxic overexposure is critical. Future research must prioritize precise dose optimization, long-term environmental monitoring, and biosafety assessments to ensure that nanotechnology applications in aromatic plant production remain both effective and sustainable [60].

While nanoparticles can positively modulate antioxidant and enzymatic activity in aromatic plants, their impact is highly dose dependent. At optimal concentrations, NPs act as stimulants of plant defense and metabolic pathways, but beyond specific thresholds, they can shift from beneficial to harmful, inducing oxidative stress and impairing cellular integrity. Thus, understanding the fine balance between bio-stimulation and toxicity is critical. This leads to the next consideration toxicological impacts and threshold concentrations which determine the safe and effective use of nanotechnology in aromatic plant systems.

6 Toxicological Implants and Threshold Concentrations

In recent decades, considerable research has been dedicated to evaluating the hazardous impacts of nanoparticles (NPs) on plants. The degree of phytotoxicity largely depends on plant species as well as the physical and chemical properties of metallic NPs, which together determine whether they stimulate or inhibit plant growth at different developmental stages [61]. Metallic NPs are absorbed primarily through roots and transported via the vascular system to aerial tissues, where they can

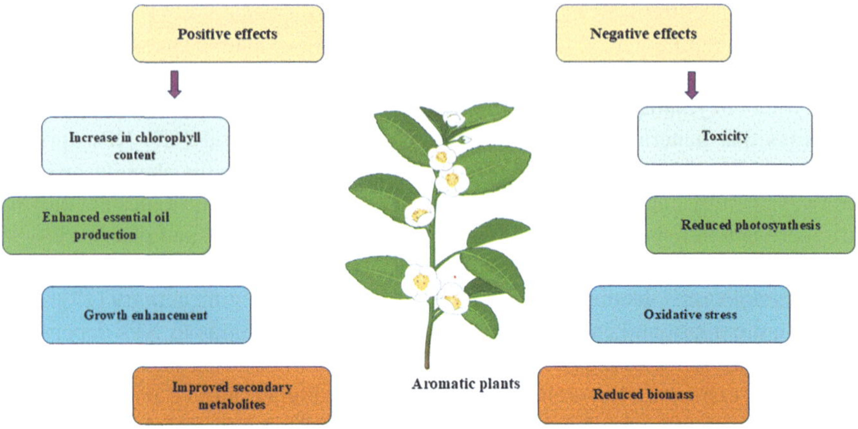

Fig. 2 Positive versus negative impacts of NPs on aromatic plants: a comparative illustration of growth and metabolic benefits against physiological stress and biomass reduction

accumulate and induce severe toxicity [62]. At elevated concentrations, even essential metal NPs such as Cu, Zn, Fe, and Ni disrupt physiological functions, leading to oxidative stress, reduced germination, impaired photosynthesis, enzyme inhibition, and decreased pigment biosynthesis [63]. Such toxic accumulation of inorganic particles and metal ions often manifests as stunted growth and weakened development [64]. Additionally, both bulk and nanoscale forms of metals and metal oxides have been associated with DNA damage, cell-cycle arrest, cytotoxicity, apoptosis, and genotoxicity [65]. Therefore, identifying safe and toxic dose ranges for each species is critical for balancing growth promotion with phytotoxicity [66] (Fig. 2).

Aromatic plants rely on secondary metabolism to produce specialized compounds essential for defense and adaptation. Metallic NPs disrupt these processes by altering nutrient balance and enzyme activity. For example, Cu NPs affected Ca, Fe, Mg, and Mn contents in oregano roots and shoots, indicating potential changes in nutritional quality and secondary metabolite biosynthesis [67, 68]. CuO NPs also induced cytological abnormalities, including disrupted mitotic and meiotic cell divisions, contributing to growth suppression [69]. Similarly, Kralova and Jampilek [13] reported Cu NP-induced nutrient imbalances in oregano, which may compromise essential oil composition and pharmacological value.

Gold nanoparticles (Au NPs) have shown both beneficial and adverse effects depending on concentration and synthesis route. Pretreatment with biogenic Au NPs synthesized from Brazilian red propolis enhanced germination in certain crops, but excessive doses reduced dry biomass in wheat and tomato seedlings, likely due to oxidative stress and hormonal imbalance [70, 71]. Comparative studies using free black rice polyphenols (BRP) and BRP-Au NPs revealed species-specific phytotoxicity, underscoring the importance of NP origin and concentration [72] (Fig. 3).

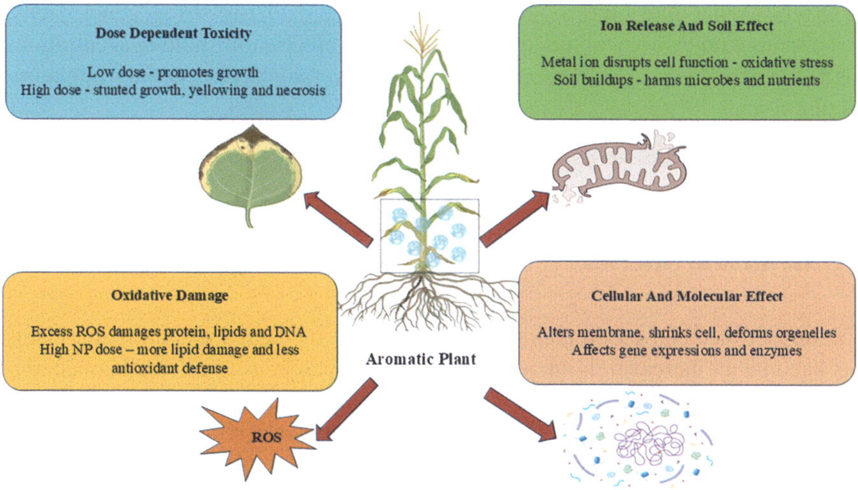

Fig. 3 Mechanisms of nanoparticle-induced toxicity in aromatic plants, including dose-dependent effects, oxidative damage, ion release with soil impact, and cellular-molecular alterations

Silver nanoparticles (Ag NPs) similarly show dual effects. In *Thymus daenensis* and *Thymus vulgaris*, Ag NPs improved drought tolerance, enhanced germination, and promoted root growth under saline conditions up to 200 mM. However, their positive influence on shoot growth was limited to 100 mM salinity [73]. Conversely, in *Allium cepa*, Ag NPs caused cytotoxic and genotoxic effects, including reduced mitotic index (MI), chromosomal aberrations, and DNA fragmentation effects along with the decreased synthesis of organosulfur compounds, such as S-alk(en)yl-cysteine sulfoxides (ACSOs) like isoalliin [74, 75].

Zinc oxide nanoparticles (ZnO NPs) also present concentration-dependent effects. At lower concentrations (40 μg/mL), they promoted callus growth in *Echinacea purpurea* and enhanced anticancer properties of extracts by increasing flavonoid content [76]. However, higher concentrations (50 μg/mL) caused cytotoxic and genotoxic effects in *Allium cepa* and other plants, including membrane damage, ROS accumulation, DNA breaks, and abnormal mitotic events [13, 77]. Additional studies confirmed their clastogenic and genotoxic potential, with mitotic index dropping from 63.6% in controls to 50% under UV-assisted ZnO NP exposure [78]. Morphological abnormalities, including chromatin bridges and anaphase irregularities, were observed at varying NP concentrations [79]. Furthermore, oxidative stress indicators such as ROS and malondialdehyde (MDA) accumulation increased significantly with exposure [80, 81].

Collectively, these findings illustrate that the toxicological impacts of Au, Ag, Zn, and Cu NPs in aromatic plants are dose-, size-, and species dependent. At sub-toxic concentrations, NPs can enhance stress tolerance and secondary metabolite biosynthesis, but above threshold levels, they induce oxidative stress, genotoxicity, and impaired physiological functions. Establishing precise threshold concentrations for different plant species is therefore critical for leveraging NP benefits while minimizing risks.

7 Comparative Effects of Au, Ag, Zn, and Cu NPs on Specific Medicinal Species

The impact of metal-based nanoparticles (NPs) on medicinal and aromatic plants has become an increasingly important focus of research, as even small concentrations can significantly alter growth, secondary metabolism, and essential oil production. Among these, Au, Ag, Zn, and Cu NPs are particularly noteworthy for their role in regulating phytochemistry and enhancing bioactive compounds.

Zinc-based nanoparticles (ZnO NPs) have shown remarkable potential in boosting essential oil biosynthesis and overall plant performance. For instance, foliar application of ZnO NPs doubled the essential oil yield in *Salvia officinalis* while enriching oxygenated monoterpenes such as camphor, thereby enhancing both quality and pharmacological potential [82, 83]. In *Ocimum sanctum* (holy basil), supplementation with ZnO NPs in combination with iron improved shoot biomass, enhanced antioxidant activity, and elevated the content of key metabolites such as stevioside [84]. Similarly, under saline conditions, ZnO NP treatment in *Ocimum basilicum* promoted growth, stabilized cell membranes, improved photosynthetic efficiency, and significantly increased essential oil yield [85]. Collectively, these studies emphasize the role of Zn-based nanofertilizers in improving stress tolerance and secondary metabolism in aromatic species.

Silver nanoparticles (Ag NPs), often synthesized via green methods using plant extracts, have demonstrated strong bioactivity and compatibility with medicinal plants. For example, Ag NPs synthesized using *Thymus vulgaris* extract exhibited outstanding antioxidant capacity, neutralizing nearly 97% of free radicals at a specific concentration [86]. These NPs also showed antimicrobial activity superior to standard antibiotics such as amoxicillin and ciprofloxacin [87]. Importantly, natural oils and phenolics from the thyme extract acted as stabilizing agents, prolonging NP stability and bioactivity for over 3 months [88]. This highlights the dual benefit of Ag NP synthesis using aromatic plant extracts: enhancing the NPs' stability while preserving their therapeutic efficacy.

Copper-based nanoparticles (Cu and CuO NPs) also demonstrate strong potential in stimulating essential oil biosynthesis. In *Dracocephalum moldavica*, application of CuO NPs increased essential oil content by 70% and overall yield by 104%, suggesting significant enhancement of metabolic pathways related to terpene synthesis [13]. Moreover, recent work has indicated that Cu NPs can modulate nutrient uptake and secondary metabolism in aromatic plants, influencing the concentration of phenolic compounds and the quality of essential oils [67]. These findings point to Cu NPs as effective tools for improving the pharmacological and aromatic profile of medicinal plants, provided that threshold concentrations are carefully regulated.

Gold nanoparticles (Au NPs) remain relatively less studied in aromatic species, but emerging evidence suggests promising applications. Au NPs are generally biocompatible and can be synthesized using aromatic plant extracts or essential oils, which not only ensures eco-friendly production but also enhances control over their size and stability [89]. In *Lavandula angustifolia*, Au NPs have been reported to enhance SOD and CAT gene expression, improving tolerance to abiotic stress and indirectly supporting secondary metabolism [56]. Recent studies also suggest that biogenic Au NPs may improve germination and seedling vigor in medicinal crops, though high concentrations can negatively affect biomass [70, 71]. These results underline the potential of Au NPs as safe and precise modulators of growth and phytochemistry, with future applications likely to expand into essential oil regulation and biofortification.

These findings demonstrate that Au, Ag, Zn, and Cu NPs exert species-specific, concentration-dependent effects on medicinal and aromatic plants. At optimal levels, they enhance growth, antioxidant defenses, and secondary metabolism, leading to improved essential oil yield and composition. However, excessive accumulation may trigger phytotoxicity, emphasizing the importance of establishing safe threshold concentrations to maximize benefits while minimizing risks (Table 1).

Table 1 Comparative effects of Au, Ag, Zn, and Cu NPs on key aromatic and medicinal plants

Nanoparticles	Aromatic plant	Positive effects (optimal dose)	Risks at higher dosage	Key references
AuNPs	*Ocimum sanctum* (holy basil)	Increased rosmarinic acid and eugenol via phenylpropanoid pathway activation	High doses reduced biomass, induced oxidative stress	[40, 70, 90]
	Lavandula angustifolia (lavender)	Enhanced SOD and CAT expression; improved abiotic stress tolerance; higher secondary metabolites	Overaccumulation led to oxidative imbalance	[56]
AgNPs	*Cymbopogon citratus* (lemongrass)	Improved plant height, leaf number, and essential oil yield	Excess reduced beneficial microbial activity in rhizosphere	[41]
	Mentha piperita (peppermint)	Increased APX and GPX activity; higher phenolic content; enhanced antioxidant potential	Cytotoxic/genotoxic effects in root meristems at higher doses	[54]
	Mentha arvensis (field mint)	Increased flavonoid content and oil yield	Oxidative stress	[91]
	Thymus vulgaris (thyme)	Strong antioxidant capacity (97% free radical neutralization); improved antimicrobial activity	High concentration caused reduced vigor and abnormal mitosis	[74, 86]

(continued)

Table 1 (continued)

Nanoparticles	Aromatic plant	Positive effects (optimal dose)	Risks at higher dosage	Key references
ZnO NPs	*Lavandula angustifolia* (lavender)	Increased biomass and photosynthesis	Chlorosis and ROS accumulation	[92]
	Ocimum basilicum (sweet basil)	Enhanced CAT and SOD activity; better membrane integrity; increased essential oil yield under salinity	ROS overproduction, DNA damage, growth inhibition	[53, 85]
	Salvia officinalis (sage)	100% increase in essential oil yield; enriched camphor and oxygenated monoterpenes	High doses impaired cell division and DNA integrity	[82, 83]
	Mentha arvensis (field mint)	Boosted oil yield; improved drought tolerance; stronger antioxidant defenses	Cytotoxicity at >50 μg/mL; chromosomal abnormalities	[13, 76]
Cu/CuO NPs	*Pelargonium graveolens* (rose geranium)	Antimicrobial, increased terpenes	Growth inhibition, ROS burst	[93]
	Rosmarinus officinalis (rosemary)	Stimulated phenolic metabolism; improved biomass and essential oil quality	Excess reduced CAT activity, elevated MDA, oxidative stress	[57]
	Dracocephalum moldavica (Moldavian balm)	Essential oil content ↑70%; yield ↑104%; enhanced terpene synthesis	Overexposure disrupted mitosis and meiotic processes	[13, 69]
	Mentha piperita (peppermint)	Increased biomass and flavonoid accumulation	ROS-mediated toxicity at higher concentrations	[49]

8 Applications in Enhancing Medicinal Value and Bioactivity

Aromatic and medicinal plants have historically been valued for their therapeutic properties, largely attributed to their essential oils and secondary metabolites. With the advent of nanotechnology, metal-based nanoparticles (NPs) such as Cu, Zn, Ag, and Au have emerged as potent modulators of plant growth, phytochemical composition, and bioactivity. Their distinct physicochemical properties allow them to interact at the molecular and cellular levels, thereby influencing phytochemical biosynthesis and enhancing the medicinal potential of these plants.

The mechanisms underlying these enhancements are multifaceted (Fig. 4). Metal-based NPs can modulate gene expression associated with phytochemical biosynthesis, stimulate secondary metabolic pathways, induce plant defense enzymes, and improve nutrient uptake and photosynthetic efficiency. Collectively, these processes contribute to elevated accumulation of phenolics, flavonoids, terpenoids, alkaloids, and other bioactive compounds that define the therapeutic efficacy of aromatic plants.

Modulation of gene expression: NPs influence the regulation of genes associated with phytochemical biosynthesis, leading to increased production of bioactive compounds. For instance, Au and Ag NPs have been shown to upregulate transcription factors involved in flavonoid and alkaloid synthesis [94].

Enhancement of nutrient uptake and photosynthetic efficiency: Zn and Cu NPs improve chlorophyll content and electron transport, thereby boosting photosynthesis and biomass accumulation [95].

Activation of secondary metabolic pathways: By triggering specific signaling cascades, NPs stimulate the accumulation of terpenoids, phenolics, and essential oils—metabolites central to the medicinal and aromatic value of plants [94].

Induction of plant defense enzymes: NPs enhance antioxidant enzyme activities such as superoxide dismutase and peroxidases, which not only improve stress resilience but also enrich phytochemicals with medicinal value [27].

Among these, Au NPs stand out due to their high biocompatibility and ability to form stable conjugates with biomolecules. They can be functionalized with peptides, nucleic acids, or phytochemicals to improve antimicrobial efficacy and

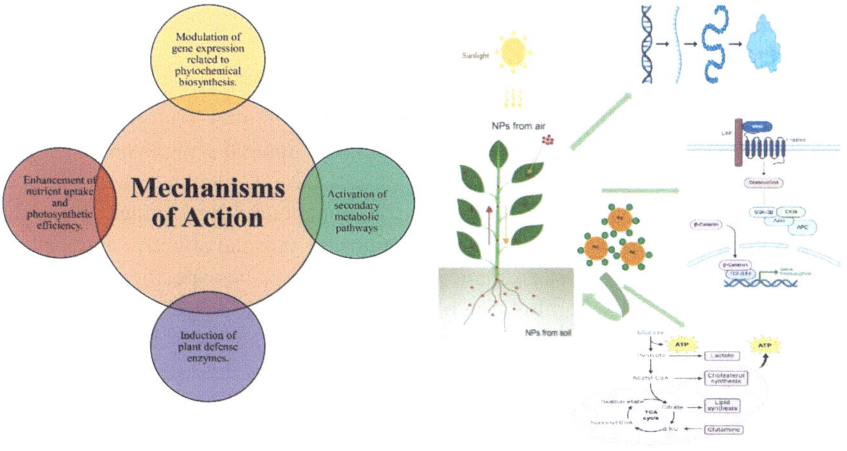

Fig. 4 Proposed mechanisms underlying the bioactivity enhancement of aromatic plants

broaden applications [96, 97]. Biofunctionalized Au NPs, which encapsulate or bind physiologically active molecules, are increasingly being explored as antibacterial agents, with the added advantage that microbial resistance is difficult to develop against them [98, 99]. Studies also indicate that Au NPs can enhance essential oil yield and modify its composition, further improving the medicinal quality of plants [100]. Similarly, Ag NPs have been reported to elevate essential oil yield and enhance antimicrobial activity in aromatic plants like *Mentha* and *Ocimum* species [97, 100]. Recent studies also demonstrate that ZnO and CuO NPs not only promote plant growth but also improve the concentration of pharmacologically important compounds such as rosmarinic acid and other essential oils in aromatic herbs [101].

In summary, the application of Cu, Zn, Ag, and Au NPs in aromatic plants enhances phytochemical biosynthesis, essential oil content, and bioactivity by modulating genetic, metabolic, and enzymatic pathways. These findings highlight the potential of nanotechnology as a novel strategy to fortify the medicinal and commercial value of aromatic plants.

9 Conclusions

Metal-based nanoparticles (Au, Ag, Zn, and Cu) represent a promising frontier in enhancing the growth, resilience, and bioactivity of aromatic and medicinal plants. Their unique nanoscale properties enable them to interact with plant tissues at molecular and cellular levels, modulating nutrient uptake, photosynthetic efficiency, secondary metabolite biosynthesis, and antioxidant defense systems. At optimal concentrations, these NPs stimulate growth, improve stress tolerance, and significantly enrich essential oil yield and composition, thereby augmenting the medicinal and commercial value of aromatic crops.

However, the dual nature of these materials is evident. While low to moderate doses act as elicitors of beneficial responses, supraoptimal concentrations can induce phytotoxicity manifested as oxidative stress, impaired photosynthesis, cytological abnormalities, and genotoxicity. The impact is highly dependent on NP type, size, surface coating, plant species, and environmental conditions, underscoring the importance of identifying threshold concentrations for safe and effective application.

Comparative studies across aromatic species reveal species-specific responses to Au, Ag, Zn, and Cu NPs, highlighting both opportunities for targeted enhancement of phytochemistry and risks of unintended toxicity. Green synthesis methods, which use plant-derived reducing and capping agents, offer a more sustainable and biocompatible alternative to conventional NP production, improving safety and compatibility for agricultural use.

Looking ahead, integrating nanotechnology into aromatic plant cultivation requires a careful balance between innovation and biosafety. Future research should focus on dose optimization, long-term ecological assessments, and mechanistic insights into NP-plant interactions at genomic and metabolomic levels. By combining precision nanomaterials with sustainable cultivation practices, it may be possible to harness the benefits of Au, Ag, Zn, and Cu NPs while minimizing risks, ultimately contributing to the production of high-value aromatic crops with enhanced medicinal efficacy.

Competing Interests The authors declare that they have no competing interest.

Declaration of generative AI and AI-assisted technologies in the writing process: During the preparation of this work, the book chapter author(s) have not taken any assistance from AI, and the book chapter author(s) reviewed and edited the content as needed and take(s) full responsibility for the content of the publication.

References

1. Abate L, Bachheti A, Bachheti RK, Husen A, Getachew M, Pandey DP (2021) Potential role of Forest-based plants in essential oil production: an approach to cosmetic and personal health care applications. In: Husen A, Bachheti RK, Bachheti A (eds) Non-timber Forest products. Springer International Publishing AG, Cham, pp 1–18. https://doi.org/10.1007/978-3-030-73077-2_1
2. Bachheti RK, Abate L, Hunde Y, Tadesse MG, Bachheti A, Pandey DP, Sharma A, Zebeaman M, Husen A (2023) Essential oils from medicinal plants and their role in nanoparticles synthesis, characterization, and applications. In: Bachheti RK, Bachheti A (eds) secondary metabolites from medicinal plants - nanoparticles synthesis and their applications. CRC Press, Boca Raton, FL, pp 177–198. https://doi.org/10.1201/9781003213727
3. Borai P, Husen A (2003) Aromatic oils from forest. In: Nautiyal S, Kaul AK (eds) Non-timber Forest products of India. Jyoti Publishers and Distributors, Dehra Dun, India, pp 314–331
4. Grigore-Gurgu L, Dumitrașcu L, Aprodu I (2025) Aromatic herbs as a source of bioactive compounds: an overview of their antioxidant capacity, antimicrobial activity, and major applications. Molecules 30:1304
5. Husen A (2022) Herbs, shrubs and trees of potential medicinal benefits. CRC press, Boca Raton, FL. https://doi.org/10.1201/9781003205067
6. Husen A (2023a) Exploring poisonous plants: medicinal values, toxicity response and other potential uses. CRC Press, Boca Raton, FL, pp 33487–32742. https://doi.org/10.1201/b23017
7. Husen A (2023b) Antidiabetic medicinal plants and herbal treatments. CRC Press, Boca Raton, FL, pp 33487–32742. https://doi.org/10.1201/b23347
8. Husen A (2024a) Antimalarial medicinal plants. Taylor & Francis Group, LLC, NW ISBN: 9781032457154. https://doi.org/10.1201/9781003378396
9. Husen A (2024b) Medicinal spices and condiment crops. Taylor & Francis Group, LLC, NW ISBN: 9781032469478. https://doi.org/10.1201/9781003387046
10. Akhilraj BC, Suresh J (2023) Implementation of plant biotechnology to the development of bioactive molecules in medicinal and aromatic plants. Arch Agric Allied Sci
11. Husen A (2023d) Nanomaterials from agricultural and horticultural products. Springer Nature Singapore Pte Ltd., Singapore, Singapore. https://link.springer.com/book/9789819934348
12. Husen A (2023e) Nanomaterials and nanocomposites exposures to plants (response, interaction, phytotoxicity and defense mechanisms). Springer Nature Singapore Pte Ltd., Singapore, Singapore. https://doi.org/10.1007/978-981-99-2419-6
13. Kralova K, Jampilek J (2021) Responses of medicinal and aromatic plants to engineered nanoparticles. Appl Sci 11:1813

14. Njuguna J, Ansari F, Sachse S, Rodriguez VM, Siqqique S, Zhu H (2021) Nanomaterials, nanofillers, and nanocomposites: types and properties. In: Health and environmental safety of nanomaterials. Woodhead Publishing, pp 3–37
15. Pirsaheb M, Gholami T, Seifi H, Dawi EA, Said EA, Hamoody AHM, Salavati-Niasari M (2024) Green synthesis of nanomaterials by using plant extracts as reducing and capping agents. Environ Sci Pollut Res 31:24768–24787
16. Shahcheraghi N, Golchin H, Sadri Z, Tabari Y, Borhanifar F, Makani S (2022) Nano-biotechnology, an applicable approach for sustainable future. 3 Biotech 12:65
17. Villagrán Z, Anaya-Esparza LM, Velázquez-Carriles CA, Silva-Jara JM, Ruvalcaba-Gómez JM, Aurora-Vigo EF, Martínez-Esquivias F (2024) Plant-based extracts as reducing, capping, and stabilizing agents for the green synthesis of inorganic nanoparticles. Resources 13:70
18. Doğaroğlu ZG, Uysal Y, Çaylalı Z, Karakulak DS (2023) Green nanotechnology advances: green manufacturing of zinc nanoparticles, characterization, and foliar application on wheat and antibacterial characteristics using *Mentha spicata* (mint) and Ocimum basilicum (basil) leaf extracts. Environ Sci Pollut Res 30:60820–60837
19. Kirubakaran D, Wahid JBA, Karmegam N, Jeevika R, Sellapillai L, Rajkumar M, SenthilKumar KJ (2025) A comprehensive review on the green synthesis of nanoparticles: advancements in biomedical and environmental applications. Biomed Mater Devices:1–26
20. Gangwar J, Kadanthottu Sebastian J, Puthukulangara Jaison J, Kurian JT (2023) Nano-technological interventions in crop production—a review. Physiol Mol Biol Plants 29:93–107
21. Khalid MF, Khan RI, Jawaid MZ, Shafqat W, Hussain S, Ahmed T, Alina Marc R (2022) Nanoparticles: the plant saviour under abiotic stresses. Nanomaterials 12:3915
22. Mosquera J, García I, Liz-Marzán LM (2018) Cellular uptake of nanoparticles versus small molecules: a matter of size. Acc Chem Res 51:2305–2313
23. Ahmed B, Rizvi A, Ali K, Lee J, Zaidi A, Khan MS, Musarrat J (2021) Nanoparticles in the soil–plant system: a review. Environ Chem Lett 19:1545–1609
24. Skiba E, Adamczyk-Szabela D, Wolf WM (2021) Metal-based nanoparticles' interactions with plants. In: Plant responses to nanomaterials: recent interventions, and physiological and biochemical responses. Springer International Publishing, Cham, pp 145–169
25. Yadav S, Nadar T, Lakkakula J, Wagh NS (2024) Biogenic synthesis of nanomaterials: bioactive compounds as reducing, and capping agents. In: Biogenic nanomaterials for environmental sustainability: principles, practices, and opportunities. Springer International Publishing, Cham, pp 147–188
26. Al-Khayri JM, Rashmi R, Surya Ulhas R, Sudheer WN, Banadka A, Nagella P, Almaghasla MI (2023) The role of nanoparticles in response of plants to abiotic stress at physiological, biochemical, and molecular levels. Plants 12:292
27. Singh R, Choudhary P, Kumar S, Daima HK (2024) Mechanistic approaches for crosstalk between nanomaterials and plants: plant immunomodulation, defense mechanisms, stress resilience, toxicity, and perspectives. Environ Sci Nano 11:2324–2351
28. Yeshi K, Crayn D, Ritmejerytė E, Wangchuk P (2022) Plant secondary metabolites produced in response to abiotic stresses has potential application in pharmaceutical product development. Molecules 27:313
29. Hong J, Wang C, Wagner DC, Gardea-Torresdey JL, He F, Rico CM (2021) Foliar application of nanoparticles: mechanisms of absorption, transfer, and multiple impacts. Environ Sci Nano 8:1196–1210
30. Mogre SS, Brown AI, Koslover EF (2020) Getting around the cell: physical transport in the intracellular world. Phys Biol 17:061003
31. Hamzah Saleem M, Usman K, Rizwan M, Al Jabri H, Alsafran M (2022) Functions and strategies for enhancing zinc availability in plants for sustainable agriculture. Front Plant Sci 13:1033092
32. Dwivedi A, Kumar S, Vajpayee P (2024) Copper oxide nanoparticles influenced ionome, reduced growth, altered pigments and stomatal morphology in *Mentha arvensis* L. Int J Plant Environ 10:53–60

33. Jadczak P, Kulpa D, Bihun M, Przewodowski W (2019) Positive effect of AgNPs and AuNPs in in vitro cultures of *Lavandula angustifolia* Mill. Plant Cell Tissue Organ Cult 139:191–197
34. Rani S, Kumari N, Sharma V (2023) Uptake, translocation, transformation and physiological effects of nanoparticles in plants. Arch Agron Soil Sci 69:1579–1599
35. Sanità G, Carrese B, Lamberti A (2020) Nanoparticle surface functionalization: how to improve biocompatibility and cellular internalization. Front Mol Biosci 7:587012
36. Djanaguiraman M, Anbazhagan V, Dhankher OP, Prasad PV (2024) Uptake, translocation, toxicity, and impact of nanoparticles on plant physiological processes. Plants 13:3137
37. Shakya P, Marslin G, Siram K, Beerhues L, Franklin G (2019) Elicitation as a tool to improve the profiles of high-value secondary metabolites and pharmacological properties of Hypericum perforatum. J Pharm Pharmacol 71:70–82
38. Ruotolo R, Maestri E, Pagano L, Marmiroli M, White JC, Marmiroli N (2018) Plant response to metal-containing engineered nanomaterials: an omics-based perspective. Environ Sci Technol 52:2451–2467
39. El-Saadony MT, Saad AM, Soliman SM, Salem HM, Desoky ESM, Babalghith AO, AbuQamar SF (2022) Role of nanoparticles in enhancing crop tolerance to abiotic stress: a comprehensive review. Front Plant Sci 13:946717
40. Mohanta YK, Biswas K, Mishra AK, Patra B, Mishra B, Panda J, Nayak D (2024) Amelioration of gold nanoparticles mediated through Ocimum oil extracts induces reactive oxygen species and mitochondrial instability against MCF-7 breast carcinoma. RSC Adv 14:27816–27830
41. Mukarram M, Khan MMA, Corpas FJ (2021) Silicon nanoparticles elicit an increase in lemongrass (*Cymbopogon flexuosus* (Steud.) Wats) agronomic parameters with a higher essential oil yield. J Hazard Mater 412:125254
42. Sobhannizadeh A, Giglou MT, Behnamian M, Estaji A, Majdi M, Szumny A (2025) The effect of plant growth regulators, FeO3-CTs nanoparticles and LEDs light on the growth and biochemical compounds of black seed (*Nigella sativa* L.) callus in vitro. BMC Plant Biol 25:539
43. Roman S, Voaides C, Babeanu N (2023) Exploring the sustainable exploitation of bioactive compounds in pelargonium sp.: beyond a fragrant plant. Plants 12:4123
44. Zhang Y, Zhang Y, Ding R, Zhang K, Guo H, Lin Y (2024) Self-assembled nanocarrier delivery systems for bioactive compounds. Small 20:2310838
45. Jasim B, Thomas R, Mathew J, Radhakrishnan EK (2017) Plant growth and diosgenin enhancement effect of silver nanoparticles in Fenugreek (Trigonella foenum-graecum L.). Saudi Pharm J 25(3):443–447
46. Al Jabri H, Saleem MH, Rizwan M, Hussain I, Usman K, Alsafran M (2022) Zinc oxide nanoparticles and their biosynthesis: overview. Life 12:594
47. Kowsalya K, Halka J, Arun M (2024) Exogenous approaches for improving the morphophysiological attributes and secondary metabolite production in medicinal plants. In: Propagation to pharmacopeia. CRC Press, pp 159–198
48. Javed R, Yucesan B, Zia M, Gurel E (2022) Nanoelicitation: a promising and emerging technology for triggering the sustainable in vitro production of secondary metabolites in medicinal plants. In: Plant and nanoparticles. Springer Nature Singapore, Singapore, pp 265–280
49. Asle-Mohammadi Z, Kharazmi M, Sheikhi H, Mohammadkhani N, Nicola S (2024) Foliar application of Fe, Zn, and Mn as a practical strategy to alleviate the soil Cu toxicity and stimulate the physiological and biochemical properties of peppermint (*Mentha piperita* L.). J Soil Sci Plant Nutr 24:371–388
50. Yang Y, Zhang T (2019) Antimicrobial activities of tea polyphenol on phytopathogens: a review. Molecules 24:816
51. Babashpour-Asl M, Farajzadeh-Memari-Tabrizi E, Yousefpour-Dokhanieh A (2022) Foliar-applied selenium nanoparticles alleviate cadmium stress through changes in physio-biochemical status and essential oil profile of coriander (*Coriandrum sativum* L.) leaves. Environ Sci Pollut Res 29:80021–80031

52. Irato P, Santovito G (2021) Enzymatic and non-enzymatic molecules with antioxidant function. Antioxidants 10:579
53. Hasan SA, Khan A, Irfan M (2024) Role of zinc oxide nanoparticles in alleviating sodium chloride-induced salt stress in sweet basil (*Ocimum basilicum* L.)
54. Ostadi A, Javanmard A, Amani Machiani M, Kakaei K (2022) Optimizing antioxidant activity and phytochemical properties of peppermint (*Mentha piperita* L.) by integrative application of biofertilizer and stress-modulating nanoparticles under drought stress conditions. Plants 12:151
55. Garcia-Caparros P, De Filippis L, Gul A, Hasanuzzaman M, Ozturk M, Altay V, Lao MT (2021) Oxidative stress and antioxidant metabolism under adverse environmental conditions: a review. Bot Rev 87:421–466
56. Hürkan K, Hürkan YK (2024) Changes in plant gene expression in response to gold nanoparticles and gold ions. In: Plant response to gold nanoparticles: plant growth, development, production, and protection. Springer, Singapore, pp 97–113
57. Iqbal J, Andleeb A, Ashraf H, Meer B, Mehmood A, Jan H, Abbasi BH (2022) Potential antimicrobial, antidiabetic, catalytic, antioxidant and ROS/RNS inhibitory activities of Silybum marianum mediated biosynthesized copper oxide nanoparticles. RSC Adv 12:14069–14083
58. Inam M, Attique I, Zahra M, Khan AK, Hahim M, Hano C, Anjum S (2023) Metal oxide nanoparticles and plant secondary metabolism: unraveling the game-changer nano-elicitors. Plant Cell Tissue Organ Cult (PCTOC) 155:327–344
59. Chrysargyris A, Mikallou M, Petropoulos S, Tzortzakis N (2020) Profiling of essential oils components and polyphenols for their antioxidant activity of medicinal and aromatic plants grown in different environmental conditions. Agronomy 10:727
60. Ur Rahim H, Qaswar M, Uddin M, Giannini C, Herrera ML, Rea G (2021) Nano-enable materials promoting sustainability and resilience in modern agriculture. Nano 11:2068
61. Dikshit PK, Kumar J, Das AK, Sadhu S, Sharma S, Singh S, Kim BS (2021) Green synthesis of metallic nanoparticles: applications and limitations. Catalysts 11:902
62. Gao M, Chang J, Wang Z, Zhang H, Wang T (2023) Advances in transport and toxicity of nanoparticles in plants. J Nanobiotechnol 21:75
63. Hatami M, Kariman K, Ghorbanpour M (2016) Engineered nanomaterial-mediated changes in the metabolism of terrestrial plants. Sci Total Environ 571:275–291
64. Goncharuk EA, Zagoskina NV (2023) Heavy metals, their phytotoxicity, and the role of phenolic antioxidants in plant stress responses with focus on cadmium. Molecules 28(9):3921
65. Attarilar S, Yang J, Ebrahimi M, Wang Q, Liu J, Tang Y, Yang J (2020) The toxicity phenomenon and the related occurrence in metal and metal oxide nanoparticles: a brief review from the biomedical perspective. Front Bioeng Biotechnol 8:822
66. Wohlmuth J, Tekielska D, Čechová J, Baránek M (2022) Interaction of the nanoparticles and plants in selective growth stages—usual effects and resulting impact on usage perspectives. Plants 11(18):2405
67. Shahhat IM, Elsheikh SYS (2024) Accumulation of secondary metabolites in the family Lamiaceae as influenced by foliar micronutrients. Bol Latinoam Caribe Plant Med Aromat 23:823–838
68. Bocso NS, Butnariu M (2022) The biological role of primary and secondary plants metabolites. J Nutr Food Process 5(3):1–7
69. Tasar N (2022) Mitotic effects of copper oxide nanoparticle on root development and root tip cells of Phaseolus vulgaris L. seeds. Microsc Res Tech 85(12):3895–3907
70. Botteon CE, Pereira ADE, de Castro LP, Justino IA, Fraceto LF, Bastos JK, Marcato PD (2025) Toxicity assessment of biogenic gold nanoparticles on crop seeds and zebrafish embryos: implications for agricultural and aquatic ecosystems. ACS Omega 10:1032–1046
71. Mancano RR, Matheus LR, LEN C, Barroso TLCT, da Rosa RG, Ferreira VC, Colpini LMS (2025) Extraction techniques for propolis and its utilization in silver nanoparticle synthesis: a comprehensive review. Eur Food Res Technol:1–37

72. Elshazly EH, Mohamed AKS, Aboelmagd HA, Gouda GA, Abdallah MH, Ewais EA, Ali GA (2022) Phytotoxicity and antimicrobial activity of green synthesized silver nanoparticles using *Nigella sativa* seeds on wheat seedlings. J Chem 2022:9609559
73. Kulak M, Yaldiz G, Camlica M (2023) Applications of some nanoparticles and responses of medicinal and aromatic plants under stress conditions. In: Sustainable Agriculture in the Era of the OMICs Revolution, pp 193–222
74. Patel J, Kumar GS, Roy H, Maddiboyina B, Leporatti S, Bohara RA (2024) From nature to nanomedicine: bioengineered metallic nanoparticles bridge the gap for medical applications. Discover Nano 19:85
75. Trzcińska-Wencel J, Mucha N, Rai M, Tyburski J, Golińska P (2025) Pre-sowing grain treatment with bio-AgNPs stimulates plant growth and affects redox homeostasis in maize. Front Plant Sci 16:1494741
76. Khan S, Al-Qurainy F, Al-Hashimi A, Nadeem M, Tarroum M, Shaikhaldein HO, Salih AM (2023) Effect of green synthesized ZnO-NPs on growth, antioxidant system response and bioactive compound accumulation in Echinops macrochaetus, a potential medicinal plant, and assessment of genome size (2C DNA content). Plants 12(8):1669
77. Sampaio GR, Guizellini GM, da Silva SA, de Almeida AP, Pinaffi-Langley ACC, Rogero MM, Torres EA (2021) Polycyclic aromatic hydrocarbons in foods: biological effects, legislation, occurrence, analytical methods, and strategies to reduce their formation. Int J Mol Sci 22:6010
78. Elenany AM, Atia MMM, Abbas EE, Moustafa M, Alshaharni MO, Negm S, Elnahal ASMA (2024) Nanoparticles and chemical inducers: a sustainable shield against onion white rot. Biology 13:219
79. Doğan Y, Öziç C, Ertaş E, Baran A, Rosic G, Selakovic D, Eftekhari A (2024) Activated carbon-coated iron oxide magnetic nanocomposite (IONPs@CtAC) loaded with morin hydrate for drug-delivery applications. Front Chem 12:1477724
80. Mahjoubian M, Naeemi AS, Moradi-Shoeili Z, Tyler CR, Mansouri B (2023) Toxicity of silver nanoparticles in the presence of zinc oxide nanoparticles differs for acute and chronic exposures in zebrafish. Arch Environ Contam Toxicol 84:1–17
81. Shaban AS, Owda ME, Basuoni MM, Mousa MA, Radwan AA, Saleh AK (2024) Punica granatum peel extract mediated green synthesis of zinc oxide nanoparticles: structure and evaluation of their biological applications. Biomass Convers Biorefinery 14:12265–12281
82. Bo C, Zhijia Z, Wenjun Z, Songsong Z, Guojun W, Qiang W, Hao W (2024) Application status of antibacterial polyurethane. J Polym Res 31:275
83. Motelica L, Vasile BS, Ficai A, Surdu AV, Ficai D, Oprea OC, Dobre AA (2023) Antibacterial activity of zinc oxide nanoparticles loaded with essential oils. Pharmaceutics 15:2470
84. Azizah NS, Irawan B, Kusmoro J, Safriansyah W, Farabi K, Oktavia D, Miranti M (2023) Sweet basil (*Ocimum basilicum* L.)—a review of its botany, phytochemistry, pharmacological activities, and biotechnological development. Plants 12:4148
85. Amirifar A, Hemati A, Asgari Lajayer B, Pandey J, Astatkie T (2022) An insight into plant nanobionics and its application to overcome the impact of environmental stress on medicinal and aromatic plants. In: Environmental challenges and medicinal plants: sustainable production solutions under adverse conditions. Springer, Cham, pp 385–408
86. Jampilek J, Kos J, Kralova K (2019) Potential of nanomaterial applications in dietary supplements and foods for special medical purposes. Nano 9:296
87. Al-Jahani GMA (2021) *Thymus vulgaris* (thyme) as a natural organic matter to biosynthesis silver nanoparticles and their antibacterial efficiency. Int J Pharm Res Allied Sci 10:118–126
88. Fakhariha M, Rafati AA, Garmakhany AD, Asl AZ (2025) Nanoencapsulation enhances stability, release behavior, and antimicrobial properties of sage and thyme essential oils. Sci Rep 15:18373
89. Antunes Filho S, Dos Santos MS, Dos Santos OAL, Backx BP, Soran ML, Opriş O, Bououdina M (2023) Biosynthesis of nanoparticles using plant extracts and essential oils. Molecules 28:3060

90. Sharma MMM, Kapoor D, Loyal A, Kumar R, Sharma P, Husen A (2024) Mitigation of drought stress in plants using gold nanoparticles. In: Plant response to gold nanoparticles: plant growth, development, production, and protection. Springer Nature Singapore, Singapore, pp 193–209
91. Haq SIU, Wali S, Sama NU, Kamran K, Ullah Z, Mohamed HI (2024) Environmentally friendly synthesis of silver nanoparticles (AgNPs) using *Mentha arvensis* plants modulates physiological and biochemical attributes and yield of sunflower (Helianthus annuus L.). J Soil Sci Plant Nutr 24:3610–3630
92. Mosaedi H, Mozafari H, Sani B, Pirbalouti AG, Rajabzadeh F (2024) Foliar-applied silicon and zinc nanoparticles improve plant growth, biochemical attributes, and essential oil profile of fennel (Foeniculum vulgare) under different irrigation regimes. Funct Plant Biol 51:10
93. Lala S (2025) Nanoparticles as elicitors and harvesters of economically important secondary metabolites in plants. In: Plant specialized metabolites: phytochemistry, ecology and biotechnology. Springer Nature Switzerland, Cham, pp 1191–1240
94. Selvakesavan RK, Kruszka D, Shakya P, Mondal D, Franklin G (2023) Impact of nanomaterials on plant secondary metabolism. In: Nanomaterial interactions with plant cellular mechanisms and macromolecules and agricultural implications, pp 133–170
95. Singh H, Desimone MF, Pandya S, Jasani S, George N, Adnan M, Alderhami SA (2023) Revisiting the green synthesis of nanoparticles: uncovering influences of plant extracts as reducing agents for enhanced synthesis efficiency and its biomedical applications. Int J Nanomed:4727–4750
96. Graczyk A, Pawlowska R, Jedrzejczyk D, Chworos A (2020) Gold nanoparticles in conjunction with nucleic acids as a modern molecular system for cellular delivery. Molecules 25:204
97. Patil T, Gambhir R, Vibhute A, Tiwari AP (2023) Gold nanoparticles: synthesis methods, functionalization and biological applications. J Clust Sci 34:705–725
98. Mba IE, Nweze EI (2021) Nanoparticles as therapeutic options for treating multidrug-resistant bacteria: research progress, challenges, and prospects. World J Microbiol Biotechnol 37:108
99. Timoszyk A, Grochowalska R (2022) Mechanism and antibacterial activity of gold nanoparticles (AuNPs) functionalized with natural compounds from plants. Pharmaceutics 14:2599
100. Lala S (2021) Nanoparticles as elicitors and harvesters of economically important secondary metabolites in higher plants: a review. IET Nanobiotechnol 15:28–57
101. Movahedi A, Taghimolaee M, Ghayebpanah S, Monajemi F, Ghanbari M, Nemat Farahzadi H (2024) A review of the effect of nanoparticles on reactive oxygen species production in different plants. J Eng Appl Res 1(1):131–163

Effects of Metal Oxide Nanoparticles (ZnO, CeO_2, Fe_2O_3) on Aromatic Plants

Shefali Singh, Smita Rai, Shifa Suhail, Gulafshan Eram, Reena Vishvakarma, Poonam Sharma, and Swati Sharma

Abstract Aromatic plants (APs) are used extensively in food, nutraceutical, medicine, and cosmetic industry owing to their bioactivity. Nanotechnology has opened up new possibilities to raise the yield of aromatic crops, more so in the era of green technology and sustainable agriculture. Metal oxide nanoparticles (MONPs), such as iron oxide (FeO_3), cerium oxide (CeO_2), and zinc oxide (ZnO), have drawn a lot of interest among nanomaterials. These nanoparticles affect plant physiology and metabolism by acting as antibacterial agents, growth promoters, stress reducers, and nanofertilizers. ZnO, CeO_2, and FeO_3 nanoparticles improve biomass output, photosynthetic efficiency, essential oil yield, seed germination, and chlorophyll synthesis when present at optimal quantities. Additionally, they are essential for regulating antioxidant defense systems, enhancing nutrient absorption, and reducing biotic as well as abiotic stress, including drought, salinity or heavy metal toxicity. Additionally, MONPs enhance the medicinal and aromatic value of plants by promoting the manufacture of beneficial secondary metabolites. The dual function of these nanoparticles as growth promoters and possible toxicants is highlighted by the fact that excessive exposure or improper application may cause oxidative stress, DNA damage, and metabolic disruptions. This chapter provides a comprehensive overview of the beneficial and adverse effects of ZnO, CeO_2, and Fe_2O_3 nanoparticles on aromatic plants, emphasizing their sustainable application for enhancing agricultural productivity.

Keywords Aromatic plants · Nanotechnology · Metal oxide nanoparticles · Agricultural · Stress tolerance

Shefali Singh and Smita Rai contributed equally with all other contributors.

S. Singh · S. Rai · S. Suhail · G. Eram · S. Sharma (✉)
Plant Microbial Interaction Laboratory, Department of Biosciences, Faculty of Science, Integral University, Lucknow, Uttar Pradesh, India

R. Vishvakarma · P. Sharma
Department of Bioengineering, Faculty of Engineering, Integral University, Lucknow, Uttar Pradesh, India

A. Husen (ed.), *Sustainable Uses of Aromatic Plants in Nanotechnology*, Smart Nanomaterials Technology, https://doi.org/10.1007/978-981-95-7409-4_20

1 Introduction

Natural compounds originating from plants, especially aromatic plants, were used for human benefit even in ancient times [1]. Since their usage dates back more than 6000 years in Indian, Chinese, Egyptian, Greek, Roman, and Persian traditions, APs are essential to traditional medicine [2, 3]. The diverse range of chemical substances that aromatic plants produce is known as secondary metabolites (SMs) [4]. These metabolites are found in trace amounts in different plant sections and are produced by plants in response to particular biological or environmental stimuli. Essential oils, terpenoids, alkaloids, phenolic compounds, glucosinolates, cyanogenic glycosides, and other SMs provide a range of therapeutic benefits, including as anti-inflammatory, anticancer, antibacterial, anti-allergy, and antidiabetic qualities [5–12]. The pharmaceutical, food, cosmetic, dye, supplement, flavoring, antimicrobial, disinfection, personal care, oil, perfume, insecticide, herbicide, plant hormone, rubber, and biofuels sectors are among those that use them widely [11, 13, 14]. An overview of potential industrial uses of such plants such as in cosmetics, medicine, or agriculture industry is given in Fig. 1.

Aromatic plants offer a variety of effects that have a health-related benefits and have nutrigenomics implications for gut and immunological growth since plant derived extracts as well as their secondary metabolites are an excellent source of bioactive compounds such phenolics, terpenoids, etc. [5, 12]. Dietary supplements known as nutraceuticals are used to maintain the proper operation of the human body, prevent disease, postpone aging, and enhance health [15]. In addition to any possible health benefits, aromatic plants are valued for their benefits to the economy and environment. The worldwide nutraceutical commercial only is huge, assessed

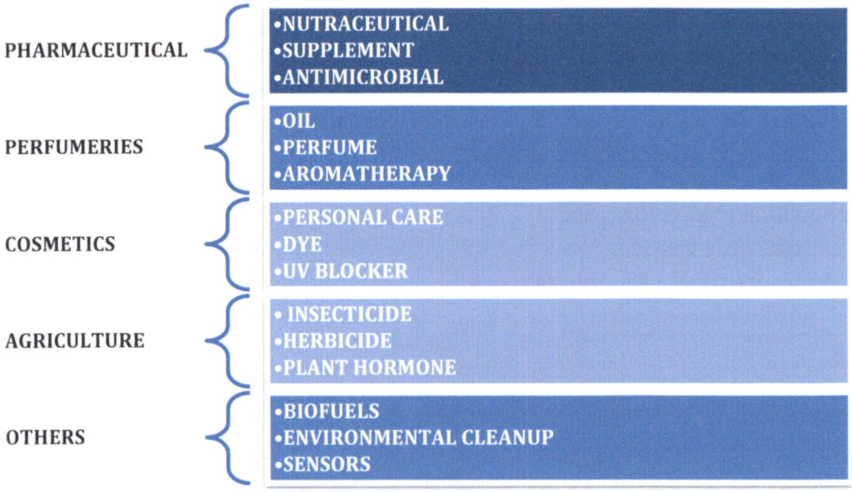

Fig. 1 Industrial sectors related to application of medicinal and aromatic plants

to be value USD 118 billion [16]. Aromatic plants need to be cultivated on a wide scale to boost local economies and reduce the negative environmental consequences of primarily chemical-based agriculture.

Nanotechnology is one of the scientific and technological fields that is expanding the fastest right now. A substance is considered a nanoparticle if at least one of its dimensions is less than 100 nm. Because of its special physicochemical characteristics, nanoparticles have been utilized in several goods [17, 18]. Nanoparticles have been widely used in medicine and cosmetics [19]. In recent years, nanoparticles have also been formulated for use as nanopesticides or nanofertilizers (including as nanocarriers for fertilizers and insecticides) in agriculture.

Metallic as well as nonmetallic both types of nanoparticles are being studied. Metal and metal oxide nanoparticles like copper, iron, aluminum, nickel, zinc oxide, copper oxide, cerium dioxide, silver, and titanium dioxide (TiO2) are primarily researched for their effects on various plants [20]. Fullerene and single-walled carbon nanotubes are two examples of nonmetal nanoparticles whose nontoxicity processes have been thoroughly investigated. On the other hand, exposure to single-walled carbon nanohorns (SWCNHs) has been shown to accelerate seed germination and enhance growth for several organs of maize, tomato, rice, and soybean [20]. When herbicide was encapsulated in chitosan nanoparticles, the herbicide's effectiveness was shown to be much increased [21]. Nanoparticles, including hydrogels (made of various polymers like chitosan and alginate), nanozeolites (basic

Fig. 2 List of metal oxide NP tested for plant growth promotion in some aromatic and medicinal plants

building blocks of silicate $[SiO^4]^-$ and aluminates $[AlO^4]^-$ tetrahedrons), and nanosensors (monitoring plants and soil health) aid in improving the quality of soil [22]. Many studies have reported the benefits of metal oxide nanoparticles in plants. The applications of metal oxide NP in some aromatic and medicinal plant growth promotion [20, 23–29] are listed in Fig. 2.

2 Zinc Oxide Nanoparticles

Zinc as an essential micronutrient plays a critical role in plant physiological processes including protein metabolism, growth regulation, enzyme activation, and chlorophyll synthesis [30]. However, zinc deficiency is a common nutritional disorder in crops worldwide, leading to reduced productivity and quality. Zinc is an essential nutrient and also has antimicrobial nature; hence, the application of zinc oxide nanoparticles (ZnO NPs) provides a novel strategy to address this challenge, by offering enhanced bioavailability and controlled release of zinc compared to conventional fertilizers thereby improving crop productivity. They have been applied across various sectors and gaining attention in agriculture [31]. ZnO NPs are extensively utilized in agriculture due to their cost-effectiveness, biocompatibility, and versatile functionality. Zinc and zinc oxide nanoparticles have been studied for promoting plant growth in plants as well as in tissue culture. ZnO NPs with aspect ratio were used with the seeds of *Capsicum annum* and *Solanum melongena* to speed up germination. This was shown by a shorter germination time and a higher mean germination rate compared to control plants. They applied 100 mg/L ZnO NPs and made sure that the transplant length, fresh weight, and dry weight were all at their highest [24]. The treatment of *Capsicum annuum* L. with ZnO NPs suspensions, respectively, improved the seed germination rate and seedling vigor. Conversely, the application of these ZnO NP concentrations led to the suppression of radical growth and the enhancement of phenolic compound accumulation [32].

The application of ZnO and other nanofertilizers has shown great potential in enhancing plant growth improving nutrient absorption and stress tolerance. These comparative effects are summarized in Table 2, which highlights the diverse roles of ZnO and other nanofertilizers in aromatic plants. Combination showed synergistic antimicrobial activity, notably stronger against various bacteria, due to essential oil loading (Table 1).

ZnO NPs have gained attention for their potential to enhance growth, yield, and stress tolerance aromatic plants such as *Dracocephalum moldavica* L. [33]. These NPs enhance both yield and fragrance in aromatic rice (*Oryza sativa*). When applied as foliar spray, they grain zinc concentration, spikelet number, and grain filling while also increasing aroma-related compounds like 2-acetyl-1-pyrroline (2-AP) thereby improving both nutritional value and aromatic quality. When applied to soil, ZnO NPs are taken up by plant roots, where they dissolve to release Zn^{2+} ions or, in some cases, enter root cells as intact nanoparticles [34]. In soil, ZnO NPs may

Table 1 Effect of ZnO NPs synthesized using aromatic plant and their application

Aromatic plant	Mode of application	Effect of ZnO NPs	Reference
Syzygium aromaticum	Applied as antifungal agent to control fungal growth postharvest	ZnO NPs synthesized using clove bud extract inhibited growth and mycotoxin production by *Fusarium graminearum* (a fungal pathogen)	Lakshmeesha et al. [69]
Melissa officinalis	Applied through irrigation with nutrient solutions at different concentrations (0, 100, and 300 mg L^{-1})	Improved nutrient uptake, biomass, lateral bud and root development, bioactive compound and antioxidant metabolites	Babajani et al. [70]
Linum usitatissimum	Applied via MS medium to seedlings and callus	Boosted antioxidants and metabolites	Zaeem et al. [71]
Dracocephalum moldavica L.	Foliar spray	Enhanced Zn biofortification, biomass, chlorophyll, essential oils, and secondary metabolites	Nekoukhou et al. [72]

Table 2 CeO_2 NPs effects on aromatic plant species

Plant name	CeO_2 NP conc.	Effects	References
Ocimum basilicum (basil)	25–100 mg/L	Increased seed germination, biomass, enhanced oil content, antioxidant enzyme activity	Kralova and Jampilek [73]
Mentha piperita (peppermint)	50 mg/L	Increased chlorophyll content, decreased oxidative stress, increased growth	Khan et al., [74]
Lavandula angustifolia (lavender)	10–100 mg/L	Modulation of terpenoid profile, increased secondary metabolites	Habán et al., [75]
Cymbopogon citratus (lemongrass)	50 mg/L	Increased germination rate, biomass, antioxidant response increased	Mirzaei et al., [76]
Coriandrum sativum (cilantro)	500 mg/kg	Increased antioxidant enzyme activity like CAT, APX and show stress tolerance response	Morales et al., [77]

remain as nanoparticle or dissolve into ion forms, and their behavior is strongly influenced by soil pH, microbial activity, and organic matter. At optimum concentration, they can improve soil plant nutrition and micronutrient availability; however, excessive use may cause zinc accumulation, alter soil chemistry, and harm beneficial microbes [31]. The use of ZnO NPs in plants has gained attention due to both source of zinc micronutrient and nanomaterial that regulate plant physiological and biochemical processes. Through their impact on enzyme activity, nutrient uptake, and metabolic regulation, ZnO NPs play a critical role in enhancing plant growth, biomass production, and overall development [35].

Several studies have shown that applying ZnO NPs at optimal concentrations can enhance plant growth and biomass production. In *Tanacetum parthenium* (feverfew), foliar application of ZnO NPs significantly improves plant height, leaf area, and dry biomass. Enhanced growth has been attributed to improved chlorophyll content, photosynthetic efficiency, and enzyme activity involved in primary metabolism [36]. ZnO NPs play a role in early plant development, especially during seed germination. At low concentration, it can improve germination percentage and promote emergence of radicle and plumule and seedling vigor. High concentration may suppress seedling growth, damage germination, and induce oxidative stress. This dose-dependent nature of ZnO NPs highlights the importance of optimal application strategies to harness benefits while minimizing phytotoxic effects [37]. A study of *Brassica juncea* demonstrated that 25 ppm ZnO NPs shows positive effects on seedling growth, while 100 ppm of these nanoparticles caused toxicity. A related study on *Brassica napus* which is an important plant of *Brassicaceae* family confirms the dose-dependent effects of ZnO NPs [38].

ZnO NPs have emerged as promising nanofertilizers and elicitors in agriculture due to their ability to modulate plant metabolism and enhance tolerance to environmental stresses. Beyond their role as a micronutrient source, ZnO NPs influence phytochemical pathways, regulate stress signaling, and impact both plant growth and defense responses [36]. ZnO NPs can act as abiotic elicitors, mimicking natural stress signals that trigger defense-related pathways in plants. By inducing mild oxidative stress or perturbing ion homeostasis, these nanoparticles activate signaling cascades involving phytohormones (jasmonic acid, salicylic acid, abscisic acid) and secondary messenger molecules (Ca^{2+}, nitric oxide). This elicitation stimulates the expression of defense genes, leading to enhanced production of protective metabolites and structural barriers such as lignin [35].

ZnO NPs play a dual role in stress management by alleviating both abiotic and biotic stresses. Under abiotic stresses such as drought, salinity, and heavy metal toxicity, ZnO NPs improve photosynthetic activity, water-use efficiency, and antioxidant enzyme activity, thereby reducing stress-induced damage [39]. Against biotic stresses, ZnO NPs enhance plant immunity and can directly inhibit the growth of pathogens through antimicrobial activity. This dual functionality makes ZnO NPs a potential tool for integrated crop management, improving yield stability under challenging environmental conditions [40].

One of the central mechanisms by which ZnO NPs influence plant physiology is through the generation of reactive oxygen species (ROS). At controlled concentrations, ROS serve as critical signaling molecules that activate antioxidant defense systems, enhance stress tolerance, and regulate secondary metabolite biosynthesis. While ZnO NPs at low to moderate doses can enhance plant productivity and stress resilience, higher concentrations pose risks of genotoxicity. Their main mechanism involves overproduction of reactive oxygen species (ROS), which at moderate levels act as beneficial signals but in excess can damage lipids, proteins, and nucleic acids, leading to DNA strand breaks, chromosomal aberrations, and impaired growth. However, excessive ROS accumulation due to high doses of ZnO NPs can cause oxidative damage to proteins, lipids, and nucleic acids, resulting in cell death and impaired plant growth. The phytotoxicity of ZnO NPs has been reported in aromatic and medicinal plants [31, 41]. The accumulation of reactive oxygen species (ROS) due to high concentrations (10–2000 mg/L) of ZnO NPs has been observed by Lee et al. [41] in buckwheat plants. There was also cell damage in the root as well as higher activity of the antioxidant defense enzymes. Thus, the balance between ROS-mediated signaling and oxidative stress determines whether ZnO NPs act as beneficial stimulants or harmful toxicants. In coriander (*Coriandrum sativum*), also soil exposure to ZnO NPs increased lipid peroxidation and stress responses, suggesting potential DNA damage [31].

These effects highlight the importance of establishing safe dose thresholds to harness the benefits of ZnO NPs without compromising genomic stability. Studies in model plants like *Allium cepa* and *Vicia faba* further confirm DNA fragmentation, altered mitotic index, and oxidative damage under ZnO NP exposure [42]. Thus, in aromatic crops, ZnO NPs act as a double-edged sword—useful nanofertilizers at low doses but possible genotoxins if overapplied. These benefits are highly concentration dependent: while optimal concentration supports growth and metabolite production, excessive levels may cause oxidative and genotoxic stress. Therefore, careful adjustment of ZnO NPs application is essential to boost quality and productivity in aromatic plants by avoiding potential risks.

3 Cerium Oxide Nanoparticles

Cerium is an irony-gray color element member of lanthanide group (atomic number, 58; atomic weight, 140.166). It dissolves in hot water easily and in icy water sluggishly because its structure at room temperature is hard ductile metal. Chemical formula of cerium oxide is CeO_2 and size ranges from 5 to 100 to 150 nm revealed through scanning electron microscope (SEM) technique. In X-ray diffraction (XRD) studies, it was found to have a highly crystalline, face-centered cubic structure. Cerium oxide (CeO_2) nanoparticles are also called as "nanoceria"; they are found in class of rare earth metal-based nanoparticles which have innovative properties like redox properties, biocompatibility, agricultural, biomedicines, cosmetics, engineering sectors, and excellent antioxidant and free radical scavengers [43]. CeO_2 NPs show a higher surface area, enhanced surface reactivity, and increased presence of oxygen vacancies. These vacancies permit the coexistence of Ce^{3+} and Ce^{4+} oxidation states and giving CeO_2 NPs exceptional redox cycling capabilities. This behavior of redox allows CeO_2 NPs to scavenge reactive oxygen species (ROS) and act as nanozymes, imitating the performance of antioxidant enzymes like superoxide dismutase (SOD) and catalase (CAT). The presence of peak at 320 nm in a sample in UV-vis spectroscopy can confirm the successful formation of nanoparticles [44]. Plant extracts mainly medicinal plants help to turn cerium salts into nanoparticles, and other natural resources also form nanoparticles such as marine oyster shell, egg white, and biopolymers (chitosan, agarose, alginate). They all are eco-friendly, nontoxic, biocompatible, and sustainable. NPs are made up of two processes:

"top-down" or "bottom-up." In top-down, large material is broken down into small pieces through various techniques like lithographic which includes milling, crushing, laser, sputtering, etc. and in bottom-up through physical, chemical, and biological processes like plant-based material, microorganisms, bio-based products, etc.

CeO_2 NPs are emerging as a valuable component in the development of advanced nanofertilizers due to large surface area, redox properties, and antioxidant properties. It also has been explored as a nanofertilizer to enhance nutrient efficiency and salt tolerance and improve growth in spearmint plants [45]. Coating of CeO_2 NPs showed protection of plants from abiotic stress like drought, salinity, and heavy metal toxicity by decreasing oxidative damage and stabilizing cellular functions. And also, studies [17, 18] show that nanoparticles coated with polyvinylpyrrolidone (PVP) (55.6 nm,- 51.8 mV) gave protection from salt stress in *Brassica napus* which exhibits anti-inflammatory and other medicinal benefits. It is attained by increasing the photosynthetic apparatus efficiency and resulted higher biomass [17, 18]. Foliar application of nanoparticles has been shown to mitigate salinity stress, enhancing chlorophyll as well as biomass, and reducing levels of malondialdehyde (MDA) and hydrogen peroxide ($H_2 O_2$) in plants. The effect of these NPs is available in Table 2. Incorporating CeO_2 NPs into nanofertilizer system holds great potential for sustainable agriculture [46].

The application of CeO_2 NPs has been found to allow the germination and early growth of many aromatic plants. The effect depends on its concentration; at low concentrations, CeO_2 NPs can improve rate of seed germination, increase root and shoot elongation, and enhance seedling vigor, which helps to attenuate oxidative stress by scavenging ROS. In many aromatic plants like *Ocimum basilicum* (basil), *Menta* spp. (mint) and *Cymbopogon citratus* (lemongrass) showed that CeO_2 NPs can promote chlorophyll synthesis, photosynthetic efficiency, and biomass accumulation [47]. CeO_2 NPs increase the percentage of seed germination and speed up by enhancing the alpha-amylase activity, leading to faster starch breakdown and energy availability and modulating phytohormones like abscisic acid and indole acetic acid [48]. Through elevation of chlorophyll and carotenoid levels, stimulation of antioxidant enzyme activities, and reduction of H_2O_2 and MDA levels, CeO_2 NP applications provided protection by salinity stress in spearmint plants. CeO_2 NPs at 100 mg L^{-1} had negative effects on physiology of plants at certain concentration [45]. An excessive accumulation of highly reactive compounds can overwhelm the plant's cellular antioxidant defenses. Specifically, more production of ROS can lead to lipid peroxidation and damage, including the oxidation of nitrogenous bases and deoxyribose backbone. CeO_2 NPs are highly effective at scavenging ROS due to their catalytic properties, making them widely utilized in industry, medical research, and plant science [49].

CeO_2 NPs are increasingly recognized for their capacity to mitigate or reduce abiotic stress in aromatic plants by promoting the physiological and biochemical resilience. In spearmint a medicinal and aromatic herb valued for its essential oils, CeO_2 NPs have been shown to attenuate the harmful impacts of salinity stress by lowering oxidative stress. They achieve this by scavenging ROS and enhancing the activity of essential antioxidant enzyme, SOD, CAT, and peroxidase (POD). This causes to reduce the accumulation of oxidative markers like H_2O_2 and MDA, preserving membrane integrity and promoting healthier growth under stress conditions. Moreover, CeO_2 NPs support improved photosynthetic efficiency and increased chlorophyll and carotenoid contents, which are often compromised under salinity or drought stress. Importantly, in spearmint, nanoparticle treatment also enhanced the biosynthesis of key secondary metabolites, including L-menthone, pulegone, and 1,8-cineole—compounds critical to plant's aromatic and medicinal qualities [45]. This dual benefit of stress mitigation and essential oil enrichment highlights the potency of a sustainable strategy for improving both yield and quality in aromatic plant production under challenging environmental conditions [50].

Its uses in plant protection from biotic stresses are also studied as it may reduce contamination of bacteria, fungi, and virus which causes disease in plants. CeO_2 NPs exhibit antimicrobial activity against a broad range of phytopathogens including bacteria, fungi, and virus. A study [51] showed that CeO_2 NPs can reduce pathogen load directly by disrupting microbial cell membrane or indirectly by strengthening plant immune responses, in *Solanum* sp. CeO_2 NPs suppress fungal growth and increase accumulation of pathogenesis-related proteins. They produce ROS that destroy microbial cells or disrupt with their metabolism. CeO_2 NPs reduce *Fusarium* wilt disease so that chlorophyll content and growth improve [51].

CeO2 microparticles and nanoparticles exert cytotoxic (cell damage) and genotoxic (DNA or chromosomal damage) effects on Allium cepa root, the cells also show lower mitotic index, chromosomal anomalies and increased DNA damage, because of nanoparticles have smaller size, more surface area, and more cellular uptake potential. The nanoparticle form was reported to exhibit a higher level of biological disruption compared to the microparticle form, underlining the importance of their size in toxicity assessments [52].

4 Iron Oxide Nanoparticle

A naturally existing substance, iron oxide, can be manufactured into nanoparticles with improved characteristics. It frequently occurs in forms such as Fe_2O_3 (hematite) and Fe_3O_4 (magnetite) [53]. Iron nanoparticles have also been used in agriculture and hydroponic settings. Iron is crucial for human and animal health as well as the growth of plants. It is essential for respiration, photosynthesis, and cell metabolism. Because iron is needed for the synthesis of certain chlorophyll protein in chloroplasts, inadequate amounts of it might result in leaf yellowing and reduced photosynthetic potential. As a vital element, iron nanoparticles scavenge reactive oxygen species (ROS), adsorb heavy metals, activate the oxidation defense mechanism, and promote the synthesis of iron films on the root surface. Many studies mention the effective implementation of different NPs that possess the special capacity to pass through plants cell wall, convey biomolecules within plant cells, and make use of their magnetic properties as a carrier for location and transportation [54]. Iron oxide has a larger surface area and enhanced chemical reactivity in its nanoparticle state, which is usually between 1 and 100 nanometers in size. Iron oxide nanoparticles (IONPs) are also recognized for their magnetic behavior, particularly in forms like magnetite and maghemite that exhibit superparamagnetic properties, which indicates that when exposed to a magnetic field, they become magnetic but lose their magnetism when the field is removed. This property makes them more functional in agricultural applications [55]. These nanoparticles are also considered biocompatible and can be readily modified at the surface for specialized uses. Analogous to iron supplements, IONPs help plants produce more chlorophyll, absorb nutrients better, and develop more quickly. However, their effects depend on their concentration. Greater levels may result in toxicity or stress, while small quantities are reported to improve plant health [56].

IONPs have demonstrated positive results when added to medicinal plants like ginger and peppermint [57]. According to reports, IONPs applied externally to *Mentha piperita* have been shown to stimulate plant growth, enhance chlorophyll content, and improve essential oil yield and quality [57]. However, a dose-dependent response may be observed since larger amounts cause the oil composition to change toward other substances such mentho-furan. Although more direct research is still required, these nanoparticles improved drought tolerance, antioxidant responses, and enzyme activity, indicating that ginger itself might benefit similarly from IONP treatment [58]. With possible applications in sustainable agriculture and the manufacturing of traditional medicines, IONPs function as a micronutrient delivery system that promotes improved growth, metabolic activity, and secondary metabolite production in plant-based medicines [59].

Fe_3O_4 NPs have been the subject of extensive study recently due to their low Curie temperature, strong coercivity, and super paramagnet. Fe_3O_4 NPs have these qualities in addition to being biocompatible and nontoxic [60]. The application of Fe_3O_4 NPs as fertilizers to replace the present conventional iron fertilizers is the subject of several investigations [61]. Magnetic Fe3O4 NPs can also be used for plant growth promotion. Because of their mesoscopic effect, small object effect, quantum impact, and surface effect, nanomaterials have better physical and chemical characteristics than their atomic or bulky alternatives. Fe_3O_4 magnetic nanoparticles (FeO_4 MNPs) demonstrated unique properties such as low toxicity, chemical stability, electromagnetic behaviors, and anti-contaminant abilities. It is commonly known that seed germination and seedling growth are stimulated by the interaction of magnetic substances, Fe_3O_4 MNPs, and cell enzymes [62]. Aquaporins in root cells open as a result of iron oxide MNPs, enhancing the absorption of minerals and water. In addition, seed germination and growth of seedlings may be affected by the concentration of magnetic nanomaterials. However, depending on the kind of plant, the amounts of these substances might have both beneficial and detrimental effects [63].

In a variety of plant systems, iron oxide nanoparticles can function as efficient elicitors. Further details of the effects of iron oxide NPs and their application on some plant species are given in Table 3. In a variety of plant systems, iron oxide nanoparticles can function as efficient elicitors. They enhance antioxidant defense, promote plant tolerance against abiotic and biotic stress, and promote the synthesis

Table 3 Iron oxide NP effects on aromatic plant species

Plant species	Application method	Effects	Reference
Mentha piperita	Foliar spray; extract used for green synthesis	Increased biomass, essential oil content; plant used for green synthesis	Khan et al., [78]
Trachyspermum ammi	Soil + plant growth promoting rhizobacteria + Fe_2O_3 NPs	Alleviated arsenic stress Enhanced antioxidants (SOD, POD, CAT, APX) Increase flavonoids and phenolics	Sun et al. [79]
Dracocephalum kotschyi	Foliar spray of Fe_2O_3 NPs	Increased growth, chlorophyll, essential oil yield, trichome density	Khanizadeh et al. [80]
Solidago virgaurea	Foliar spray (Fe NPs)	Increased growth Enhanced mineral content Improve secondary metabolites (phenols, flavonoids)	Ahmed et al., [81]
Moringa oleifera	Foliar spray (Fe_3O_4 NPs)	Increased biomass under salinity Enhance photosynthesis Improve N, P, K uptake	Tawfik et al., [82]
Sageretia thea	Root extract used in green synthesis	Synthesized FeO NPs Had antioxidant, antifungal, antibacterial effects	Israeel et al., [83]
Ocimum basilicum	Foliar application of FeO NPs	Enhance chlorophyll, essential oil content, stress tolerance	Khan [84]
Argemone mexicana	Leaf extract used for green synthesis	Synthesized FeO NPs showed strong antioxidant potential	Arokiyaraj et al., [85]

of essential metabolites (such as flavonoids, alkaloids, and phenolics). The size of the nanoparticles and the variety of plant have an impact on these dose-dependent effects [64]. They enhance antioxidant defense, promote plant tolerance against abiotic and biotic stress, and promote the synthesis of essential metabolites (such as flavonoids, alkaloids, and phenolics). Iron in nanoform is an ordinary part of micronutrient nanofertilizers, and in plants, it is considered to be a dual action micronutrient. Although it is an essential substance that regulates many physiologic processes, iron may also be harmful to plants in its ionic form by generating hazardous reactive oxygen species (ROS) (e.g., when Fe2+ reacts with H_2O_2 to create hydroxyl radicals via Fenton reaction). The toxicity of nanoparticles is mainly affected by their incredibly tiny particle size and capacity to generate ROS. We employed big Fe_3O_4 nanoparticles, around 95 nm in size, to mitigate the possible toxicity [65]. Significant genotoxic effects have been identified in studies on garden cress treated with Fe_3O_4 nanoparticles. The study found abnormalities in chromosomes and increased micronuclei formation in root cells, which are signs of DNA damage. These results bring attention to the possible dangers of using Fe_3O_4 nanoparticles to aromatic plant-based agricultural applications [66]. As antibacterial agents and plant growth inducers, IONPs convey a variety of benefits in the fields of agriculture. By restricting the

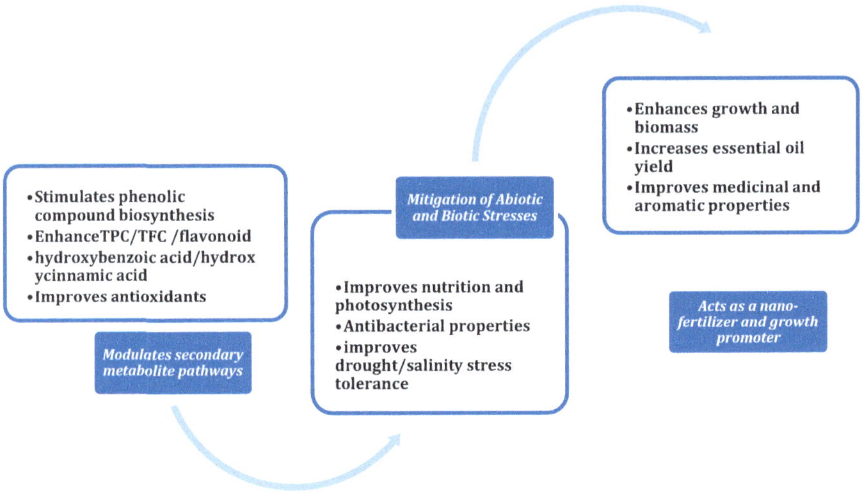

Fig. 3 Mechanism of action of metal oxide nanoformulations in promoting growth of medicinal plants

amount of oxygen required for respiration, NPs great surface-area-to-volume ratio and compact dimensions adequately sheath the microbe while exhibiting antimicrobial activity features [67].

Compared to conventional fertilizers, these nanofertilizers release nutrients more slowly and sustainably with the specific needs throughout the growing season [68]. Collectively, nanofertilizers represent a promising approach for sustainable agriculture by boosting productivity and minimizing nutrient losses and environmental damage. Studies suggest that nanofertilizers can enhance nutrient use efficiency by 20–30% compared to conventional fertilizers while greatly reducing nutrient losses to the environment. All the three nanoparticles have the characteristics required for plant growth promotion in medicinal and aromatic plants. They use direct (nutrient acquisition) as well as indirect (protection from biotic and abiotic stress as control of pathogens) mechanisms or both to promote plant growth. They also increase the yield and medicinal and aromatic properties of essential oil or extracts (Fig. 3).

5 Conclusion

Aromatic plants have direct application in pharmaceutical, cosmetic, food, agriculture, and other industries. As a slight change in the environment directly affects them, there is a need to study and develop new mechanisms to sustain their growth. Nanofertilizers are nutrient formulations integrated with nanomaterials, allowing for controlled release and gradual distribution in the soil. Iron nanoparticles have also been used in promoting plant growth in soil as well as in hydroponic settings.

Studies can be conducted to make use of their magnetic properties as an optimized carrier in precision agriculture. Although nanoparticles may be made of a wide variety of materials, only a few of them are widely utilized due to safety concerns. They tend to be toxic in lower dimensions, and they can also cross blood-brain barrier. Moreover, they tend to bioaccumulate in plants; as a result, their exposure to the environment can be dangerous. When deciding on health and environmental laws for novel materials, one must take into account the size, shape, and surface reactivity of the nanoparticles as well as their agglomeration. Therefore, metal oxide NPs offer great potential for improving aromatic plant productivity and quality and long-term assessment to maximize their benefits while minimizing environmental and phytotoxic risks. Henceforth, there is a need of comprehensive overall research in the field before their commercial application.

Competing Interests The authors declare that they have no competing interest.

Declaration of generative AI and AI-assisted technologies in the writing process During the preparation of this work, the book chapter author(s) have not taken any assistance from AI, and the book chapter author(s) reviewed and edited the content as needed and take(s) full responsibility for the content of the publication.

References

1. Chaachouay N, Zidane L (2024) Plant-derived natural products: a source for drug discovery and development. Drugs and Drug Candidates 3(1):184–207
2. Boukhatem MN, Setzer WN (2020) Aromatic herbs, medicinal plant-derived essential oils, and phytochemical extracts as potential therapies for coronaviruses: future perspectives. Plants 9(6):800
3. Gunther RT (1934) The Greek herbal of Dioscorides. Oxford University Press, Oxford, UK
4. Dafni A, Petanidou T, Vallianatou I, Kozhuharova E, Blanche C, Pacini E, Benitez G (2020) Myrtle, basil, rosemary, and three-lobed sage as ritual plants in the monotheistic religions: an historical–ethnobotanical comparison. Econ Bot 74(3):330–355
5. Elshafie HS, Camele I, Mohamed AA (2023) A comprehensive review on the biological, agricultural and pharmaceutical properties of secondary metabolites based-plant origin. Int J Mol Sci 24(4):3266
6. Husen A (2024a) Antimalarial Medicinal Plants. Taylor & Francis Group, LLC., NW. https://doi.org/10.1201/9781003378396
7. Husen A (2023a) Antidiabetic medicinal plants and herbal treatments. CRC Press, Boca Raton, FL, pp 33487–32742. https://doi.org/10.1201/b23347
8. Husen A (2022) Herbs, shrubs and trees of potential medicinal benefits. CRC Press, Boca Raton, FL, pp 33487–32742. https://doi.org/10.1201/9781003205067
9. Husen A (2023b) Exploring poisonous plants: medicinal values, toxicity response and other potential uses. CRC Press, Boca Raton, FL, pp 33487–32742. https://doi.org/10.1201/b23017
10. Husen A (2023c) Secondary metabolites based green synthesis of nanomaterials and their applications. Springer Nature Singapore Pte Ltd., Singapore, Singapore. https://doi.org/10.1007/978-981-99-0927-8
11. Husen A (2024b) Medicinal spices and condiment crops. Taylor & Francis Group, LLC., NW. https://doi.org/10.1201/9781003387046

12. Khan A, Kanwal F, Ullah S, Fahad M, Tariq L, Altaf MT, Zhang G (2025) Plant secondary metabolites—central regulators against abiotic and biotic stresses. Meta 15(4):276
13. Prado JM, Rostagno MA (eds) (2022) Natural product extraction: principles and applications. Royal Society of Chemistry
14. Yeşilyurt S, Gürgan M, Sertkahya M (2024) Biologically active compounds from medicinal and aromatic plants for industrial applications. In: Medicinal and aromatic plants: current research status, value-addition to their waste, and agro-industrial potential (Vol I). Springer Nature Switzerland, Cham, pp 1–11
15. Rajat S, Manisha S, Robin S, Sunil K (2012) Nutraceuticals: a review. Int Res J Pharm 3(4):95–99
16. Keservani RK, Sharma AK, Kesharwani RK (2020) Nutraceuticals and dietary supplements: applications in health improvement and disease management. CRC Press
17. Khan MN, Li Y, Fu C, Hu J, Chen L, Yan J et al (2022a) CeO2 nanoparticles seed priming increases salicylic acid level and ROS scavenging ability to improve rapeseed salt tolerance. Global Chall 6(7):2200025
18. Khan Y, Sadia H, Ali Shah SZ, Khan MN, Shah AA, Ullah N et al (2022b) Classification, synthetic, and characterization approaches to nanoparticles, and their applications in various fields of nanotechnology: a review. Catalysts 12(11):1386
19. Gupta V, Mohapatra S, Mishra H, Farooq U, Kumar K, Ansari MJ et al (2022) Nanotechnology in cosmetics and cosmeceuticals—a review of latest advancements. Gels 8(3):173
20. Banerjee S (2024) Introduction to ethnobotany and traditional medicine. In: Traditional resources and tools for modern drug discovery: ethnomedicine and pharmacology. Springer Nature Singapore, Singapore, pp 1–30
21. Singh H, Sharma A, Bhardwaj SK, Arya SK, Bhardwaj N, Khatri M (2021) Recent advances in the applications of nano-agrochemicals for sustainable agricultural development. Environ Sci: Processes Impacts 23(2):213–239
22. Pandya T, Patel S, Kulkarni M, Singh YR, Khodakiya A, Bhattacharya S, Prajapati BG (2024) Zeolite-based nanoparticles drug delivery systems in modern pharmaceutical research and environmental remediation. Heliyon 10(16)
23. Afrouz M, Ahmadi-Nouraldinvand F, Elias SG, Alebrahim MT, Tseng TM, Zahedian H (2023) Green synthesis of spermine coated iron nanoparticles and its effect on biochemical properties of Rosmarinus officinalis. Sci Rep 13(1):775
24. Alkasir M, Samadi N, Sabouri Z, Mardani Z, Khatami M, Darroudi M (2020) Evaluation cytotoxicity effects of biosynthesized zinc oxide nanoparticles using aqueous Linum Usitatissimum extract and investigation of their photocatalytic activityackn. Inorg Chem Commun 119:108066
25. Din MI, Tariq M, Hussain Z, Khalid R (2020) Single step green synthesis of nickel and nickel oxide nanoparticles from Hordeum vulgare for photocatalytic degradation of methylene blue dye. Inorg Nano-Metal Chem 50(4):292–297
26. Lovanh N, Agga G, Ruiz-Aguilar G, Loughrin J, Sistani K (2025) Synthesis of silver nanoparticles from bitter melon (Momordica charantia) extracts and their antibacterial effect. Microorganisms 13(8):1809
27. Naghavi F, Khoshroo SMR, Kazemipour M, Zarandi MM (2023) Potassium copper sulfate hydrate nanoparticles modulated salinity stress through improving germination, growth, and biochemical attributes of common bean (Phaseolus vulgaris L.). Russ J Plant Physiol 70(6):152
28. Nayak A, Sahoo JK, Sahoo SK, Sahu D (2022) Removal of Congo red dye from aqueous solution using zinc oxide nanoparticles synthesised from Ocimum sanctum (Tulsi leaf): a green approach. Int J Environ Anal Chem 102(19):7889–7910
29. Tombuloglu H, Ercan I, Alshammari T, Tombuloglu G, Slimani Y, Almessiere M, Baykal A (2020) Incorporation of micro-nutrients (nickel, copper, zinc, and iron) into plant body through nanoparticles. J Soil Sci Plant Nutr 20(4):1872–1881

30. Rai-Kalal P, Jajoo A (2021) Priming with zinc oxide nanoparticles improve germination and photosynthetic performance in wheat. Plant Physiol Biochem 160:341–351. https://doi.org/10.1016/j.plaphy.2021.01.032
31. Ruiz-Torres N, Flores-Naveda A, Barriga-Castro ED, Camposeco-Montejo N, Ramírez-Barrón S, Borrego-Escalante F, Niño-Medina G, Hernández-Juárez A, Garza-Alonso C, Rodríguez-Salinas P, García-López JI (2021) Zinc oxide nanoparticles and zinc sulfate impact physiological parameters and boosts lipid peroxidation in soil grown coriander plants (Coriandrum sativum). Molecules 26(7):1998. https://doi.org/10.3390/molecules26071998
32. García-López JI, Zavala-García F, Olivares-Sáenz E, Lira-Saldívar RH, Díaz Barriga-Castro E, Ruiz-Torres NA et al (2018) Zinc oxide nanoparticles boosts phenolic compounds and antioxidant activity of Capsicum annuum L. during germination. Agronomy 8(10):215
33. Ghaffari Yaichi Z, Hassanpouraghdam MB, Rasouli F, Aazami MA, Vojodi Mehrabani L, Jabbari SF, Asadi M, Esfandiari E, Jimenez-Becker S (2025) Zinc oxide nanoparticles foliar use and arbuscular mycorrhiza inoculation retrieved salinity tolerance in Dracocephalum moldavica L. by modulating growth responses and essential oil constituents. Sci Rep 15(1):492. https://doi.org/10.1038/s41598-024-84198-2
34. Afzal S, Aftab T, Singh NK (2022) Impact of zinc oxide and iron oxide nanoparticles on uptake, translocation, and physiological effects in Oryza sativa L. J Plant Growth Regul 41(4):1445–1461. https://doi.org/10.1007/s00344-021-10388-1
35. Anjum S, Anjum I, Hano C, Kousar S (2019) Advances in nanomaterials as novel elicitors of pharmacologically active plant specialized metabolites: current status and future outlooks. RSC Adv 9(69):40404–40423. https://doi.org/10.1039/c9ra08457f
36. Kráľová K, Jampílek J (2022) Metal- and metalloid-based Nanofertilizers and Nanopesticides for advanced agriculture. In: Fernandes Fraceto L, Pereira de Carvalho HW, de Lima R, Ghoshal S, Santaella C (eds) Inorganic Nanopesticides and Nanofertilizers: a view from the mechanisms of action to field applications. Springer International Publishing, pp 295–361. https://doi.org/10.1007/978-3-030-94155-0_10
37. Sánchez-Pérez DM, Márquez-Guerrero SY, Ramírez-Moreno A, Rodríguez-Sifuentes L, Galindo-Guzmán M, Flores-Loyola E, Marszalek JE (2023) Impact of biologically and chemically synthesized zinc oxide nanoparticles on seed germination and seedlings' growth. Horticulturae 9(11):1201. https://doi.org/10.3390/horticulturae9111201
38. Sarkhosh S, Kahrizi D, Darvishi E, Tourang M, Haghighi-Mood S, Vahedi P, Ercisli S (2022) Effect of zinc oxide nanoparticles (ZnO-NPs) on seed germination characteristics in two Brassicaceae family species: Camelina sativa and Brassica napus L. J Nanomater 2022(1):1892759. https://doi.org/10.1155/2022/1892759
39. Shoala T (2023) 3—the dual role of nanotechnology in the management of biotic and abiotic stresses in plants. In: Ingle AP (ed) Nanotechnology in agriculture and agroecosystems. Elsevier, pp 51–70. https://doi.org/10.1016/B978-0-323-99446-0.00011-8
40. Chanu Thounaojam T, Thounaojam TM, Upadhyaya H (2021) Chapter 16—role of zinc oxide nanoparticles in mediating abiotic stress responses in plant. In: Abd-Elsalam KA (ed) Zinc-based nanostructures for environmental and agricultural applications. Elsevier, pp 323–337. https://doi.org/10.1016/B978-0-12-822836-4.00027-6
41. Lee S, Kim S, Kim S, Lee I (2013) Assessment of phytotoxicity of ZnO NPs on a medicinal plant, Fagopyrum esculentum. Environ Sci Pollut Res 20(2):848–854. https://doi.org/10.1007/s11356-012-1069-8
42. Ghosh M, Jana A, Sinha S, Jothiramajayam M, Nag A, Chakraborty A, Mukherjee A, Mukherjee A (2016) Effects of ZnO nanoparticles in plants: cytotoxicity, genotoxicity, deregulation of antioxidant defenses, and cell-cycle arrest. Mutat Res Genet Toxicol Environ Mutagen 807:25–32. https://doi.org/10.1016/j.mrgentox.2016.07.006
43. Sadidi H, Hooshmand S, Ahmadabadi A, Javad Hoseini S, Baino F, Vatanpour M, Kargozar S (2020) Cerium oxide nanoparticles (Nanoceria): hopes in soft tissue engineering. Molecules 25(19):4559

44. Choudary MRP, Surya M, Saravanan M (2024) Green synthesis of cerium oxide nanoparticles using Tribulus terrestris: characterization and evaluation of antioxidant, anti-inflammatory and antibacterial efficacy against wound isolates. Biomed Phys Eng Express 10(6):065033
45. Haghmadad Milani M, Mohammadi A, Panahirad S, Farhadi H, Labib P, Kulak M et al (2024) Cerium oxide nanoparticles (CeO2 NPs) enhance salt tolerance in spearmint (Mentha spicata L.) by boosting the antioxidant system and increasing essential oil composition. Plants 13(20):2934
46. Hassanpouraghdam MB, Vojodi Mehrabani L, Bonabian Z, Aazami MA, Rasouli F, Feldo M et al (2022) Foliar application of cerium oxide-salicylic acid nanoparticles (CeO2: SA nanoparticles) influences the growth and physiological responses of portulaca oleracea l. under salinity. Int J Mol Sci 23(9):5093
47. Ukwattage NL, Zhiyong Z (2025) Impacts of cerium dioxide nanoparticles on the soil–plant system and their potential agricultural applications. Nano 15(12):950
48. Rout A, Jalan K, Banerjee P (2025) Effect of nanomaterials on the physiological status of crop plants. Nanotechnology-based sustainable agriculture, 187–230
49. Pietrzak M, Skiba E, Wolf WM (2024) Root-applied cerium oxide nanoparticles and their specific effects on plants: a review. Int J Mol Sci 25(7):4018
50. Mohammadi MHZ, Panahirad S, Navai A, Bahrami MK, Kulak M, Gohari G (2021) Cerium oxide nanoparticles (CeO2-NPs) improve growth parameters and antioxidant defense system in Moldavian balm (Dracocephalum moldavica L.) under salinity stress. Plant Stress 1:100006
51. Wang Q (2014) Chronic impact of cerium oxide nanoparticles on Solanum lycopersicum L. and Brassica rapa L. Southern Illinois University at Carbondale
52. Liman R, Acikbas Y, Ciğerci İH (2019) Cytotoxicity and genotoxicity of cerium oxide micro and nanoparticles by allium and comet tests. Ecotoxicol Environ Saf 168:408–414
53. Panda J, Das D (2025) Superparamagnetic iron oxide nanoparticle-based nanosystems for cancer theranostics. Glob Transl Med 4(2):31–50
54. Spanos A, Athanasiou K, Ioannou A, Fotopoulos V, Krasia-Christoforou T (2021) Functionalized magnetic nanomaterials in agricultural applications. Nano 11(11):3106
55. Ajinkya N, Yu X, Kaithal P, Luo H, Somani P, Ramakrishna S (2020) Magnetic iron oxide nanoparticle (IONP) synthesis to applications: present and future. Materials 13(20):4644
56. Waszkowska K, Lesiak A (2025) Application of iron oxide nanoparticles for promoting eucalyptus growth under salt stress. In: Nanomaterials in agroforestry systems. Springer Nature Singapore, Singapore, pp 291–311
57. Abou El-Nasr MK, Hassan KM, Abd-Elhalim BT, Kucher DE, Rebouh NY, Ansabayeva A et al (2025) The emerging roles of nanoparticles in managing the environmental stressors in horticulture crops—a review. Plants 14(14):2192
58. Ali BS, Hassan SM, Abbas G, Aslam T, Tanveer R, Arshad MIM et al (2024) Synthesis, characterization and biological applications of green synthesized iron oxide nanostructures using Zingiber officinale. Pak J Sci 76(3):372
59. Chowdhury SS, Ibnat N, Hasan M, Ghosh A (2024) Medicinal plant-based nanoparticle synthesis and their diverse applications. In: Ethnopharmacology and OMICS advances in medicinal plants volume 2: revealing the secrets of medicinal plants. Springer Nature Singapore, Singapore, pp 213–250
60. Ganapathe LS, Mohamed MA, Mohamad Yunus R, Berhanuddin DD (2020) Magnetite (Fe3O4) nanoparticles in biomedical application: from synthesis to surface functionalisation. Magnetochemistry 6(4):68
61. Rui M, Ma C, Hao Y, Guo J, Rui Y, Tang X et al (2016) Iron oxide nanoparticles as a potential iron fertilizer for peanut (Arachis hypogaea). Front Plant Sci 7:815
62. Bahrami MK, Movafeghi A, Mahdavinia GR, Hassanpouraghdam MB, Gohari G (2018) Effects of bare and chitosan-coated Fe_3O_4 magnetic nanoparticles on seed germination and seedling growth of Capsicum annuum L. Biointerface Res Appl Chem 8(5):3552–3559

63. Kornarzyński K, Sujak A, Czernel G, Wiącek D (2020) Effect of Fe3O4 nanoparticles on germination of seeds and concentration of elements in Helianthus annuus L. under constant magnetic field. Sci Rep 10(1):8068
64. Nourozi E, Hosseini B, Maleki R, Abdollahi Mandoulakani B (2019) Iron oxide nanoparticles: a novel elicitor to enhance anticancer flavonoid production and gene expression in Dracocephalum kotschyi hairy-root cultures. J Sci Food Agric 99(14):6418–6430
65. Feng Y, Kreslavski VD, Shmarev AN, Ivanov AA, Zharmukhamedov SK, Kosobryukhov A et al (2022) Effects of iron oxide nanoparticles (Fe3O4) on growth, photosynthesis, antioxidant activity and distribution of mineral elements in wheat (Triticum aestivum) plants. Plants 11(14):1894
66. Mošenoka A, Kokina I, Plaksenkova I, Jermaļonoka M, Sledevskis E, Krasovska M (2024) Effects of metal oxide nanoparticles on the growth and genotoxicity of garden cress (Lepidium sativum L.). Agronomy 14(10):2324
67. Ashraf H, Batool T, Anjum T, Illyas A, Li G, Naseem S, Riaz S (2022) Antifungal potential of green synthesized magnetite nanoparticles black coffee–magnetite nanoparticles against wilt infection by ameliorating enzymatic activity and gene expression in Solanum lycopersicum L. Front Microbiol 13:754292
68. Jithendar B, Kumar R, Rana N (2024) Revolutionizing crop nutrition: exploring nano fertilizers in agriculture. Int J Plant Soil Sci 36(6):327–339
69. Lakshmeesha TR, Kalagatur NK, Mudili V, Mohan CD, Rangappa S, Prasad BD et al (2019) Biofabrication of zinc oxide nanoparticles with Syzygium aromaticum flower buds extract and finding its novel application in controlling the growth and mycotoxins of Fusarium graminearum. Front Microbiol 10:1244
70. Babajani A, Iranbakhsh A, Oraghi Ardebili Z, Eslami B (2019) Differential growth, nutrition, physiology, and gene expression in Melissa officinalis mediated by zinc oxide and elemental selenium nanoparticles. Environ Sci Pollut Res 26(24):24430–24444
71. Zaeem A, Drouet S, Anjum S, Khurshid R, Younas M, Blondeau JP et al (2020) Effects of biogenic zinc oxide nanoparticles on growth and oxidative stress response in flax seedlings vs. in vitro cultures: a comparative analysis. Biomolecules 10(6):918
72. Nekoukhou M, Fallah S, Abbasi-Surki A, Pokhrel LR, Rostamnejadi A (2022) Improved efficacy of foliar application of zinc oxide nanoparticles on zinc biofortification, primary productivity and secondary metabolite production in dragonhead. J Clean Prod 379:134803
73. Kralova K, Jampilek J (2021) Responses of medicinal and aromatic plants to engineered nanoparticles. Appl Sci 11(4):1813
74. Khan M, Wali R, Mashwani ZUR, Raja NI, Ullah R, Bari A et al (2024) Nanowarriors from mentha: unleashing nature's antimicrobial arsenal with cerium oxide nanoparticles. ACS Omega 9(13):15449–15462
75. Habán M, Korczyk-Szabó J, Čerteková S, Ražná K (2023) Lavandula species, their bioactive phytochemicals, and their biosynthetic regulation. Int J Mol Sci 24(10):8831
76. Mirzaei M, Ladan Moghadam A, Hakimi L, Danaee E (2020) Plant growth promoting rhizobacteria (PGPR) improve plant growth, antioxidant capacity, and essential oil properties of lemongrass (Cymbopogon citratus) under water stress. Iran J Plant Physiol 10(2):3155–3166
77. Morales MI, Rico CM, Hernandez-Viezcas JA, Nunez JE, Barrios AC, Tafoya A et al (2013) Toxicity assessment of cerium oxide nanoparticles in cilantro (Coriandrum sativum L.) plants grown in organic soil. J Agric Food Chem 61(26):6224–6230
78. Khan S, Akhtar N, Rehman SU, Jamil M (2023) Iron oxide nanoparticles: plant response, interaction, phytotoxicity and defense mechanisms. In: Nanomaterials and nanocomposites exposures to plants: response, interaction, phytotoxicity and defense mechanisms. Springer Nature Singapore, Singapore, pp 227–245
79. Sun Y, Ma L, Ma J, Li B, Zhu Y, Chen F (2022) Combined application of plant growth-promoting bacteria and iron oxide nanoparticles ameliorates the toxic effects of arsenic in Ajwain (*Trachyspermum ammi* L.). Front Plant Sci 13:1098755

80. Khanizadeh P, Mumivand H, Morshedloo MR, Maggi F (2024) Application of Fe2O3 nanoparticles improves the growth, antioxidant power, flavonoid content, and essential oil yield and composition of Dracocephalum kotschyi Boiss. Front Plant Sci 15:1475284
81. Ahmed MA, Shafiei-Masouleh SS, Mohsin RM, Salih ZK (2023) Foliar application of iron oxide nanoparticles promotes growth, mineral contents, and medicinal qualities of Solidago virgaurea L. J Soil Sci Plant Nutr 23(2):2610–2624
82. Tawfik MM, Mohamed MH, Sadak MS, Thalooth AT (2021) Iron oxide nanoparticles effect on growth, physiological traits and nutritional contents of Moringa oleifera grown in saline environment. Bull Natl Res Cent 45(1):177
83. Israeel M, Iqbal J, Abbasi BA, Ijaz S, Ullah R, Zarshan F et al (2024) Potential biological applications of environment friendly synthesized iron oxide nanoparticles using Sageretia thea root extract. Sci Rep 14(1):28310
84. Khan AAH (2021) Plant physiological responses to engineered nanoparticles. In: Plant-microbes-engineered nano-particles (PM-ENPs) nexus in agro-ecosystems: understanding the interaction of plant, microbes and engineered Nano-particles (ENPS). Springer International Publishing, Cham, pp 85–99
85. Arokiyaraj S, Saravanan M, Prakash NU, Arasu MV, Vijayakumar B, Vincent S (2013) Enhanced antibacterial activity of iron oxide magnetic nanoparticles treated with Argemone mexicana L. leaf extract: an in vitro study. Mater Res Bull 48(9):3323–3327

Challenges, Opportunity, Safety, and Regulatory Issues Associated with Sustainable Uses of Aromatic Plants in Nanotechnology

Bhupal Bhattacharya, Amit Kumar Mandal, Debasish Biswas, Ahmet Kati, and Sarmistha Bhattacharya

Abstract Metal nanoparticles (especially Au, Ag, Zn, Cu) hold great promise for enhancing yield and quality in aromatic plant cultivation, but they also pose serious environmental and health risks if unchecked. This study summarizes key mechanisms by which metal-based nanomaterials interact with aromatic plants, promoting growth or oil synthesis at low doses but causing oxidative stress and DNA damage at higher doses and then traces the downstream pathways of exposure. We highlight how nano-agrochemicals can persist and bioaccumulate in soils, water, and food chains, imperiling ecosystems and human health. Crucially, we focus on the legal and ethical challenges of governing this technology globally. Current regulatory frameworks are fragmented and reactive: most jurisdictions apply old chemical laws to novel nanoforms, leaving gaps in labeling, worker protection, and public transparency. Developing countries often lack any nano-specific rules, risking a "regulatory dumping" of untested products. We examine how these gaps violate basic principles of environmental justice and intergenerational equity: vulnerable farmworkers and future generations face the burdens of nanotechnology's unknown harms, while benefits accrue elsewhere. We argue that nanotechnology policy must be reoriented by precaution, transparency, and equity: from requiring premarket risk assessments and clear product labels to forging international treaties that harmonize standards and ensure inclusive, multi-stakeholder governance.

B. Bhattacharya
Department of Law, Raiganj University, Raiganj, West Bengal, India

A. K. Mandal
Department of Sericulture, Raiganj University, Raiganj, West Bengal, India

D. Biswas
Department of Economics, Raiganj University, Raiganj, West Bengal, India

A. Kati
Institute of Health Sciences, Biotechnology, University of Health Sciences, Istanbul, Turkey

S. Bhattacharya (✉)
Amex Law College, Burdwan, West Bengal, India

A. Husen (ed.), *Sustainable Uses of Aromatic Plants in Nanotechnology*, Smart Nanomaterials Technology, https://doi.org/10.1007/978-981-95-7409-4_21

Keywords Nanotechnology · Aromatic plant cultivation · Environmental justice · Regulatory dumping · Legal and ethical challenges

1 Introduction

Nanotechnology is transforming plant cultivation by enabling targeted delivery of nutrients, pesticides, and growth regulators at the molecular level. In aromatic plant production involving herbs, spices, and medicinal crops, metallic nanoparticles (NPs) like gold (Au), silver (Ag), zinc (Zn), and copper (Cu) can boost seed germination, oil yield, and disease resistance [1, 2]. For example, AuNPs at low concentrations have been reported to improve chlorophyll content and essential oil biosynthesis in basil [1], and zinc oxide NPs can act as nanofertilizers enhancing mint oil yield [3]. These advances promise higher productivity and resource efficiency.

However, the very properties that make NPs useful, extremely small size, high reactivity, and large surface area also create new hazards. NPs can cross biological barriers, interact unpredictably with cellular chemistry, and remain in the environment for long periods. Importantly, innovation in nano-enabled horticulture is racing ahead of governance [4]. Commercial products are entering markets globally even as risk assessments lag behind. This "techno-regulatory gap" means aromatic-crop growers may unwittingly apply materials whose long-term effects are unknown. The social consequences are profound: rural farmworkers and consumers face exposure risks, while the benefits (improved yields, patentable formulations) accrue mainly to companies and wealthier consumers [5, 6].

2 Mechanisms and Phytotoxicity of Metal-Based Nanoparticles

Metal nanoparticles exhibit dual effects on aromatic plants beneficial at low doses and harmful at high doses driven by their unique physicochemical properties. In general, small (1–100 nm) particles can enter plants through stomata or root pores, slowly dissolving to release ions (a "Trojan horse" effect).

The above diagram depicts that the controlled release of metal nanoparticles underlies their utility as nanofertilizers or nanopesticides (Fig. 1). However, uncontrolled uptake or excessive ion release can trigger phytotoxicity [7].

(a) Gold NPs are largely inert but can stimulate growth signaling. Studies have found that low concentrations of AuNPs improve seed germination and essential oil production in herbs like basil. The particles can modulate stress-response pathways and increase phenolic compounds, enhancing plant vigor. Yet at higher levels (e.g., >100 mg/L) or with very small (<5 nm) AuNPs, toxic effects

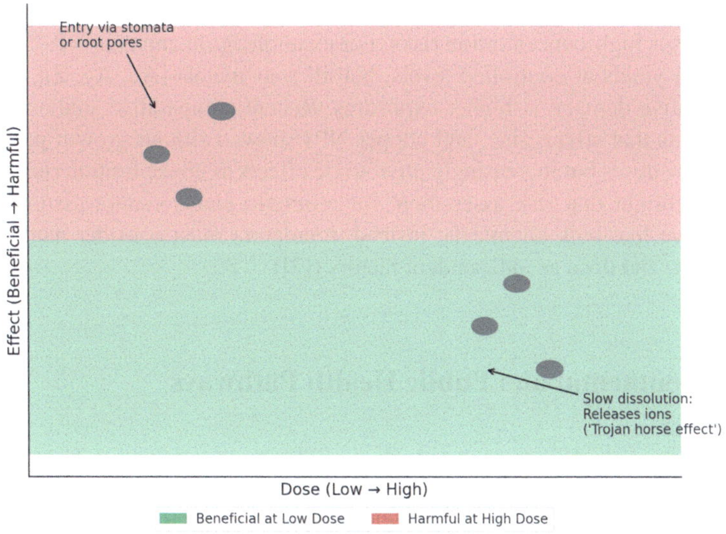

Fig. 1 Effects of metal nanoparticles on aromatic plants

emerge: they induce oxidative stress and damage cellular organelles [8]. This illustrates the narrow window between benefit and harm for AuNP use.

(b) Silver NPs are potent antimicrobial agents. In aromatic crops such as lavender and rosemary, AgNPs have been used to suppress fungal pathogens and improve cutting viability. However, AgNPs readily dissolve to release Ag^{+}, which generates reactive oxygen species (ROS). This causes lipid peroxidation, DNA strand breaks, and enzyme inhibition in plant cells [9]. Even trace AgNPs applied to soil can bioaccumulate and later enter the human food chain [10]. In short, AgNPs at low doses can protect plants but can also cause severe phytotoxicity through ROS and genotoxic mechanisms.

(c) Zinc is an essential micronutrient. ZnO NPs used as nanofertilizers can greatly improve nutrient delivery and oil yield in mint and basil ([3]). By gradually releasing Zn^{2+}, they enhance cell division, photosynthesis, and nitrogen metabolism. This can raise essential oil content and overall growth. However, excessive ZnNPs lead to root inhibition and nutrient imbalances [7, 10]. Since Zn is reactive, overloaded ZnNPs trigger oxidative stress similar to silver, interfering with plant redox homeostasis.

(d) Copper NPs serve as fungicides in crops. They improve growth under pathogen pressure and alleviate some stresses [11]. Yet CuNPs oxidize readily to release Cu^{2+}, which can disturb the mitochondria and electron transport in plant cells. Like silver, high CuNP exposure induces ROS formation and stunts growth [7]. Interestingly, some studies report that very low CuNP doses mitigate salt stress better than bulk Cu^{2+} salts, underscoring that *nanoform matters* ([10]. The complexity is that CuNPs may help or harm depending on dose and context.

Each nanoparticle type displays a dose-dependent trade-off: low-concentration benefits versus high-concentration risks. For example, gold and zinc NPs can boost aromatic oil yields at controlled levels, but all four metals (Au, Ag, Zn, Cu) can cause oxidative damage at higher exposures. Recent comparative studies confirm these patterns that silver, zinc, and copper NPs showed similar growth-promoting effects at low doses but increasingly phytotoxic effects as concentration rises. These findings highlight that risk assessment for aromatic plants cannot assume nanoforms behave like bulk chemicals; instead, regulators must consider nanoparticle size, coating, and dose as independent factors ([7]).

3 Environmental and Public Health Pathways

Nanoparticles applied to aromatic plants do not remain confined to treated fields. They enter soils, water, and biota, creating a web of exposure that challenges conventional governance. Key pathways include the following:

(a) Bioaccumulation and Persistence: Metal NPs are chemically stable and persistent. Once in soil or water, particles or ions can accumulate in organisms. For example, studies have shown that AgNPs introduced to marine or freshwater systems accumulate in fish and shellfish [12]. Even if edible tissues have low nanoparticle levels, accumulation in organs like liver or muscle raises food-safety concerns. In soil, NPs can alter microbial communities and then move up the food web. Recent work using model ecosystems found that silver and zinc NPs can be transferred from plants to herbivores, potentially magnifying exposure with each trophic step. Because of these dynamics, the true risk is not just eating a single herb leaf but the cumulative ecosystem effect. Nanomaterials could subtly disrupt pollinators, soil fungi, or insects that sustain aromatic crops, leading to long-term ecological degradation.

(b) Occupational Exposure: Workers who cultivate, harvest, or process aromatic plants may inhale or absorb NPs. Dermal contact is often the largest route: one study of nanopesticide applicators found that most exposure occurred through skin on clothing-covered areas. Respiratory exposure is also possible during powder handling or spraying. Crucially, many farmworkers lack guidance on nanomaterial hazards. Major safety agencies (OSHA, NIOSH) have general statements about nanoparticles, but no binding standards for farm settings [13]. In practice, farmers often receive no extra training or PPE when using nano-formulations. This is an ethical issue of workplace justice: the people at greatest risk (often low-income rural laborers) have the least power to demand safety, while agribusiness reaps productivity gains. As one review warns, handling nanoparticles without adequate controls "implicitly shifts from a 'safety-first' to a 'market-first' approach" [7]. The regulatory vacuum for occupational exposure violates the basic right of workers to a safe work environment [13].

(c) Consumer Exposure and Transparency: Aromatic plant products (teas, extracts, oils, herbs) may contain trace nanoparticles if fields were treated. The health impact of chronic low-dose ingestion is unknown. This uncertainty is compounded by a policy vacuum on transparency. Currently, neither the USA nor the EU requires explicit labeling of nano-ingredients in foods or herbal productseprints.lse.ac.uk. Major agencies (EPA, FDA, EFSA) generally treat nanomaterials under existing laws without calling them out by name. Consumer advocacy groups argue this leaves a "nano-blind" space: people cannot tell if their chamomile tea or essential oil tablets contain engineered nanoparticleseprints.lse.ac.uk. The LSE briefing on transatlantic regulation notes that governments on both sides of the Atlantic have so far "indicated no intention to change the status quo" [14] regarding labeling. This lack of transparency undermines informed choice. Surveys show that many consumers, especially women and the elderly, are already wary of novel technologies in food, and this caution is rational given the missing information. Ethically, failing to disclose nano-content violates principles of autonomy and trust [5]. Without mandatory labels or public risk communication, the introduction of nano-aromatic products creates an involuntary "experiment" on consumers.

(d) Ecological Disruptions: Field applications of NPs can disturb nontarget organisms. Soil microbes, earthworms, pollinators, and other beneficials may be affected. For instance, high concentrations of AgNPs have been shown to shift soil bacterial communities, while lower levels of ZnO or CuO sometimes even enhance soil enzyme activity [15]. In some cases, nanoparticles can boost nutrient absorption by increasing beneficial rhizosphere microbes. However, these effects are highly context specific. Without comprehensive, case-by-case data, unintended ecosystem imbalances are likely. Over time, contaminated soils may yield aromatic plants with altered chemical profiles, and groundwater runoff could carry NPs to distant habitats. These environmental impacts raise an ethical concern of *intergenerational equity*: future generations may inherit degraded soils and water quality from today's nano-uses. In short, environmental pathways of nanopesticides and nanofertilizers underscore the long-tail risks that standard regulatory tests often overlook ([12].

4 Legal and Regulatory Landscape

To date, no country has a fully comprehensive framework for nano-enabled aromatic cultivation. Instead, governance relies on a patchwork of sectoral laws, voluntary guidelines, and ad hoc responses. This fragmented approach creates regulatory gaps in at least four key areas:

(a) A fundamental challenge is that most chemical statutes do not define "nanomaterials." For example, in the USA, the Environmental Protection Agency (EPA) regulates new substances under TSCA but uses the same law for nanoforms as for bulk chemicals. The FDA and USDA likewise apply existing statutes to foods and organic agriculture, issuing only guidance on nanotech [16]. Without a clear legal definition, products may slip through oversight. The EU has taken a more proactive stance: the 2011 EU Recommendation defines a nanomaterial by size (10–100 nm), and since 2020, REACH requires companies to report nanoform data [17]. However, enforcement has lagged and the EU chemicals agency found most safety data sheets still omit required nano-details. In practice, then, neither US nor EU laws explicitly regulate "nanofertilizers" or "nanopesticides" as such. New ingredient approvals often happen under traditional pesticide or fertilizer law without nano-specific analysis. In short, legal ambiguity reigns: products are generally presumed safe unless proven harmful, flipping the burden of proof onto regulators ([18]).

(b) In principle, US pesticide law (FIFRA) and EU plant-protection regulations could catch some nanopesticides by requiring registration of new active ingredients. Yet in practice, many nano-agrochemicals reach the market labeled simply as "enhanced formulations" or "proprietary blends." A notable enforcement case came in 2008 when the EPA fined a company for falsely claiming a regular pesticide was a "nano-coating" with nanotech benefits. This highlights that agencies rely on enforcement actions rather than clear rules. Without mandatory nano-specific testing or labeling, nano-products may hide behind conventional claims. For consumers, this means that products from "nanotreated" aromatic farms could enter the food chain with no on-package indication of nano-use [5, 14] eprints.lse.ac.uk. Many food-safety authorities have stated that existing regulations suffice for nanomaterials. But this reactive stance means regulators learn about potential problems only after products are already in use.

(c) Neither chemical workplace laws nor pesticide handling rules specifically address nanoparticle hazards. The US Occupational Safety and Health Administration has issued general guidance on engineered nanomaterials, but it is geared to factories, not open farms. Likewise, international labor guidelines (e.g., ILO codes) barely mention nanoparticles. This gap shifts risk onto farmworkers and processors who use herbal extracts and oils [13]. For example, a worker grinding nano-augmented turmeric powder or spraying nano-insecticide receives no new training. This lack of enforceable PPE or exposure limits is an ethical failure. By contrast, some countries have begun to integrate nanosafety: for instance, India's 2020 guidelines require data on nano-inputs for pesticides, and Brazil's recent agrochemical law (2023) mandates risk assessments [5, 19]. However, these efforts are piecemeal. Until comprehensive occupational standards are adopted, farm laborers remain vulnerable by default.

(d) No binding international treaty governs nanomaterials. Some multilateral agreements touch on relevant issues. Existing chemical conventions (Rotterdam, Stockholm, Basel) could in theory list certain metal NPs as hazardous wastes or toxic chemicals, but to date, few nano-specific amendments exist. There have been global initiatives promoting safe nano-use: the OECD has published voluntary principles on nanotechnology [4], and UN agencies like FAO/WHO have convened expert meetings [5]. Yet these are not enforceable. Experts argue that piecemeal guidance will not address the systemic risks of nanomaterials. In fact, analysts warn that emerging economies might become "testing grounds" for nano-agrochemicals if wealthier nations impose stricter rules [6]. This "regulatory dumping" phenomenon parallels debates in genetic engineering: without an international framework, companies may market nano-enabled fertilizers and pesticides preferentially in countries with laxer oversight. As Picecchi [6] observes, no single treaty covers all nano-risks, so a dedicated global instrument or at least an amendment to a major convention is needed (Table 1).

The above table reviews the fragmentation of global regulation. In the USA, the EPA and FDA rely on old statutes and voluntary notices [16], yielding "legal uncertainty" for nano [20]. The EU, grounded in precaution, at least requires nano-specific data under REACH [17], but enforcement is spotty and no separate nano-regulator exists. In Brazil and India, new rules are emerging [19] but remain sectoral. Japan and Australia have taken more proactive stances: Australia's APVMA requires nano-definitions in agri-chem cases. No country yet harmonizes nano-specific labeling, safety testing, and worker protections under one regime. This decentralization means accountability gaps, for instance, a fertilizer-maker may only have to reveal nanoparticle content if a country's law explicitly asks, leaving citizens in other nations in the dark.

The current legal landscape is chaotic and reactive. Governments largely treat aromatic plant nanoproducts as conventional chemicals, creating huge blind spots. Where laws exist, they often rely on "soft" measures (voluntary reporting, vague guidance) instead of binding standards. And enforcement is weak: studies have found that many nano-labeled products on the market lack safety data [21]. The upshot is that innovation has outpaced governance, allowing potential harm to slip through. From a justice perspective, this is problematic: technologies with uncertain

Table 1 Comparative overview of global nanotechnology regulation

Region	Regulatory basis	Strengths	Weaknesses
United States	Old statutes (EPA, FDA), voluntary notices	Flexible use of existing laws; agencies adapt to nano	Legal uncertainty; no nano-specific definitions
European Union	Precautionary principle; REACH requires nano-specific data	Nano-specific reporting required; grounded in precaution	Weak enforcement; no separate nano-regulator
Brazil	New rules emerging; sectoral approach	Risk analysis mandatory for pesticides	Fragmented across sectors; not fully cohesive
India	New rules emerging; sectoral approach	Guidelines for agri-inputs/food nano	Gap between policy and practice; enforcement weak
Japan	Proactive stance; emerging nano-rules	Proactive and clear definitions emerging	Still evolving; not comprehensive
Australia	Proactive stance; APVMA requires nano-definitions	Dedicated nano-regulator (APVMA); case-by-case guidance	Reliance on industry-funded data; safety vs. market tension

risks are being deployed without ensuring that affected communities can consent or even know what is happening on their farms or in their foods.

5 Ethical and Social Considerations

The technical issues discussed above raise profound ethical questions. Three overlapping principles should guide nano-governance: precaution, justice (including intergenerational equity), and responsible innovation.

5.1 The Precautionary Principle

At the core of the precautionary principle lies a simple yet profound assertion: when scientific uncertainty surrounds a new technology, society has a responsibility to prevent potential harm rather than wait for conclusive proof of danger. This principle, firmly embedded in international environmental law through instruments such as the Rio Declaration and the UNFCCC, has particular resonance for nanotechnology applications in aromatic plants [21]. Unlike traditional agricultural inputs, nanoparticles behave unpredictably in biological and ecological systems, and once released, they cannot be easily retrieved or neutralized. Applying precaution in this context means that no new nano-formulation, whether a fertilizer, pesticide, or growth stimulant, should be commercialized without comprehensive and independent safety testing. In practice, this may require temporary moratoria on certain nanopesticides until their long-term environmental and health effects are rigorously studied and transparently reported. The ethical significance of precaution is that it shifts the burden of proof from regulators to developers. Innovators must demonstrate safety proactively, instead of expecting governments to prove harm after exposure has already occurred. Current regulatory practice often allows companies to rely on their own in-house data, with oversight bodies responding reactively only once adverse outcomes become visible. This approach undermines public trust and exposes ecosystems to irreversible damage. A precaution-first framework is not anti-innovation or anti-science; rather, it ensures that risk is anticipated and mitigated before harm materializes. Given the well-documented ability of nanoparticles to persist in soils, bioaccumulate in food chains, and disrupt ecosystem functions [12], adopting precautionary governance is a moral imperative as much as a legal necessity. It demands that governments codify mandatory premarket assessments specific to nanoscale materials, create adaptive regulatory mechanisms that update standards as new evidence emerges, and prevent situations where untested nanoproducts are effectively trialed in open environments with farmers and ecosystems as unwitting participants.

5.2 Environmental Justice and Intergenerational Equity

While precaution addresses uncertainty, questions of justice focus on fairness in the distribution of benefits and burdens. In the case of nanotechnology in aromatic plants, the benefits, such as increased yields, patentable innovations, and higher-value essential oils, flow largely to biotechnology firms, agribusinesses, and affluent consumers of high-end herbal products. Conversely, the risks are disproportionately borne by those least equipped to manage them: rural farmworkers, marginalized communities, and future generations. Rural workers, often with limited economic or social power, face direct exposure to nanoparticles during cultivation, spraying, or postharvest handling. They frequently do so without proper training, protective equipment, or legal safeguards, a clear violation of environmental justice [5, 13]. Such inequity subjects vulnerable populations to involuntary risks, echoing Schlosberg's [22] argument that justice must encompass not only distributional fairness but also recognition and participation. Beyond the immediate, the issue of intergenerational equity looms large. Nanoparticles can persist in soils for decades, degrade into toxic ions, and disrupt ecological balance in ways that future generations will be forced to confront. The contamination of soils, biodiversity loss, and potential reduction in soil fertility represent legacies of harm that today's decisions may leave behind. From a fairness perspective, it is ethically indefensible for the present generation to externalize unknown harms onto those yet to come. Communities most dependent on aromatic crops for livelihood subsistence or smallholder farmers are also the least likely to be compensated for damages caused by nano-pollution. Addressing these inequities requires concrete safeguards. First, testing of risky nano-products should never be conducted on vulnerable communities without informed consent. Second, the polluter-pays principle should be rigorously enforced so that corporations, not farmers, bear the financial responsibility for environmental remediation. Finally, regulatory frameworks must expand beyond environmental impact assessments to include mandatory social impact assessments, ensuring that questions of justice and fairness are integrated into decision-making about nanotechnology deployment [22].

5.3 Responsible Innovation and Stakeholder Engagement

While precaution and justice offer guiding principles, governance also requires operational mechanisms for responsible innovation. The framework of responsible research and innovation (RRI) emphasizes anticipation, inclusiveness, reflexivity, and responsiveness in the development of emerging technologies [23]. Applied to nanotechnology in aromatic plants, RRI would mean involving all key stakeholders, i.e., farmers, workers, consumers, nongovernmental organizations, and local communities, at the earliest stages of product design and regulatory evaluation. Evidence shows that public attitudes toward novel food and plant technologies are shaped less by the technologies themselves than by the transparency and inclusiveness of decision-making processes. Where dialogue is open and responsive, acceptance is higher; where information is withheld and decisions are top-down, distrust and resistance grow. Indeed, grassroots actors have already demonstrated their willingness to act in the absence of formal governance: some organic certification bodies, such as the Soil Association in the UK and the Biological Farmers of Australia, have preemptively banned engineered nanomaterials from certified products, reflecting consumer skepticism and precautionary ethics. This bottom-up pressure underscores the importance of building participatory governance structures. Mechanisms for responsible innovation might include advisory committees composed of ethicists, consumer advocates, and farmer representatives to review proposed nanoproducts; requirements for public consultation and comment before regulatory approvals; and full disclosure of funding sources and potential conflicts of interest in nanotechnology research. Such measures democratize governance and ensure that decisions reflect societal values rather than narrow corporate interests. Ultimately, the goal is to reframe nanotechnology governance not as a purely technical or economic issue but as a democratic one rooted in transparency, accountability, and inclusivity. By embedding stakeholder engagement in regulatory practice, societies can ensure that the development and deployment of nano-enabled aromatic plant technologies align with ethical principles and public trust.

6 Conclusion and Recommendations

Metal-based nanomaterials offer real benefits to aromatic plant cultivation: higher yields, reduced pesticide use, and novel product properties, but these come with unaccounted risks. This review has shown that every major class of nano-agrochemical can generate reactive intermediates or persistent residues that conventional regulation does not fully address. We found that global law remains a patchwork: existing statutes and guidelines only partially capture nano-specific hazards, and enforcement is weak. In this setting, ethical lapses are almost inevitable: the burdens of uncertainty fall on workers, the environment, and future generations, while the economic upsides flow elsewhere.

Addressing nanotechnology in aromatic plant production demands more than tweaking pesticide laws. It requires a paradigm shift: treating emerging nano-products as public goods subject to the highest safety standards, not mere market novelties. Ethically, policies must be recalibrated so that "innovation" does not override "protection." Integrating the precautionary principle with environmental justice and participatory innovation means, for example, formalizing the right of consumers to know (and refuse) nano-treated foods, legally empowering farmworkers to demand safe practices, and ensuring that developing nations are not left to clean up someone else's experiment.

To align nanotechnology with the goals of sustainability and justice, we propose the following strategic actions:

(a) Governments should enact laws designed for the nanoscale era, rather than shoehorning nano into outdated chemical rules. New statutes must require pre-market risk assessments for all nanoforms, not just their bulk analogues. For example, a garlic oil pesticide containing 10 nm CuNPs should be evaluated separately from one with bulk $CuSO_4$. Nano-legislation should demand detailed physicochemical characterization (size, shape, and coating) and long-term fate studies as part of approval. Mandatory environmental and toxicological testing, including effects on nontarget plants, soil microbes, and animal models, should be instituted. By moving from reactive consent orders to proactive registration, regulators can catch hazards before products reach farms.
(b) Full transparency is critical for public trust. Food and consumer products derived from nano-treated aromatic crops should carry clear labels identifying the nanomaterials used (type, size range). This information must also be publicly reported in a registry. Labeling serves multiple purposes: it informs consumers' choices, alerts sensitive individuals, and drives companies to be accountable. The EU's new Novel Food Regulation already requires notifying the presence of engineered nanomaterials; this approach should be expanded globally. In addition, workers and applicators must be informed about nano-ingredients in the products they handle. Without such disclosure, markets will perpetuate an "information vacuum" that erodes both safety and confidence.
(c) Specific worker-protection rules for nano-agrochemicals are urgently needed. Regulators should define exposure limits for common agricultural nanomaterials, analogous to existing thresholds for pesticides or dust. Personal protective equipment standards such as nanoparticle-grade respirators or gloves must be mandated for anyone applying or synthesizing these materials. Governments should require nano-handling training for farmworkers (in all languages spoken) and periodic health monitoring. International bodies like the ILO and WHO can develop model guidelines tailored to farming contexts. Ensuring safe working conditions is not just a technical fix but a justice issue: it upholds the rights of vulnerable workers [13].
(d) The international community should launch a centralized registry for all nano-agrochemicals. This database, overseen by an appropriate UN agency (e.g., UNEP or FAO), would catalogue each nanoparticle-containing product by country, detailing its composition, intended use, and safety data. Such a registry would illuminate cross-border trade and prevent duplicate or secret registrations. Alongside the registry, nations should collaborate on unified testing protocols and labeling norms to avoid the current "patchwork" of rules. Ideally, the G20 or UN Environment Programme would convene a treaty process or framework convention on nanomaterials, incorporating the precautionary principle and capacity building for developing countries [6]. This could mirror agreements on hazardous chemicals, ensuring no single nation becomes a dumping ground.
(e) Policymakers must explicitly incorporate environmental justice and intergenerational equity into decision-making. Regulatory agencies should assess not only the scientific risk but also the distribution of that risk across different communities and future generations. For instance, risk-benefit analyses of a new

nanopesticide should consider its impact on rural communities in the Global South, not just aggregate yield gains. Formal stakeholder councils including farmers' unions, indigenous leaders, consumer advocates and ethicists should advise on novel approvals. This inclusive approach [23] ensures that technology serves societal needs fairly.

By enacting these reforms, nanoparticle-tailored laws, mandated transparency, rigorous worker protections, international cooperation, and ethical oversight, nanotechnology can be steered toward truly sustainable uses in aromatic plants.

Competing Interests The authors declare that they have no competing interest.

Declaration of generative AI and AI-assisted technologies in the writing process: During the preparation of this work, the book chapter author(s) have not taken any assistance from AI, and the book chapter author(s) reviewed and edited the content as needed and take(s) full responsibility for the content of the publication.

References

1. Kralova K, Jampilek J (2021) Responses of medicinal and aromatic plants to engineered nanoparticles. Appl Sci 11(4):1813
2. Kah, M., Tufenkji, N. and White, J.C., 2019. Nano-enabled strategies to enhance crop nutrition and protection. Nature nanotechnology, 14(6), pp.532–540
3. Raliya R, Tarafdar JC, Biswas P (2016) Zinc oxide nanoparticle biosynthesis and its effect on improving growth and yield of agricultural crops. J Crop Improv 30(3):256–277
4. Organisation for Economic Co-operation and Development (2010) OECD principles for the oversight of nanotechnologies and nanomaterials. OECD Publishing, Paris
5. Food and Agriculture Organization of the United Nations & World Health Organization (2012) Application of nanotechnologies in the food and agriculture sectors: joint FAO/WHO meeting report (27 March 2012). FAO, Rome
6. Picecchi D (2018) Tiny things with a huge impact: the international regulation of nanomaterials. Mich J Environ Admin L 7(2):447–473
7. Bondarenko O, Juganson K, Ivask A, Kasemets K, Mortimer M, Kahru A (2013) Nanotoxicology: a review of the toxicological properties of CuO and ZnO nanoparticles in aquatic and terrestrial organisms. Nanoscale Res Lett 8:387
8. Siddiqi, K.S. and Husen, A., 2017. Plant response to engineered metal oxide nanoparticles. Nanoscale research letters, 12(1), p.92
9. Gorka, S.M., Phan, K.L., Lyons, M., Mori, S., Angstadt, M. and Rabinak, C.A., 2016. Cannabinoid modulation of frontolimbic activation and connectivity during volitional regulation of negative affect. Neuropsychopharmacology, 41(7), pp.1888–1896
10. Dang F, Huang Y, Wang Y, Zhou D, Xing B (2021) Transfer and toxicity of silver nanoparticles in the food chain. Environ Sci Nano 8(6):1519–1535
11. Nair, R. and Kumar, D.S., 2012. Plant diseases—control and remedy through nanotechnology. In Crop improvement under adverse conditions (pp. 231–243). New York, NY: Springer New York
12. Wijnhoven SWP, Peijnenburg WJ, Herberts CA, Hagens WI, Oomen AG, Heugens EH et al (2009) Nano-silver; a review of available data and knowledge gaps in human and environmental risk assessment. Nanotoxicology 3(2):109–138

13. Maynard AD, Aitken RJ, Butz T, Colvin V, Donaldson K, Oberdörster G et al (2006) Safe handling of nanotechnology. Nature 444(7117):267–269
14. Falkner R, Breggin L, Jaspers N, Pendergrass J, Porter R (2009) Consumer labelling of nanomaterials in the EU and US: convergence or divergence? Royal Institute of International Affairs (Chatham House), London
15. Simonin, M. and Richaume, A., 2015. Impact of engineered nanoparticles on the activity, abundance, and diversity of soil microbial communities: a review. Environmental Science and Pollution Research, 22(18), pp.13710–13723
16. U.S. Food and Drug Administration (FDA) (2014) Guidance for industry: considering whether an FDA-regulated product involves the application of nanotechnology. FDA, Washington, D.C.
17. Lövestam G, Rauscher H, Roebben G, Klüttgen BS, Gibson N, Putaud JP, Stamm H (2010) Considerations on a definition of nanomaterial for regulatory purposes. JRC Reference Report
18. Schummer J (2009) The ethics of nanotechnology: risk, uncertainty, and precaution. In: ten Have H (ed) Nanotechnology and ethics. Routledge, London, pp 25–52
19. Wacker MG, Proykova A, Santos GML (2016) Dealing with nanosafety around the globe; regulation vs. innovation. Int J Pharm 509(1–2):95–106
20. U.S. Environmental Protection Agency (EPA) (2015) TSCA framework for nanomaterials regulation. EPA, Washington, D.C.
21. Foster KR, Vecchia P, Repacholi MH (2000) Science and the precautionary principle. Science 288(5468):979–981
22. Schlosberg D (2007) Defining environmental justice: theories, movements, and nature. Oxford University Press
23. Stilgoe J, Owen R, Macnaghten P (2013) Developing a framework for responsible innovation. Res Policy 42(9):1568–1580

The manufacturer's authorised representative in the EU is Springer Nature Customer Service Centre GmbH, Europaplatz 3, 69115 Heidelberg, Germany. If you have any concerns regarding our products, please contact ProductSafety@springernature.com

Printed and bound by CPI Group (UK) Ltd, Croydon, CR0 4YY
20/10/2026
02236467-0018